Aspects of Terpenoid Chemistry and Biochemistry

Aspects of Terpenoid Chemistry and Biochemistry

PROCEEDINGS OF THE
PHYTOCHEMICAL SOCIETY SYMPOSIUM
LIVERPOOL, APRIL 1970

Edited by

T. W. GOODWIN

Department of Biochemistry
University of Liverpool, Liverpool, England

1971

ACADEMIC PRESS
LONDON AND NEW YORK

ACADEMIC PRESS INC. (LONDON) LTD
Berkeley Square House
Berkeley Square,
London, W1X 6BA

U.S. Edition published by
ACADEMIC PRESS INC.
111 Fifth Avenue,
New York, New York 10003

Library of Congress Catalog Card Number: 70-141738

ISBN: 0-12-289840-0

PRINTED IN GREAT BRITAIN BY
WILLIAM CLOWES & SONS LIMITED
LONDON, COLCHESTER AND BECCLES

Contributors

G. BRITTON, *Department of Biochemistry, The University of Liverpool, Liverpool, England*

R. B. CLAYTON, *Department of Psychiatry, Stanford University School of Medicine, Stanford, California, U.S.A.*

M. J. O. FRANCIS, *Nuffield Department of Orthopaedic Surgery, University of Oxford, Nuffield Orthopaedic Centre, Headington, Oxford, England*

T. W. GOODWIN, *Department of Biochemistry, The University of Liverpool, Liverpool, England*

V. HEROUT, *Institute of Chemistry and Biochemistry, Czechoslovak Academy of Science, Prague, Czechoslovakia*

S. LIAAEN-JENSEN, *Organic Chemistry Laboratories, Norwegian Institute of Technology, University of Trondheim, Trondheim, Norway*

J. MACMILLAN, *Department of Chemistry, The University of Bristol, Bristol, England*

B. V. MILBORROW, *Shell Research Ltd., Milstead Laboratory of Chemical Enzymology, Sittingbourne, Kent, England*

A. PFIFFNER, *Forschungsabeilung der F. Hoffmann-La Roche & Co. A.G., Basel, Switzerland*

H. H. REES, *Department of Biochemistry, The University of Liverpool, Liverpool, England*

D. R. THRELFALL, *Department of Biochemistry and Agricultural Biochemistry, University College of Wales, Aberystwyth, Wales*

O. B. WEEKS, *New Mexico State University Arts and Science Research Centre, Las Cruces, New Mexico, U.S.A.*

G. R. WHISTANCE, *Department of Biochemistry and Agricultural Biochemistry, University College of Wales, Aberystwyth, Wales*

Preface

In 1966 the Phytochemical Society held a Symposium on "Terpenoids in Plants" in Aberystwyth and the Proceedings were published under that title (Ed. J. B. Pridham, Academic Press, 1967). Only four years later the Committee of the Phytochemical Society found itself making a carefully considered decision to hold another Symposium on essentially the same topic. We hope that the present book justifies that decision. A comparison of the two proceedings shows that there is virtually no overlap and emphasizes the enormous strides which terpene chemistry, biochemistry and biology have taken in the past four years. No doubt this will not be the last symposium on terpenoids which the Phytochemical Society will hold, and I am sure future volumes will record equally fascinating developments.

I am grateful to the various contributors for providing such up to date reviews in their extremely active areas and to the Academic Press for preparing it for publication with a minimum of delay.

January, 1971

T. W. GOODWIN

Contents

CHAPTER 1

The Biological Significance of the Terpenoid Pathway of Biosynthesis

R. B. Clayton

CHAPTER 2

Monoterpene Biosynthesis

M. J. O. Francis

CHAPTER 3

Biochemistry of Sesquiterpenoids

V. Herout

CHAPTER 4

Juvenile Hormones

A. Pfiffner

CHAPTER 5

Abscisic Acid

B. V. Milborrow

CHAPTER 6

Diterpenes—The Gibberellins

J. Macmillan

CHAPTER 7

Ecdysones

H. H. Rees

CHAPTER 8

Recent Progress in Carotenoid Chemistry

S. Liaaen-Jensen

CHAPTER 9

General Aspects of Carotenoid Biosynthesis

G. Britton

CHAPTER 10

Biosynthesis of C_{50} Carotenoids

Owen B. Weeks

CHAPTER 11

Algal Carotenoids

T. W. Goodwin

CHAPTER 12

Biosynthesis of Isoprenoid Quinones and Chromanols

D. R. Threlfall and G. R. Whistance

CHAPTER 1

The Biological Significance of the Terpenoid Pathway of Biosynthesis

R. B. CLAYTON

Department of Psychiatry, Stanford University School of Medicine, Stanford, California U.S.A.

I. Introduction

In the introductory lecture in an earlier symposium in this series, Goodwin (1967) gave an authoritative survey of the functions of terpenes in plants and it is obviously unnecessary to reiterate this material. It is noteworthy, however, that terpenoid compounds of physiological importance for the growth and reproductive functions of higher and lower plant forms are still being characterized and no doubt will continue to be characterized for a long time to come. It is unnecessary to belabor the fact that steroids, also terpenoid in origin, fulfill hormonal functions in vertebrates that are essential, either to the life of the individual (adrenal steroids) or of the species (gonadal hormones). The physiological and biochemical roles of these compounds and their metabolic pathways are well documented and the many similarities between steroid metabolism in higher plants and animals are now clearly appreciated (Heftmann, 1963, 1967, 1968; Tschesche, 1965).

There are, to be sure, enormous numbers of steroids and terpenoid products, for the most part of plant origin, whose functions (aside from stimulating the genius of some of the world's greatest organic chemists and contributing to the despair of their graduate students) are at present unknown. We have, by now, however, a well established view of terpenoid and steroid compounds as primary regulators of cellular processes. In many instances, moreover, very similar features of these control mechanisms have been revealed, to show that control is exerted at some level between the reservoir of genetic information and the synthesis of cellular protein (Varner and Ram Chandra, 1964; Overbeek, 1968; Wareing *et al.*, 1968; Williams-Ashman, 1965; Tata, 1966; Sekeris, 1965; Mueller, 1968; Ilan *et al.*, 1970).

Rather than dwelling further upon these functions of terpenes and steroids that have been so frequently discussed, I would like to review some aspects of the function of terpenoid compounds at the ecological or inter-organismic level. While in some cases these remain speculative, in others they are clearly defined and have profound effects upon species distribution, social organization and behavior. It may not be inappropriate to point out that this emphasis does not reflect the ecological panic that is currently sweeping the United States, but stems from an earlier involvement with insect biochemistry, especially in relation to their sterol requirements. From here it was a short step to the behavior and communication of insects and to behavioral and social interactions in a more general sense. Some of the topics and conclusions to be discussed here have been presented elsewhere (Clayton, 1970). In assembling my material I have assumed that most categories of compounds that I shall discuss will be treated from the biochemical and structural viewpoint in great detail by other authors in this symposium and these aspects will therefore not be strongly emphasized.

I would like to begin by drawing attention to some terpenoid compounds that function as vitamins. The categorization of compounds as vitamins implies their absent or limited synthesis in the organism that requires them, coupled with their indispensable functional role in that species. The distribution of such compounds in potential food sources therefore has obvious ecological implications. It is interesting to note that at least two terpenoid vitamins required by mammals serve similar essential roles in much more primitive organisms: Vitamin A, (Fig. 1) or its carotenoid precursor (Olson

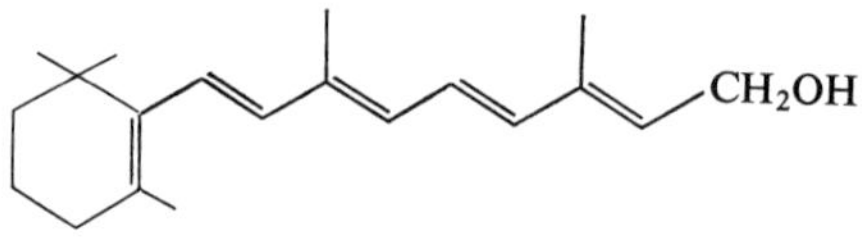

CH_2OH

Retinol

FIG. 1. Retinol (vitamin A).

and Hayaishi, 1965; Olson, 1967) is essential for visual function in insects (Goldsmith and Warner, 1964; Goldsmith *et al.*, 1964) just as in mammals, (Wald, 1968) though no parallels seem to have been found in these lower organisms for its other actions (Olson, 1967, 1969) in mammals (for example, upon reproductive function). Olson (1969) has recently discussed the biochemistry and physiology of vitamin A with particular reference to the sociological problems of its deficiency in human populations in South East Asia. Vitamin E which fulfills a vital role in reproductive and muscular function in mammals (Roels, 1967; Draper and Saari Csallany, 1969) is also an essential factor in the reproduction of several species of lower organisms though the manifestations of its role differ according to the species. An interesting example is its control of sexuality and morphological characteristics of rotifers of the genus *Asplanchna* (Fig. 2) (Gilbert, 1967, 1968; Gilbert and Thompson, 1968). The females of these organisms are of two types, "amictic",

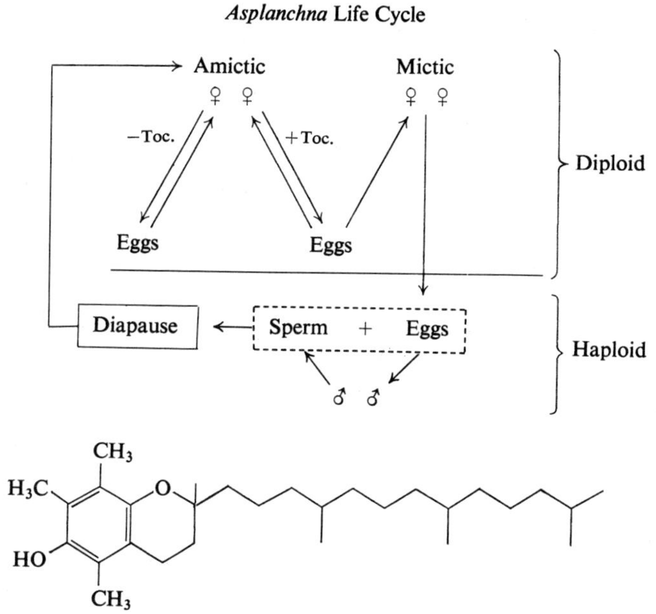

FIG. 2. α-Tocopherol (vitamin E) and the effects of its presence (+Toc) or absence (–Toc) in the diet of the rotifer, *Asplanchna*.

which lay eggs that are diploid and always hatch out parthenogenetically to give females, and "mictic", which give haploid eggs that develop into males if unfertilized and into diapausing amictic females if fertilized. The intervention of tocopherol occurs at the amictic female stage. In its presence the amictic female lays some eggs which develop parthenogenetically into mictic females, i.e. females which will lay haploid eggs, capable of fertilization. Thus, tocopherol acts here as a key that unlocks the door to sexual reproduction with its concomitant possibilities for the reorganization of genetic material and consequent evolutionary change. It seems that in Nature the sources of tocopherol are green algae that are eaten by protozoons such as *Parameceum*, or by other small rotifers, which in turn are eaten by *Asplanchna*. Interestingly, the blue green algae which are devoid of tocopherol, do not serve as adequate members of the food chain (Gilbert, 1967).

Sterols are vitamins for many lower organisms (Table I), including, apparently, all insects, none of which, it seems, can perform the synthesis of these compounds (Clayton, 1964; Clark and Bloch, 1959). In insects sterols perform a dual role as components of membrane structures (Lasser *et al.*, 1966; Lasser and Clayton, 1966; Roeske and Clayton, 1968), and as precursors of the molting hormone ecdysone (Karlson and Hoffmeister, 1963; Vroman

TABLE I
Sterol requirements of lower organisms

PROTOZOA
Paramecium[a]
Tetrahymena setifera and *paravorax*[b,c,d,e]
T. pyriformis (partial requirement)[f]
MYCOPLASMAS (PPLO) (Saprophytic but not non-saprophytic types)[g]
PURPLE PHOTOSYNTHETIC BACTERIA (*Rhodopseudomonas palustris*)[h]
FUNGUS—*Phytophthora cactorum*—29-isofucosterol or sitosterol required for sexual reproduction.[i,j,k]
YEAST (anaerobically grown)[l]
COELENTERATES—*Rhizostoma* sp.[m] and *Paracentrotus*[t]
NEMATODES—*Turbatrix aceti* and *Caenorhabditis briggsae*, free living forms utilizing *E. coli.*—did well on phytosterols and ergosterol, but poorly on cholesterol.[n,o,p]
TAPE WORM (*Spirometra mansonoides*)[q]
ANNELID—*Lumbricus terrestris.*[r]
MOLLUSCS (cuttlefish) *Sepia officinalis,*[s] *Ostrea*[t]
CRUSTACEANS[u,v]
INSECTS[w,x]

References to Table I.

[a] Conner and Van Wagtendonk (1955); [b] Wagner and Erwin (1961); [c] Holz *et al.* (1961a); [d] Holz *et al.* (1961b); [e] Hutner and Holz (1962); [f] Conner and Ungar (1964); [g] Smith and Rothblat (1960); [h] Aaronson (1964); [i] Elliott *et al.* (1964); [j] Hendrix (1964); [k] Elliott *et al.* (1966); [l] Andreasen and Stier (1953); [m] Van Aarem *et al.* (1964); [n] Hieb and Rothstein (1968); [o] Rothstein (1968); [p] Cole and Krusberg (1968); [q] Meyer *et al.* (1966); [r] Wooton and Wright (1962); [s] Zandee (1967); [t] Salaque *et al.* (1966); [u] Zandee (1966); [v] Van der Oord (1964); [w] Clark and Bloch (1959); [x] Clayton (1964).

et al., 1964). With respect to the former function at least, their quantitative importance as a component of the insect's diet exceeds that of any mammalian "vitamin" (Clayton, 1964), and the insects have extraordinarily efficient mechanisms for the retention of dietary sterols when once incorporated (Lasser *et al.*, 1966; Vroman *et al.*, 1964). There is evidence that ecological restrictions are placed on certain carnivorous insects by their highly specific requirement for cholesterol and their inability to produce it from phytosterols (Clayton, 1964; Fraenkel *et al.*, 1941; Levinson, 1955, 1962). Conversely, for some phytophagous species plant sterols may be nutritionally more effective than cholesterol (Clayton, 1964). Many species can carry out the transformation of phytosterols to cholesterol, probably by way of $\Delta^{24(28)}$ and Δ^{24} sterols as intermediates (Fig. 3) (Ritter and Wientjens, 1967; Svoboda and Robbins, 1968; Svoboda *et al.*, 1969). While these observations recall the fact that these same types of intermediates are involved in the reverse process of phytosterol biosynthesis (Lederer, 1969), there is so far no evidence to indicate a mechanistic relationship between the two processes and indeed such a relationship seems unlikely. Again, other species cannot degrade phytosterols to cholestane derivatives in measurable amounts, yet apparently can utilize phytosterols *per se*, at least for structural purposes (Kaplanis *et al.*, 1963).

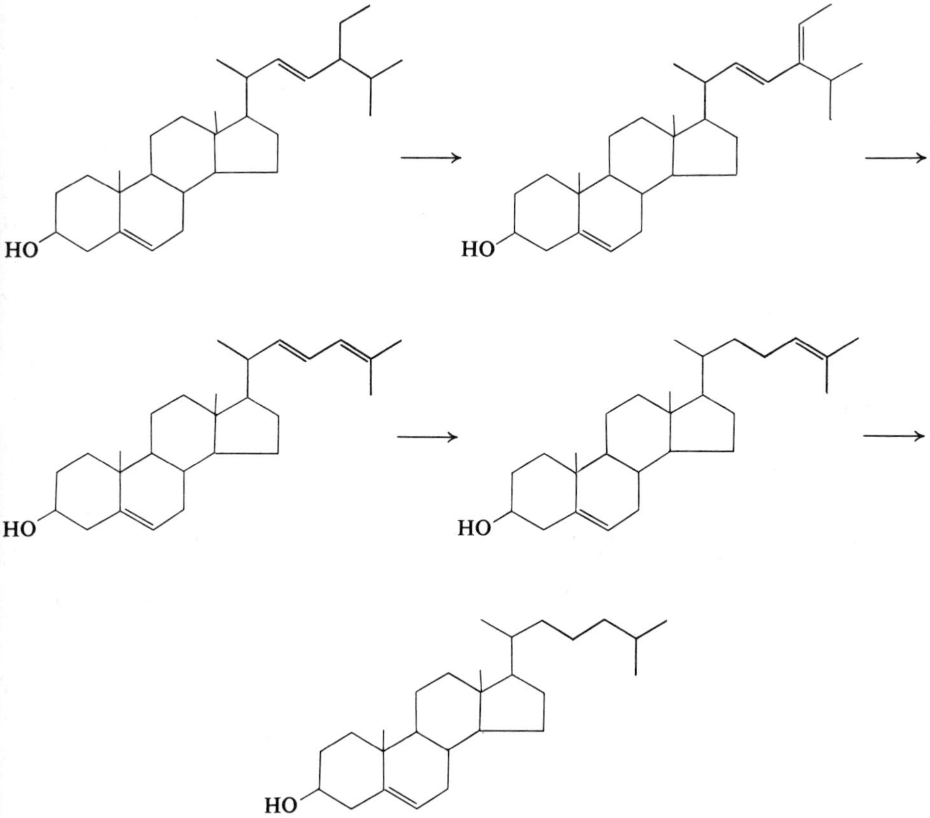

HO → HO →

HO → HO →

HO

FIG. 3. Possible stages in the conversion of stigmasterol to cholesterol in insects.

Until recently, the only reported case, (Heed and Kircher, 1965) for which cholesterol was found inadequate to support growth and development was *Drosophila pachea*. A Δ^7-sterol is required by this insect and is obtained in a restricted habitat in association with the Senita cactus whose sterol products are largely of this type. Earlier this year there appeared a report (Kok *et al.*, 1970)

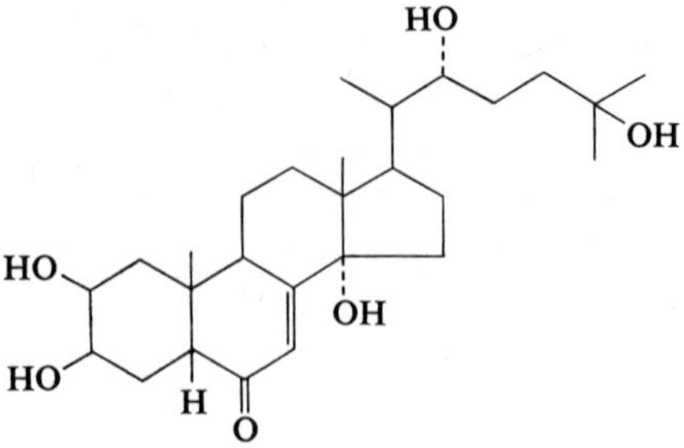

α-Ecdysone[1]

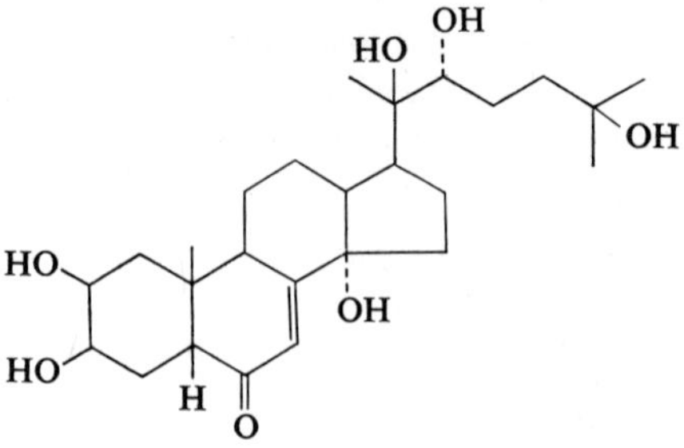

Ecdysterone[2]

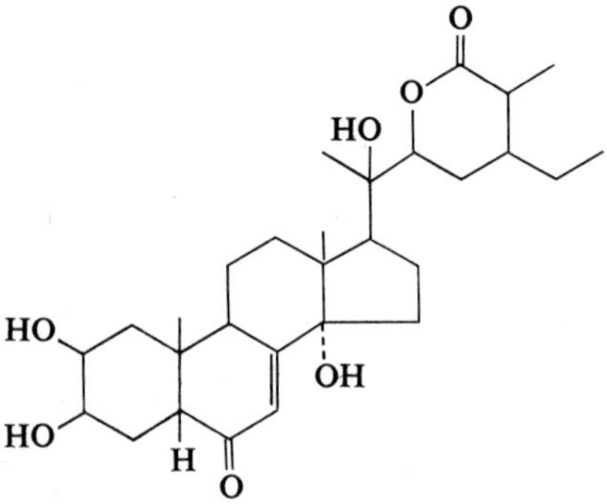

Capitasterone[3]

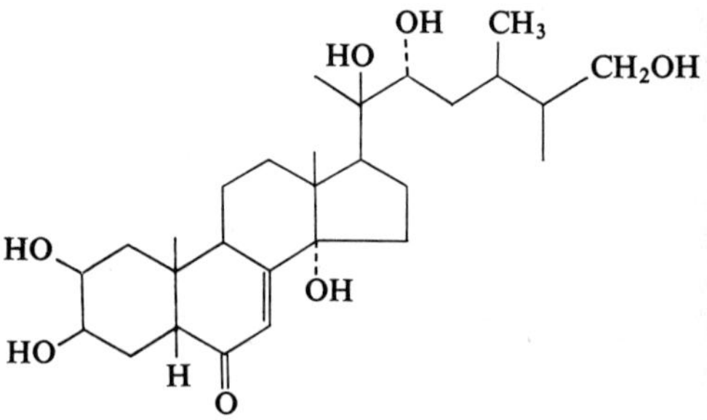

Callinecdysone-B[4]

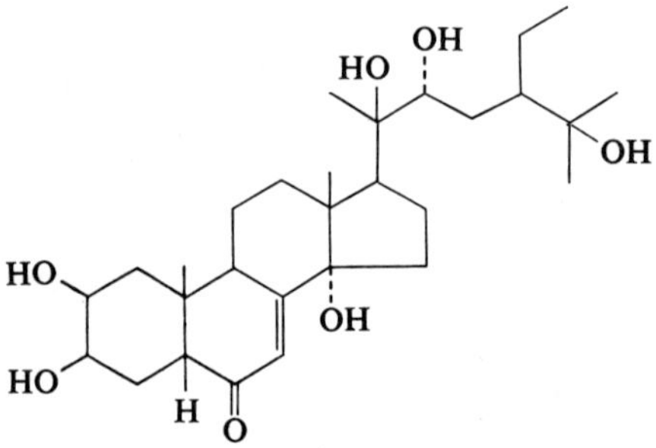

Podecdysone[5]

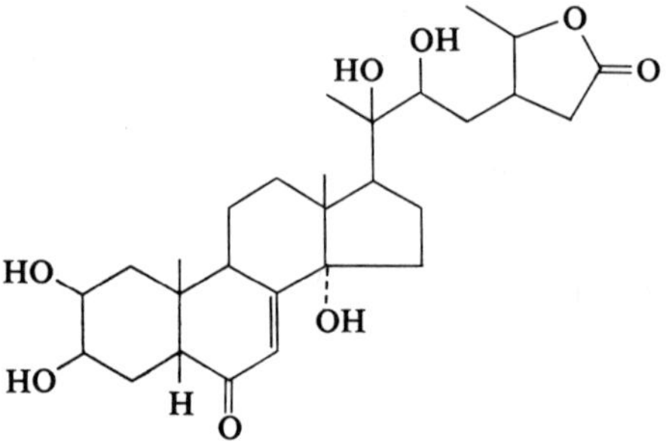

Cyasterone[6]

FIG. 4. Some examples of ecdysones and their sources. Those isolated from plants may have either a cholestane or phytosterol skeleton. Those obtained from animal sources are generally cholestane derivatives but callinecdysone-B is an exception.

References:
[1] From *Bombyx mori*—Butenandt and Karlson (1964); Huber and Hoppe (1965). From bracken fern, *Pteridium aquilinum*—Kaplanis *et al.* (1967). [2] From *B. mori*—Hocks and Wiechert (1966). From crayfish, *Jasus lalandii*—Hampshire and Horn (1966). From *P. aquilinum*—Kaplanis *et al.* (1967). From *Podocarpus elatus* (Taxaceae)—Galbraith and Horn (1966). [3] From *Cyathula capitata* (Amaranthaceae)—Takemoto *et al.* (1968). [4] From soft shell crab, *Callinectes sapidus*—Faux *et al.* (1969). [5] From *P. elatus*—Galbraith *et al.* (1968). [6] From *C. capitata*—Takemoto *et al.* (1967).

of a study of the relationship between the beetle *Xyleborus ferrugineus* and a fungal symbiont of its gut, *Fusarium solani*. It was shown that, like many other insects (Clayton, 1964) this beetle depends upon the symbiotic fungus for its supply of sterol—specifically in this case, for the $\Delta^{5,7}$-sterol ergosterol. In experiments with defined diets and aposymbiotic insects it was found that pupation required the fungal sterol, ergosterol, and that although egg laying and growth could occur in the presence of cholesterol this sterol failed to allow the insects to pupate. However, 7-dehydrocholesterol could replace ergosterol as a dietary factor that allowed completion of the life cycle. The most reasonable explanation for this observation would seem to be that these insects lack the enzymic mechanism for introduction of the Δ^{7}-bond which is a universal feature of the structure of ecdysone, the maturation hormone.

Turning to the role of steroids and terpenoids as hormones in insects, we find, again, considerable evidence for interactions between plants and insects that may be important ecologically. There are now numerous instances of the isolation from various plant species of ecdysones (Fig. 4) (Siddall, 1970) of the C_{27} or C_{29} series which suggest a defensive interference on the part of the plant in the endocrinology of predatory insects, although this remains a matter of controversy (Carlisle and Ellis, 1968; Robbins *et al.*, 1968). This

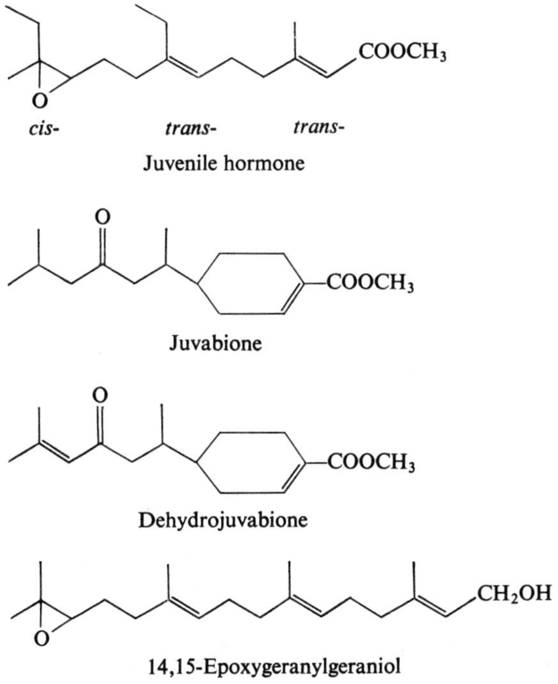

COOCH$_3$
O
cis- *trans-* *trans-*

Juvenile hormone

O
COOCH$_3$

Juvabione

O
COOCH$_3$

Dehydrojuvabione

CH$_2$OH
O

14,15-Epoxygeranylgeraniol

FIG. 5. Juvenile hormone and some analogues.

principle extends to the elaboration by certain plant species of compounds with juvenile hormone activity (Fig. 5) in particular insects (Siddall, 1970; Slama and Williams, 1965; Bowers *et al.*, 1966; Roller and Dahm, 1968) and, as indicated by Pfiffner (1971, p. 95) it has excited considerable interest on the part of the chemical industry as a new approach to insect control. Incidentally, when considering the possible defensive role of plant products with endocrine function in insects one should probably view the question in

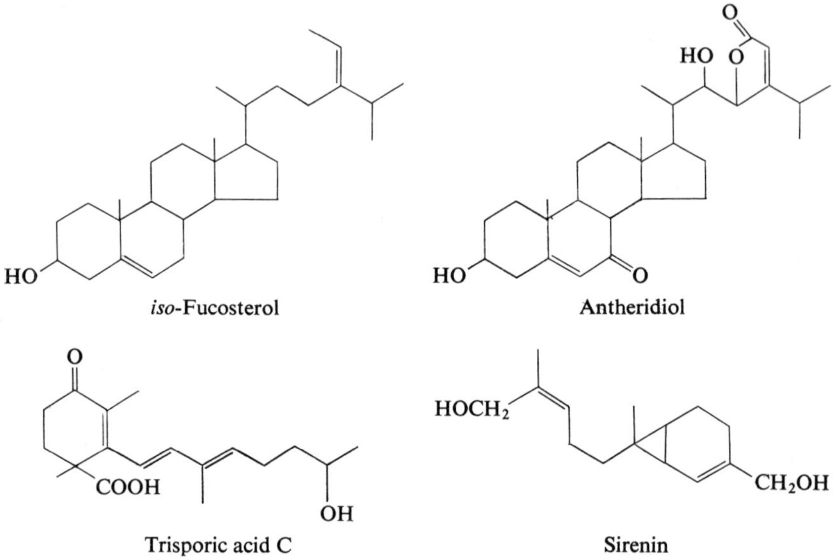

iso-Fucosterol

Antheridiol

Trisporic acid C

Sirenin

FIG. 6. Steroids and terpenoids with reproductive functions in fungi.

an evolutionary context. It may not be entirely coincidental that the ferns, which have recently proved to be so rich a source of "phytoecdysones", and the orthopteran insects apparently experienced a simultaneous surge of evolution during early carboniferous times. If many modern descendants of these primitive insects have metabolic mechanisms by which they can defend themselves against hormonally active plant materials, we should perhaps not assume that this was always true for their ancestors.

In relation to the action of insect hormones, it is interesting to recall that the life cycles of protozoa that inhabit the gut of termites and wood-eating cockroaches are apparently under close endocrine control by the host (Cleveland and Nutting, 1955; Cleveland *et al.*, 1960). Recent evidence for the inhibition of protozoal development by 2,3-epoxygeranylgeraniol (Fig. 5) (Mors *et al.*, 1967) and by a preparation with juvenile hormone activity (Ilan *et al.*, 1969) suggests that these earlier observations reflect the direct action of the host hormones on the symbionts and implies a broad spectrum of action of insect hormones and closely related compounds among lower organisms. In view of this possibility one must hope that the use of insecticides of the juvenile hormone type, especially of those analogues of more bizarre structure, such as the halogenated and phosphonylated derivatives (Siddall, 1970; Pfiffner, 1971), will not be undertaken without exhaustive studies of their possible effects on species which, while less obtrusive than the insects, are nevertheless vital to the maintenance of proper ecological balance.

The sterol requirement of the parasitic fungi, *Phytophthora cactorum* and *Pithium periplocum* which require isofucosterol or sitosterol for reproduction (Elliott *et al.*, 1966) was discussed by Professor Goodwin in 1966. Although this relationship has not been established, it seems likely that this requirement reflects a need for these sterols as starting materials for synthesis of steroids with reproductive significance, such as antheridiol (Fig. 6) (Arsenault *et al.*, 1968). This compound is secreted by the female hyphae of the aquatic fungus *Achlya ambisexualis* and initiates formation of antheridial branches in the male growing close by (McMorris and Barksdale, 1967). We may, in passing, note that the trisporic acids that were also discussed at some length in Goodwin's presentation in 1966 are now established as sexual factors in the fungi, *Blakeslea trispora* and *Mucor mucedo* (Austin *et al.*, 1969), and that sirenin (Machlis *et al.*, 1966), the agent released by female gametes of Allomyces, which exerts a powerful attraction upon the male gametes, has also recently been characterized as a terpene (Nutting *et al.*, 1968). This newly developed area of terpenoids with significance for fungal reproduction has recently been reviewed by Barksdale (1969).

There are many other ways in which terpenoid compounds are involved in the control of population distribution and behavior. Alkaloids and cardiac-active steroid glycosides of certain plants, undoubtedly act as deterrents to grazing animals, and while some of these compounds are known to repell some insects (Fraenkel, 1959) others are exploited at second hand for defense against predators by some insect species whose own physiological processes are immune to their effects (von Euw *et al.*, 1967; Reichstein, 1967; Aplin *et al.*, 1968). The use of heart-active steroids in defense by amphibians has long been recognized and the structure of the most potent naturally occurring neurotoxic agent known, batrachotoxin, of the Colombian arrow frog, has recently been elucidated by Witkop and coworkers (Fig. 7) (Tokuyama *et al.*,

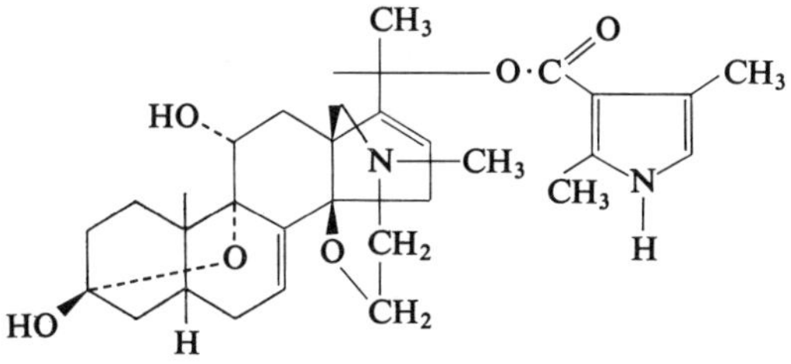

Batrachotoxin

FIG. 7. Batrachotoxin: the arrow poison of the Colombian frog *Phyllobates aurotoenia*.

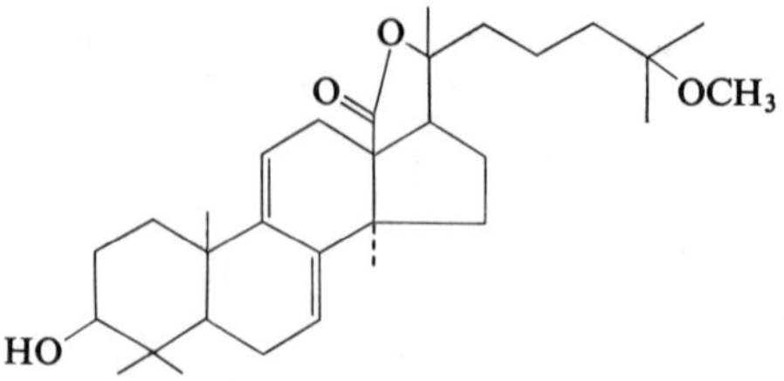

Ternaygenin

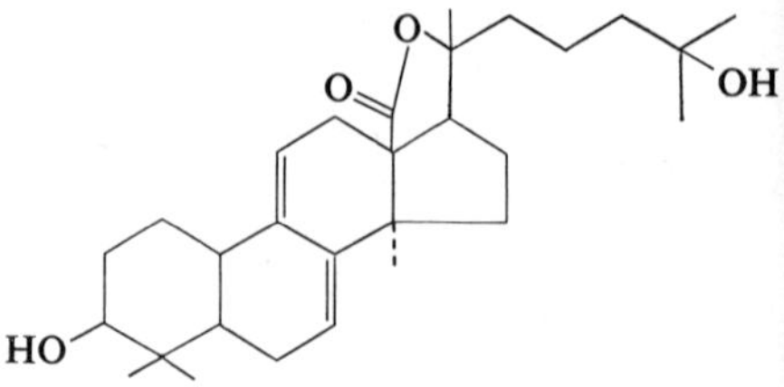

Koellikerigenin

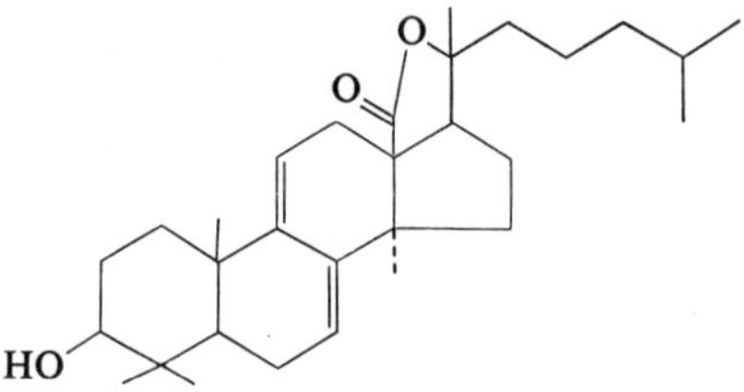

Seychellogenin

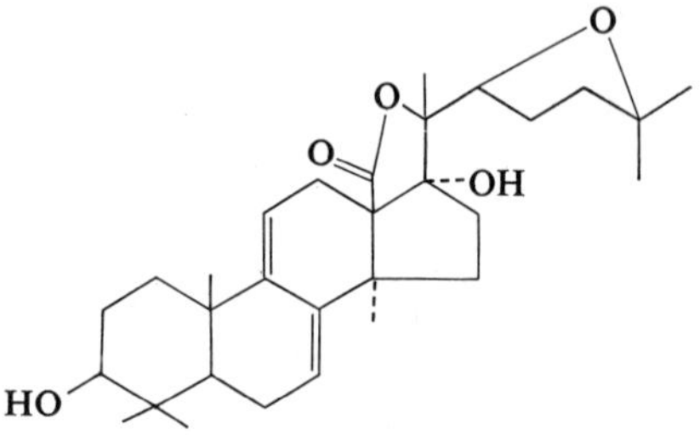

22,25-Oxidholothurinogenin

FIG. 8. Toxins of sea cucumbers.

1969). It is a highly unusual C_{21} steroidal alkaloid. Similarly, recent work by Djerassi and coworkers (Roller *et al.*, 1969) has revealed the structures of several newly recognized toxic agents of the sea cucumbers (Fig. 8). These are apparently based on lanosterol as a precursor and are closely related to the holothurinogenin structure first recognized by Sobotka *et al.* (1964) as occurring in these organisms.

Within recent years, simpler terpenes have gained increasing attention as agents of communication and defense among insects and as attractants to food sources. Several detailed reviews of this field have been published (Weatherston, 1967; Regnier and Law, 1968; Cavill and Robertson, 1965; also M. Beroza, In press) and I will not attempt to cover the same ground again here. I will, however, illustrate the kind of subtle biological interaction that may be involved by reference to one particularly fascinating case, that is the infestation of pine trees by the pine bark beetles of the genus *Dendroctonus*. The initial attack of these beetles upon the tree is stimulated by the terpene exudates from the bark (Fig. 9) (Rudinsky, 1966). The question as to which component is most potent in this respect is in dispute, α-pinene (Renwick and Vite, 1969), and 3-carene (Pitman, 1969) have been proposed and experimental evidence has been more fully described which implicates myrcene (Bedard *et al.*, 1969, 1970). In any event it seems that this substance attracts more females than males, thereby assuring a rapid establishment of the infesting population. As the insects attacking the tree increase in numbers an interesting change in the proportions of males and females attracted to the site takes place. Now, depending upon the species, two to three times more males than females are attracted in a massive attack, but at a certain point this attack becomes much reduced in scale and virtually ceases at those sites that have become most heavily infested. Thus, the chance of fertilization of the initially attracted females is increased and the population growth, that results not only directly from the first wave of attack but also from the egg-laying by female colonizers that mate after having penetrated the bark, is kept within bounds (Renwick and Vite, 1969; Rudinsky, 1969).

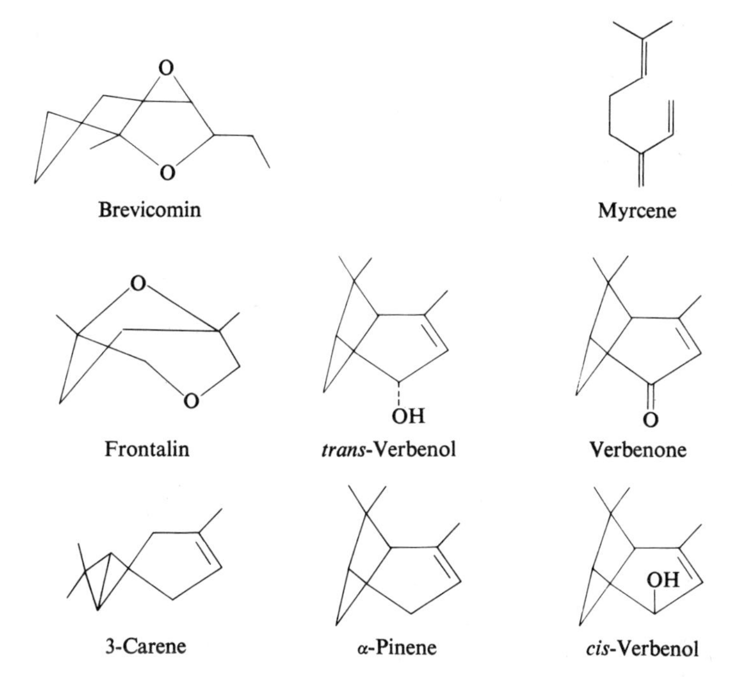

FIG. 9. Compounds implicated in the colonization of pine trees by beetles of the genus *Dendroctonus*.

The details of the mechanisms controlling these events apparently differ somewhat from species to species and are controversial. In one species that has been studied, however (*D. brevicomis*) two attractant pheromones (Fig. 9) are apparently involved. One of them, brevicomin, was identified by Silverstein *et al.* (1968) and a second compound of related structure, designated frontalin on account of its attractive activity in the related species, *D. frontalis*, was isolated by a group at the Boyce Thompson Institute (Kinzer *et al.*, 1969). The latter group of workers have also implicated *trans*-verbenol, a known oxidation product of α-pinene, as an attractant produced by females of *D. frontalis* (Renwick, 1967). The males of both species are reported to produce a third pheromone, verbenone (Renwick, 1967) which acts as a repellent, first to other males, then to other females, and it is suggested that the control of colonization is accomplished through the interaction of these attractive and repulsive agents.

In a third species, *D. pseudotsugae*, which has been subjected to close study (Rudinsky, 1969) the mechanism of "shutting off" the attack seems to be different and even more interesting. It seems that in this species the initial events of the attack and the sex ratios involved are similar to those just described, but the signal for stopping the attack is the sound of the wing-beating of frustrated male insects who cannot find a female. This auditory signal apparently stimulates the female to produce a "masking" pheromone which nullifies the effect of her attractant secretions. It is interesting that *cis*-verbenol, but not *trans*-verbenol is used as a sex attractant and aggregation pheromone by another insect predator of pine trees, the bark beetle, *Ips confusus* (Wood *et al.*, 1968). These are selected examples of what is probably a commonly occurring type of chemical communication among insects in which terpenoid compounds act to control mating and population growth in relation to available food sources. It is interesting to note that synergism often plays an important role in these interactions, as exemplified by the case of *Ips confusus* (Wood *et al.*, 1968) and more recently in the case of the bollweevil, *Anthoromus grandis* (Tumlinson *et al.*, 1969; Zurfluh *et al.*, 1970) (Fig. 10). The male of this species produces two alcohols and two aldehydes and the presence of both alcohols with either of the aldehydes is necessary for attraction of the female.

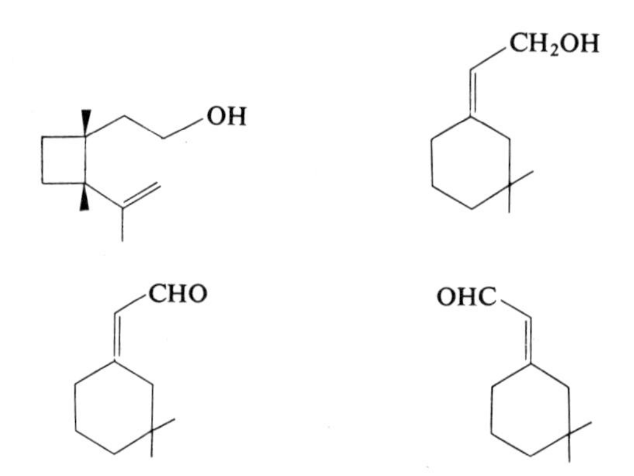

FIG. 10. Sex attractants of the bollweevil *Anthoromus grandis*.

In case one imagines that only the lower organisms utilize these forms of communication one should recall that sexual excitement in female pigs is elicited by the odor of urinary Δ^{16}-C_{19} steroids of the boar (Fig. 11) (Patterson, 1968). It has, of course been known for some time, that estrus and reproductive success in mice are strongly influenced by pheromones (Whitten, 1956; Whitten *et al.*, 1968; Bruce, 1960; Dominic, 1966) and in recent publications evidence has been presented for the existence of a powerful sex attractant in the vaginal secretions of estrogen-stimulated rhesus monkeys (Michael and Keverne, 1970). One is bound to speculate that these physiologically active scents are terpenoid in origin—possibly steroids.

Finally, I would like to draw your attention to an effect of androgenic steroids which has profound behavioral and social significance for all vertebrates, most probably including man. This is in causing sexual differentiation of the brain during a specific phase of embryonic or early postnatal development, the result of which is to lay the foundations, not only of later male

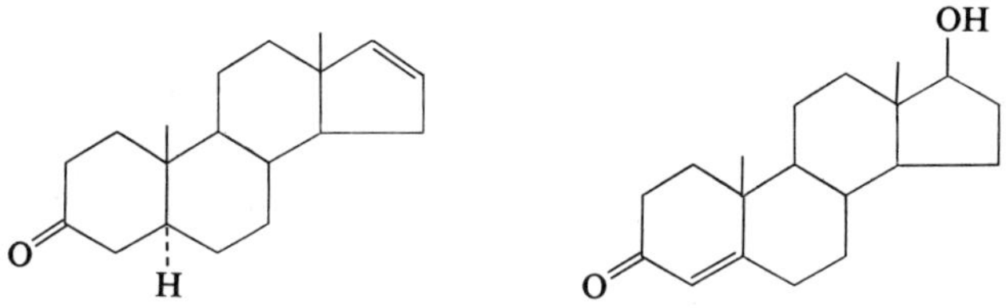

5α-Androst-16-en-3-one

Testosterone

FIG. 11. A Δ^{16}-androstane sex attractant of the boar, and testosterone, whose functions include the sexual differentiation of the brain in mammals.

patterns of sexual behavior, but of those aggressive characteristics which typify maleness of behavior in general. A great deal of behavioral work has been done in these areas, especially with rodents and it has been frequently reviewed (Young *et al.*, 1964; Harris, 1964; Phoenix *et al.*, 1967). I will merely report that recent work in our own laboratory (Clayton *et al.*, 1970) on the biochemical events underlying this process strongly suggest that it involves influences upon RNA metabolism in specific highly localized brain areas known from the work of others to be involved in sexual and sex-related behavior. This surely is one of the most striking examples of the intervention of a terpenoid compound at a point which directly interrelates the genetic endowment of the individual with his behavioral role as a member of the species.

II. Importance of Pathways of Terpenoid Biosynthesis

I have touched quite superficially on a considerable number of topics to emphasize the wide-ranging significance of the terpenoid compounds at all levels of biological organization: intracellular, intercellular, interorganismic and interspecific. I want to devote the remainder of this article to a brief discussion of some features of the pathway of terpene biosynthesis that seem to make it specially suitable for providing these numerous compounds that have such diverse functions.

First of all we should note that all the functions that we have discussed involve a high degree of structural specificity for the compounds in question.

Whatever the particular biological function we are concerned with, whether it is the stimulation of nervous pathways via olfactory receptors, the initiation of a developmental process by catalysing the release of genetic information, or the inhibition of Na^+-K^+-activated ATPase; in a particular animal species, it is clear that there is only a narrow range of molecular structures capable of eliciting any one of these effects. We are in fact dealing with an area of molecular biology which is fundamental to the organization of living systems despite the fact that it has been overshadowed in recent years by other areas of biochemistry to which this title has been exclusively applied. In a recently published article Kendrew (1968) examines the development and current status of molecular biology with particular emphasis upon the "interplay between information and conformation, between genotype and phenotype, between nucleic acid and protein". He observes that, "it has become apparent that both the genetic apparatus and the metabolic chains require a system of control; genes must be switched on and off in sympathy with the needs of the individual cell for particular proteins, and metabolic pathways must be closed down or rendered more active in accordance with demand for their products", and he goes on to indicate that these controls are dependent upon subtle conformational changes of genetic repressors and allosteric enzymes. In discussing controlling factors at the other extreme of biological complexity, Darnell (1970) has pointed out the fundamental role of the ecosystem as a unit of "co-evolution" and has emphasized the importance of communication within the ecosystem between members of the same and different species. Again, in a recent article, Brown *et al.* (1970) have pointed out that the field of chemical interactions between species has become so highly developed that two new terms are called for, the one "allomones" to connote a chemical messenger released by one species and eliciting a reaction on the part of a second species that is favorable to the first (releasing) species; and the second "kairomones" to refer to such a substance the *recipient* of which, however, rather than the releaser, gains the advantage. After noting various examples, some of which have also been cited here, they conclude that they "only emphasize the versatility with which evolution has exploited the use of chemical agents as messages relating to one another the parts of an organism, individual organisms of a species and different species of a community". Clearly, the terpenoid compounds play a major role in all of these processes which essentially involve the transmission of biologically significant information, and the evolutionary importance of the pathway by which they are formed almost certainly rests upon the facility with which it yields molecules of high structural specificity.

Let us consider this point a little further. In general terms the number of possible conformations of a molecule can be regarded as an exponential function of the number of conformationally significant bonds. If one equates the structural specificity of a molecule comprising a given number of carbon atoms with the inverse of its number of conformational possibilities, one can easily calculate that the structural specificity of an open chain monofunctional terpenoid with 10 carbon atoms, of the usual isoprenoid structure:

$$\begin{matrix} CH_3 \\ CH_3 \end{matrix}\!\!>\!C{=}CH{-}CH_2{-}CH_2{-}\overset{\overset{CH_3}{|}}{C}{=}CH{-}CH_2{-}X$$

is four orders of magnitude greater than that of the corresponding saturated monofunctional C_{10} compound: $CH_3{-}(CH_2)_9{-}X$. It is interesting to compare the terpenoids with a series of chemical communication agents synthesized (presumably) via the fatty acid pathway. Such a family of compounds comprises the majority of insect sex attractants of known structure (Fig. 12) (Regnier and Law, 1968). It can be seen that specificity depends heavily upon the presence and position of one or more double bonds and upon their geometry,

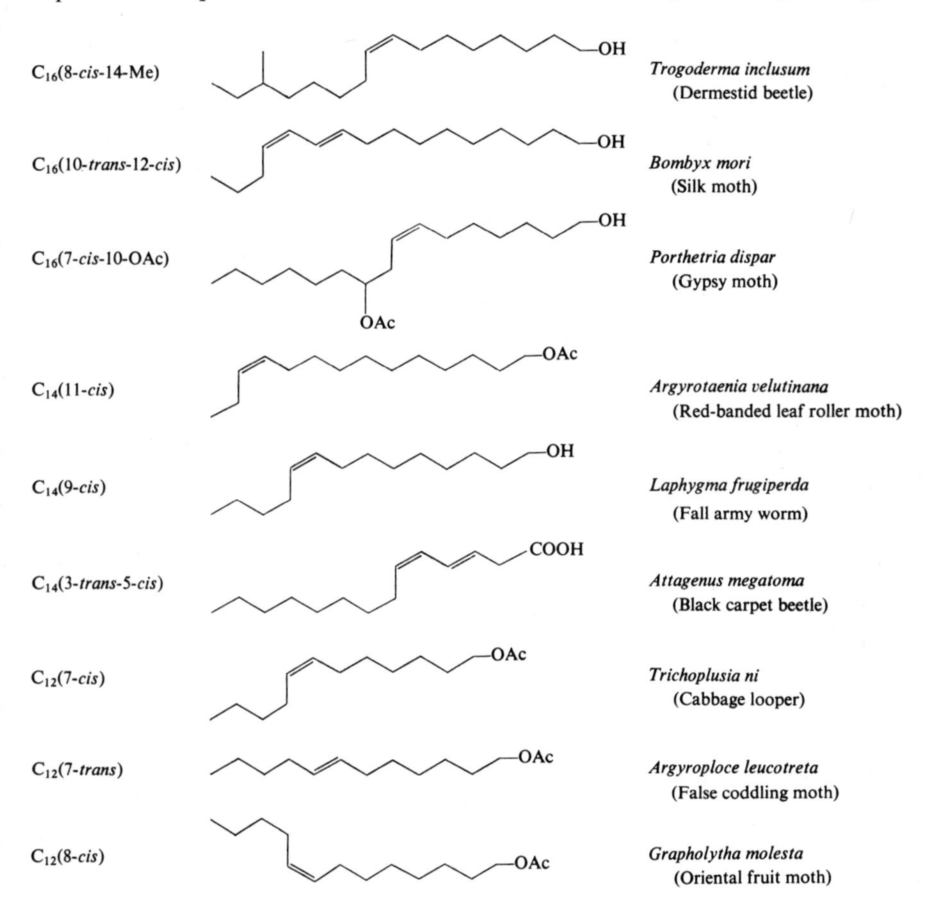

FIG. 12. Insect sex attractants with structures related to the fatty acids.

and upon the presence of substituent groups. These are reminiscent of the characteristics of terpenoid compounds and, indeed, in one case the resulting compound is (formally) a partial terpenoid, though its structure makes it seem unlikely that it arises from mevalonic acid.

The simple terpenoid structures may be oxidized selectively or modified in other ways, to increase their informational content. In particular, they may be converted to cyclic compounds in which the structural specificity is maximized. Functional groups of informational significance may also be introduced into the cyclic structures, thus fitting them for specific regulatory functions—as in the case of the steroid hormones.

It is true that the so-called acetate pathway of biosynthesis (Birch, 1957; Richards and Hendrickson, 1964) that operates in plants generates compounds of high structural specificity comparable to that of many cyclic terpenes and steroids. It is still unclear how this pathway relates to the pathway of fatty acid synthesis, but in any case it seems in one important respect to be less versatile and flexible than the terpenoid pathway, in that it most frequently yields products bearing multiple oxygen functions. These functional groups contribute to the polarity and reactivity of these products and no doubt predetermine their volatility and the types of molecular interactions that they can enter into. The terpenoid pathway, on the other hand, primarily yields hydrocarbon or monofunctional structures the derivatives of which encompass a wider range of chemical characteristics.

From such considerations, one must conclude that the terpenoid pathway of biosynthesis is the most versatile and economical route to compounds capable of carrying significant biological information that is exploited in control mechanisms in an almost unlimited range of contexts. The most striking feature of terpenoid biochemistry, especially in relation to the formation of cyclized derivatives, is, of course, the close analogy that it bears to the organic chemistry of this series of compounds. This aspect of terpenoid biochemistry has often been stressed (Ruzicka, 1953, 1956; Cornforth, 1959; Clayton, 1965; Stork and Burgstahler, 1955; Eschenmoser *et al.*, 1955; Oehlschlager and Ourisson, 1967; Johnson, 1967; van Tamelen, 1968) and has been so extensively and repeatedly reviewed (Richards and Hendrickson, 1964; Ruzicka, 1956; Cornforth, 1959; Clayton, 1965) that another survey along these lines would be redundant. The most detailed and influential of these bio-organic chemical rationalizations is the well known postulated derivation of the polycyclic triterpenes, including lanosterol, from squalene (Stork and Burgstahler, 1955; Eschenmoser *et al.*, 1955). I would therefore like to close, by summarizing briefly some of the results of recent work on the cyclization of squalene oxide analogues, carried out both in our own laboratories at Stanford, in collaboration with van Tamelen and coworkers, and in Corey's laboratory at Harvard, which reinforcę the view, so often expressed or implied, that the biochemistry of the terpenoids may be considered as an extension of their organic chemical properties, and suggests that this, together with its great versatility, is the underlying reason for the evolutionary selection of this pathway.

III. Cyclization of Squalene

The now well established biological relationship between squalene, squalene 2,3-oxide (Corey and Russey, 1966; van Tamelen *et al.*, 1966; Yamamoto *et al.*, 1969) and two very similar cyclization products, lanosterol and cycloartenol, (Eppenburger *et al.*, 1969; Hewlins *et al.*, 1969) is shown in Fig. 13. The study of this biological role of squalene oxide was prompted by the stereochemical analogy between the initial phases of the chemical cyclization (van Tamelen *et al.*, 1966) of this compound and its enzymic cyclization, as well as the similar close analogy with the steric course of chemical cyclization of the corresponding oxide of a farnesol derivative (van Tamelen *et al.*, 1963) (Fig. 14). In our studies of the behavior of various synthetic analogues of squalene 2,3-oxide in the presence of the rat liver cyclase that normally converts this compound into lanosterol, we have attempted to examine discrete phases of the cyclization process by making selective modifications in the structure of the substrate.

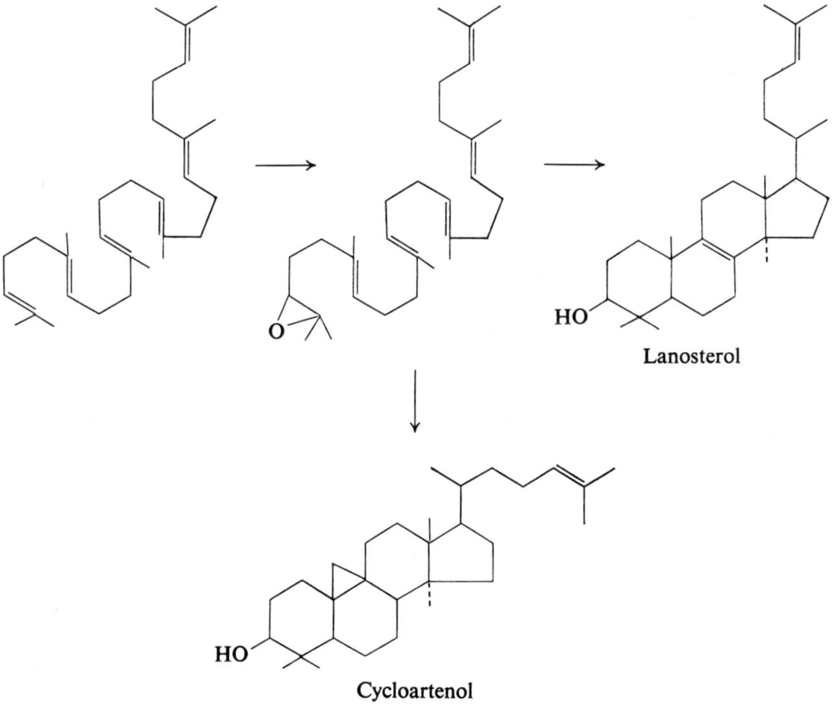

Lanosterol

Cycloartenol

FIG. 13. Biological relationship between squalene, squalene 2,3-oxide and the sterol cyclization products, lanosterol (in animal tissues) and cycloartenol (in plants).

I want to focus primarily on those experiments that bear first of all on the initiation of cyclization and secondly upon the rearrangement phase of the reaction. The importance of the isopropylidene carbon for the initiation of acid-catalyzed cyclizations in a series of geranic acid analogues was established long ago in experiments by Schinz and coworkers (Helg and Schinz, 1952; Gamboni *et al.*, 1954; Helg *et al.*, 1956) (Fig. 15). The result, that indicated that the gem-dimethyl group of naturally occurring terpenes was essential for cyclization, was interpreted as a clarification of the characteristic structure of terpenoid compounds (Helg *et al.*, 1956). We asked a similar question as to the importance of the tertiary center at C-2 of squalene 2,3-oxide in its enzymic cyclization and with strikingly similar results (Fig. 16). No cyclization could be detected with the bisdesmethyl compound and of the two geometrical isomers of the monomethyl analogue, only one, that in which the terminal methyl is *trans*- to the main chain, was cyclized and this in extremely low yield (Clayton *et al.*, 1968). On the other hand, when the methyl-ethyl derivatives were tried (Corey *et al.*, 1968; Crosby *et al.*, 1969) the efficiency of cyclization was restored to a level approaching that for the natural substrate, but now the cyclization only of the *trans*-ethyl compound took place (Crosby *et al.*, 1969).

The results indicate the overriding importance of the tertiary center at C-2 in the initiation of the cyclization but also point to the operation of steric factors on the β-surface of the developing cyclized product that work against

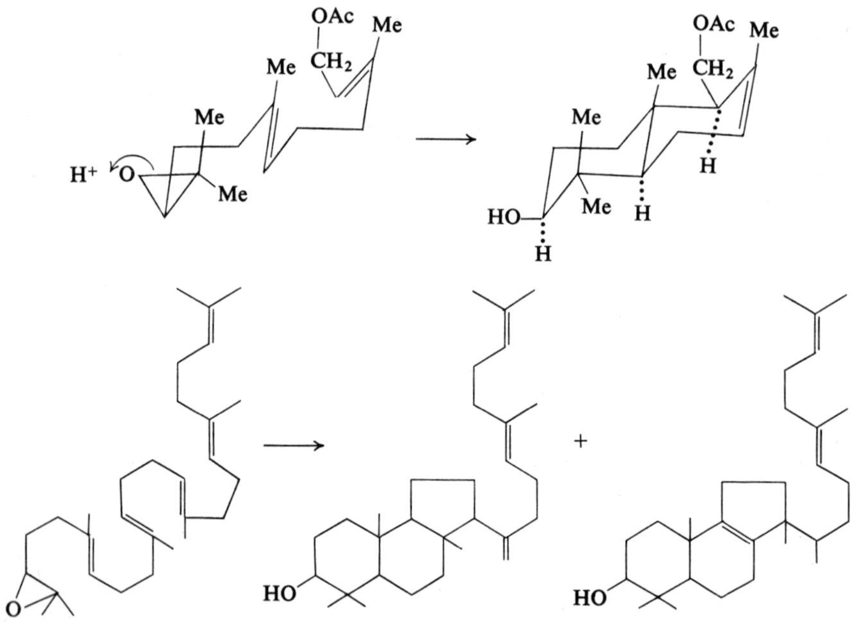

OAc
Me
Me CH_2
Me
H^+ O
Me
OAc
Me
Me CH_2
Me
H
HO
Me H
H
HO
HO
O

FIG. 14. Chemical cyclization reactions of terpenoid oxides that provide a partial analogy with the enzymic cyclization of squalene 2,3-oxide.

FIG. 15. Early studies of the significance of the isopropylidene group in the acid-catalysed cyclization of terpenes. Geranic acid (1) and its 3-desmethyl analogue (2), were cyclized but the structures (3–5) were not.

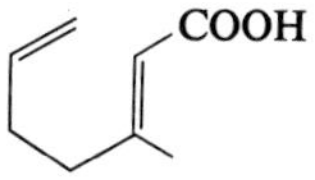

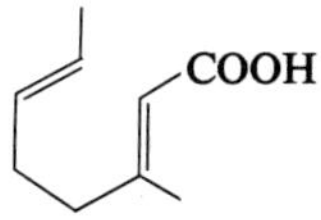

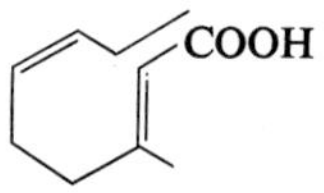

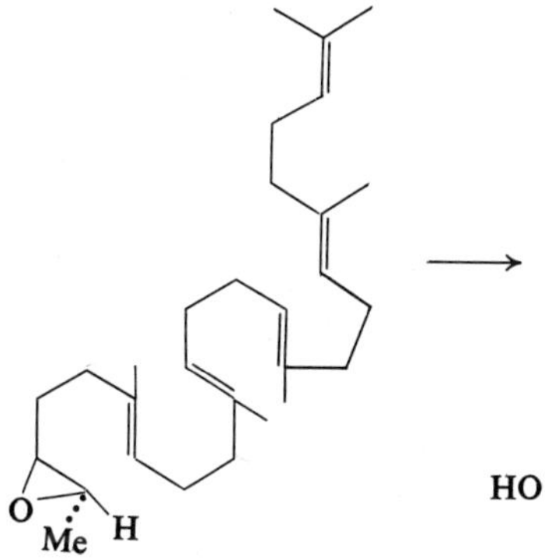

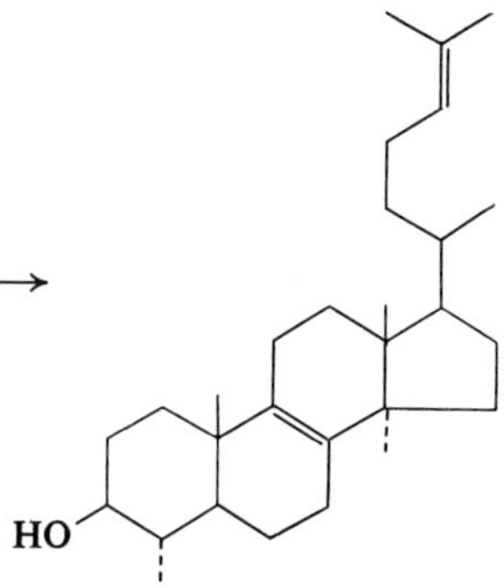

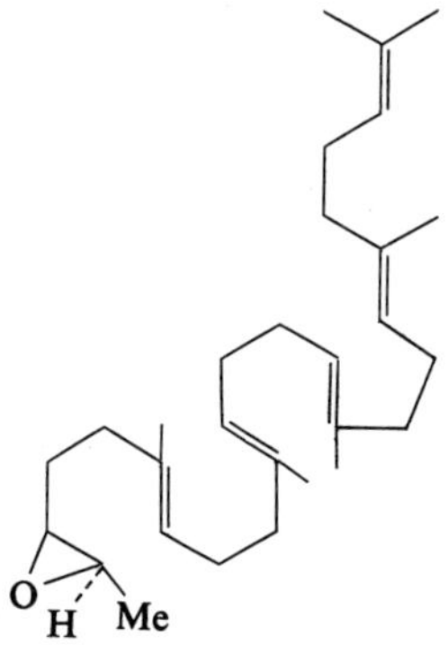

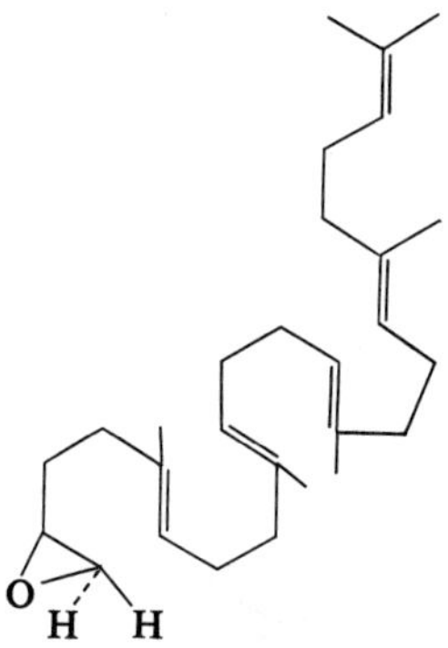

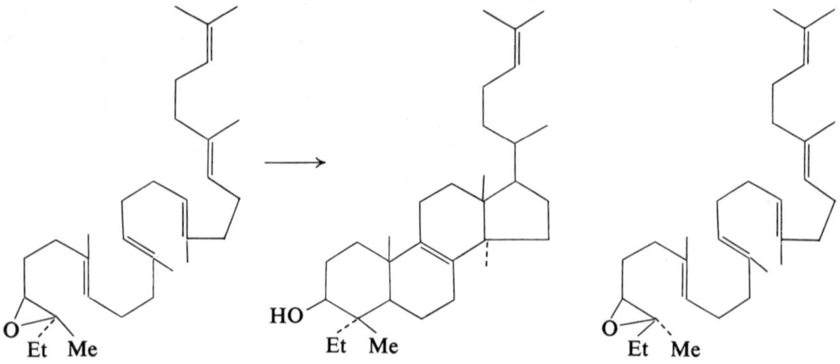

(5) (6) (7)

FIG. 16. Effects of varying the alkyl substitution at C_2 of squalene 2,3-oxide upon the course of enzymic cyclization. The disecondary oxide (1), with a *trans*-methyl substituent is poorly cyclized to the 4α-methyl sterol (2), but its *cis*-methyl analogue (3) and the bisdesmethyl oxide (4) fail to cyclize. The *trans*-ethyl, *cis*-methyl substrate (5) gives (6) in good yield, but the epimer (7) does not cyclize.

cyclization. This could imply an unfavorable interaction between the 4β-alkyl substituent and the 10β-methyl, an interaction within the active site of the enzyme or both factors operating simultaneously.

In considering the later phases of the cyclization process one may ask the general question whether the reaction is indeed properly to be regarded as a sequential series of steps, and if so, which parts of the process are concerted and which can occur independently of each other. Results obtained in our laboratory (van Tamelen *et al.*, 1967) (Fig. 17) gave the first indication of the sequential nature of the reaction, and an experiment reported by Corey *et al.* (1968) showed in the case of a 10,15-bisdesmethyl substrate, that cyclization could go to completion in the absence of rearrangement. In another experiment in Corey's laboratory (Corey *et al.*, 1969) an oxide substrate with a conjugated diene terminal portion of the chain gave an unrearranged enzymic product in which stabilization had been achieved presumably by capture of OH^-. Again,

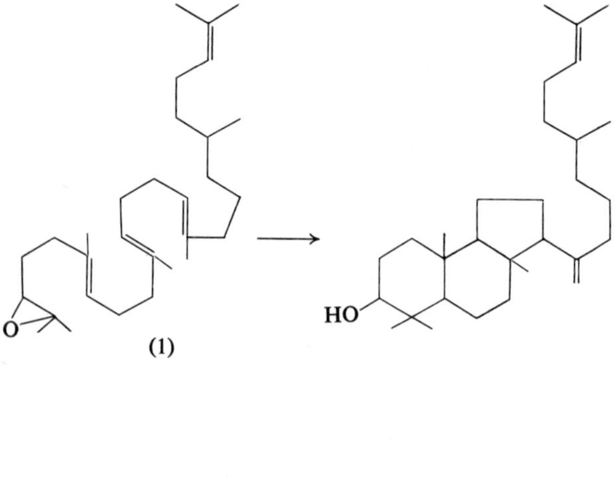

(1)

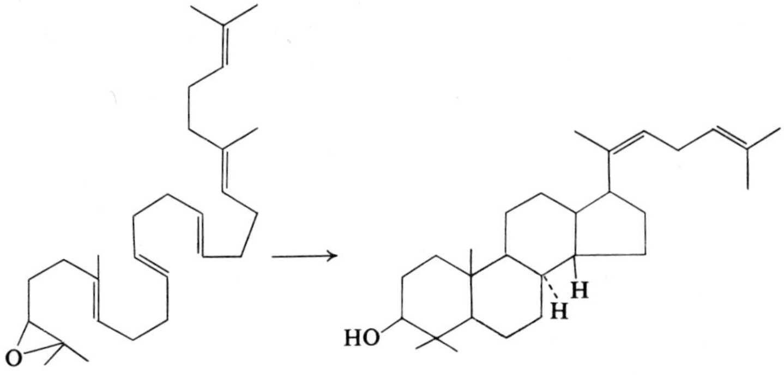

FIG. 17. Experiments with the 18,19-dihydroanalogue (1) of squalene 2,3-oxide (van Tamelen *et al.*, 1967) and with the 10,15-bisdesmethyl analogue (2) (Corey *et al.*, 1968) which indicate the sequential nature of the enzymic cyclization and rearrangement of squalene 2,3-oxide. The 10-desmethyl and 15-desmethyl analogues (3) and (4), respectively, yield rearranged products.

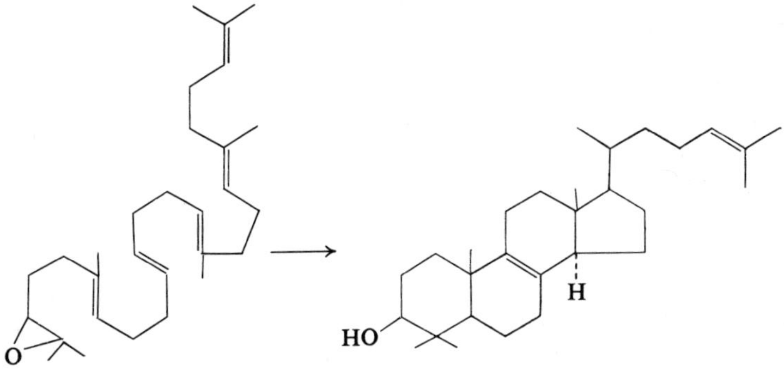

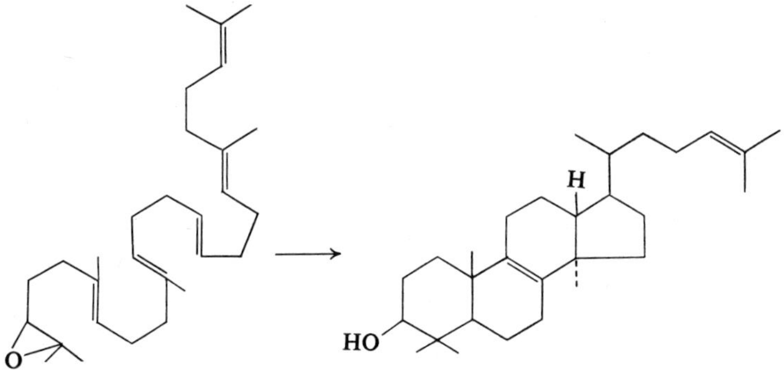

cyclization divorced from rearrangement was demonstrated and while the significance of this finding is uncertain, it may have relevance to a recent proposal by Cornforth (Cornforth, 1968).

The experiment with the 10,15-bisdesmethyl substrate (Corey *et al.*, 1968) shows that these methyl groups are important for rearrangement to proceed. However, analysing this problem further, in our own laboratory, we have been able to show that only one (either one) of the methyl groups at C_{10} or C_{15} (van Tamelen *et al.*, 1968, 1970) is required for rearrangement to take place, though in these cases the yields of product are lower than of the normal product from squalene 2,3-oxide. The results can be rationalized (Hanzlik, 1970) on the basis of the relief of unfavorable non-bonding interactions in a transition state that has the configuration postulated according to the Stork–Eschenmoser scheme, (Stork and Burgstahler, 1955; Eschenmoser *et al.*, 1955) without invoking any specific form of enzymic intervention. It is nevertheless probable that interactions with the enzyme active site intrude upon the intra-molecular steric interactions within the substrate molecule either to facilitate or limit rearrangement. It was noted by Oehlschlager and Ourisson (1967) in the 1966 symposium of this society, in discussing the diterpenes, that "Obviously, all of the rearrangement pathways exhibited in nature do not proceed to products of the greatest thermodynamic stability." Many examples might be cited from the triterpene series to substantiate this view. It remains true, nevertheless, on the basis of the evidence at hand that stereoelectronic factors inherent in the substrate molecule profoundly influence its reactivity and in a manner that is readily rationalized, assuming that the primary role of the enzyme (apart from acting as a proton donor) is to impose a specified conformation upon the substrate.* Perhaps, this conformational role having been discharged, the enzyme fulfills what is essentially a "permissive function" in which rearrangements proceed spontaneously under thermodynamic control up to a point at which a specific interaction with the active site intervenes.

However this may be, all these experiments further strengthen the often expressed view that the enzymic transformations of terpenoid compounds can be largely regarded as an exploitation of their inherent chemical reactivities. They make all the more plausible our view of this pathway as nature's most economical and evolutionarily favored route to the generation of a wide range of compounds of high structural specificity (Clayton, 1970). This conclusion as to the place of the terpenoids in biochemical and biological

* We have speculated elsewhere (Clayton, 1970) that one function of the active site of the cyclase may be to provide a hydrophobic environment which shields the cyclizing substrate against interaction with ions in the medium. While this may be the case, it is interesting that a non-enzymic acid-catalysed polyene cyclization which leads to a tricyclic product has now been shown unequivocally to take place by a concerted mechanism (Johnson and Schaaf, 1969). The necessity for a protected environment in which to achieve such a transformation, even in the more complex case of the cyclization and rearrangement of squalene 2,3-oxide, must therefore be questioned.

evolution is in complete accord with the concept of "biochemical predestination", recently discussed by Kenyon and Steinman (1969). These authors point to the conceptual gap which, tantalizingly, still exists in our attempts to understand the evolutionary link between the inherent chemical reactivities of molecules and their exploitation by enzymes in organized metabolic systems. Calvin (1969) has recently noted evidence for the prebiotic formation of isoprenoids which may require a reassessment of the view that the finding of such compounds in ancient geological specimens is evidence for the presence of early forms of life. Again, this observation stimulates our curiosity about the precise evolutionary relationship between the enzymic and non-enzymic reactivities of the terpenes and tempts us to wonder whether these relatively simple but informationally rich compounds, probably present from an early stage of chemical evolution, have contributed from the beginning to the organization and integration of living material.

ACKNOWLEDGEMENTS

Work from the author's laboratory that is cited in this article has been supported by U.S.P.H.S. Grants Nos. GM-12493, GM-10421, HD-00801, by grants in aid from the American Heart Association and, between 1960 and 1965, by an Established Investigatorship of the American Heart Association.

REFERENCES

Aaronson, S. (1964). *J. Gen. Microbiol.* **37**, 225.
Andreasen, A. A. and Stier, T. J. B. (1953). *J. Cell. Comp. Physiol.* **41**, 23.
Aplin, R. T., Benn, M. H. and Rothschild, M. (1968). *Nature, Lond.* **219**, 747.
Arsenault, G. P., Biemann, K., Barksdale, A. W. and McMorris, T. C. (1968). *J. Am. chem. Soc.* **90**, 5635.
Austin, D. J., Bu'Lock, J. D. and Gooday, G. W. (1969). *Nature, Lond.* **223**, 1178.
Barksdale, A. W. (1969). *Science, N.Y.* **166**, 831.
Bedard, W. D., Tilden, P. E., Wood, D. L., Silverstein, R. M., Brownlee, R. G. and Rodin, J. O. (1969). *Science, N.Y.* **164**, 1284.
Bedard, W. D., Silverstein, R. M. and Wood, D. L. (1970). *Science, N.Y.* **167**, 1638.
Beroza, M. (ed.) (1971). "Chemicals Controlling Insect Behavior", Academic Press, New York and London. (In press.)
Birch, A. J. (1957). *Fortschr. Chem. Org. Naturst.* **14**, 186.
Bowers, W. S., Fales, H. M., Thompson, M. J. and Uebel, E. C. (1966). *Science, N.Y.* **154**, 1024.
Brown, W. L. Jr., Eisner, T. and Whittaker, R. H. (1970). *BioScience* **20**, 21.
Bruce, H. M. (1960). *J. Reprod. Fertil.* **1**, 96.
Butenandt, A. and Karlson, P. (1964). *Z. Naturf.* **9b**, 389.
Calvin, M. (1969). "Chemical Evolution, Molecular Evolution Towards the Origin of Living Systems on the Earth and Elsewhere", Oxford University Press, New York and Oxford.
Carlisle, D. B. and Ellis, P. E. (1968). *Science, N.Y.* **159**, 1472.
Cavill, G. W. K. and Robertson, P. L. (1965). *Science, N.Y.* **149**, 1337.
Clark, A. J. and Bloch, K. (1959). *J. biol. Chem.* **234**, 2578.
Clayton, R. B. (1964). *J. Lipid Res.* **5**, 3.
Clayton, R. B. (1965). *Q. Rev. (Lond.)* **19**, 168, 201.

Clayton, R. B. (1970). *In* "Chemical Ecology" (E. Sondheimer and J. B. Simeone, eds), p. 235, Academic Press, London and New York.
Clayton, R. B., van Tamelen, E. E. and Nadeau, R. G. (1968). *J. Am. chem. Soc.* **90**, 820.
Clayton, R. B., Kogura, J. and Kraemer, H. C. (1970). *Nature, Lond.* **226**, 810.
Cleveland, L. R. and Nutting, W. L. (1955). *J. Exp. Zool.* **130**, 485.
Cleveland, L. R., Burke, A. W. and Karlson, P. (1960). *J. Protozool.* **7**, 229.
Cole, R. J. and Krusberg, L. R. (1968). *Life Sci.* **7**, 713.
Conner, R. L. and Van Wagtendonk, W. J. (1955). *J. gen. Microbiol.* **12**, 31.
Conner, R. L. and Ungar, F. (1964). *Expl. Cell Res.* **36**, 134.
Corey, E. J. and Russey, W. E. (1966). *J. Am. chem. Soc.* **88**, 4750.
Corey, E. J., Ortiz de Montellano, P. R. and Yamamoto, H. (1968). *J. Am. chem. Soc.* **90**, 6254.
Corey, E. J., Lin, K. and Yamamoto, H. (1969). *J. Am. chem. Soc.* **91**, 2132.
Corey, E. J., Lin, K. and Jautelat, M. (1968). *J. Am. chem. Soc.* **88**, 2727.
Cornforth, J. W. (1959). *J. Lipid Res.* **1**, 3.
Cornforth, J. W. (1968). *Angew. Chem.* (Intl. Edn.) **7**, 903.
Crosby, L. O., van Tamelen, E. E. and Clayton, R. B. (1969). *Chem. Commun.* 532.
Darnell, R. M. (1970). *Am. Zool.* **10**, 9.
Dominic, C. J. (1966). *J. Reprod. Fertil.* **11**, 407, 415.
Draper, H. H. and Saari Csallany, A. (1969). *Fedn. Proc.* **28**, 1690.
Elliott, C. G., Hendrie, M. R., Knights, B. A. and Parker, W. (1964). *Nature, Lond.* **203**, 427.
Elliott, C. G., Hendrie, M. R. and Knights, B. A. (1966). *J. Gen. Microbiol.* **42**, 425.
Eppenberger, U., Ourisson, G. and Hirth, L. (1969). *Europ. J. Biochem.* **8**, 180.
Eschenmoser, A., Ruzicka, L., Jeger, O. and Arigoni, D. (1955). *Helv. chim. Acta* **38**, 1890.
Faux, A., Horn, D. H. S. and Middleton, E. J. (1969). *Chem. Commun.* 175.
Fraenkel, G. S. (1959). *Science, N.Y.* **129**, 1466.
Fraenkel, G., Reed, J. A. and Blewett, M. (1941). *Biochem. J.* **35**, 712.
Galbraith, M. N. and Horn, D. H. S. (1966). *Chem. Commun.* 905.
Galbraith, M. N., Horn, D. H. S., Porter, Q. N. and Hackney, R. J. (1968). *Chem. Commun.* 971.
Gamboni, G., Schinz, H. and Eschenmoser, A. (1954). *Helv. chim. Acta* **37**, 964.
Gilbert, J. J. (1967). *Proc. Natn. Acad. Sci. U.S.A.* **57**, 1218.
Gilbert, J. J. (1968). *Physiol. Zool.* **41**, 14.
Gilbert, J. J. and Thompson, G. A., Jr. (1968). *Science, N.Y.* **159**, 734.
Goldsmith, T. H. and Warner, L. T. (1964). *J. gen. Physiol.* **47**, 433.
Goldsmith, T. H., Barker, R. J. and Cohen, C. F. (1964). *Science, N.Y.* **146**, 65.
Goodwin, T. W. (1967). *In* "Terpenoids in Plants" (Pridham, J. B., ed.), p. 1, Academic Press, New York and London.
Hampshire, F. and Horn, D. H. S. (1966). *Chem. Commun.* 37.
Hanzlik, R. P. (1970). Ph.D. Thesis, Stanford University.
Harris, G. W. (1964). *Endocrinology* **75**, 627.
Heed, W. B. and Kircher, H. W. (1965). *Science, N.Y.* **149**, 758.
Heftmann, E. (1963). *Ann. Rev. Pl. Physiol.* **14**, 225.
Heftmann, E. (1967). *Lloydia* **30**, 209.
Heftmann, E. (1968). *Lloydia* **31**, 293.
Helg, R. and Schinz, H. (1952). *Helv. chim. Acta* **35**, 2406.
Helg, R., Zobrist, F. Lauchenauer, A., Brack, K., Caliezi, A., Stauffacher, D., Zweifel, E. and Schinz, H. (1956). *Helv. chim. Acta* **39**, 1269.
Hendrix, J. W. (1964). *Science, N.Y.* **144**, 1028.

Hewlins, M. J. E., Ehrhardt, J. D., Hirth, L. and Ourisson, G. (1969). *Europ. J. Biochem.* **8**, 184.

Hieb, W. F. and Rothstein, M. (1968). *Science, N.Y.* **160**, 778.

Hocks, P. and Wiechert, R. (1966). *Tetrahedron Letters* 2989.

Holz, G. G., Wagner, B., Erwin, J. A., Britt, J. J. and Bloch, K. (1961a). *Comp. Biochem. Physiol.* **2**, 202.

Holz, G. G., Erwin, J. A. and Wagner, B. (1961b). *J. Protozool.* **8**, 297.

Holz, G. G., Erwin, J. A. and Wagner, B. (1962). *J. Protozool.* **9**, 359.

Huber, R. and Hoppe, W. (1965). *Chem. Ber.* **98**, 2403.

Hutner, S. H. and Holz, G. G. (1962). *Ann. Rev. Microbiol.* **16**, 189.

Ilan, J., Ilan, J. and Ricklis, S. (1969). *Nature, Lond.* **224**, 179.

Ilan, J., Ilan, J. and Patel, N. (1970). *J. biol. Chem.* **245**, 1275.

Johnson, W. S. (1967). *Trans. N.Y. Acad. Sci.* **29**, 1001.

Johnson, W. S. and Schaaf, T. K. (1969). *Chem. Commun.* 611.

Kaplanis, J. N., Monroe, R. E., Robbins, W. E. and Louloudes, S. J. (1963). *Ann. Entomol. Soc. Am.* **56**, 198.

Kaplanis, J. N., Thompson, M. J., Robbins, W. E. and Bryce, B. M. (1967). *Science, N.Y.* **157**, 1436.

Karlson, P. and Hoffmeister, H. (1963). *Z. physiol. Chem.* **331**, 298.

Kendrew, J. C. (1968). *In* "Structural Chemistry in Molecular Biology" (A. Rich and N. Davidson, eds.), p. 187, Freeman and Co., New York.

Kenyon, D. H. and Steinman, G. (1969). "Biochemical Predestination", McGraw-Hill, New York.

Kinzer, G. W., Fentiman, A. F., Page, T. F. Jun, Foltz, R. L., Vite, J. P. and Pitman, G. B. (1969). *Nature, Lond.* **221**, 477.

Kok, L. T., Norris, D. M. and Chu, H. M. (1970). *Nature, Lond.* **225**, 661.

Lasser, N. L. and Clayton, R. B. (1966). *J. Lipid Res.* **7**, 413.

Lasser, N. L., Edwards, A. M. and Clayton, R. B. (1966). *J. Lipid Res.* **7**, 403.

Lederer, E. (1969). *Q. Rev. (Lond.)* **23**, 453.

Levinson, Z. H. (1955). *Riv. Parassitol.* **16**, 183.

Levinson, Z. H. (1962). *J. Insect Physiol.* **8**, 191.

Machlis, L., Nutting, W. H., Williams, M. W. and Rapoport, H. (1966). *Biochemistry* **5**, 2147.

McMorris, T. C. and Barksdale, A. W. (1967). *Nature, Lond.* **215**, 320.

Meyer, F., Kimura, S. and Mueller, J. F. (1966). *J. biol. Chem.* **241**, 4224.

Michael, R. P. and Keverne, E. B. (1970). *Nature, Lond.* **225**, 84.

Mors, W. B., dos Santos fo, M. F., Monteiro, H. J., Gilbert, B. and Pellegrino, J. (1967). *Science, N.Y.* **157**, 950.

Mueller, G. C. (1968). *In* "Biogenesis and Action of Steroid Hormones" (Dorfman, R. I., Yamasaki, K. and Dorfman, M., eds.), p. 1, Geron-X, Inc. Los Altos, California.

Nutting, W. H., Rapoport, H. and Machlis, L. (1968). *J. Am. chem. Soc.* **90**, 6434.

Oehlschlager, A. C. and Ourisson, G. (1967). *In* "Terpenoids in Plants" (Pridham, J. B., ed.), p. 83, Academic Press, London and New York.

Olson, J. A. (1967). *Pharmacol. Rev.* **19**, 559.

Olson, J. A. (1969). *Fedn. Proc.* **28**, 1670.

Olson, J. A. and Hayaishi, O. (1965). *Proc. Natn. Acad. Sci. U.S.A.* **54**, 1364.

Overbeek, J. van (1968). *In* "Plant Growth Regulators", S.C.I. Monograph No. 31, p. 181, Society of Chemical Industry, London.

Patterson, R. L. S. (1968). *J. Sci. Fd. Agric.* **19**, 31, 434.

Pfiffner, A. (1971). *In* "Aspects of Terpenoid Chemistry and Biochemistry" (T. W. Goodwin, ed.), p. 95, Academic Press, London and New York.

Pitman, G. B. (1969). *Science, N.Y.* **166**, 905.
Phoenix, C. H., Goy, R. W. and Young, W. C. (1967). *In* "Neuroendocrinology" (Martini, L. and Ganong, F., eds.), Vol. II, p. 163, Academic Press, London and New York.
Regnier, F. E. and Law, J. H. (1968). *J. Lipid Res.* **9**, 541.
Reichstein, T. (1967). *Naturwiss. Rundsch.* **20**, 499.
Renwick, J. A. A. (1967). *Contrib. Boyce Thompson Inst.* **23**, 355.
Renwick, J. A. A. and Vite, J. P. (1969). *Nature, Lond.* **224**, 1222.
Richards, J. H. and Hendrickson, J. B. (1964). "The Biosynthesis of Steroids, Terpenes and Acetogenins", W. A. Benjamin, New York and Amsterdam.
Ritter, F. J. and Wientjens, W. H. J. M. (1967). *T.N.O. Nieuws* **22**, 381.
Robbins, W. E., Kaplanis, J. N., Thompson, M. J., Shortino, T. J., Cohen, C. F. and Joyner, S. C. (1968). *Science, N.Y.* **161**, 1158.
Roels, O. A. (1967). *Nutr. Rev.* **25**, 33.
Roeske, W. R. and Clayton, R. B. (1968). *J. Lipid Res.* **9**, 276.
Roller, H. and Dahm, K. H. (1968). *Rec. Prog. Horm. Res.* **24**, 651.
Roller, P., Djerassi, C., Cloetens, R. and Tursch, B. (1969). *J. Am. chem. Soc.* **91**, 4918.
Rothstein, M. (1968). *Comp. Biochem. Physiol.* **27**, 309.
Rudinsky, J. A. (1966). *Science, N.Y.* **152**, 218.
Rudinsky, J. A. (1969). *Science, N.Y.* **166**, 884.
Ruzicka, L. (1953). *Experientia* **9**, 357, 362.
Ruzicka, L. (1956). *In* "Perspectives in Organic Chemistry" (Sir A. Todd, ed.), p. 265, Interscience, New York.
Salaque, A., Barbier, M. and Lederer, E. (1966). *Comp. Biochem. Physiol.* **19**, 45.
Sekeris, C. E. (1965). *In* "Mechanisms of Hormone Action" (Karlson, P., ed.), p. 149, Thieme, Stuttgart.
Siddall, J. B. (1970). *In* "Chemical Ecology" (E. Sondheimer and J. B. Simeone, eds.), p. 281, Academic Press, New York and London.
Silverstein, R. M., Brownlee, R. G., Bellas, T. E., Wood, D. L. and Browne, L. E. (1968). *Science, N.Y.* **159**, 889.
Slama, K. and Williams, C. M. (1965). *Proc. Soc. Natn. Acad. Sci. U.S.A.* **54**, 411.
Smith, P. F. and Rothblat, G. H. (1960). *J. Bact.* **80**, 842.
Sobotka, H., Friess, S. L. and Chanley, J. D. (1964). *In* "Comparative Neurochemistry" (D. Richter, ed.), p. 471, Macmillan (Pergamon), New York.
Stork, G. and Burgstahler, A. (1955). *J. Am. chem. Soc.* **77**, 5068.
Svoboda, J. A. and Robbins, W. E. (1968). *Experientia* **24**, 1131.
Svoboda, J. A., Hutchins, R. F. N., Thompson, M. J. and Robbins, W. E. (1969). *Steroids* **14**, 469.
Takemoto, T., Hikino, Y., Nomoto, K. and Kikino, H. (1967). *Tetrahedron Letters* 3191.
Takemoto, T., Nomoto, K., Hikino, Y. and Kikino, H. (1968). *Tetrahedron Letters* 4929.
Tata, J. R. (1966). *In* "Progress in Nucleic Acid Research" (Davidson, J. N., and Cohn, W. E. eds.), Vol. 5, p. 191, Academic Press, New York and London.
Tokuyama, T., Daly, L. and Witkop, B. (1969). *J. Am. chem. Soc.* **91**, 3931.
Tschesche, R. (1965). *Bull. Soc. Chim. Fr.* 1219.
Tumlinson, J. H., Hardee, D. D., Gueldner, R. C., Thompson, A. C., Hedin, P. A. and Minyard, J. P. (1969). *Science, N.Y.* **166**, 1010.
Van Aarem, H. E., Vonk, H. J. and Zandee, D. I. (1964). *Arch. Int. Physiol. Biochem.* **72**, 606.
Van der Oord, A. (1964). *Comp. Biochem. Physiol.* **13**, 461.

van Tamelen, E. E. (1968). *Acc. chem. Res.* **1**, 111.
van Tamelen, E. E., Storni, A., Hessler, E. J. and Schwartz, M. (1963). *J. Am. chem. Soc.* **85**, 3195.
van Tamelen, E. E., Willett, J. D., Clayton, R. B. and Lord, K. E. (1966a). *J. Am. chem. Soc.* **88**, 4752.
van Tamelen, E. E., Willett, J. D., Schwartz, M. and Nadeau, R. (1966b). *J. Am. chem. Soc.* **88**, 5937.
van Tamelen, E. E., Sharpless, K. B., Hanzlik, R., Clayton, R. B., Burlingame, A. L. and Wszolek, P. C. (1967). *J. Am. chem. Soc.* **89**, 7150.
van Tamelen, E. E., Hanzlik, R. P., Sharpless, K. B., Clayton, R. B., Richter, W. J. and Burlingame, A. L. (1968). *J. Am. chem. Soc.* **90**, 3284.
van Tamelen, E. E., Hanzlik, R. P., Clayton, R. B. and Burlingame, A. L. (1970). *J. Am. chem. Soc.* **92**, 2137.
Varner, J. E. and Ram Chandra, G. (1964). *Proc. Natn. Acad. Sci. U.S.A.* **52**, 100.
vonEuw, J., Fishelson, L., Parsons, J. A., Reichstein, T. and Rothschild, M. (1967). *Nature, Lond.* **214**, 35.
Vroman, H. E., Kaplanis, J. N. and Robbins, W. E. (1964). *J. Lipid Res.* **5**, 418.
Wagner, B. and Erwin, J. A. (1961). *Comp. Biochem. Physiol.* **2**, 202.
Wald, G. (1968). *Nature, Lond.* **219**, 800.
Wareing, P. F., Good, J., Potter, H. and Pearson, A. (1968). *In* "Plant Growth Regulators", S.C.I. Monograph No. 31, p. 191, Society of Chemical Industry, London.
Weatherston, J. (1967). *Q. Rev. (Lond.)* **21**, 287.
Whitten, W. K. (1956). *J. Endocrinol.* **13**, 399.
Whitten, W. K., Bronson, F. H. and Greenstein, J. A. (1968). *Science, N.Y.* **161**, 584.
Williams-Ashman, H. G. (1965). *Cancer Res.* **25**, 1096.
Wood, D. L., Browne, L. E., Bedard, W. D., Tilden, P. E., Silverstein, R. M. and Rodin, J. O. (1968). *Science, N.Y.* **159**, 1373.
Wooton, J. A. M. and Wright, L. D. (1962). *Comp. Biochem. Physiol.* **5**, 253.
Yamamoto, S., Lin, K. and Bloch, K. (1969). *Proc. Natn. Acad. Sci. U.S.A.* **63**, 110.
Young, W. C., Goy, R. W. and Phoenix, C. H. (1964). *Science, N.Y.* **143**, 212.
Zandee, D. I. (1966). *Arch. Internat. Physiol. Biochem.* **74**, 435.
Zandee, D. I. (1967). *Arch. Internat. Physiol. Biochem.* **75**, 487.
Zurfluh, R., Dunham, L. L., Spain, V. L. and Siddall, J. B. (1970). *J. Am. chem. Soc.* **92**, 425.

CHAPTER 2

Monoterpene Biosynthesis

M. J. O. FRANCIS

Nuffield Department of Orthopaedic Surgery, University of Oxford, Nuffield Orthopaedic Centre, Headington, Oxford, England

I. Introduction

The biosynthesis and metabolism of monoterpenes was discussed in detail by Loomis (1967) and in general this review will be confined to more recent work. Various specialized aspects of monoterpene chemistry and biochemistry have also been reviewed recently (Battersby, 1967, 1970; Taylor and Battersby, 1969; Sticher, 1969; Waller, 1969; Weatherston, 1967; Wood, 1969).

Monoterpenes are predominantly plant products and are most often isolated as the major components of the oils obtained by steam distillation of plant material. Monoterpenes may also be present in non-steam distillable forms. For example the family of methylcyclopentanoid monoterpenes (the iridoids) are normally present in plant tissues as their β-D-glucosides; geraniol has also been found as its β-D-glucoside (Francis and Allcock, 1969) and as its fatty acid ester (Dunphy, 1970). In addition monoterpenes are formed by certain arthropods either as components of their defensive secretions or of their pheromones (Weatherston, 1967). Thus irodial (26, Fig. 3) or its derivatives is secreted by various species of ant, and verbenol (16, Fig. 1) and verbinone by males and females respectively of the beetle *Dendroctonus frontalis* (Renwick, 1967; Silverstein *et al.*, 1966). Species ranging from mammals to micro-organisms metabolize monoterpenes, though in these organisms *de novo*

ACYCLIC

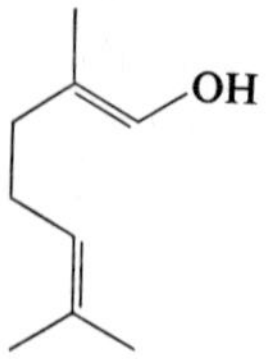

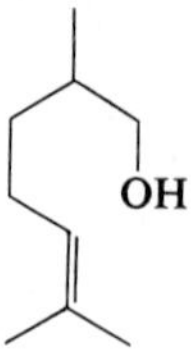

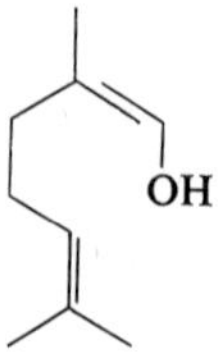

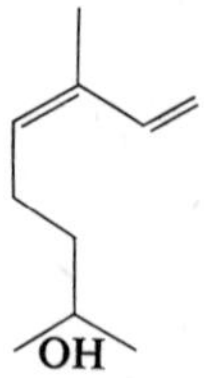

(1) Geraniol (2) Citronellol (3) Nerol (4) Ocimenol

MONOCYCLIC

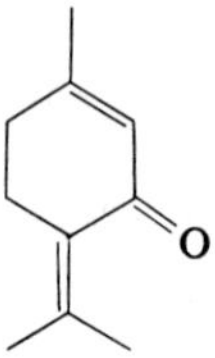

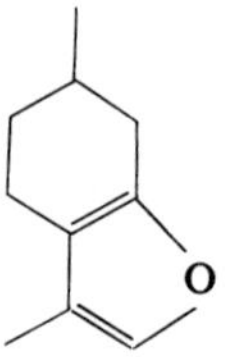

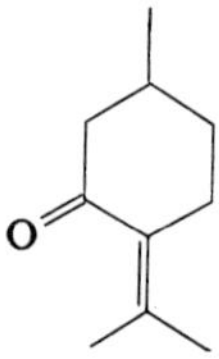

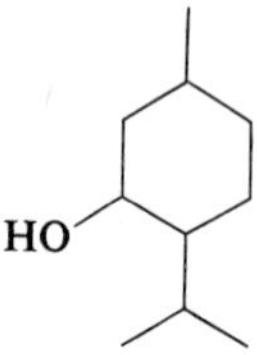

(5) Piperitenone (6) Menthofuran (7) Pulegone (8) Menthol

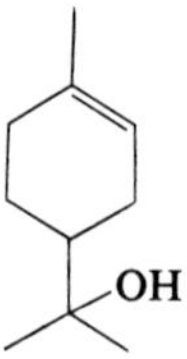

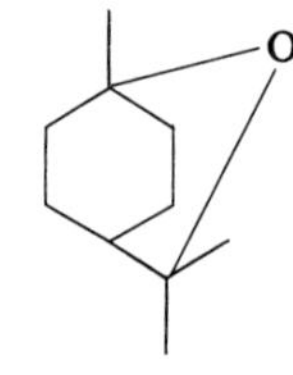

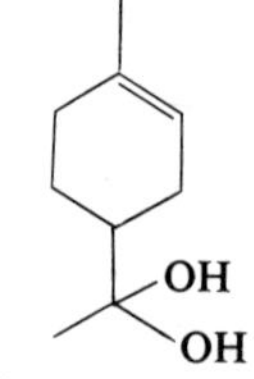

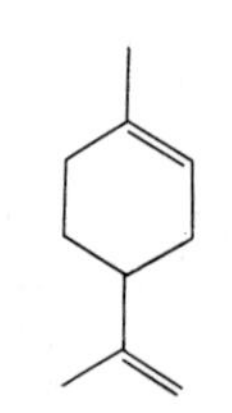

(9) α-Terpineol (10) 1,8-Cineole (11) Uroterpenol (12) Limonene

BICYCLIC

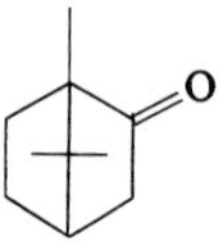

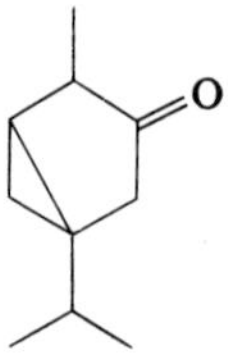

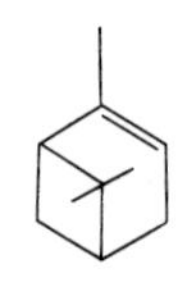

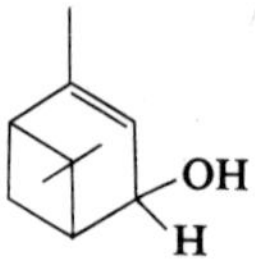

(13) Camphor (14) Thujone (15) α-Pinene (16) Verbenol

IRREGULAR

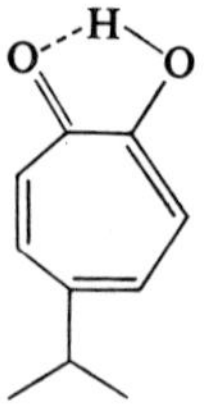

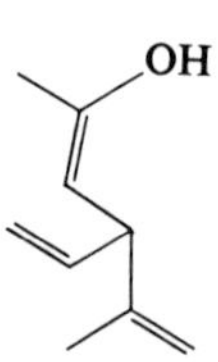

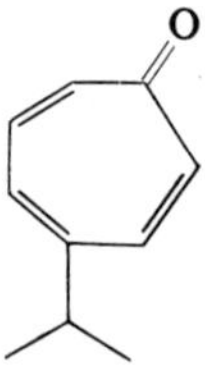

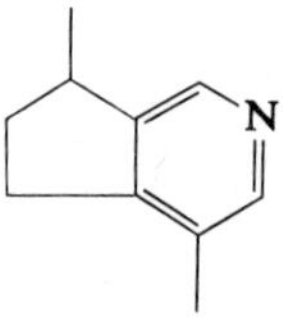

(17) γ-Thujaplicin (18) Lyratol (19) Nezukone (20) Actinidine

FIG. 1.

synthesis of the monoterpenoid C-10 carbon skeleton may not occur save as an intermediate in steroid biosynthesis. Several bacterial species degrade monoterpenes to CO_2 and water via a pathway analogous to that used in the catabolism of leucine (Wood, 1969; Waller, 1969). In man dietary limonene (12, Fig. 1) is converted into uroterpenol (11, Fig. 1), which is excreted as its β-D-glucuronide (Dean *et al.*, 1967).

Monoterpenes are compounds whose structures can be derived, either directly or indirectly, from the condensation of two isoprene units. A wide range of such compounds is now recognized and a selection of those referred to in this review is given in Fig. 1. The discovery of mevalonic acid (MVA; 21, Fig. 2) as a precursor, first of the steroids and then of the terpenoids

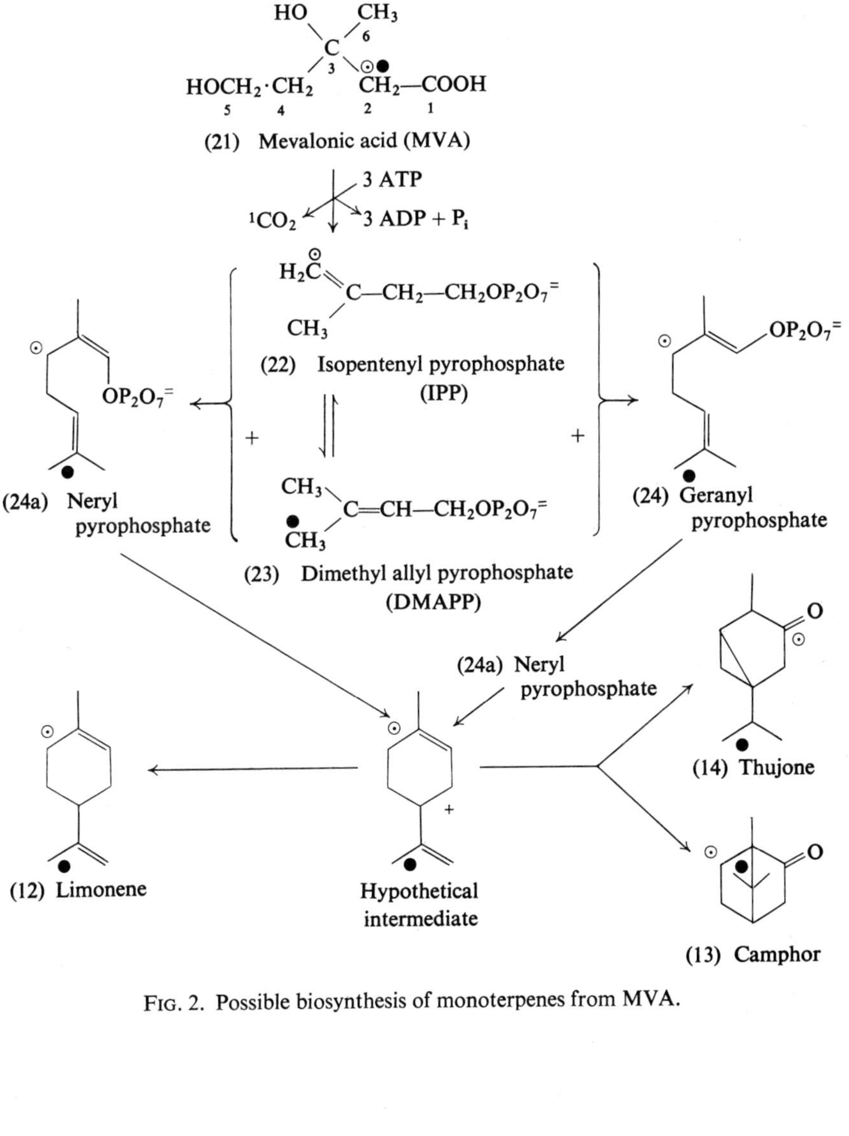

FIG. 2. Possible biosynthesis of monoterpenes from MVA.

generally has naturally focused attention on MVA as a precursor of the mono-terpenoid family. Monoterpenes are often assumed to be synthesized from geranyl pyrophosphate (24, Fig. 2) the C-10 monoterpene intermediate in the synthesis of higher terpenes. Geranyl pyrophosphate is a product of the head-to-tail condensation of isopentenyl pyrophospate (IPP; 22, Fig. 2) and dimethylallyl pyrophosphate (DMAPP; 23, Fig. 2), both formed *in vivo* from MVA. The simplest alicyclic monoterpenes [e.g. nerol (3) and citronellol (2), Fig. 1] can be formed directly from geranyl pyrophosphate (24) or geraniol (1). Monocyclic and bicyclic monoterpenes may, in turn, be derived from one or more of the alicyclic monoterpenes [e.g. α-terpineol (9) and α-pinene (15), Fig. 1]. Plausible chemical schemes have been proposed for such re-arrangements of the monoterpene carbon skeleton. Perhaps the best known is Ruzicka's scheme based on the rearrangements of the C-10 carbonium ion derived from nerol or neryl pyrophosphate (24a) (see Loomis, 1967 and Fig. 2). It is, however, unfortunate that the chemistry of the monoterpenes

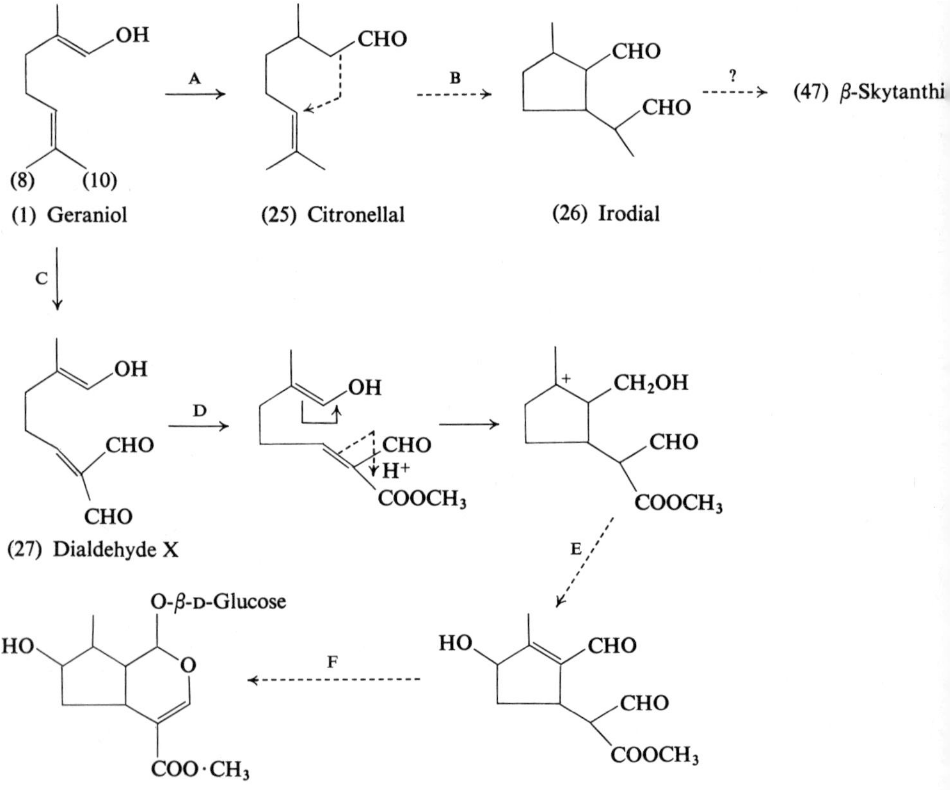

OH
A
CHO
B
CHO
CHO
?
(47) β-Skytanthi
(8) (10)
(1) Geraniol
(25) Citronellal
(26) Irodial
C
OH
D
OH
CH_2OH
CHO
CHO
H^+
CHO
$COOCH_3$
$COOCH_3$
CHO
(27) Dialdehyde X
E
O-β-D-Glucose
HO
O
F
HO
CHO
CHO
$COO \cdot CH_3$
$COOCH_3$
(28) Loganin

FIG. 3. Possible biosynthesis of loganin from geraniol. (Adapted from Bowman and Leete, 1969, and reproduced with permission.)

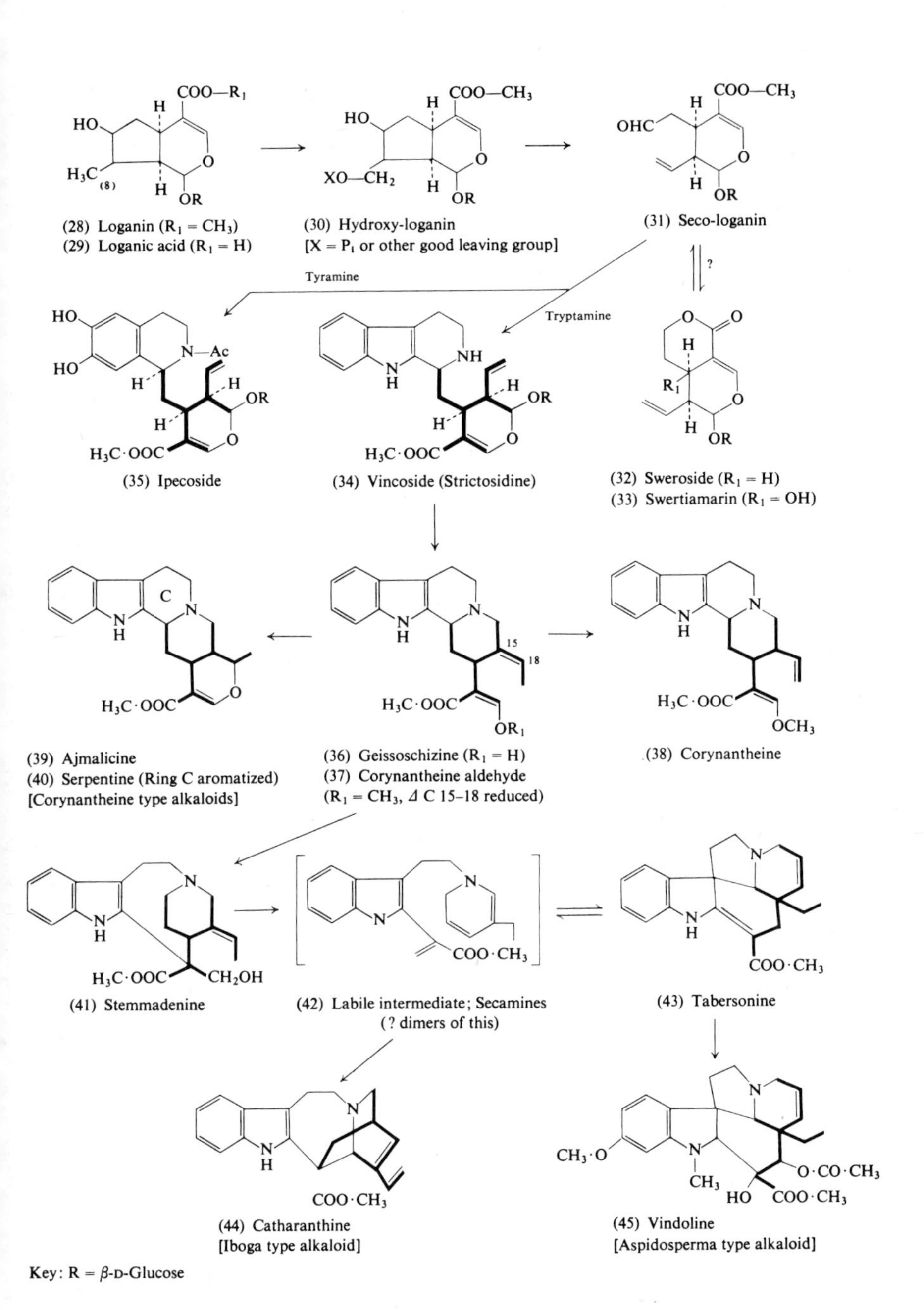

COO—R₁
HO
H₃C
(8)
OR
(28) Loganin (R₁ = CH₃)
(29) Loganic acid (R₁ = H)
COO—CH₃
HO
XO—CH₂
OR
(30) Hydroxy-loganin
[X = Pᵢ or other good leaving group]
OHC
(31) Seco-loganin
Tyramine
Tryptamine
HO
HO
N—Ac
H₃C·OOC
(35) Ipecoside
NH
(34) Vincoside (Strictosidine)
(32) Sweroside (R₁ = H)
(33) Swertiamarin (R₁ = OH)
C
(39) Ajmalicine
(40) Serpentine (Ring C aromatized)
[Corynantheine type alkaloids]
15
18
OR₁
(36) Geissoschizine (R₁ = H)
(37) Corynantheine aldehyde
(R₁ = CH₃, Δ C 15–18 reduced)
OCH₃
(38) Corynantheine
CH₂OH
(41) Stemmadenine
COO·CH₃
(42) Labile intermediate; Secamines
(? dimers of this)
(43) Tabersonine
(44) Catharanthine
[Iboga type alkaloid]
CH₃·O
O·CO·CH₃
HO
(45) Vindoline
[Aspidosperma type alkaloid]
Key: R = β-D-Glucose

FIG. 4. Possible biosynthesis of indole alkaloids from loganin.

lends itself so readily to the construction of hypothetical biosynthetic pathways. For the biochemical evidence is limited and the design and interpretation of experimental results complicated by this multiplicity of feasible chemical mechanisms. Furthermore, though it might be logical to suppose that a plant which synthesizes geranyl pyrophosphate would also use it as a precursor of the monoterpenes, this need not necessarily be so. For example, plant species can synthesize the pyridine ring in three separate ways, by: (a) the condensation of a three-carbon compound related to glycerol and a four-carbon compound related to aspartic acid (Yang *et al.*, 1965); (b) the use of lysine as a precursor (Hylin, 1964); (c) the use of MVA and geranyl pyrophosphate (24) as precursors (Auda *et al.*, 1967a).

Much recent work on monoterpene biosynthesis has been concentrated on the iridoids (e.g. 28, Fig. 3) and the indole (e.g. 38, 44 and 45, Fig. 4) and isoquinoline alkaloids, which have unexpectedly been shown to be biosynthesized from the iridoid loganin (28) (Battersby, 1967). Here the derivation from MVA is more difficult to envisage, yet paradoxically enough the evidence for biosynthesis of this group of monoterpenes from MVA and geranyl pyrophosphate (24) is the most rigorously established (see Figs. 3 and 4 for outline pathways).

The structures of some other irregular monoterpenes are given in Fig. 1. These are assumed to be derived, if indeed they are monoterpenes, from either abnormal condensations of two isoprene units or secondarily by shifts of carbon atoms in a normal monoterpene C-10 skeleton [e.g. geraniol (1)].

II. General and Chemical Considerations

The biosynthesis of monoterpenes is often considered as an offshoot of the well-known pathways to the higher terpenes. Yet, as Loomis (1967) has pointed out, the evidence supporting this assumption is slight. The incorporation of MVA into monoterpenes by whole plants or plant tissues is generally low, in the range of 0·01–0·1% or, in some cases, negligible. Such results cannot be ascribed to poor translocation of MVA from the sites of injection to the active tissues as MVA is rapidly translocated (Regnier *et al.*, 1968; Horodysky *et al.*, 1969). They may, however, be a reflection of the low permeability of externally applied MVA to the intracellular sites of biosynthesis as has been demonstrated in the biosynthesis of some higher terpenoids (Rogers *et al.*, 1967).

The low incorporation of MVA introduces complicating factors that are often overlooked. Label from [2-^{14}C] MVA can enter the general metabolic pathways of the plant. In *Mentha piperita* up to 4·2% of the label from [2-^{14}C] MVA was present in the respired CO_2 (Battu and Youngken, 1966). A reflection of these pathways is seen in the biosynthesis of plumieride (49, Fig. 5; Yeowell and Schmid, 1964) from [2-^{14}C] MVA where 10% of the label was present in the glucoside moiety of the molecule. Sometimes, even, there is more label derived from MVA in the non-monoterpenoid portion of the compound under study than in the monoterpene moiety. The acetyl residue of sabinyl acetate isolated from *Juniperus sabina* L. plants fed [2-^{14}C] MVA contained thirty times as much activity as the sabinyl moiety (Banthorpe, personal communication).

The routes leading from [2-^{14}C] MVA to a general pool of labelled carbon intermediates are unknown. The biosynthesis of L-MVA from 3-hydroxy-3-methylglutaryl-CoA is considered to be essentially irreversible and it is unlikely that a reversal of L-MVA biosynthesis is a major route for MVA catabolism. D-MVA isomer may, however, be catabolized. In contrast recent work has emphasized both the metabolic lability of monoterpenes and the complicated relationships of monoterpene biosynthesis to the physiological condition of the plants under study (Burbott and Loomis, 1967, 1969; Francis and Allcock, 1969). Monoterpenes are also precursors of the higher terpenes (Baisted, 1967; Banthorpe and Wirz-Justice, 1969); these higher terpenes can be catabolized (e.g. Dicks and Friend, 1968). MVA-carbon may thus enter the general metabolic pools via biosynthesis to monoterpenes and higher terpenoids and then by the further catabolism of these compounds.

Biosynthetic experiments using MVA are often carried out using several days incubation. Such long incubations may lead to significant randomization of the tracer in the MVA skeleton. In *Mentha* sp. $^{14}CO_2$ is a much more effective precursor than [2-^{14}C] MVA (Loomis, 1967; Hefendehl *et al.*, 1967) and since MVA is degraded to CO_2 in these plants, specific incorporation of MVA may not occur. The low incorporations, such as are obtained with [2-^{14}C] MVA make determinations of complete specific labelling patterns essential before results can be used to justify a biosynthetic mechanism. The rapid appearance of label from $^{14}CO_2$ in monoterpenes does, however, provide a means whereby the interrelationships between different monoterpenes may be studied, even though the use of $^{14}CO_2$ may not distinguish between different routes of biosynthesis of the monoterpene skeleton. $^{14}CO_2$ has the additional advantage that it is the only precursor that can be introduced under entirely physiological conditions. Nevertheless even $^{14}CO_2$ may not always be an effective monoterpene precursor (Charlwood, 1970). The rapid turnover of monoterpenes suggests that several days incubation period may be too long. Labelled monoterpenes may be further metabolized before the extraction procedures are begun. In rose flowers, label from [2-^{14}C] MVA found in the free and β-D-glucosyl monoterpene pools had reached less than 10% of its maximal levels 3 h after exposure of the flowers to [2-^{14}C] MVA (Francis and O'Connell, 1969). Experiments on the time course of appearance of label in the desired products should always be undertaken as a first priority.

The low incorporation of label from [2-^{14}C] MVA into monoterpenes may result from MVA not being the sole precursor of the two isoprene units found in the monoterpenes. One aspect of this possibility is that the MVA synthesized *in vivo* from acetyl-CoA never appears as free MVA but only as enzyme-bound MVA. Enzyme bound intermediates in the synthesis of MVA have been described in the synthesis of MVA from acetyl-CoA by enzyme preparations from pigeon liver (Brodie *et al.*, 1964). Fatty acid synthetases which use acetyl-CoA are multienzyme complexes (Lynen, 1961) and the same may be true for the terpene biosynthetic enzymes. Whether externally added MVA is acted on by the synthetase would depend on the equilibrium between free and enzyme-bound MVA. This is a parameter that may vary between synthetases from different plant species or plant tissues (viz. the reactions of different fatty acid synthetases (Wakil and Ganguly, 1959; Lynen, 1961). Such variability would account for the low or negligible incorporations of MVA into monoterpenes and the difficulties of isolating monoterpene-synthesizing enzymes (see p. 45). Even if free MVA is an intermediate, its concentration in tissues is low and addition of excess MVA, 50% of which is normally the unphysiological D-isomer, may inhibit monoterpene biosynthesis thus accounting for the observed results.

Cyclic monoterpenes are related stereochemically to nerol (3) rather than to geraniol (1) (Fig. 1). Mild acid hydrolysis of neryl phosphate yields cyclic monoterpenes as products whereas acid hydrolysis of geranyl phosphate yields unsaturated acyclic monoterpenes (Rittersdorf and Cramer, 1968). Neryl pyrophosphate (24a), rather than geranyl pyrophosphate (24), is therefore the more likely immediate intermediate in the synthesis of the cyclic monoterpenes. Two biosynthetic routes to neryl pyrophosphate are possible: either by isomerization of geranyl pyrophosphate or by condensation of IPP and DMAPP with opposite stereochemistry to that found in the synthesis of geranyl pyrophosphate (see Fig. 2). These two pathways may be distinguished by carrying out monoterpene biosynthetic studies using 4*S* and 4*R*[^{3}H]-2-^{14}C-MVA as precursors. Condensations to give *cis* rather than *trans* double bonds [e.g. nerol (3) rather than geraniol (1)] are a feature of the biosynthesis of rubber (Archer and Audley, 1967). Cyclic monoterpenes may also be derived from linaloyl pyrophosphate (56, Fig. 7) since mild acid hydrolysis of linaloyl esters gives the same carbonium ion as do neryl esters (Rittersdorf and Cramer, 1968). The direct formation of linaloyl pyrophosphate (56, Fig. 7) from IPP (22) and DMAPP (23) (Fig. 2) has been proposed by Attaway *et al.* (1967) to account for changes in composition of orange and tangerine peel monoterpenes during fruit maturation. It is possible that all three pathways will be found. The evolution of monoterpene biosynthesis could be related either to the biosynthesis of higher terpenes—that is, to geranyl pyrophosphate (24), or to the biosynthesis of rubber—that is, to neryl pyrophosphate (24a). The presence of monoterpenes either in discrete oil glands (e.g. *Mentha* or *Pelargonium* sp.) or in resin ducts (e.g. *Pinus* sp.) may be a reflection of such a relationship.

Finally the trace monoterpene constituents of plant essential oils could arise secondarily in the isolation processes rather than be formed biosynthetically. These oils are often produced by steam distillation of plant products under primitive conditions in reactive, such as iron, containers. The pH of such distilling mixtures can be as low as 2·8 (M. J. O. Francis, unpublished observations on *Pelargonium graveolens*). Such hot acid conditions favour monoterpene isomerization and could account for some of the minor monoterpene constituents found in essential oils.

III. Biosynthesis in Whole Plants or Plant Tissues

A. NORMAL MONOTERPENES

The further work on *Mentha* sp. has supported the main features of Reitsema's scheme (see Loomis, 1967) for the formation of *Mentha* monoterpenes. However, piperitenone(s) does not appear to be an important precursor of menthone, and pulegone (7) may not be the only precursor of menthofuran (6) (Fig. 1) (Loomis, 1967; Hefendehl *et al.*, 1967; Battu and Youngken, 1966). In general the *de novo* synthesis of reduced compounds such as menthofuran (6) and pulegone (7) occurs before that of oxidized compounds such as menthone. The proportions of oxidized and reduced monoterpenes vary according to the metabolic status of the plant and monoterpenes may act as an important store of physiological reducing power. In *Mentha* sp. this metabolic lability of monoterpenes seems to be controlled by the balance between daytime photosynthesis and night time utilization of photosynthate (Burbott and Loomis, 1967). Monoterpenes would be good substrates for energy metabolism. The aerobic oxidation of menthone via pathways similar to those described for the microbial oxidation of geraniol (Wood, 1969) could yield 11–12 moles of ATP/C-2 unit. This compares with a yield of 16–17 moles of ATP/C-2 unit for fatty acid oxidation and 10–11 moles of ATP/C-2 unit for glucose oxidation. Intermediates produced during the microbial oxidation of geraniol include dimethylacryl-CoA. Dimethylacryl-CoA can be converted to β-hydroxy-β-methylglutarul-CoA and thence to MVA. Partial breakdown and resynthesis of monoterpenes could occur via such a pathway.

Appreciable quantities of simple acylic monoterpenol β-D-glucosides (e.g. geranyl-β-D-glucoside) have been isolated from rose flowers (Francis and Allcock, 1969). *Pelargonium graveolens* also contains monoterpene β-D-glucosides (M. J. O. Francis, unpublished observations). Previously the only monoterpene glucosides to be characterized have been amongst the iridoids and related monoterpenes.

Geranyl-β-D-glucoside could be formed directly from geranyl-pyrophosphate (24) and thus provide the means whereby a monoterpene pool is formed by diversion of the pathways to the higher terpenes. Direct hydrolysis of geranyl pyrophosphate to geranol (1) might not be so effective partly because geraniol is insoluble in water and partly because geraniol is readily rephosphorylated (van Aller and Nes, 1968). Geranyl-β-D-glucoside is, however, equally soluble in organic and aqueous phases and could, in addition, represent a transport form of monoterpene or indeed of glucose. In general, glycoside moieties could act as protecting groups stabilizing otherwise reactive intermediates, for example α-terpineol (9). This seems to be the case in the biosynthesis of the indole alkaloids from loganin (28) (Figs. 3 and 4) where a potentially reactive hydroxyl group is combined with glucose until a late stage in the biosynthesis. Monoterpene glycosides could also play a role in plant cell wall biosynthesis similar to that of mannosyl-L-phosphoryl polyisoprenol in bacterial mannan synthesis (Scher *et al.*, 1968).

Rose flowers at the stage of opening when the concentrations of monoterpenes were increasing at a maximal rate, incorporated label from [2-^{14}C] MVA into monoterpenes rapidly (Francis and O'Connell, 1969), as did isolated rose petals (Paseshnichenko and Guseva, 1967). Maximum incorporation of label into monoterpenes occurred 45–60 min after introduction of a pulse dose of [2-^{14}C] MVA and had fallen to low levels after 3 h (Francis and O'Connell, 1969). These experiments provide further evidence that monoterpenes can be in a state of rapid metabolic flux. The maximum incorporation into free monoterpenes was of the order of 1% but into the monoterpenol β-D-glucoside pool was 11%. These levels of incorporation are ten to a hundred times the incorporation levels of [2-^{14}C] MVA previously reported for the biosynthesis of non-methylcyclopentane monoterpenes from MVA. Specific degradation of the geraniol (1) and nerol (3) residues (Fig. 1) present both as such and as their β-D-glucosides, showed that the MVA had been incorporated specifically (Francis *et al.*, 1970). The isoprene units of geraniol and nerol derived respectively from IPP (22) and DMAPP (23) (see Fig. 2) were labelled in the ratio of 1:1, as would be expected if geraniol was biosynthesized by the normal MVA pathway. The ^{14}C/^{3}H isotope ratios of geraniol and nerol, biosynthesized from 4*S* and 4*R* [$^{3}H_1$]-[2-^{14}C] MVA respectively, showed that geraniol or its glucoside, rather than nerol or its glucoside was the first product of condensation of IPP (22) and DMAPP (23) (Fig. 2). Geraniol has also been shown to be a biosynthetic precursor of α-pinene (15), thujone (14) (Fig. 1) and isothujone (Ayrey *et al.*, in preparation). These results also confirmed that the [4*S*]-H of MVA is lost in the isomerization of IPP to DMAPP in higher plants, as in yeast (Agranoff *et al.*, 1960) and mammalian tissue (Shah *et al.*, 1965). The specific incorporation of label from [2-^{14}C] MVA, together with the high levels and rapid rate of incorporation, provide convincing evidence that, in rose petals, monoterpenes are indeed synthesized by the MVA pathway.

Other labelling work on whole plants or tissues has included that of Scora and Mann (1967) on *Monarda punctata*, Attaway *et al.* (1967, 1968, 1969) on citrus, Waller *et al.* (1968) on *Santolina chamaecyparissus* L., Achilladelis and Hanson (1968) on *Rosmarinus officinalis* and Banthorpe *et al.* (1966, 1968) on a number of plant species. The time sequence of incorporation of label from [2-^{14}C] malonate and [2-^{14}C] MVA into the monoterpenes of *M. punctata* led Scora and Mann to postulate the operation of two monoterpene biosynthetic pathways. In *S. chamaecyparissus* [2-^{14}C] MVA was incorporated into α-pinene and three unidentified monoterpenes in 0·5–1·2% yield (2–7 days incubation). Interestingly one of these monoterpenes tended to become more extensively labelled after time lapses in older plants whereas the opposite occurred in younger plants. The work on citrus fruit peel monoterpenes (Attaway *et al.*, 1967) suggested that linaloyl pyrophosphate (56, Fig. 7) rather than geranyl pyrophosphate might be the first monoterpene intermediate. Further work on leaves and fruit peel from the same species (Attaway and Buslig, 1968 and 1969) showed that [3-^{14}C] linaloyl was converted over a two-day period to a spectrum of monoterpenes of which α-terpineol (9), terpenen-4-ol and *trans*-2,8-*p*-menthadien-1-ol were quantitatively the most important. [1-^{14}C] Geranyl pyrophosphate (24) is incorporated in 0·22% yield into 1–8 cineole (10) by *R. officinalis* (see Fig. 1). As expected from the symmetrical structure of cineole the label was approximately equally distributed between the two C-5 units.

The work of Banthorpe and his colleagues has, however, provided the most puzzling results on monoterpene biosynthesis. They have shown that several thujone derivatives (e.g. 14) (Banthorpe and Turnbull, 1966), camphor (13) (Banthorpe and Baxendale, 1968), α-pinene (15) (Ayrey *et al.*, in preparation) and artemesia ketone (57) (Fig. 7) (Charlwood, 1970) were formed specifically from [2-^{14}C] MVA. Yet the bulk of the tracer is contained in one ring position such that IPP (22) but not DMAPP (23) was the tracer containing precursor (see Fig. 2). In these experiments the [2-^{14}C] MVA was stem fed over a period of 4–5 days with incorporations of *ca.* 0·02%. The authors suggested that monoterpene biosynthesis from MVA in higher plants involves condensation of IPP (22) derived from MVA with DMAPP (23) that is mainly present in a metabolic pool. As discussed above this pool could represent MVA [and so DMAPP (23)] derived from the catabolism of monoterpenes or higher terpenes. If monoterpenes are synthesized at an intra-cellular surface, the two halves of the monoterpene could be formed from two MVA molecules situated on each side of an impermeable membrane only one side of which was accessible to added [2-^{14}C] MVA. A similar non-uniform labelling pattern of isoprene units has been found in the biosynthesis from [2-^{14}C] MVA of the sesquiterpenes, coriamyrtin and tutin (Biollaz and Arigoni, 1969). Further studies on this intriguing phenomenon are awaited with interest. It should be noted that earlier work on the specific incorporation of [2-^{14}C] MVA into leaf or stem monoterpenes was mainly based on incomplete degradation of the monoterpene carbon skeletons. The results of Banthorpe and his colleagues suggest therefore that the pattern of monoterpene biosynthesis (other than of methycyclopentane monoterpenes) they have described may be a universal one for the biosynthesis of leaf and stem monoterpenes. This is in contrast to the results on monoterpene biosynthesis in flowers (Francis *et al.*, 1970; Godin *et al.*, 1963) where label from [2-^{14}C] MVA is distributed equally in the two isoprene residues.

B. METHYL-CYCLOPENTANOID MONOTERPENES AND DERIVED ALKALOIDS

Modern biosynthetic studies on the methyl-cyclopentanoid monoterpenes (iridoids) began with the demonstration that [2-^{14}C] MVA was a specific precursor of plumieride (49) (Yeowell and Schmid, 1964). These studies were given great impetus by the suggestions that the indole alkaloids are related to the iridoid family of monoterpenes (Thomas, 1961; Wenkert, 1962); suggestions that have been amply confirmed by the fact that MVA (21), geraniol (1), geranyl pyrophosphate (24) and the iridoid loganin (28) are specific precursors of the three main classes of indole alkaloid, for example ajmalicine (39), vindoline (45) and catharanthine (44), (see Figs. 1–4). This earlier work has been summarized by Battersby (1967, 1970), Ramage (1967, 1968), Leete (1969) and Taylor and Battersby (1969). A range of compounds related to the iridoids and indole alkaloids have also been described. These include sweroside (32), swertiamarin (33), oleoropein and gakankosin (Wieffering, 1966) and the valepotriates (Thies, 1968).

The biosynthesis of the iridoid carbon skeleton from geraniol has not been established. The experiments of Yeowell and Schmid (1964) showed that in *Plumiera acutifolia* [2-^{14}C] MVA gave rise to labelled plumieride (49) in 1·03% yield. Specific degradation of the aglucone gave the labelling pattern for plumieride shown in Fig. 5. This distribution of label confirms the mevalonoid origin of the methyl-cyclopentanetetrahydropyran ring portions of plumieride and is characteristic of the labelling pattern found in the biosynthesis of these iridoids and of the indole alkaloids from [2-^{14}C] MVA. The labelling pattern suggests that at one stage in the biosynthesis of the iridoids the C-8 and C-10 methyl groups of geraniol become biosynthetically equivalent (see Fig. 3). Yeowell and Schmid (1964) suggested that irodial (26) formed from geraniol via reactions A–B (Fig. 3), was an intermediate in the biosynthesis of plumieride, thus accounting for the observed randomization of the C-3 and C-15 of plumieride (49) (Fig. 5). However, Bowman and Leete (1969) have been unable to show that irodial (26) is a precursor of loganin (28). They suggest the alternative pathway for loganin biosynthesis from geraniol shown as reactions C–F (Fig. 3). Randomization in this pathway could occur at the level of the dialdehyde X (27). It may be that compounds such as the skytanthines (e.g. 47, Fig. 5) or actinidine (20) are biosynthesized from geraniol via irodial (26) whereas those which have loganin (28) as a precursor are formed via the dialdehyde X (27) [but see Waller (1969) for alternative proposals]. Yet negative findings of this kind may reflect the spatial isolation of a biosynthetic pathway rather than the nature of the pathway itself. For example MVA has been demonstrated as a precursor of loganin in only some plant species (see discussion on loganin biosynthesis).

Eisenbraun and Waller and their colleagues have studied the biosynthesis of several irodoids or related monoterpenes. The skytanthus alkaloids (e.g. 47) (Auda *et al.*, 1967), were shown to be monoterpenoid; there are therefore three separate routes to the piperidine nucleus as there are to the pyridine nucleus, the other routes having lysine or a poly-β-keto acid derived from four acetate units as precursors. Four days after feeding [2-^{14}C] MVA to *Skytanthus acutus* M. 0·6% of the activity was present in the isolated alkaloids. Alkaloids with the highest specific activities were isolated from the green stems of the plant, the site of MVA injection. The most interesting

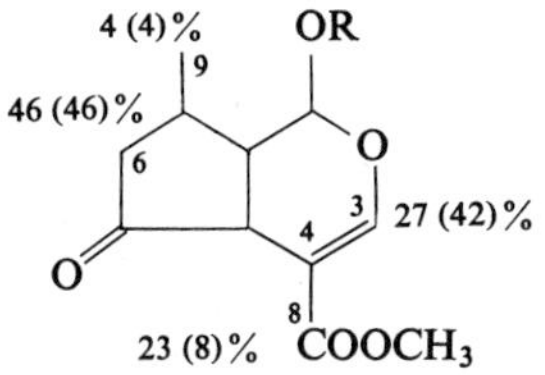

(46) Verbenalin
1–2 Month old plants
(4 Month old plants)

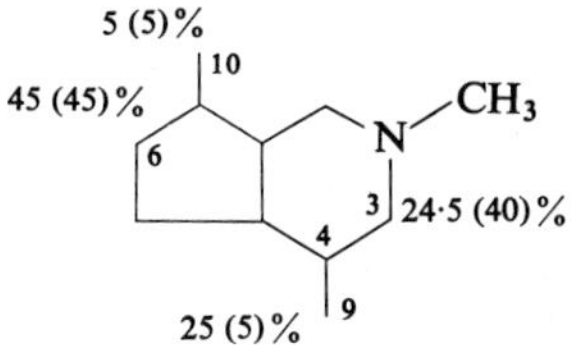

(47) β-Skytanthine
1·3 Year old plants
(3 Year old plants)

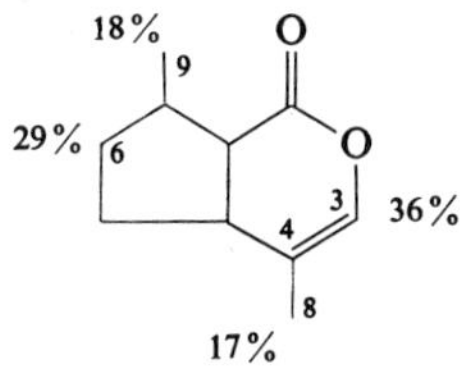

(48) Nepetalactone

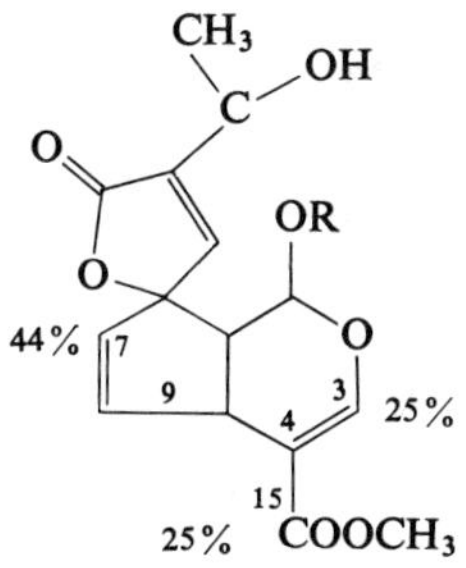

(49) Plumieride

R = β-D-Glucose

FIG. 5. Labelling patterns of methyl-cyclopentane monoterpenes biosynthesized from [2-^{14}C] mevalonate.

findings concerned the differences in labelling pattern of β-skytanthine (47) isolated after feeding [2-^{14}C] MVA to 1-, 2- and 3-year-old plants respectively. In the young flowering plants 25% of the label was present in C-3 and 25% in C-9 and in the old plants 40% in C-3 and 5% in C-9 (see 47) (Fig. 5). This change in labelling pattern with age of the plant is also apparent in the biosynthesis of verbenalin (46) from [2-^{14}C] MVA [see (46), Fig. 5] by *Verbena officinalis* L. (Horodysky *et al.*, 1969). This lack, or otherwise, of randomization may be a reflection of the relative pool sizes of the methylcyclopentanoid monoterpenes and of their monoterpene precursors or perhaps the enzymatic mechanism of synthesis changes with age (Horodysky *et al.*, 1969). The pattern of randomization is, however, similar to that found in the biosynthesis of plumieride (49) from [2-^{14}C] MVA (see Fig. 5). The labelling pattern found in β-skytanthine (47) and verbenalin (46) biosynthesized from [2-^{14}C] MVA provide additional support for the mechanism of isomerization of IPP (22) and DMAPP (23) (Fig. 2) first proposed by Agranoff *et al.* (1960) and later established by Shah *et al.* (1965). In contrast the labelling pattern found in nepetalactone (48) (Fig. 5, Regnier *et al.*, 1968) biosynthesized from [2-^{14}C] MVA by *Nepeta cataria* L. suggests that some randomization of label may take place at the IPP as well as at the monoterpenoid stage. This would not

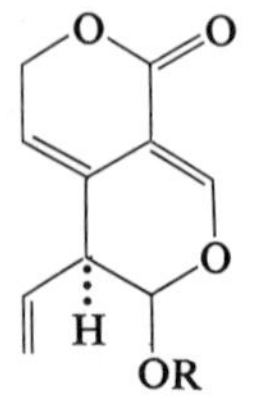

(50) Gentiopicroside

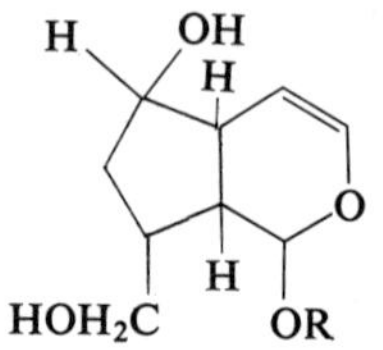

(51) Aucubin

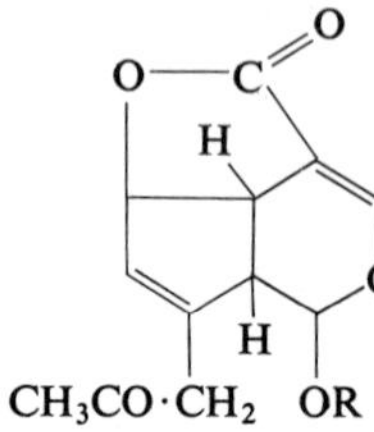

(52) Asperuloside

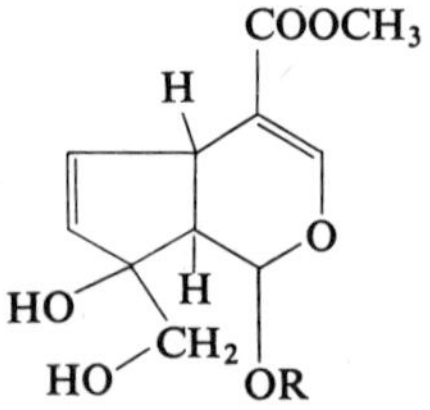

(53) Monotropeine methyl ester

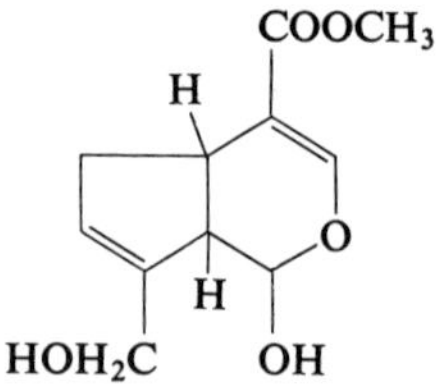

(54) Genepin

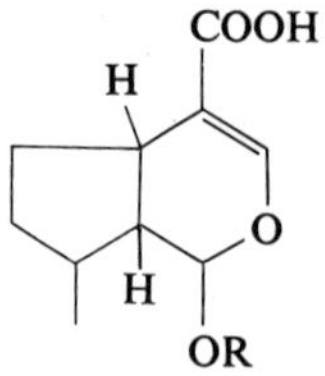

(55) Compound Z

R = β-D-Glucose

FIG. 6. The structures of some iridoids.

be in accord with the proposed mechanism of IPP isomerization. The randomization of label in nepetalactone may, however, be a reflection of nepetalactone and MVA catabolism since the incubation periods were of long duration. Evidence for MVA catabolism (or of substances derived from MVA) in *N. cataria* plants was obtained by the demonstration that 50 h after administration of [2-^{14}C] MVA 0·15% of the label had been evolved as respiratory CO_2. Additional evidence for a monoterpenoid origin of the iridoids (46–48) was obtained by the incorporation of geraniol/geranyl pyrophosphate into verbenalin (46), nepetalactone (48) (Fig. 5) and the related pyridine monoterpene alkaloid, actinidine (20) (Fig. 1) (Auda *et al.*, 1967a).

Loganin (28) is an important intermediate in the biosynthesis both of other iridoids and of the indole alkaloids (Battersby, 1967, 1970). Loganin was first isolated in 1961 (Sheth *et al.*, 1961) and its stereochemistry is now known in detail (Brechbuhler-Bader *et al.*, 1968; Battersby *et al.*, 1968a). Experiments with *Vinca rosea* plants have shown that MVA is a specific precursor of loganin (Loew and Arigoni, 1968). In contrast whole and sliced rhizomes of *Menyanthes trifoliata* did not incorporate [2-^{14}C] MVA into loganin though the triterpene and sterol fractions were well labelled (Brechbuhler-Bader *et al.*, 1968). Yet [1-^{3}H] geraniol, [2-^{14}C] and [3-^{14}C] geraniol were incorporated in 0·2%, 0·25% and 0·1% yields respectively (Battersby *et al.*, 1968a; Brechbuhler-Bader *et al.*, 1968). With [3-^{14}C] geraniol 85% of the label was in the expected position. Geranyl pyrophosphate (24) has also been shown to be a precursor of loganin. Compound Z (55) (Fig. 6) is converted into loganin in 0·27% yield as well as into verbenalin (46; 11% yield), asperuloside

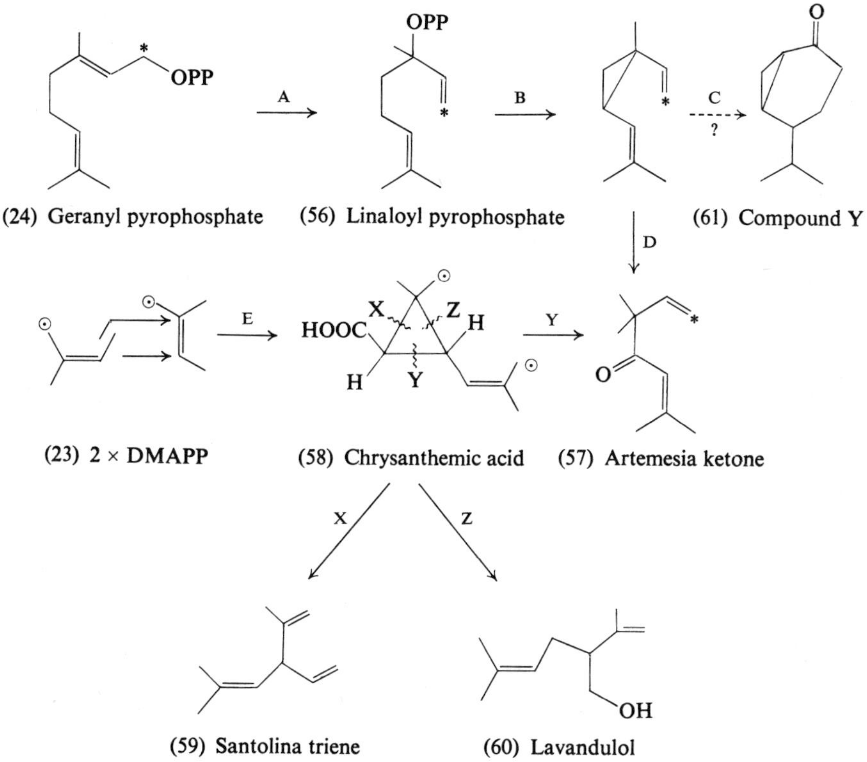

(24) Geranyl pyrophosphate (56) Linaloyl pyrophosphate (61) Compound Y

(23) 2 × DMAPP (58) Chrysanthemic acid (57) Artemesia ketone

(59) Santolina triene (60) Lavandulol

FIG. 7. Possible biosynthesis of irregular monoterpenes.

(52; 0·60% yield) and aucubin (51; 0·5% yield); compound Z (55) could be formed from irodial (26) (Inouye *et al.*, 1969b). The hydroxylation of desoxyloganin to form loganin has been shown to be the final stage of loganin biosynthesis (Battersby, 1970).

Figure 4 outlines current ideas on the biosynthesis of the indole alkaloids and related compounds from loganin. Experiments with *V. rosea* have shown that O-methyl-[^{3}H] loganin is incorporated into ajmalicine (39), serpentine (40) and catharanthine (44) (Fig. 4) (Battersby *et al.*, 1966). In contrast the related iridoids, monotropeine methyl ester (52, Fig. 6), verbenalin (46, Fig. 5) and genepin (54, Fig. 6) were not incorporated. That loganin is a specific precursor of the three types of indole alkaloid was confirmed by isolating labelled ajmalicine (39), catharanthine (44) and vindoline (45) (Fig. 4) when loganin (mostly labelled in the C-8 methyl group) was fed to young shoots of *V. rosea* (Loew and Arigoni, 1968). This incorporation was specific (Battersby *et al.*, 1968b).

Labelled loganin gives rise to gentiopicroside (50) and asperuloside (51) (Fig. 6) in 4·5% and 0·45% yields respectively (Inouye *et al.*, 1969b), to strictosidine (34) in 5·2% yield (Brown *et al.*, 1968) and to ipecoside (35) (Fig. 4) in 1·7% yield (Battersby and Gregory, 1968). This last compound is a seco-cyclopentane monoterpene derivative and is an example of a mixed isoquinoline-monoterpene alkaloid (Battersby, 1967; Battersby and Gregory, 1968). Not only did the determination of the structure of ipecoside suggest loganin as a key indole alkaloid precursor but also increased the likelihood that a seco-loganin (31) derived from loganin gives rise to such varied compounds as gentiopicroside (50), swertiamarin (32), sweroside (33) (Fig. 4) and bankankosin in addition to the indole alkaloids. The isolation of compounds such as foliamenthin (Loew *et al.*, 1968; Battersby *et al.*, 1968d) and the tetrahydro-Δ-carboline monoterpenoid glucoside strictosidine (34) [vincoside (Smith, 1968)] lend further weight to this suggestion, as do the results of Coscia *et al.* (1969) on gentiopicroside (50) biosynthesis. Seco-loganin has now been detected in *V. rosea* plants. It is formed from loganin and gives rise to representative compounds of the three major classes of indole alkaloid (Battersby *et al.*, 1968c; Battersby *et al.*, 1969a). Seco-loganin could be formed via a hydroxyloganin (30, Fig. 5). Sweroside (32) which is related to seco-loganin (31) (see Fig. 4), gives rise to the indole alkaloid vindoline (45) in 11% yield (Inouye *et al.*, 1968) and to gentiopicroside (50) (Inouye *et al.*, 1967). *In vitro* seco-loganin and tryptamine condense to yield vincoside (34) (probably equivalent to strictosidine) and isovincoside (Battersby *et al.*, 1968e and 1969b, and see Fig. 4). Vincoside (34) gives rise to geissoschizine (36), and the indole alkaloids (Battersby *et al.*, 1968e; Battersby and Hall, 1969). Swertimarin (33) does not appear to be an intermediate as proposed by Inouye *et al.* (1968). Present evidence suggests that the rearrangement of the C-10 monoterpenoid skeleton to give the three classes of indole alkaloid takes place only after condensation of the C-10 unit, probably seco-loganin (31), with tryptamine. Several studies including a novel approach which uses the appearance of indole alkaloids in the germination of *V. rosea* seeds (Qureshi and Scott, 1968b) have suggested that the indole alkaloid skeletons arise in the order: corynanthe type, aspidosperma type, iboga type with stemmadenine (41) and tabersonine (43) as important intermediates (Ramage, 1968; Battersby *et al.*, 1968f; Kutney *et al.*, 1968). Geissoschizine (36) rather than corynantheine aldehyde (37) may also be an intermediate in these transformations (Battersby *et al.*, 1968c). These biosynthetic pathways, summarized in Fig. 4, are given further support by the *in vitro* conversions of the aspidosperma alkaloid (−) tabersonine (43) to the iboga alkaloid (±) catharanthine (44), and of (+) stemmadenine (41) to tabersonine (44) (Qureshi and Scott, 1968a), though Brown *et al.* (1969) have not been able to repeat these experiments.

The biosynthesis of quinine, reserpine and cephaeline is also related to those of the methyl-cyclopentane monoterpenes. Thus sweroside is a specific precursor of quinine and reserpine (Inouye *et al.*, 1969a) and loganin of cephaeline (Battersby and Gregory, 1968) and of quinine (Battersby and Hall, 1970). An interesting aspect of the biosynthesis of cephaeline (Garg and Gear, 1969a) as of ajmalicine (39; Garg and Gear, 1969b) is that [2-^{14}C] glycine is a specific precursor of the C-9/10 unit of these compounds whereas [2-^{14}C] acetate is not, though [2-^{14}C] acetate is a specific precursor of β-sitosterol.

C. OTHER IRREGULAR MONOTERPENES

Artemesia ketone (57) biosynthesis has been studied. [2-^{14}C] MVA did not give rise to labelled artemesia ketone in *S. chaemaecyparissus* under conditions where other monoterpenes were labelled (Waller *et al.*, 1968). This finding has been confirmed by Charlwood (1970) but in *Artemesia annua* [1-^{14}C] acetate, [2-^{14}C] MVA and [1-^{14}C] geranyl pyrophosphate were all precursors of artemesia ketone. Here again only the isoprene units derived from IPP in the molecule became labelled. The labelling pattern was consistent with Reactions A, B and D in Fig. 7.

The carbon skeleton of artemesia ketone may also be derived from fission of bond Y of chrysanthemic acid (58). In chemical experiments careful choice of conditions leads to cleavage of any of the three C–C bonds of the chrysanthemic acid to give respectively the santolinyl (59), artemesyl (57) and lavandulyl (60) carbon skeletons (Crombie *et al.*, 1967) (Fig. 7). Chrysanthemyl alcohol, or its pyrophosphate, was therefore suggested as a biosynthetic precursor of these irregular monoterpenes. Chrysanthemic acid is formed from [2-^{14}C] MVA (Crowley *et al.*, 1962). The labelling pattern suggested that it might be synthesized by direct condensation of two DMAPP units (Reaction E, Fig. 7) (Godin *et al.*, 1963).

The structure of nezukone (19) is that of an isopropyl ketone (Hirose *et al.*, 1966). Nezukone may be related to the thujaplicins (17) (Fig. 1). The occurrence of compound Y (61, Fig. 7) in association with related tropones in a *Thuja* species known to produce monoterpenes of the pinene (15) or thujone type (14) (Birch and Keeton, 1968) reinforces an earlier suggestion by Birch (1950) that monoterpenes are intermediates in tropone biosynthesis.

IV. BIOSYNTHESIS IN CELL-FREE SYSTEMS

Plant enzyme systems have proved notoriously difficult to work with but recent advances in methodology seem to have eased these difficulties (Loomis and Battaile, 1966; Anderson, 1968). Cell-free systems capable of *de novo* terpene synthesis can now be prepared (e.g. Jungalwala and Porter, 1967).

Such systems synthesize the higher terpenes predominantly. Furthermore, though Graebe (1968) and Oster and West (1968) have obtained cell-free extracts which will utilize MVA with efficiency, most workers have found it necessary to study the polymerization of IPP (23) or its condensations with geranyl and farnesyl pyrophosphates. This apparent sensitivity of mevalonate phosphokinase to the isolation procedures may be due to its ready inhibition by geranyl and farnesyl pyrophosphates as has been found for the kinase isolated from pig liver (Dorsey and Porter, 1968) or to the factors discussed in Section II above. Monoterpenes are not synthesized by these systems except as minor by-products (e.g. Graebe, 1968).

However, Cori and his colleagues have described an extract from *Pinus radiata* seedlings that synthesizes monoterpenes from MVA (Beytia *et al.*, 1969; Valenzuela *et al.*, 1966). The enzyme system was soluble and required ATP, mercaptoethanol and bivalent ions for activity. The ratio of allylic phosphates to IPP (22) formed by this system could be altered by changes in Mn^{2+} and Mg^{2+} ion concentrations and ratios. Such an alternation in product ratio dependent on Mn^{2+}/Mg^{2+} ratios has also been found by Graebe (1968) in his study of kaurene, squalene and phytoene synthesis from MVA by extracts of pea seedlings. In a large scale experiment 0·4% of the label from [2-^{14}C] MVA was present in monoterpenes, 2·0% in C-5 alcohols and 2·4% in allylic alcohols. A certain proportion of the activity remained protein-bound. This activity could be extracted with n-butanol and comprised labelled farnesol, nerolidol and geranylgeraniol. The protein-bound activity may be related to the synthesis of higher terpenes in *Pinus* seedlings. A similar n-butanol extractable activity has been demonstrated by a bacterial prenyl synthetase (Allen *et al.*, 1967). Specific activities were not measured and hence no conclusions could be drawn on whether geranyl pyrophosphate (24) or neryl pyrophosphate (24a) was the first product of IPP (22) and DMAPP (23) condensation. Further experiments (Cori, 1969) have shown that [^{14}C] neryl pyrophosphate (24a) is a better precursor of α- and β-pinene (15) than [^{14}C] geranyl pyrophosphate (24). An interesting property was the stimulation of incorporation of label from [2-^{14}C] MVA into both C-5 and C-10 prenols by unsaturated cephalins and lecithins (George-Nascimento *et al.*, 1969).

A crude homogenate of *N. cataria* leaves has been shown to be effective in incorporating radioactivity from [2-^{14}C] MVA into nepetalactone (48) and other monoterpenes (Regnier *et al.*, 1968).

Battaile *et al.* (1968) have isolated a cell-free system from *M. piperita* which will convert [^{14}C] pulegone (7) in *ca.* 30% yield to products tentatively identified as menthone, isomenthone and menthol (8). NADPH was an essential co-factor. Enzymes catalysing the further metabolism of [^{14}C] pulegone (7) are also present in these extracts. Successful isolation of this enzyme system required removal of endogenous substrate and phenols as well as anaerobic conditions. The enzyme activities were very labile to storage. These results, together with those of Cori's group demonstrate the difficulties that will have to be overcome before isolation of other monoterpene biosynthetic enzymes can be expected.

V. Localization of Biosynthetic Sites

The accumulation of monoterpenes appears to be associated in many cases with glandular structures ("oil glands"). The development of these structures has been described in *Mentha* sp. (see Loomis, 1967) and in *N. cataria* (Regnier *et al.*, 1968). Studies on the secretory structures (glands) of *Pogostemmon cablin* (a sesquiterpene-producing plant) have shown an apparent developmental sequence (Hart and Henderson, 1970). Though glands at all stages of development were seen in the youngest leaves, young leaves contained higher proportions of the juvenile forms. *De novo* synthesis of monoterpenes (or sesquiterpenes) from externally added precursors may occur predominantly in juvenile glands while in the older glands only endogenous precursors give rise to monoterpenes (or sesquiterpenes). This would be in accord with the suggestion of Loomis (1967) to explain his results on monoterpene biosynthesis and accumulation in *Mentha* sp. leaves (Battaile and Loomis, 1961; Burbott and Loomis, 1969) and would also account for the results of Regnier *et al.* (1968) on nepetalactone biosynthesis in *N. cataria*. However, the relationships of these structures to monoterpene biosynthesis await elucidation.

Analyses of the contents of oil-glands are now possible using direct injection capillary gas–liquid chromatography. The studies of Hart and Henderson on *P. cablin*, *Pelargonium graveolens* and *Mentha* sp. using this technique have confirmed for the first time that these glands contain the mono and sesquiterpene mixtures characteristic of the plant species examined. This method provides an important technique for studies on the interrelationships between monoterpene biosynthesis and leaf and gland development.

De novo biosynthesis of monoterpenes can also be carried out by isolated flowers. In both rose and *Chrysanthemum cinerariaefolium* flowers, [2-^{14}C] MVA was a specific precursor of the characteristic flower monoterpenes (Francis and O'Connell, 1969; Crowley *et al.*, 1962). In *C. cinerariaefolium* the period for which biosynthesis occurred depended on the state of opening of the flower when picked, i.e. upon the ripeness and freshness of the ovules, which are the major site of monoterpene biosynthesis. In rose flowers, the maximum increase in rate of monoterpene accumulation occurred during the period when the petals were unfurling (Francis and Allcock, 1969). A study of the anatomy of the rose flowers showed a series of glandular structures on the upper epidermal surfaces of the petals (Stubbs and Francis, 1970). The cells comprising these upper surfaces contained unusual intra-cellular inclusions which could have a relationship to monoterpene biosynthesis. These structures themselves are suggestive of degraded chloroplasts. Fruits metabolize monoterpenes (Attaway and Buslig, 1968, 1969) and recently Potty and Bruemmer (1970) have shown that fruits contain mevalonate-activating enzymes. Roots do not appear to be required for monoterpene biosynthesis (e.g. Regnier *et al.*, 1968). Thus composition of the leaf monoterpenes of *Pelargonium* sp. grafted onto different *Pelargonium* sp. root stocks was determined by the scion (M. J. O. Francis, unpublished observations).

VI. Conclusions

Monoterpenes in plants are now firmly established as metabolically labile compounds whose function and relationship to the overall physiology of the plant still awaits elucidation. Moreover, their biosynthesis from MVA has been established only in the special cases of the iridoids and of the flower monoterpenes. The pathway of biosynthesis of leaf monoterpenes from MVA still awaits detailed elucidation and there remains a possibility that MVA may not be a precursor of the whole monoterpene molecule. These difficulties seem to stem from the apparent isolation of monoterpenes in specialized cells and should be overcome both by the increasing use of the recently described cell-free preparations and of direct analyses of individual monoterpene-containing cells. The subtle variations of monoterpene structure suggest that studies on the enzymes that interconvert monoterpenes would be a particularly valuable exercise on the detailed factors determining structure/activity relationships. The pathways of monoterpene catabolism in higher plants are unknown. Monoterpene metabolism in species (other than plants) which appear to synthesize them remains an unexplored field.

ACKNOWLEDGEMENTS

I wish to thank Dr D. V. Banthorpe of University College, London, and Drs B. W. Nichols and P. Dunphy of the Unilever Research Laboratories, Colworth House, Sharnbrook, Bedfordshire, for their help in the preparation of this manuscript.

REFERENCES

Achilladelis, B. and Hanson, J. R. (1968). *Phytochemistry* **7**, 1317.
Allen, C. M., Allworth, W., Macrae, A. and Bloch, K. (1967). *J. biol. Chem.* **242**, 1895.
Anderson, J. W. (1968). *Phytochemistry* **7**, 1973.
Archer, B. L. and Audley, B. G. (1967). *Adv. Enzymol.* **29**, 221.
Agranoff, B. W., Eggerer, H., Henning, U. and Lynen, F. (1960). *J. biol. Chem.* **235**, 326.
Attaway, J. A. and Buslig, B. S. (1968). *Biochim. biophys. Acta* **164**, 609.
Attaway, J. A. and Buslig, B. S. (1969). *Phytochemistry* **8**, 1671.
Attaway, J. A., Pieringer, A. P. and Barabas, L. J. (1967). *Phytochemistry* **6**, 25.
Auda, H., Waller, G. R. and Eisenbraun, E. J. (1967a). *J. biol. Chem.* **242**, 4157.
Auda, H., Juneja, H. R., Eisenbraun, E. J., Waller, G. R., Kays, W. R. and Appel, H. H. (1967b). *J. Am. chem. Soc.* **89**, 2476.
Ayrey, G., Banthorpe, D. V., Barnard, D. and Le Patourel, G. N. J. (In Preparation.)
Baisted, D. L. (1967). *Phytochemistry* **6**, 93.
Banthorpe, D. V. and Baxendale, D. (1968). *Chem. Commun.* 1553.
Banthorpe, D. V. and Turnbull, K. W. (1966). *Chem. Commun.* 177.

Banthorpe, D. V. and Wirz-Justice, A. (1969). *J. chem. Soc.* (C) 541.
Battaile, J. and Loomis, W. D. (1961). *Biochim. biophys. Acta* **51**, 545.
Battaile, J., Burbott, A. J. and Loomis, W. D. (1968). *Phytochemistry* **7**, 1159.
Battersby, A. R. (1967). *Pure Appl. Chem.* **14**, 117.
Battersby, A. R. (1970). *In* "Natural Substances formed Biologically from Mevalonic Acid" (T. W. Goodwin, ed.), p. 157, Academic Press, New York and London.
Battersby, A. R. and Gregory, B. (1968). *Chem. Commun.* 134.
Battersby, A. R. and Hall, E. S. (1969). *Chem. Commun.* 793.
Battersby, A. R. and Hall, E. S. (1970). *Chem. Commun.* 194.
Battersby, A. R., Brown, R. T., Kapil, R. S., Martin, J. A. and Plunkett, A. O. (1966). *Chem. Commun.* 812.
Battersby, A. R., Kapil, R. S. and Southgate, R. (1968a). *Chem. Commun.* 131.
Battersby, A. R., Kapil, R. S., Martin, J. A. and Mo, L. (1968b). *Chem. Commun.* 133.
Battersby, A. R., Byrne, J. C., Kapil, R. S., Martin, J. A., Payne, T. G., Arigoni, D. and Loew, P. (1968c). *Chem. Commun.* 951.
Battersby, A. R., Burnett, A. R., Knowles, G. D. and Parsons, P. G. (1968d). *Chem. Commun.* 1277.
Battersby, A. R., Burnett, A. R. and Parsons, P. G. (1968e). *Chem. Commun.* 1282.
Battersby, A. R., Burnett, A. R., Hall, E. S. and Parsons, P. G. (1968f). *Chem. Commun.* 1582.
Battersby, A. R., Burnett, A. R. and Parsons, P. G. (1969a). *J. chem. Soc.* (C) 1187.
Battersby, A. R., Burnett, A. R. and Parsons, P. G. (1969b). *J. chem. Soc.* (C) 1193.
Battu, R. G. and Youngken, H. W. (1966). *Lloydia* **29**, 360.
Beytia, E., Valenzuela, P. and Cori, O. (1969). *Archs Biochem. Biophys.* **129**, 346.
Biollaz, M. and Arigoni, D. (1969). *Chem. Commun.* 633.
Birch, A. J. (1950). *Ann. Rep. Chem. Soc. (Lond.)* **47**, 193.
Birch, A. J. and Keeton, R. (1968). *J. chem. Soc.* (C) 109.
Bowman, R. M. and Leete, E. (1969). *Phytochemistry* **8**, 1003.
Brechbühler-Bader, S., Coscia, C. J., Loew, P., von Szczepanski, Ch. and Arigoni, D. (1968). *Chem. Commun.* 136.
Brodie, J. D., Wasson, G. and Porter, J. W. (1964). *J. biol. Chem.* **239**, 1346.
Brown, R. T., Smith, G. N. and Stapleford, K. S. J. (1968). *Tetrahedron Letters* 4349.
Brown, R. T., Hill, J. S., Smith, G. N., Stapleford, K. S. J., Poisson, J., Muquet, M. and Kunesch, N. (1969). *Chem. Commun.* 1475.
Burbott, A. J. and Loomis, W. D. (1967). *Pl. Physiol.* **42**, 20.
Burbott, A. J. and Loomis, W. D. (1969). *Pl. Physiol.* **44**, 173.
Charlwood, B. V. (1970). Ph.D. Thesis, University of London.
Cori, O. (1969). *Archs Biochem. Biophys.* **135**, 416.
Coscia, C. J., Guarnacia, R. and Botta, L. (1969). *Biochemistry, Easton* **8**, 5036.
Crombie, L., Houghton, R. P. and Woods, D. K. (1967). *Tetrahedron Letters* 4553.
Crowley, M. P., Godin, P. J., Inglis, H. S., Snarey, M. and Thain, E. M. (1962). *Biochim. biophys. Acta* **60**, 312.
Dean, F. M., Price, A. W., Wade, A. P. and Wilkinson, G. S. (1967). *J. chem. Soc.* (C) 1893.
Dicks, J. W. and Friend, J. (1968). *Phytochemistry* **7**, 1933.
Dorsey, J. K. and Porter, J. W. (1968). *J. biol. Chem.* **243**, 4667.
Francis, M. J. O. and Allcock, C. (1969). *Phytochemistry* **8**, 1339.
Francis, M. J. O. and O'Connell, M. (1969). *Phytochemistry* **8**, 1705.
Francis, M. J. O., Banthorpe, D. V. and Le Patourel, G. N. J. (1970). *Nature, Lond.* **228**, 1005.

Garg, A. K. and Gear, J. R. (1969a). *Tetrahedron Letters* 4377.
Garg, A. K. and Gear, J. R. (1969b). *Chem. Commun.* 1447.
Godin, P. J., Inglis, H. S., Snarey, M. and Thain, E. M. (1963). *J. chem. Soc.* 5878.
George-Nascimento, C., Beytia, E., Aedo, R. and Cori, O. (1969). *Archs Biochem. Biophys.* **132**, 470.
Graebe, J. E. (1968). *Phytochemistry* **7**, 2003.
Hart, J. W. and Henderson, W. (1970). *Phytochemistry* **9**, 1219.
Hefendehl, F. W., Underhill, E. W. and von Rudloff, E. (1967). *Phytochemistry* **6**, 823.
Hirose, Y., Tomita, B. and Nakatsuka, T. (1966). *Tetrahedron Letters* 5875.
Holloway, P. W. and Popják, G. (1968). *Biochem. J.* **106**, 835.
Horodysky, A. G., Waller, G. R. and Eisenbraun, E. J. (1969). *J. biol. Chem.* **244**, 3110.
Hylin, J. W. (1964). *Phytochemistry* **3**, 161.
Inouye, H., Ueda, S. and Nakamura, Y. (1967). *Tetrahedron Letters* 3221.
Inouye, H., Ueda, S. and Takeda, Y. (1968). *Tetrahedron Letters* 3453.
Inouye, H., Ueda, S. and Takeda, Y. (1969a). *Tetrahedron Letters* 407.
Inouye, H., Ueda, S., Aoki, Y. and Takeda, Y. (1969b). *Tetrahedron Letters* 2351.
Jungalwala, F. B. and Porter, J. W. (1967). *Archs Biochem. Biophys.* **119**, 209.
Kutney, J. P., Ehret, C., Nelson, V. R. and Wigfield, D. C. (1968). *J. Am. chem. Soc.* **90**, 5929.
Leete, E. (1969). *Adv. Enzymol.* **32**, 373.
Loew, P. and Arigoni, D. (1968). *Chem. Commun.* 137.
Loew, P., von Szczepanski, Ch., Coscia, C. J. and Arigoni, D. (1968). *Chem. Commun.* 1276.
Loomis, W. D. (1967). *In* "Terpenoids in Plants" (J. B. Pridham, ed.), p. 59, Academic Press, New York and London.
Loomis, W. D. and Battaile, J. (1966). *Phytochemistry* **5**, 423.
Lynen, F. (1961). *Fedn. Proc. Fedn. Am. Socs. exp. Biol.* **20**, 941.
Oster, M. O. and West, C. A. (1968). *Archs Biochem. Biophys.* **127**, 112.
Paseshnichenko, V. A. and Guseva, A. R. (1967). *Biokhimiya* **32**, 1020.
Potty, V. H. and Bruemmer, J. H. (1970). *Phytochemistry*, **9**, 99.
Qureshi, A. A. and Scott, A. I. (1968a). *Chem. Commun.* 945.
Qureshi, A. A. and Scott, A. I. (1968b). *Chem. Commun.* 948.
Ramage, R. (1967). *Ann. Rep. Chem. Soc.* (*Lond.*) **64B**, 511.
Ramage, R. (1968). *Ann. Rep. Chem. Soc.* (*Lond.*) **65B**, 577.
Regnier, F. E., Waller, G. R., Eisenbraun, E. J. and Auda, H. (1968). *Phytochemistry* **7**, 221.
Renwick, J. A. A. (1967). *Contribut. Boyce Thompson Inst.* **23**, 355.
Rittersdorf, W. and Cramer, F. (1968). *Tetrahedron* **24**, 43.
Rogers, L. J., Shah, S. P. J. and Goodwin, T. W. (1967). *In* "Biochemistry of Chloroplasts" (T. W. Goodwin, ed.), Vol. 2, p. 283, Academic Press, London and New York.
Scher, M., Lennarz, W. J. and Sweeley, C. C. (1968). *Proc. natn. Acad. Sci. U.S.A.* **59**, 1313.
Scora, R. W. and Mann, J. D. (1967). *Lloydia* **30**, 236.
Shah, D. H., Cleland, W. W. and Porter, J. W. (1965). *J. biol. Chem.* **240**, 1946.
Sheth, K., Ramstad, E. and Wolinsky, K. (1961). *Tetrahedron Letters* 394.
Silverstein, R. M., Rodin, J. O. and Wood, D. L. (1966). *Science, N.Y.* **154**, 509.
Smith, G. N. (1968). *Chem. Commun.* 912.
Sticher, O. (1969). *Pharm. Acta Helv.* **44**, 453.

Stubbs, J. M. and Francis, M. J. O. (1970). *Planta Medica*. (In press.)
Taylor, W. I. (1966). *Science, N.Y.* **153**, 954.
Taylor, W. I. and Battersby, A. R. (1969). "Cyclopentanoid Terpene Derivatives", M. Dekker, New York.
Thies, P. W. (1968). *Tetrahedron* **24**, 313.
Thomas, R. (1961). *Tetrahedron Letters* 544.
Valenzuela, P., Cori, O. and Yudelevich, A. (1966). *Phytochemistry* **5**, 1005.
van Aller, W. T. and Nes, W. R. (1968). *Phytochemistry* **7**, 85.
Wakil, S. J. and Ganguly, J. (1959). *J. Am. chem. Soc.* **81**, 2597.
Waller, G. R. (1969). *Prog. Chem. Fat and other Lipids.* **10**, 151.
Waller, G. R., Frost, G. M., Burleson, D., Brannon, D. and Zalkow, L. H. (1968). *Phytochemistry* **7**, 213.
Weatherston, J. (1967). *Q. Rev. chem. Soc.* **21**, 287.
Wenkert, E. (1962). *J. Am. chem. Soc.* **84**, 98.
Wieffering, J. H. (1966). *Phytochemistry* **5**, 1053.
Wood, B. J. B. (1969). *Process Biochemistry* **4**, 50.
Yang, K. S., Gholson, R. K. and Waller, G. R. (1965). *J. Am. chem. Soc.* **87**, 4184.
Yeowell, D. A. and Schmid, H. (1964). *Experientia* **20**, 250.

CHAPTER 3

Biochemistry of Sesquiterpenoids

V. HEROUT

Institute of Chemistry and Biochemistry, Czechoslovak Academy of Sciences, Prague, Czechoslovakia

I. Review of Sesquiterpenoid Types

Sesquiterpenoids are usually characterized as compounds having a basic skeleton of fifteen carbon atoms, formed, similarly to other terpenoids, by a regular repetition of the basic isoprene unit, and occurring mainly in plant material. As regards the last point, the definition of sesquiterpenoids had to be appreciably extended in recent years (cf. also Bryant, 1969). Such substances have been often encountered in the animal kingdom, especially in insects, but also in Coelenterata, Mollusca, etc. Not less numerous are sesquiterpenoids, often specifically transformed, in the third evolutionary branch of living organisms, the Fungi (cf. Whittaker, 1969). The occurrence of sesquiterpenoid derivatives is not limited to moulds, but is even more frequent in the so-called higher fungi. In contrast to this the presence of sesquiterpenoids in the members of the kingdom of Monera and Protista (see Whittaker, 1969), i.e. in organisms of low levels of organization, has not yet been observed.

Parallel with the broad occurrence of substances of sesquiterpenic character, there is also a rich variety of structures of these compounds. In fact, it is much richer than had been supposed earlier. First of all I should like to mention the existence of compounds containing elements which were quite unusual in the natural molecules of sesquiterpenoids. Some derivatives contain nitrogen, as for example some alkaloids of clearly sesquiterpenoid origin, others also contain sulphur, bromine or chlorine. It is true that sulphur is usually bound in supernumerary chemical groupings, as for example in the form of β-methylthioacrylic acid in S-petasin (1) or S-petasitolide A or B (2, 3) (see Aebi *et al.*, 1958; Novotný *et al.*, 1962, 1964). Brominated sesquiterpenoids represent the products of marine flora and fauna, e.g. laurenisol (4) or aplysin (5) (see Irie *et al.*, 1969; Yamamura and Hirata, 1963). The occurrence of chlorine is also interesting. It was found by Kupchan *et al.* (1968) in some sesquiterpenoid lactones from *Eupatorium rotundifolium*, for example in eupachlorin acetate (6). Groupings formerly unknown in sesquiterpenoids

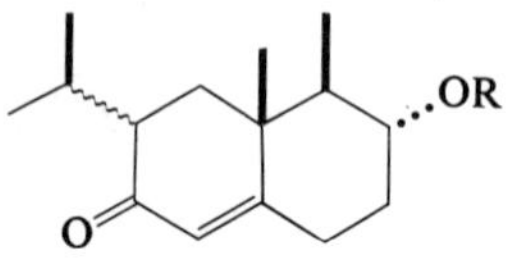

(1) Petasin R = A

β-Me-thioacrylic acid
(*cis*) = A
or
(*trans*) = B
(not decided)

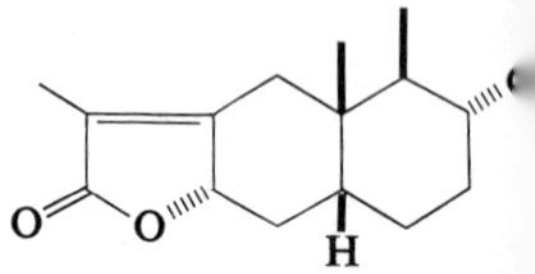

(2) Petasitolide A R =
(3) Petasitolide B R =

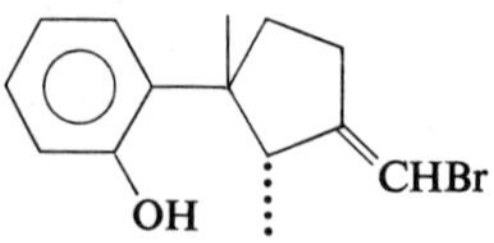

(4) Laurenisol

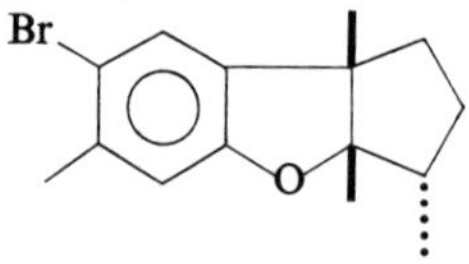

(5) Aplysin

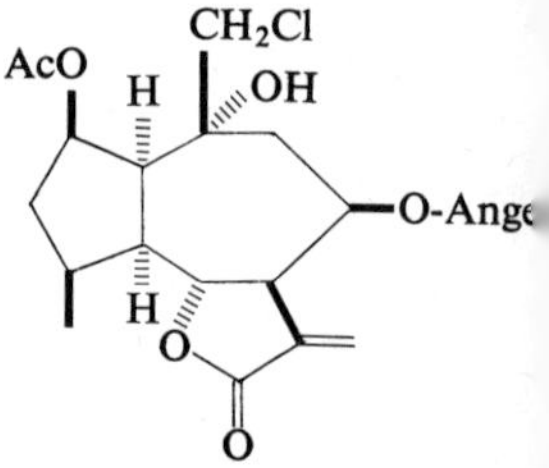

(6) Eupachlorin acetate

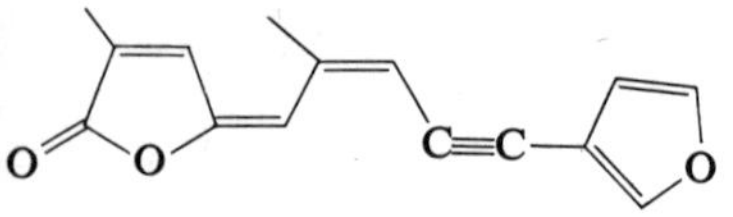

(7) Freelingyne

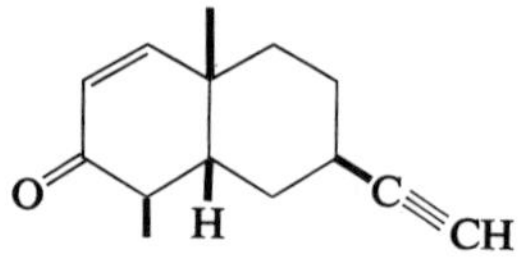

(8) Chamaecynone

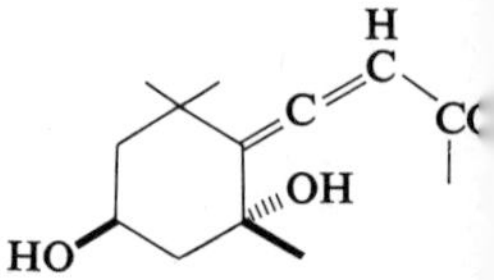

(9) Allenic sesquiterpenoi

have also been recently detected, as for example an acetylenic bond in freelingyne (7) (Massy-Westrop *et al.*, 1966) or in compounds of chamaecynone (8) type (Nozoe *et al.*, 1966). The only allene type grouping was observed in a sesquiterpenoid (9) from the flightless grasshopper *Romalea microptera* (Meinwald *et al.*, 1968).

The very rapidly increasing number of sesquiterpenoid compounds found in nature is a striking feature of this group. From the time of the structural determination of farnesol in 1913, Barton in 1953 listed the structures of some 30 sesquiterpenoids, subdivided into 16 types, but by 1964 well over 200 derivatives belonging to 40 types (e.g. Bryant, 1969) were reviewed. Towards the end of 1969 I was able to count well over 850 naturally occurring compounds with well-defined structures belonging to almost one hundred skeletal types, though this number is rather incomplete. Interrelationships between these sesquiterpenoids are quite well understood and the probable biosynthetic pathways leading to them have already been reviewed. The importance of the isoprene biogenetic rule in these studies can be mentioned. This rule represents one of the most important achievements of Professor Ruzicka in the field of the chemistry of terpenoids (cf. Ruzicka, 1956, 1959). When new types were discovered possible biosynthetic pathways were proposed by Hendrickson (1959), Richards and Hendrickson (1964), and partly also by Teisseire (1969). They were reviewed in great detail also by Parker *et al.* (1967). However, it remains true that very few of the biogenetic schemes considered today have been corroborated by ^{14}C tracer studies. From our own experience we can confirm that one reason for the paucity of such corroborations is the difficulty of feeding tracers to higher plants. Therefore up to now the main results with tracer experiments have been obtained with fungi.

For the classification of sesquiterpenoid compounds their carbon skeletons were mainly made use of. Originally these were grouped according to chemical principles, mainly the number of rings in the molecule. Nowadays it is possible to summarize in a very condensed form our present knowledge of their biogenetical relations. However, in view of our present state of knowledge this attempt, presented in Tables I–VII must necessarily be incomplete and, possibly in some cases inaccurate; one important reason is that it is frequently impossible to decide with certainty whether a single biosynthetic pathway exists. For single substances several pathways can be considered, but usually no single mechanism was clearly proved. In biological material ring formation need not be a gradual process, although it is usually presented as such implicating hypothetical intermediates. Sometimes a concerted reaction plays a role, when the whole sequence of reactions takes place on the surface of the enzymes. Another difficulty is that for the sake of ease of presentation the situation in this representation is oversimplified, for example the steric situation is not taken into account. It is indeed the differing steric situation which suggests the varying biogenetical course of the changes and cyclizations; it is clearly different in *trans*-decalin derivatives such as compounds with a basic cadinane or very probably also of bulgarane skeleton, than in substances with the same gross structure but with *cis*-fused rings, as for example muurolane and amorphane derivatives. Many similar examples are already known; in the accompanying tables the names of the known stereoisomeric skeletons are given in brackets under the most common name of the skeletal type. In broad outline this genetic dependence is based on the review by Parker *et al.* (1967), but it has been completed by the addition of recent data. Therefore, in the following discussion only the more recent information will be fully documented.

In Table I a sesquiterpene of the sesquicarane type (sesquicarene, 10) is mentioned (Ohta and Hirose, 1968) the biosynthesis of which is undoubtedly connected with the biosynthesis of substances of the bisabolane type. Its name

TABLE I
Bisabolane class

Farnesane

Cuparane

Trichothecane

Laurinterisane

Bisabolane

Bazzanane

Mayurane

Laurane

Sesquicarane

Chamigrane

Widdrane

Thujopsane

Acoriane

Bergamotane

α-Santalane

β-Santalane

Acorane

Cedrane

Anisatane

reflects the fact that this hydrocarbon is practically a carene of the sesquiterpene series. The mode of formation of substances of the cedrane type, proposed a long time ago, was recently checked experimentally *in vitro* by transforming the newly isolated hydrocarbon acoradiene (11) and acoronol (12) (Tomita and Hirose, 1970) by acid catalysed cyclization into α- or β-cedrene. A fully new skeleton is represented by bazzanane type, which is derived from a hydrocarbon isolated from the liverwort species *Bazzania pompeana* (Lac.) Mitt. (Hayashi *et al.*, 1969). The expression "acoriane type" is inferred from the name of acoric acid (Birch *et al.*, 1964); it can be clearly derived by oxidative

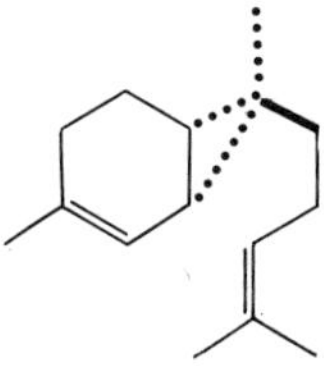

α- and β-Cedrene

(10) Sesquicarene

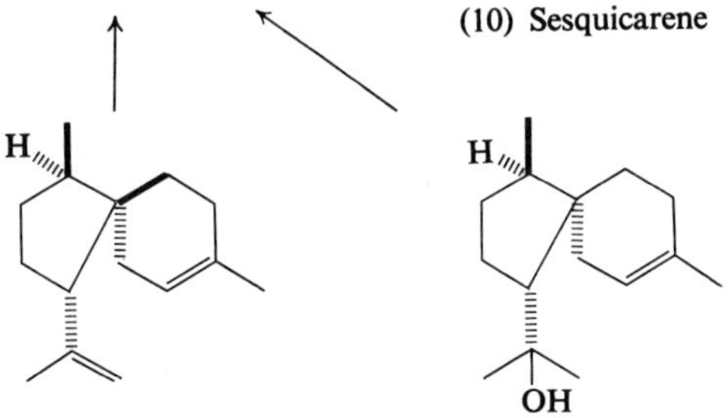

(11) Acoradiene

(12) Acorenol

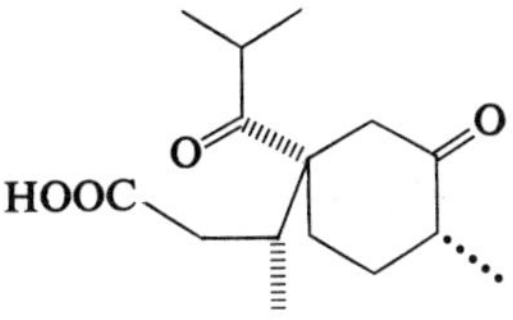

(13) Acoric acid

fission of the five-membered ring in acorone. Exceptionally the presence of an ether-bound oxygen atom is indicated in the skeleton of trichothecane, because the authors (Godfredsen *et al.*, 1967) proposed the name and the numbering of the skeleton in this form.

In Table II a group of types is given, which belong to an interesting group of sesquiterpenoid derivatives from higher fungi. They are represented by derivatives of the illudin group of which we currently know three types differing by the transformation of the cyclopropane and cyclobutane grouping localized on the six-membered ring; the recently studied illudalic acid (14a) or the alkaloid illudinine (14b) (Nair *et al.*, 1969) can be derived formally from derivatives with a cyclobutane ring by opening the ring. The supposition of a biogenetic pathway involving an intermediary stage of the humulane type was expressed both for the derivatives just mentioned and for marasmic and hirsutic acids, (15) and (16), respectively (Dugan *et al.*, 1965; Corner *et al.*, 1965). The newly described fomannosin (17) isolated recently from *Fomes annosus* (Kepler *et al.*, 1967) has an irregular skeleton, but important features of its molecule are strikingly similar to the structures just described. Its assignment to this group has, however, a speculative character, because no detailed basis for its biosynthesis has yet been proposed.

The basic skeleton of the cadinane type given in Table III represents one type of sesquiterpenoid substances which occurs frequently in nature. For the genesis of substances of this type three possibilities exist. However, recent data suggest that the biosynthetic pathway, involving a ten-membered germacrane type compound as an intermediate, seems the most probable. According to Yoshihara *et al.* (1969) germacrene-D (18a), found in nature,

TABLE II
Humulane class

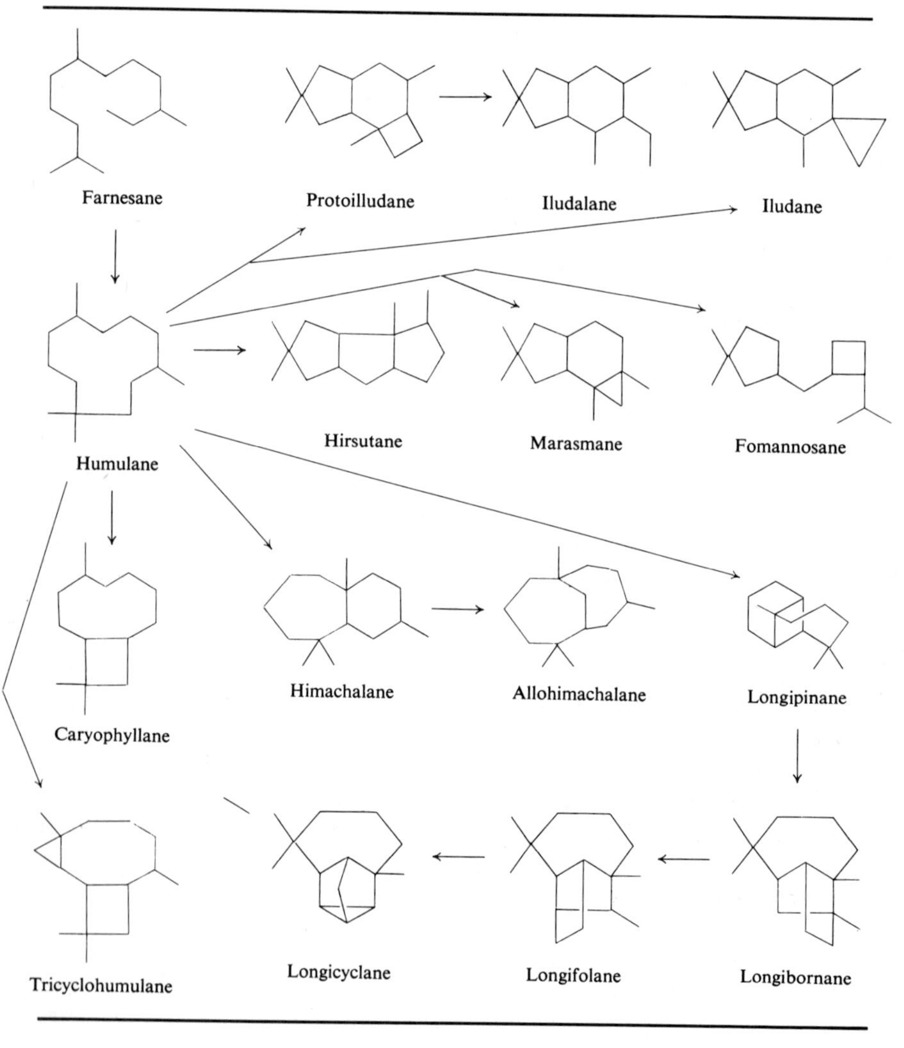

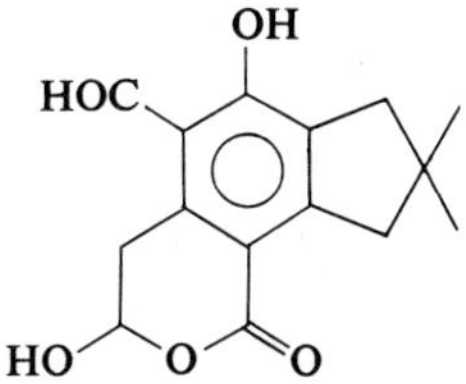

(14a) Illudalic acid

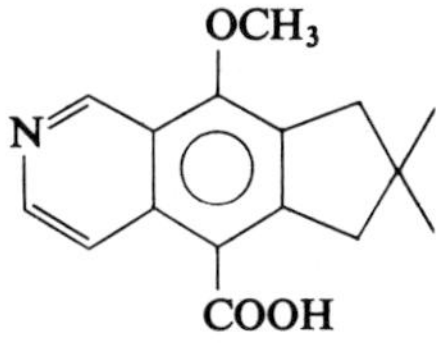

(14b) Illudinine

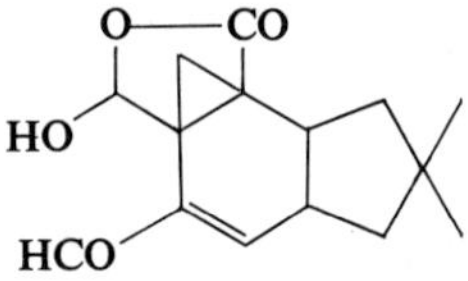

(15) Marasmic acid

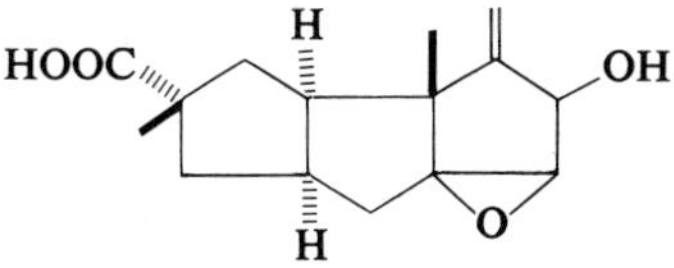

(16) Hirsutic acid

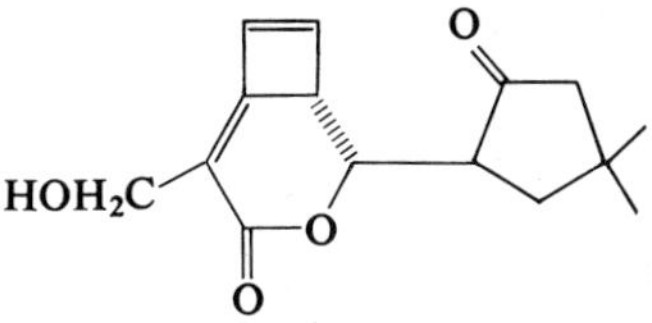

(17) Fomannosin

is a key intermediate in the biosynthesis of the cadinene type sesquiterpenoids; at least, it is highly probable that the γ-isomers of cadinene (18b) originates solely from the D-isomer. α- and β-bourbonenes (18c) may be formed very easily by a photoreaction, as indicated in Scheme 1.

An interesting group of substances derived from a cadinane skeleton was recently revealed by the discovery of copaborneol (19) (Kolbe and Westfelt, 1967) and cyclosativene (20b) (Smedman and Zavarin, 1968). A series of pairs of substances differing only by the configuration on the carbon substituted with an isopropyl group is thus represented by copaene and ylangene (21a,b), copacamphene and sativene (22a,b), and cyclocopacamphene and cyclosativene (20a,b).

In Table IV a review of substances of the selinane type is given. In the literature the name eudesmane type is also quite current. Their formation from germacrane precursors is common, but confusion was caused by the fact that such types of substances were found in nature which have a different configuration on some centres (see for example intermedeol and neointermedeol (23, 24) (Zalkow *et al.*, 1963, 1964). Recently the selinane group was

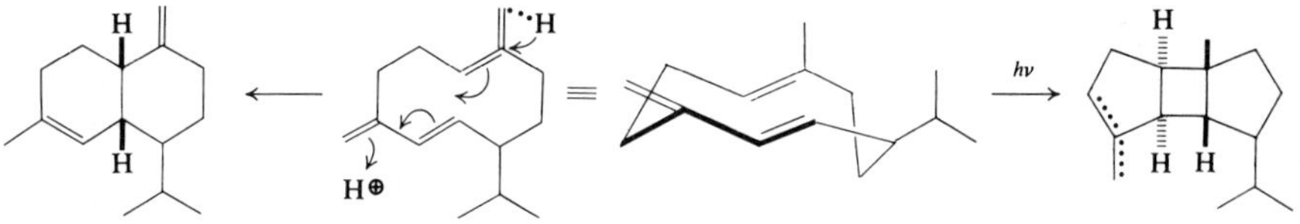

(18b) γ-Cadinene

(18a) Germacrene D

(18c) Bourbonenes

Scheme 1 (Yoshihara *et al.*, 1969).

TABLE III
Germacrane class (Cadinane group)

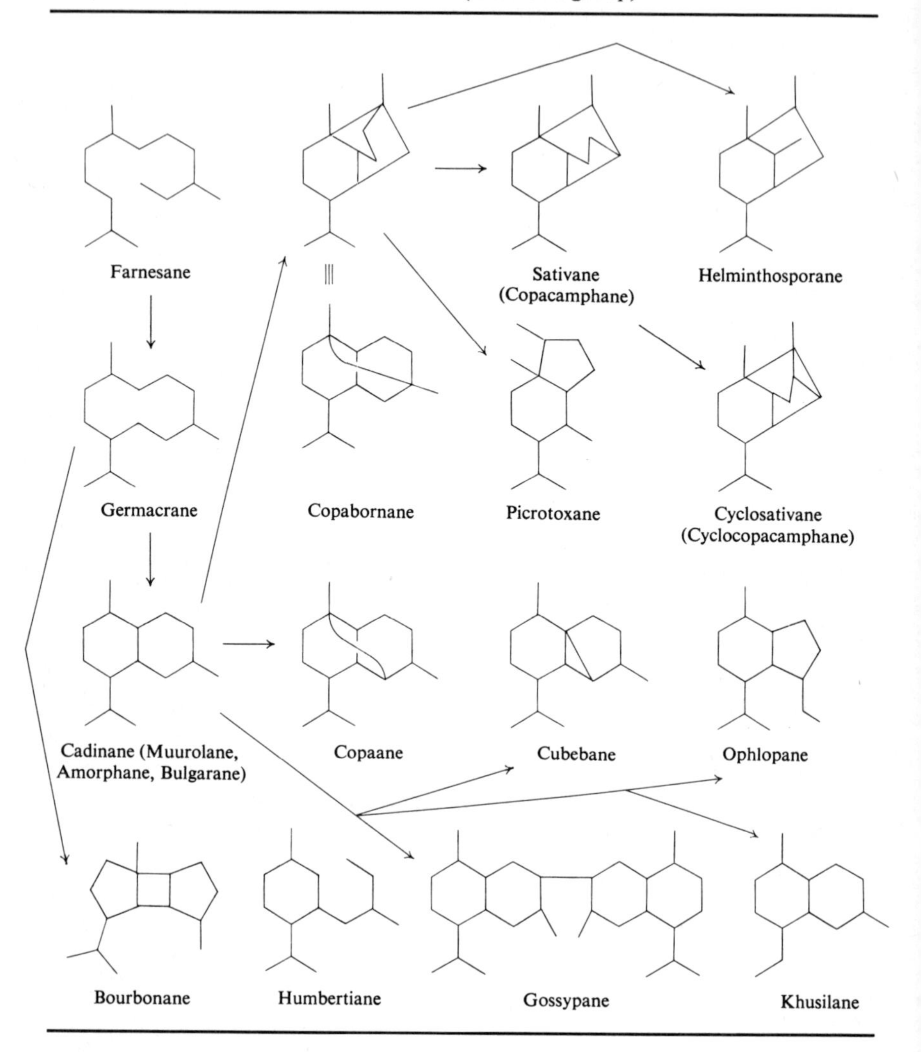

completed by a series of types, as for example the carbon skeleton of ivangulin (25) (Herz *et al.*, 1967), representing in fact a seco derivative, and that of lindenone (26) which has a carbon skeleton containing an additional, three-membered ring (Takeda *et al.*, 1967). The formation of compounds with the cyperane skeleton is explained by the contraction of one of the rings in the selinane precursor; cyperolone (27) for example belongs to this group (Hikino *et al.*, 1966). Among derivatives belonging to the selinane type substances with a chamaecynane skeleton can be found, representing non-derivatives

OH

(19) Copaborneol (20a) Cyclocopacamphene (20b) Cyclosativene

(21a) Copaene (21b) Ylangene

(22a) Copacamphene (22b) Sativene

created by the loss of a methyl group from the original isopropyl group (see chamaecynone 8), or substances with a rishitane skeleton, where the methyl group in position 10 was eliminated; rishitin (28) is a product of mould metabolism in infected potato tubers (Katsui *et al.*, 1968).

For derivatives of the tricyclovetivane type (syn.: zizaane type) only very recently a biosynthetic pathway was proposed which enables the simple correlation of all types of sesquiterpenoids identified up to the present time in vetiver oil (*Vetiveria zizanoides*) (see MacSweeney *et al.*, 1970). On the basis of two different conformers of the germacrane type an easy combination can be made of the selinane types with different configurations at C-4 and C-10

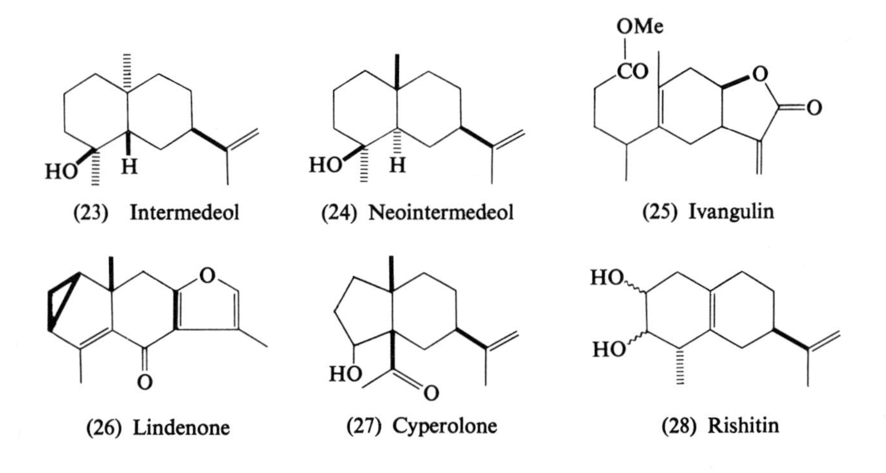

(23) Intermedeol (24) Neointermedeol (25) Ivangulin

(26) Lindenone (27) Cyperolone (28) Rishitin

TABLE IV
Germacrane class (Selinane = Eudesmane group)

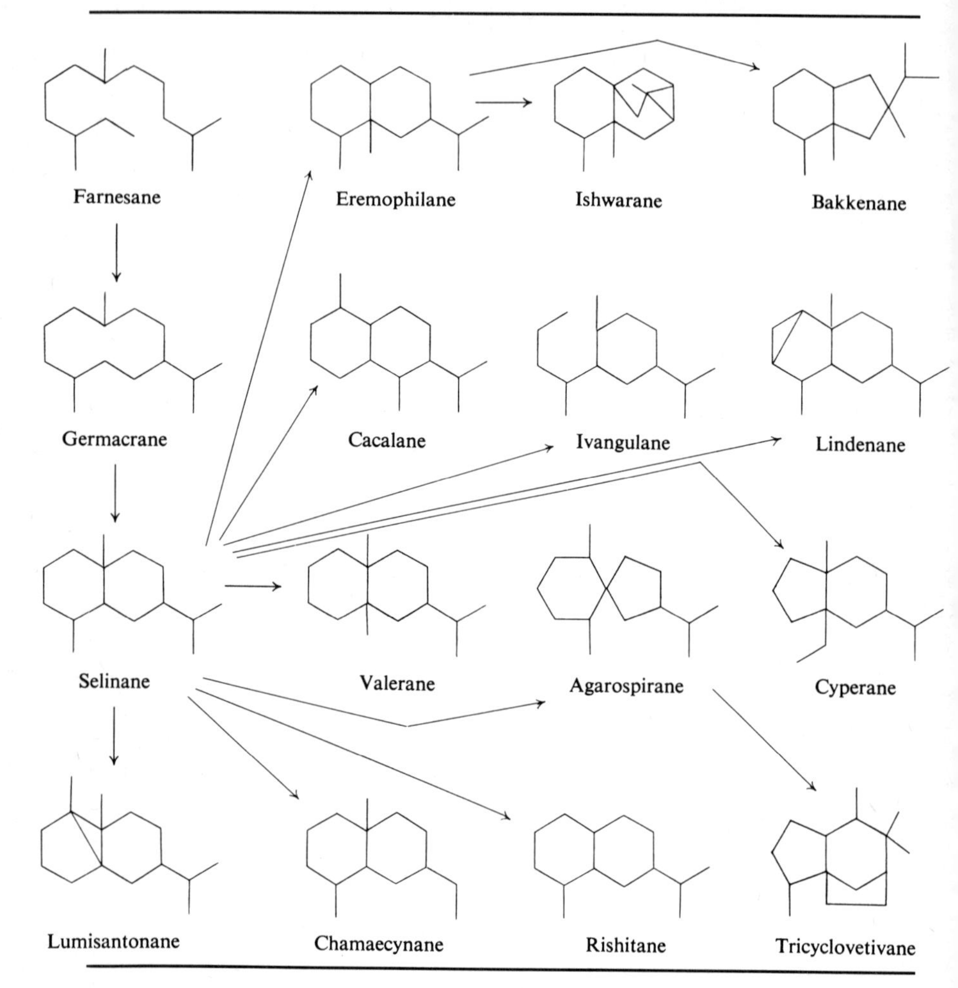

with such types as aristolane (see Table VI), eremophilane (with a configuration like that known in α-vetivone (29) for example; see Endo and de Mayo, 1967), agarospirane (in this case in β-vetivone (30); see Marshall and Johnson, 1967), and eventually also tricyclovetivane (for example in tricyclovetivene 31a; see Sakuma and Yoshikoshi, 1968; or in zizanoic acid 31b, Kido *et al.*, 1967). However, for details the original paper should be consulted. The conclusion of the latter papers is that substances with a perhydrovetivazulene skeleton probably do not occur in nature.

The dimeric absinthane type given in Table V is created, supposedly, by diene addition from precursors (Vokáč *et al.*, 1968), presumably the ses-

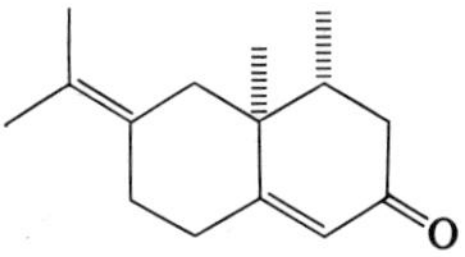

(29) α-Vetivone

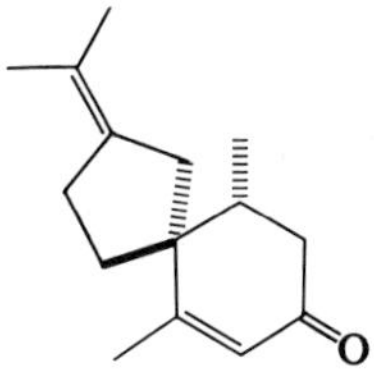

(30) β-Vetivone

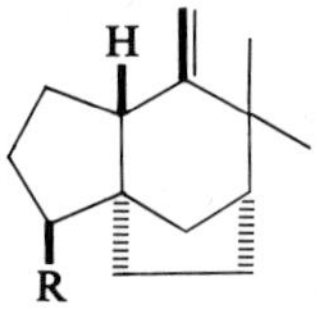

(31a) Tricyclovetivene $R = CH_3$
(31b) Zizanoic acid $R = COOH$

quiterpenoid lactone artabsin (32) or substances differing from artabsin only by the position of the double bonds in the five-membered ring (Vokáč *et al.*, 1968). Hence, the course of the dimerization is different from that which can be supposed for the formation of the second known type of sesquiterpenoid dimer, gossypol.

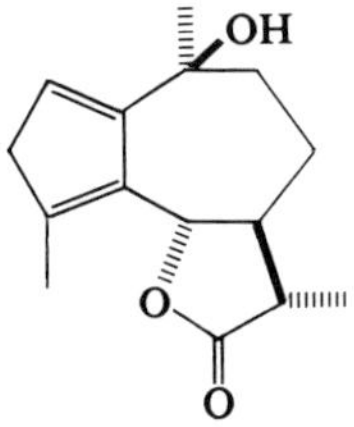

(32) Artabsin

The three related types, represented for example by patchoulenone, β-patchoulene, and patchouli alcohol (33–35), i.e. the isopatchoulane, patchoulane, and patchouliane types—were recently increased by a fourth type, seychelane. Its existence in seychellene (36) was proved by Wolff and Ourisson (1968) who also suggested the possibility (Wolff and Ourisson, 1969) of its biosynthesis by the migration of one of the gem-methyl groups in a precursor of patchouliane type according to Scheme 2.

The bicyclogermacrane type mentioned in Table VI has now been proved in nature by the discovery of bicyclogermacrene (37a) in cold-pressed *Citrus junos* oil (Nishimura *et al.*, 1969). This suggests that the hydrocarbon not only is a precursor of bicycloelemene (38, Vlahov *et al.*, 1967) but a key intermediate of sesquiterpenes of the aromadendrane and alloaromadendrane type. Its direct acid-catalysed cyclization to γ-cadinene (see Scheme 3) further supports the formation of cadinene derivatives via a germacrane intermediate. The recently described occurrence in nature of a substance with a gorgonane skeleton [β-gorgonene (39)], isolated from the gorgonian *Pseudopterogorgia americana* Gmelin (Weinheimer *et al.*, 1968), can be explained from the point of view of biosynthesis probably in an analogous manner to the formation of the zierane skeleton, i.e. by the opening of a three-membered

TABLE V
Germacrane class (Guaiane group)

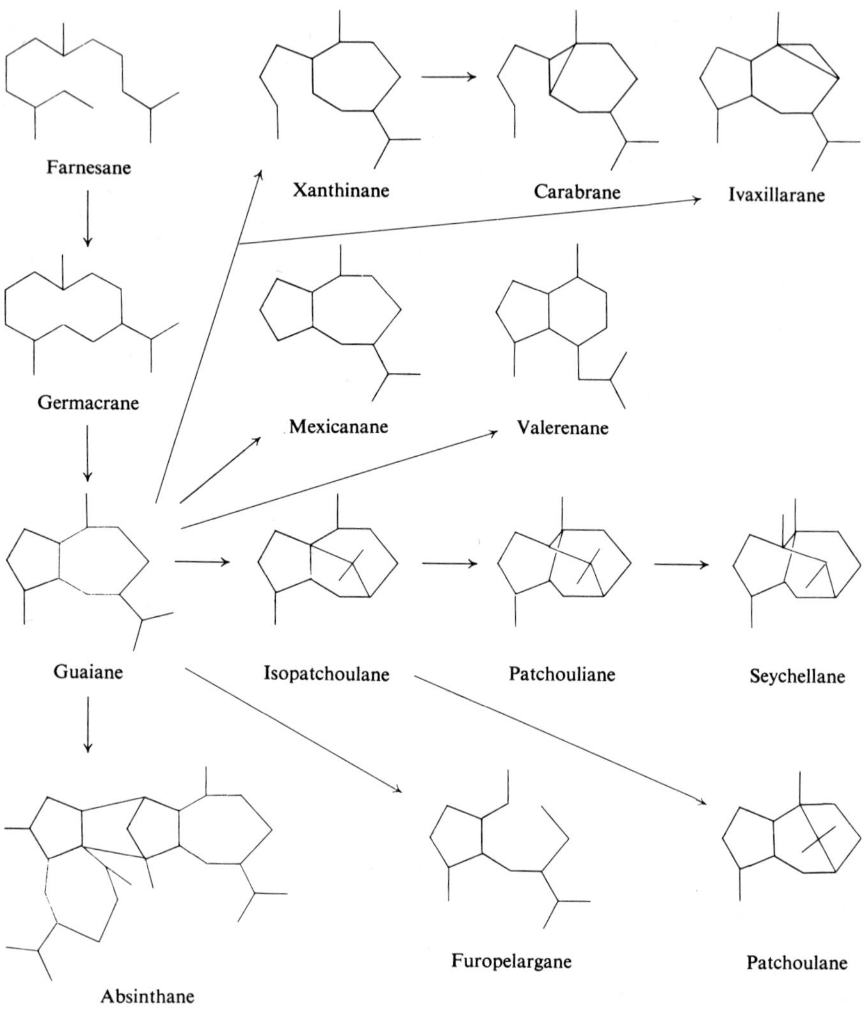

Farnesane

Xanthinane

Carabrane

Ivaxillarane

Germacrane

Mexicanane

Valerenane

Guaiane

Isopatchoulane

Patchouliane

Seychellane

Absinthane

Furopelargane

Patchoulane

ring, for example in the precursor of the bicyclogermacrane type, between the carbon atoms 7 and 11. The nardosinane type also fits into a similar biosynthetic scheme. It was first described in nardosinone (40) (for structure see Rücker, 1968) and it is evidently derived formally from a precursor with a eremophilane skeleton. The easy formation of substances of the elemane type, for example from precursors of the germacrane type at not too elevated temperatures, led to the conclusion that the substances derived from elemane

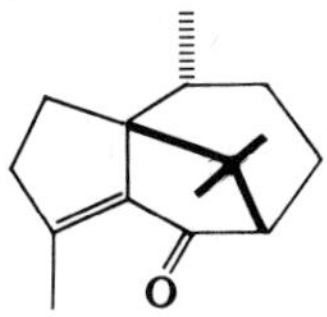

(33) Patchoulenone

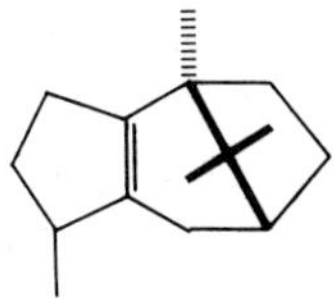

(34) β-Patchoulene

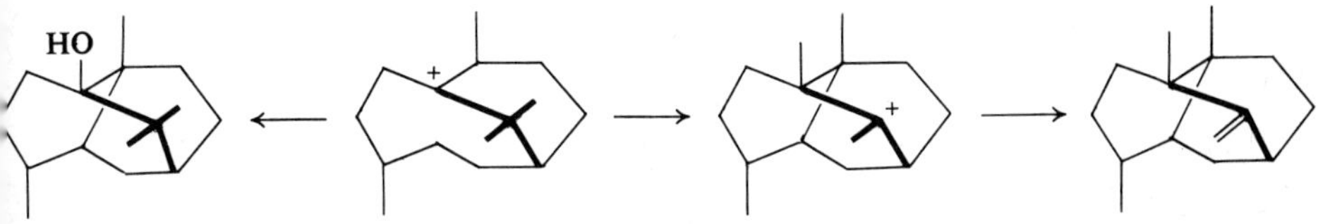

HO

+

+

) Patchouli alcohol

(36) Seychellene

Scheme 2 (Wolff and Ourisson, 1969).

are probably not present in nature; hence, they are only artifacts as is stated for example in the study on the occurrence of hedycaryol (41) in place of the previously reported elemol (Jones and Sutherland, 1968). However, the clear existence of vernolepin (42) (Kupchan *et al.*, 1968) representing a rather oxidized elemane sesquiterpenoid speaks against this conclusion.

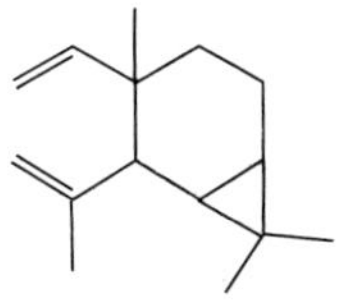

(38) Bicycloelemene

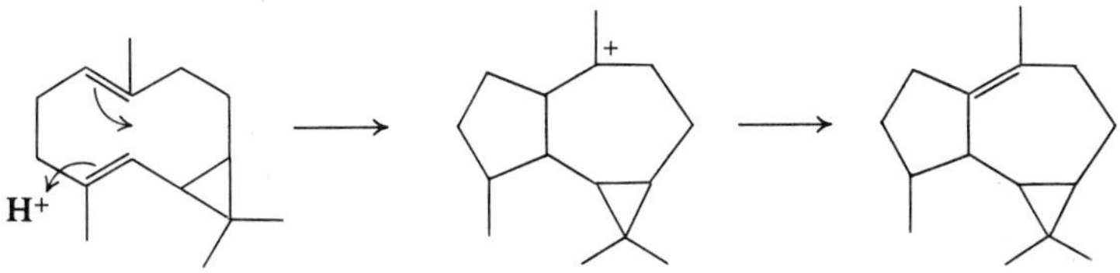

H+

(37a) Bicyclogermacrene

(37b) Ledene

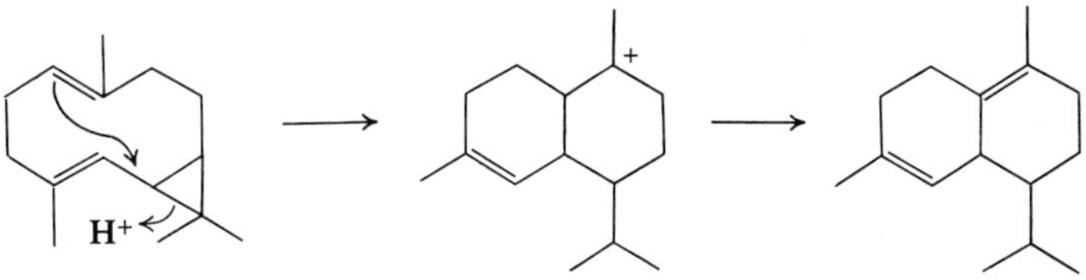

H+

(37a)

(37c) δ-Cadinene

Scheme 3 (Nishimura *et al.*, 1969).

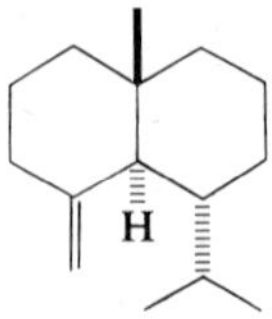

(39) β-Gorgonene

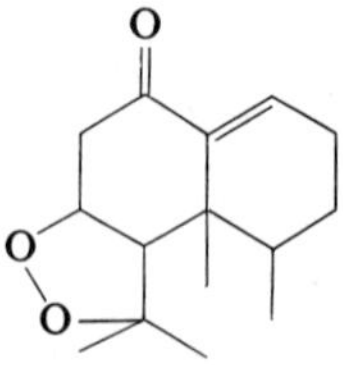

(40) Nardosinone

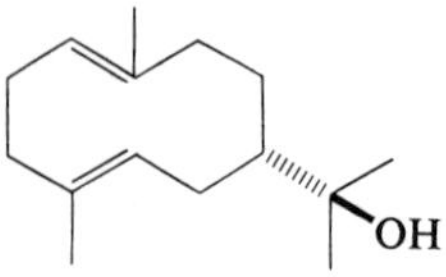

(41) Hedycaryol

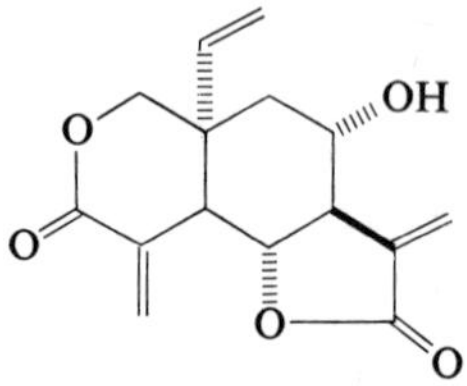

(42) Vernolepin

Nothing sure is known about geosmin (43) (Safferman *et al.*, 1967) and its biosynthesis. Its skeleton is identical to that of cogeijerene (44) the biosynthesis of which can be explained by the cyclization of pregeijerene (45) by analogy with the formation of elemane derivatives (Gough *et al.*, 1961).

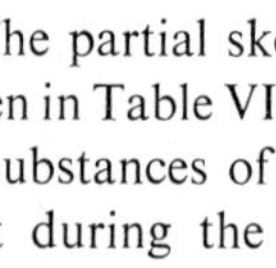

(43) Geosmin

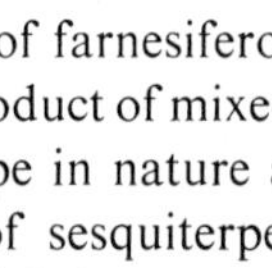

(44) Cogeijerene

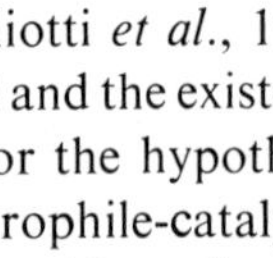

(45) Pregeijerene

The partial skeletal formula of farnesiferol B (46b) (Cagliotti *et al.*, 1958) given in Table VII (which is a product of mixed biosynthesis!) and the existence of substances of the iresane type in nature are evidences for the hypothesis that during the biosynthesis of sesquiterpenoids an electrophile-catalysed cyclization is not excluded, which is very typical of diterpenoids and triterpenoids. The mixed course of the biosynthesis, i.e. the combination of a part of sesquiterpenic origin with another part with a different biogenetic origin, is not unusual. As a classic example the already mentioned farnesiferols A–C (46a,b,c) (Cagliotti *et al.*, 1959) may be mentioned. More recently a series of other examples was found especially among mould metabolites, as for example tauranin, grifolin, siccanin and its precursors. These substances will be discussed later.

The abscisane skeleton discovered first in abscisic acid (47) can easily be derived from farnesol by cyclization. It is interesting to note that the abscisane skeleton had been synthesized already in 1950 as a possible sesquiterpenoid skeleton (Šorm and Dolejš, 1950). The substance which is considered to be

TABLE VI
Germacrane class (Elemane and Bicycloelemane groups)

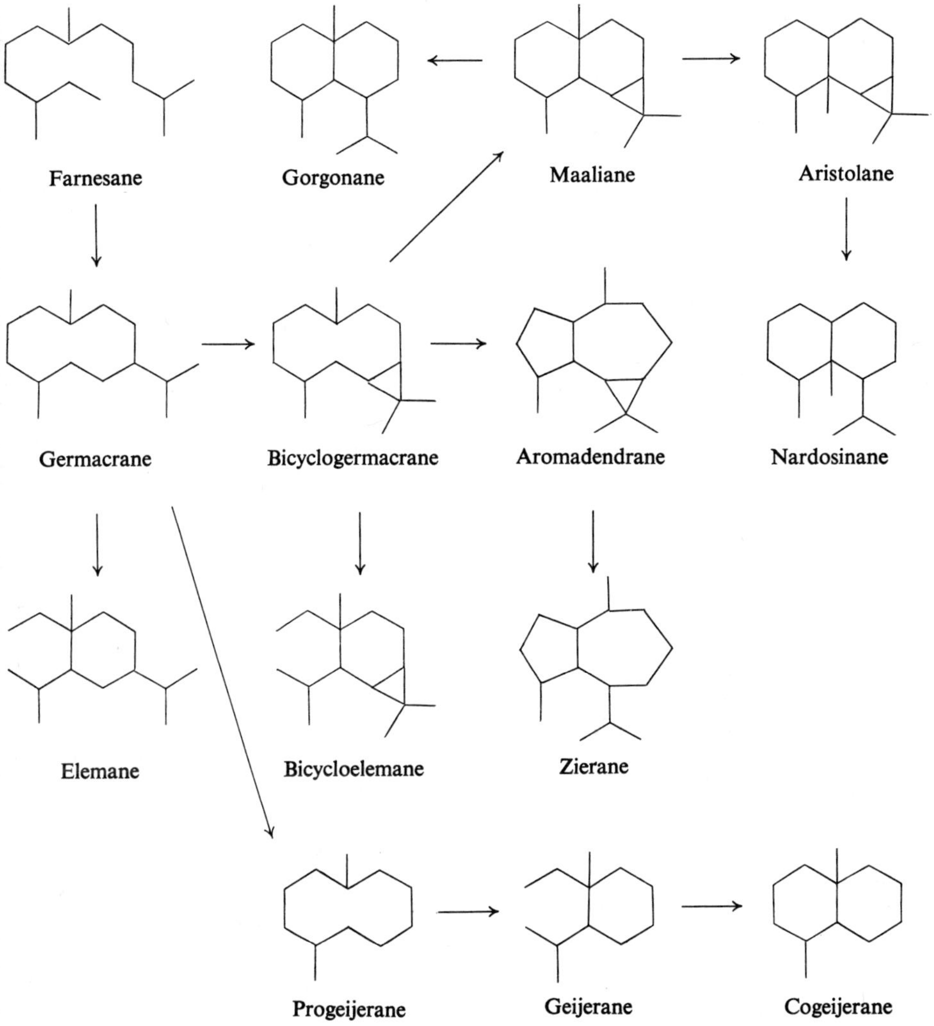

Farnesane Gorgonane Maaliane Aristolane

Germacrane Bicyclogermacrane Aromadendrane Nardosinane

Elemane Bicycloelemane Zierane

Progeijerane Geijerane Cogeijerane

one of the principal components of the aroma of tea—theaspirone (48) (Ina *et al.*, 1968)—is formally a bisnor-derivative of the abscisane type. However, its biosynthetic origin could be quite different in view of the existence of the above mentioned allenic sesquiterpene from *R. microptera* (9); the origin of this substance is regarded as the biooxidation of a precursor of polyenic origin, for example neoxanthin (Meinwald *et al.*, 1968).

TABLE VII

A. Iresane class

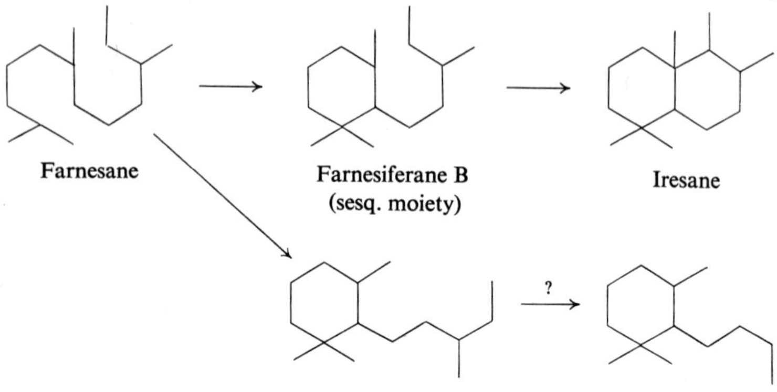

Farnesane

Farnesiferane B
(sesq. moiety)

Iresane

?

Abscisane

Theaspirane

B. Sesquiterpenoid types of non-farnesane or unknown origin

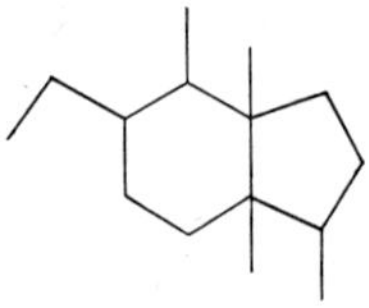

Pinguisane

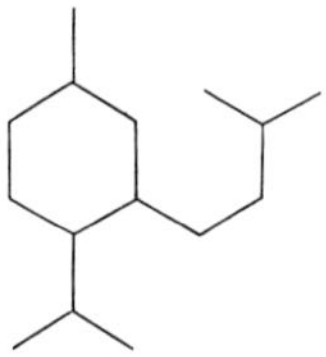

Calacane

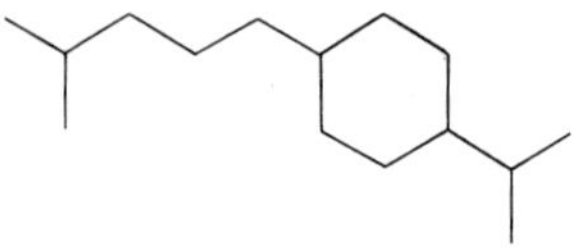

Furoventalane

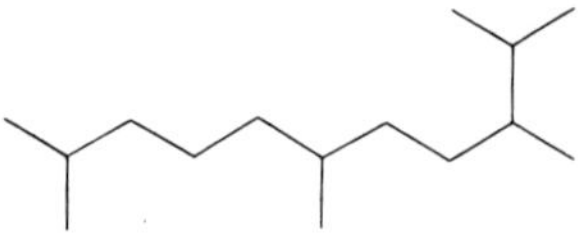

Sesq. from *Anthemis cotula*

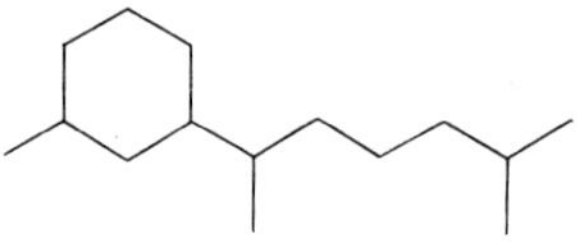

Sesq. from *Elvira biflora*

Nothing is known as yet about the biosynthesis of a substance from the liverwort *Aneura pinguis* (L.) Dum., called pinguisone (49) (Benešová *et al.*, 1969). Its skeleton has a unique character, consisting of the presence of four vicinally located methyl groups. The only well established fact is that ^{14}C-mevalonic acid is incorporated in good yield in this compound, which at the same time seems to prove that it is of sesquiterpenoid nature.

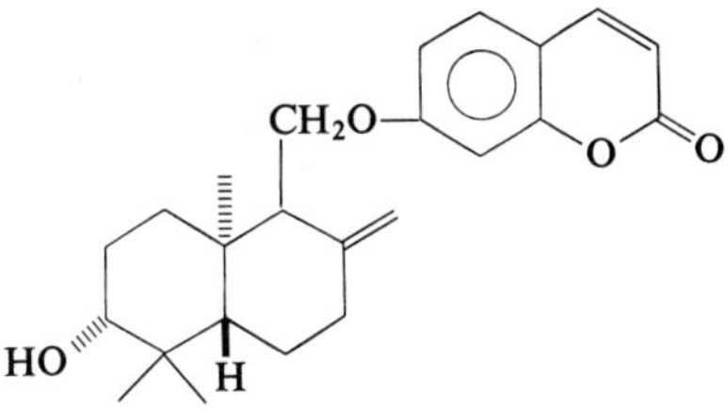

A

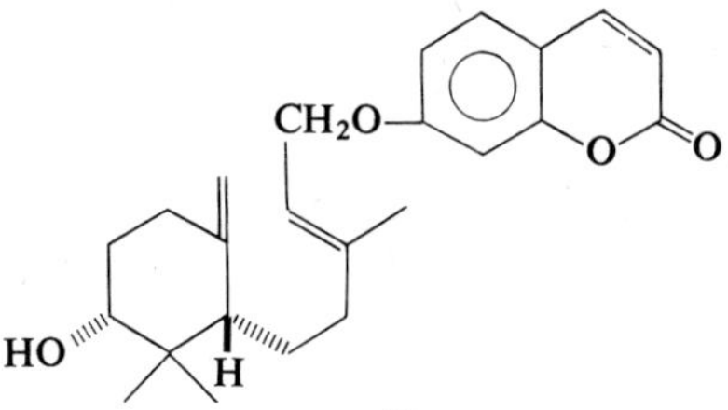

B

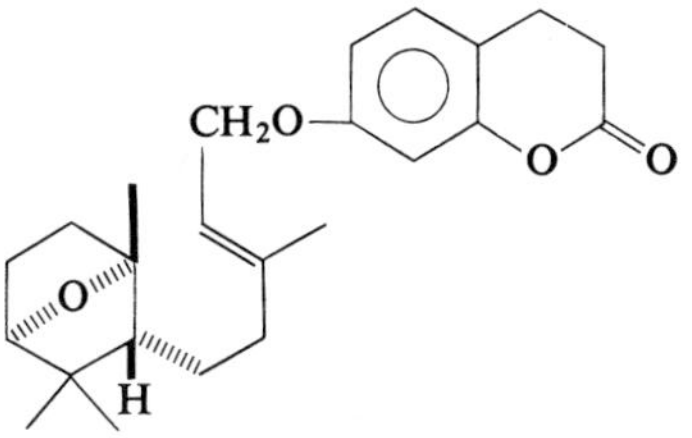

C

(46) Farnesiferols A–C

Some of the substances described up to now, which according to their molecular structure are evidently terpenoids and according to the number of carbon atoms belong to the sesquiterpenoids, cannot be derived from a precursor of the farnesyl pyrophosphate type. In some cases the explanation is plausible that the monoterpenic basis was further alkylated by another isoprene unit, as can be supposed, for example, in the case of calacone (50) (Vrkoč *et al.*, 1961), furoventalene (51) (Weinheimer and Washecheck, 1969), and an unnamed sesquiterpene from *Anthemis cotula* (52) (Bohlmann *et al.*, 1969). Another so far unnamed sesquiterpene obtained by the same group (Bohlmann and Zdero, 1969) from *Elvira biflora* (Compositae) evidently has a different generic origin: for example, structure (53) can be deduced formally by the opening of the three-membered ring in a precursor of the sesquicarene type.

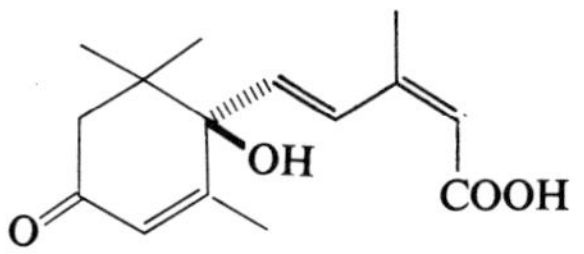

(47) Abscisic acid

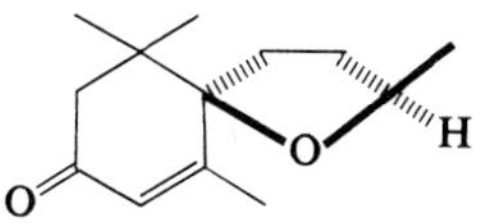

(48) Theaspirone

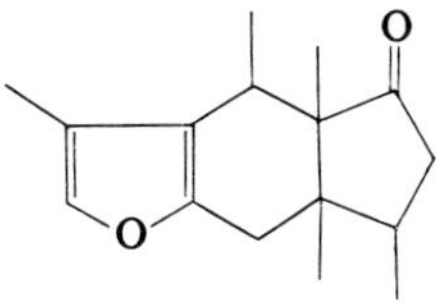

(49) Pinguisone

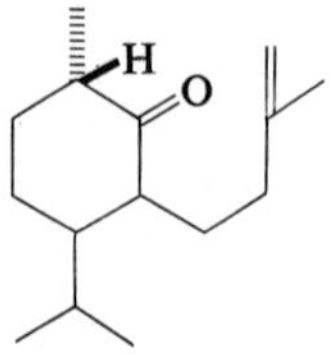

H O

(50) Calacone

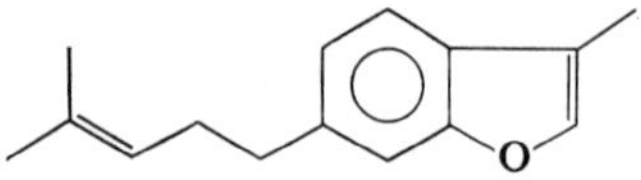

O

(51) Furoventalene

O O O

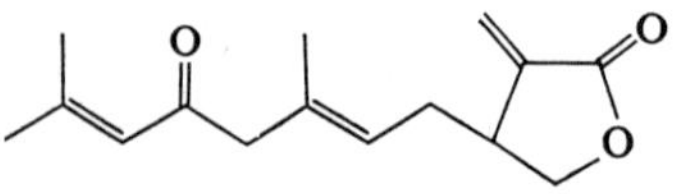

(52) Sesq. from *Anthemis cotula*

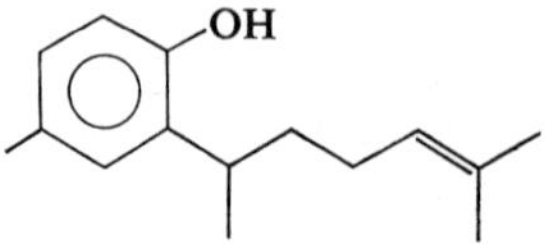

OH

(53) Sesq. from *Elvira biflora*

The group of tropolonoid compounds with 15 carbon atoms in the molecule has been omitted from the sesquiterpenoid types; in the case of nootkatin, chanootin, and nootkatinone we have substances the origin of which is evidently not taken from farnesol.

II. Biological Activity of Sesquiterpenoids

At the moment older theories are being revised which state that terpenoids in plants are only some kind of "waste products" and that being secondary products they have no practical meaning for the plant organism. It has been demonstrated in a number of examples that even some sesquiterpenoids—to which the following few examples will be limited—must be now regarded as one of the important groups of biologically active compounds. The essential functions of various individual representatives, as, e.g. hormones, phago-repellents, or cytotoxic substances are clear enough, as is also the incidental pharmacological activity exhibited in animals by a wide variety of terpenoids (Martin-Smith and Sneader, 1969).

In the following short review I shall try to enumerate some of the observed biological activities in plants, fungi and animals.

A. Plant Kingdom

1. Phytohormones

Hormonal regulation in higher plants is brought about by interaction between promotive and inhibitory hormones. Here it should be stressed that two well-known groups of these substances are represented by terpenoids: diterpene derivatives called gibberellins (functioning as "promoting" agents) and abscisic acid (representing in principle an "inhibitory" hormone). This substance will be dealt with in greater detail elsewhere (see p. 137), but the fact should be stressed here that a sesquiterpenoid, which abscisic acid is (47), has such an outstanding function. It was discovered some years ago and it was

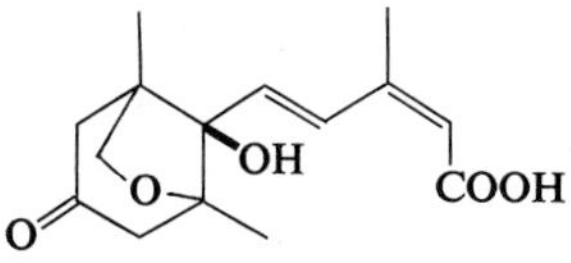

a 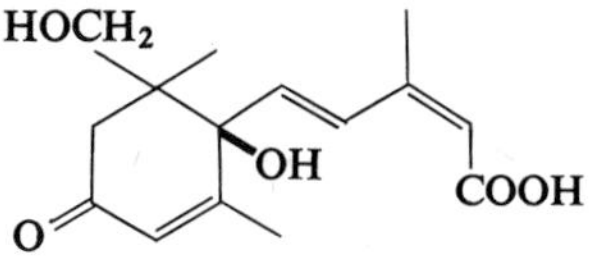 b

(54a) Phaseic acid
(54b) "Metabolite C"

previously known as both dormin or abscisin II (Eagles and Wareing, 1963; Ohkuma *et al.*, 1963). Its structure was elucidated mainly by Cornforth and his co-workers (1965, 1967). From that time the function of abscisic acid has been discussed in the literature several times, especially from the point of view of plant growth control and its complex relationships with other plant hormones (cf., Overbeek, 1966; Galston and Davies, 1969). Among the degradation products of abscisic acid phaseic acid and "metabolite C" are known (54a,b) (Millborrow, 1968, 1969; see also p. 142 of this volume).

A role in plant growth regulation has also been ascribed to some other sesquiterpenoids, nevertheless, their importance and distribution is much more limited than those of abscisic acid. An example is xanthinin (55) present in

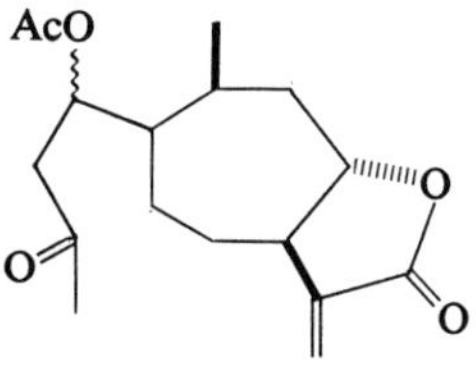

(55) Xanthinin

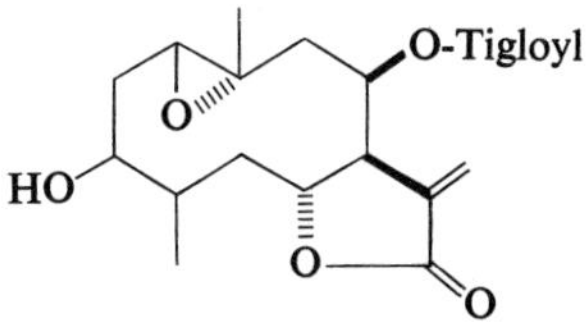

(56) Heliangin

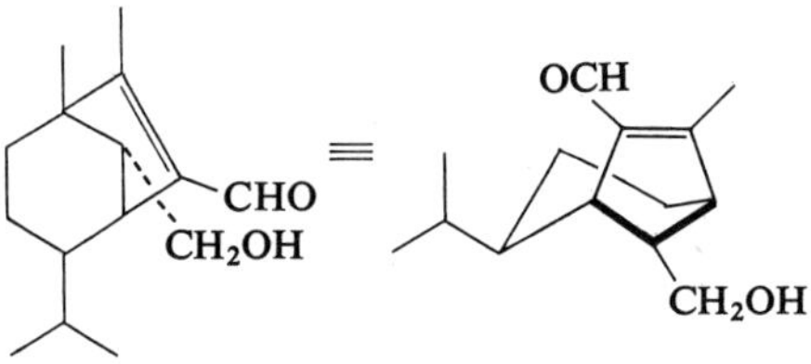

(57) Helminthosporol

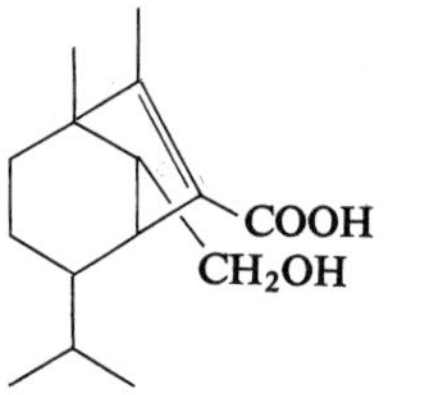

(58) Helminthosporic acid

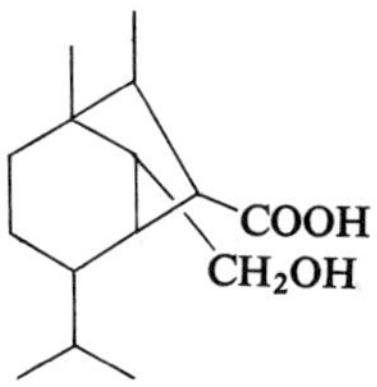

(59) Dihydrohelminthosporic acid

many species of *Xanthium* (*X. strumarium*, *X. spinosum*, *X. commune*, *X. chinense*, *X. chassei*; Winters *et al.*, 1969); this compound shows an antagonism toward auxin. Similarly, heliangin (56) has been isolated from *Helianthus tuberosus* on the basis of an inhibitory effect in the Avena curvature of straight growth test (Morimoto *et al.*, 1966). Helminthosporol (57) and recently its derivatives, helminthosporic acid (58) and dihydrohelminthosporic acid (59) (Kato *et al.*, 1964; Sakurai and Tamura, 1965) have growth-regulating properties. In the oat mesocotyl test they are all active, even at about 10^{-7} M. This activity is low and their properties exclude them from the group of genuine gibberellins. The problem remains whether these compounds are growth regulators *per se* or are active only after transformation into gibberellins.

2. *Antifeeding Effects of Sesquiterpenoids*

From the leaves of *Parabenzoin trilobum* (L.) Wada *et al.* (1968) and Wada and Munakata (1968) isolated two substances, to which they gave the names shiromodiol acetate and shiromodiol diacetate (60a,b). In bioassays using a polyphagous insect *Prodenia litura* Fabricius and an oligophagous insect *Trimeresia miranda* Butler, respectively, they proved that the substances have pronounced anti-feedant effects on insects. A series of other sesquiterpenoids has been tested and the results justify the hypothesis that other secondary metabolites of a sesquiterpenoid character present in higher plants can also have a similar effect to that of the two germacrane derivatives of the shiromodiol type. This effect was proved for pinguisone (49), a sesquiterpenoid from the liverwort *Aneura pinguis* (Benešová *et al.*, 1969) (for the biotest we thank Professor Wada).

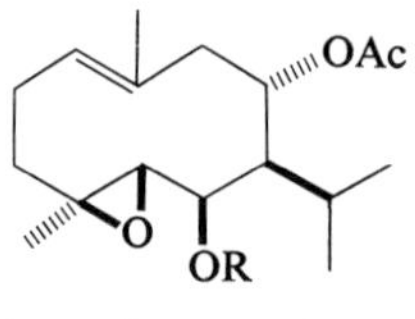

OAc

O OR

(60a) Shiromodiol monoacetate, R = H
(60b) Shiromodiol diacetate, R = Ac

3. *Juvenile Hormone Mimics*

The discovery of the function, the methods of isolation, the structural proof, and several syntheses of one of the bioregulators of the life processes in insects—the juvenile hormone (61)—are the subject of a large series of communications and reviews (see for example Berkoff, 1969; Röller and Dahm, 1968). Elsewhere in this book (p. 95) more is said on the juvenile hormone. I should like to point out that from the biochemical point of view the structure of the juvenile hormone can be deduced formally from farnesane-type compounds by two methylations. However, such a type of biomethylation is not common. As regards the true biosynthesis little is known of it. The fact that farnesol and its derivatives have to a certain extent juvenile hormone activity may represent evidence that the insect organism is readily able to synthesize the necessary hormone from this precursor.

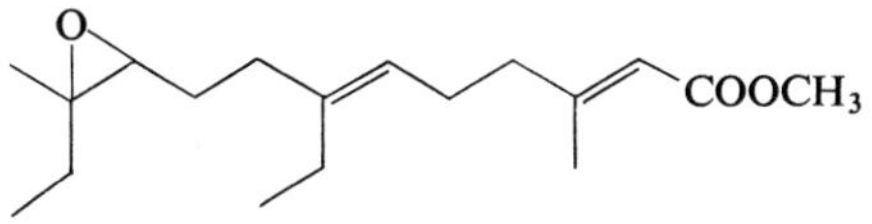

O COOCH$_3$

(61) Juvenile hormone

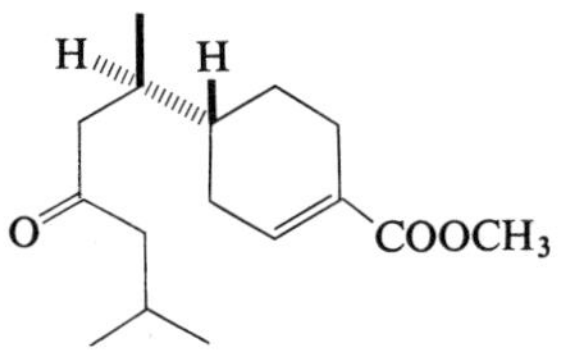

H H O COOCH$_3$

(62) Juvabione

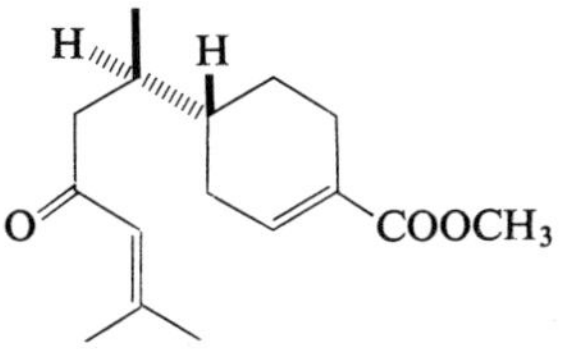

H H O COOCH$_3$

(63) Dehydrojuvabione

In contrast to this a long series of compounds is known, the structure of which is quite different from that of farnesane, which in effect mimic to various degrees the function of juvenile hormone in various arthropod families (Gilbert and Goodfelow, 1965; Bowers and Thompson, 1963). It is worth mentioning here one type of substances present as native components in certain conifers. The discovery of the so-called "paper factor" is fascinating (Sláma and Williams, 1965). It led to the discovery of juvabione (todomatuic acid methyl ester, 62; Bowers *et al.*, 1966) and of dehydrojuvabione (63) (Černý *et al.*, 1967) as the active principles. The presence of this substance in some trees excludes completely certain potential pests, especially certain bugs (Hemiptera). If observed from a somewhat unscientific aspect, its occurrence in trees represents their incredibly sophisticated self-defence against insects.

4. *Heartwood Preserving Sesquiterpenoids*

A similar "self-defence" against insects of wood-decaying fungi is exerted by substances occurring in the inactive part of tree trunks, i.e. in heartwood. An enormous number of secondary metabolites has already been isolated from numerous species of trees. In many cases it was established (e.g. Erdtman, 1939) that they represent a more or less active protective agent against infection by parasitic fungi. Such compounds are particularly substances of phenolic nature, tropolones, and frequently quinones, which have a pronounced fungistatic effect (Cruickshank and Perrin, 1964). Thus it appears that heartwood decay resistance can seldom be ascribed specifically to one substance; rather it depends on the presence of a complex mixture of polyphenols and other factors. The high content of sesquiterpenoid compounds has certainly an appreciable influence on the water-repellent properties and on the physical nature of the timber. A good example is the appreciable stability of root wood of *Fokenia hodginsii* which is stable for long periods even when buried in the soil under tropical conditions in Vietnam. The stability is apparently caused by the unusually high content of ether-extractives (about 28%), mostly of a sesquiterpenoid nature (Dolejš and Herout, 1961). Many sesquiterpenoids are the components of heartwoods of trees of various evolutionary stages; from very ancient ones (*Ginkgo biloba*; bilobanone; 64; Irie *et al.*, 1967), through strikingly rich coniferous trees, up to certain, evolutionary higher trees as for example *Ulmus rubra* and *U. thomasii*, which contain a group of substances of the cadalene type (Frachebond *et al.*, 1968; Rowe and Toda, 1969).

5. *Toxic Sesquiterpenoids from Plants*

The idea of "toxicity" can be interpreted in a very broad sense and therefore the classification of substances into this group is rather arbitrary. However, for some sesquiterpenoids their toxic properties are so pronounced that the reason for their isolation from the vegetable material originated from their toxicity (e.g. picrotoxine group). Many substances of phenolic character can be considered as toxic, as for example the dimeric sesquiterpenoid gossypol (65) from cotton seeds, or a series of quinones, as for example mansonones from the wood of *Mansonia altissima* (Marini-Bettolo *et al.*, 1965). Similarly as toxins causing mass poisoning among sheep in Africa, sesquiterpene lactones from some *Geigeria* species were also studied (Compositae; *G. aspera*, *G. africana*). It was found that the true cause of the illness is geigerin (66) and particularly vermeerin (67) (Watt and Breyer-Branwijk, 1962). The toxic principle of the Japanese star anis (*Illicium anisatum* L.) is anisatin (68) the rather complicated structure of which was elucidated only recently (Yamada *et al.*, 1965).

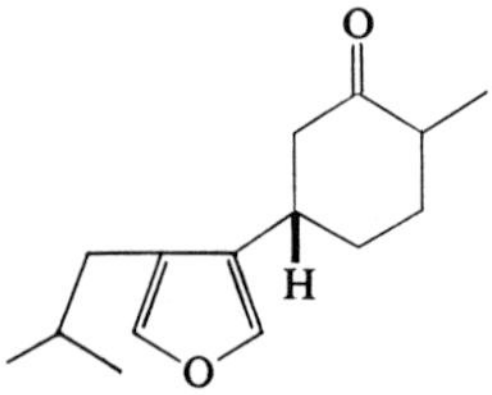

(64) Bilobanone

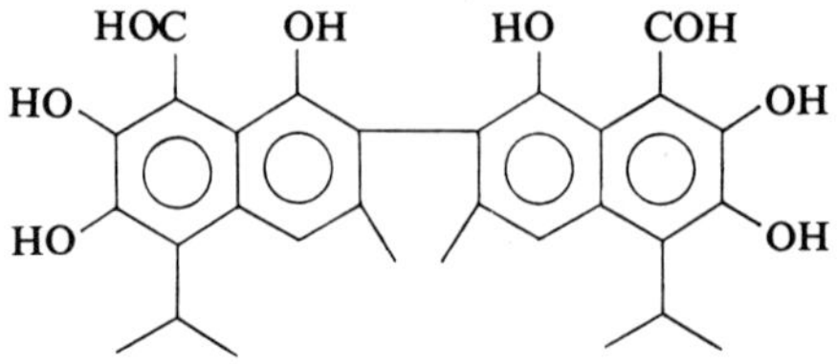

(65) Gossypol

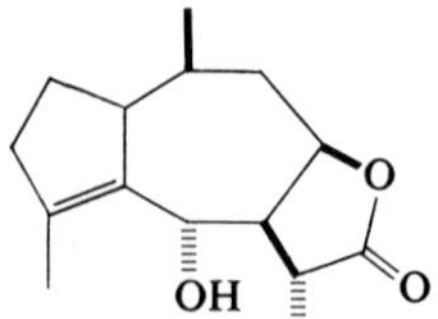

(66) Geigerin

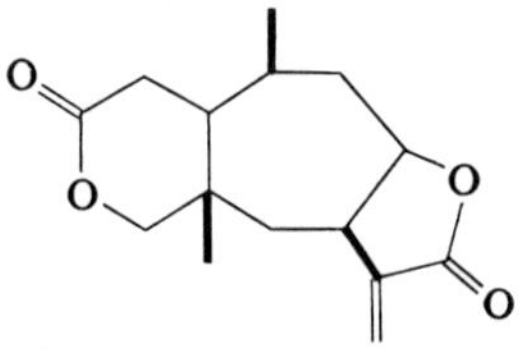

(67) Veermerin

(68) Anisatin

However, among the most interesting sesquiterpenoid toxins are the group of picrotoxin and coriamyrtin, which have been known for a long time (see surveys of Porter, 1967; Coscia, 1969). Picrotoxinin, one of the main representatives of these amaroids of typical cage-like structure is expressed by the complicated formula (69a) which has been demonstrated only recently by modern methods of organic chemistry (see Coscia, 1969). Picrotoxinin and picrotin (69b) give a molecular compound known originally under the name of picrotoxin and representing a typical component of a number of plants belonging to Menispermaceae (*Menispermum cocculus*, *Cocculus indica*, *Anamirta cocculus*). Their original practical use was to stun fish. It causes extreme central nervous system excitation and coriamyrtin and its hydroxy-derivative tutin (70a,b) from *Coriaria* sp. (Coriariaceae) are similarly active. These two toxins are responsible for the death of many cattle brought to New Zealand. They cause the stimulation of the respiratory, vasomotor, and cardioinhibitory centres in the brain. In view of their interesting origin other substances of this group may be mentioned, as for example mellitoxin (71). This was first discovered in honey as its toxic principle; it became apparent that it is a metabolic product of an insect called the "passion vine hopper" (*Scolypopa australis*) which feeds on *Coriaria arborea* Lindsey and leaves anal excretions containing some sugar. For this reason, bees incorporate the excretion into their honey. It can be assumed, therefore, that it is an insect metabolite of tutin. The same substance was also later found in vegetable material, *Hyenache globosa* (Jommi *et al.*, 1965).

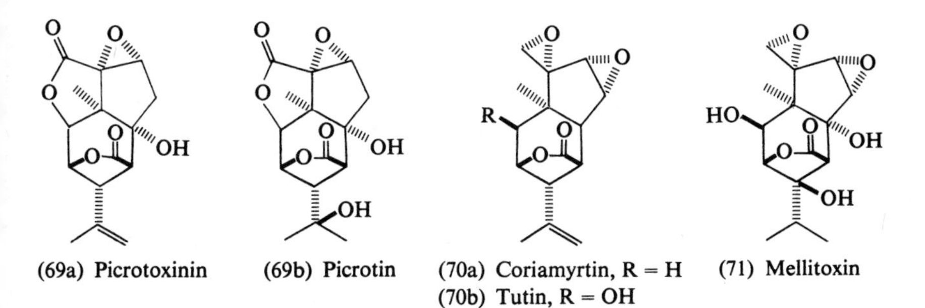

(69a) Picrotoxinin (69b) Picrotin (70a) Coriamyrtin, R = H (70b) Tutin, R = OH (71) Mellitoxin

6. *Sesquiterpenic Alkaloids*

Dendrobium alkaloids are obviously structurally related to the type of picrotoxane sesquiterpenoids. These substances were found in the thirties in some species of *Dendrobium* which are used in popular Chinese medicine (*D. nobile*, mainly). However, their structures were elucidated only recently together with the structure of a series of minor alkaloids (Yamamura and Hirata, 1964; Inubushi *et al.*, 1964; Onaka *et al.*, 1964). Their structures are so complicated that they could not be determined by classical methods and surrendered only after an attack with NMR spectroscopy. As the main components dendrobine (72) and nobilonine (73) should be mentioned, the pharmacological effect of which is nearly identical with that of picrotoxin; in small doses dendramine has analgesic and antipyretic activity and it also diminishes cardiac activity. In large doses it causes convulsions originating in the central nervous system.

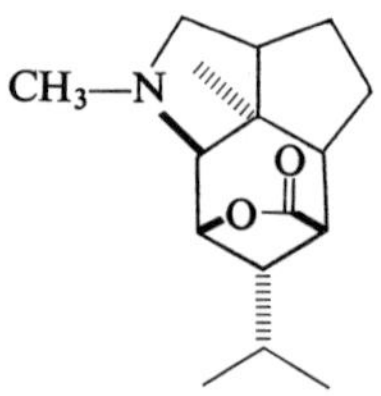

CH_3—N O O

(72) Dendrobine

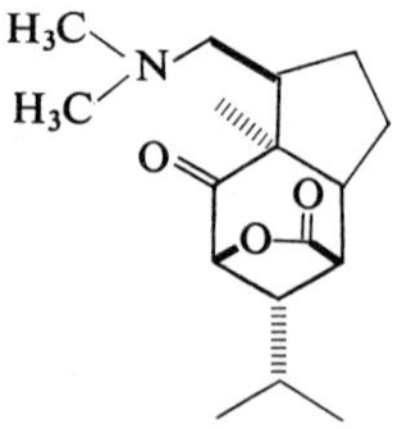

H_3C N H_3C O O O

(73) Nobilonine

Some other groups of alkaloids of sesquiterpenoid type have already been described, for a review see Edwards (1969). There is a large group isolated from various water lilies (*Nuphar* sp.). Here belongs nupharidine (74) and nupharamine (75) from *N. japonicum* (see Kotake *et al.*, 1960, 1962; Ohashi, 1959; Kawasaki *et al.*, 1963). From *N. luteum* substances were isolated of evidently dimeric character which contained sulphur in the molecule, i.e. thiobinupharidine and neothiobinupharidine (see Birnbaum, 1965).

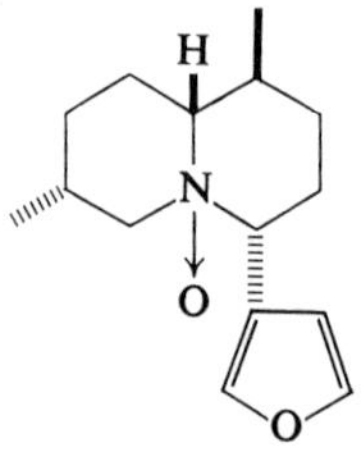

H N O O

(74) Nupharidine

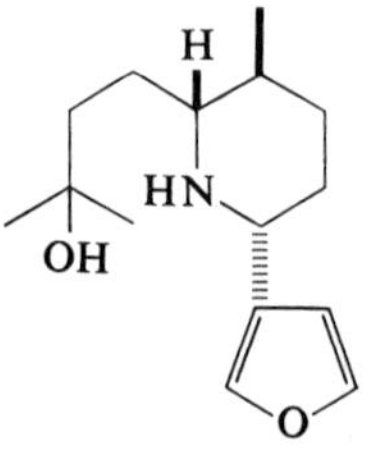

H HN OH O

(75) Nupharamine

The carbon skeleton of nupharidine and related bases is of the farnesane type, containing one furan ring and a quinolizidine grouping in the molecule. Two other alkaloids, present in *Pogostemon patchouli* (Labiates), contain a pyridine nucleus formed formally by the extension of the five-membered ring of the sesquiterpene basis. Thus in epiguaipyridine (76) the basis is a guaiane carbon skeleton, while in patchoulipyridine (77) the basis can be derived from the patchoulane type; it was also prepared synthetically from β-patchoulene (for a study of both bases see Büchi *et al.*, 1966; see further the data from Edwards, 1969). Somewhat questionable is the origin of another alkaloid, fabianine (78) from *Fabiana imbricata* (Loganaceae) (Edwards and Elmore, 1962). The view has also been expressed that it arises by oxidative cleavage through the loss of one carbon from a precursor with a cadinane skeleton (Edwards, 1969).

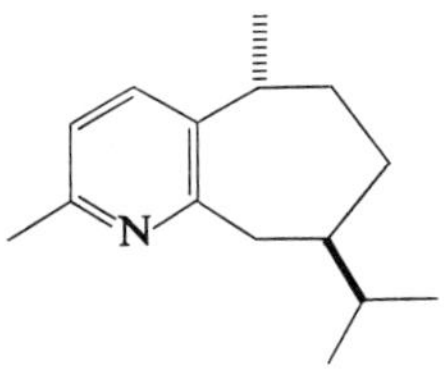

N

(76) Epiguaipyridine

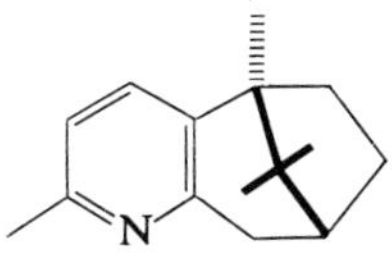

N

(77) Patchoulipyridine

N

OH

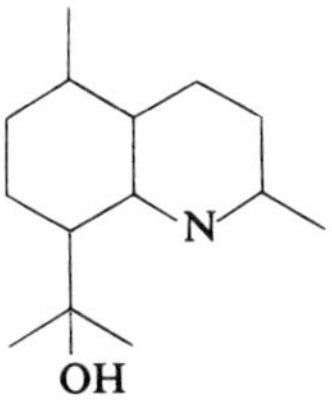

(78) Fabianine

7. *Pharmacological Properties and the Utilization of Plant Sesquiterpenoids*

Some time ago great interest was paid by pharmacologists to the anti-inflammatory action of some azulenes, especially of guaiazulene and chamazulene isolated from plant material (see Treibs *et al.*, 1955). It seems that the rush of extravagant claims for this property of azulene has died away and that their application remains restricted to cosmetics rather than to real pharmacology. However, the increased interest in these substances enabled the discovery to be made that "natural" azulenes practically do not exist and regularly are formed as artifacts from sesquiterpenic precursors. Therefore, it is worth mentioning that in the liverwort *Calypogeia trichomanis* Corda 1,4-dimethylazulene (79) is deposited in visually observable oily droplets in the cells, as shown by Menche and Huneck (1966).

β-Santonin (80) from Flores Artemisia, which has long been known and utilized as a vermifuge has been excluded from official use in many countries because of its toxicity.

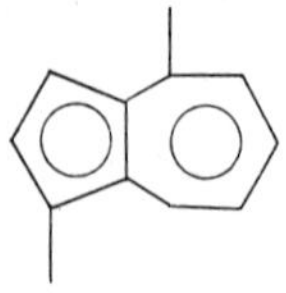

(79) 1,4-Dimethylazulene

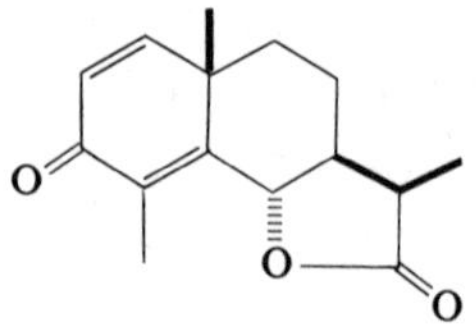

(80) β-Santonin

Recently, however, renewed interest was aroused by the fact that certain sesquiterpene lactones, especially those substituted with many oxygen functions, display a significant inhibitory activity *in vitro* against cells derived from human carcinoma of the nasopharynx (KB) and also against the Walker 256 intramuscular carcinosarcoma *in vivo* in rats. Kupchan's group (Kupchan *et al.*, 1966, 1968, 1969) also attributes this activity to elephanthin (81) from *Elephanthopus elatus*, eupachlorinacetate (6) from *Eupatorium rotundum*, and vernodalin (82) from *Vernonia amygdalina*. All the mentioned plants are members of the Compositae family. Doskotch and Hufford (1969) also attribute similar properties to damsin (83), obtained from *Ambrosia maritima*, or *A. ambrosioides*. Similarly as in the case of many other cancerostats the problem here is whether these compounds will prove meaningful in terms of potential clinical desirability.

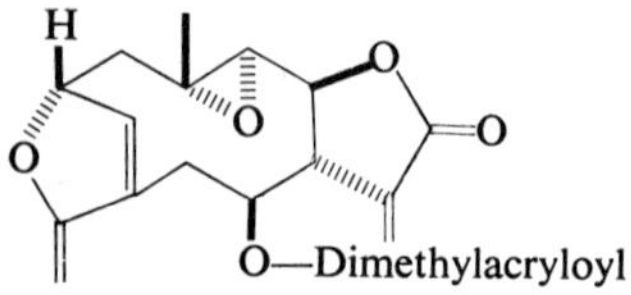

O—Dimethylacryloyl

(81) Elephanthin

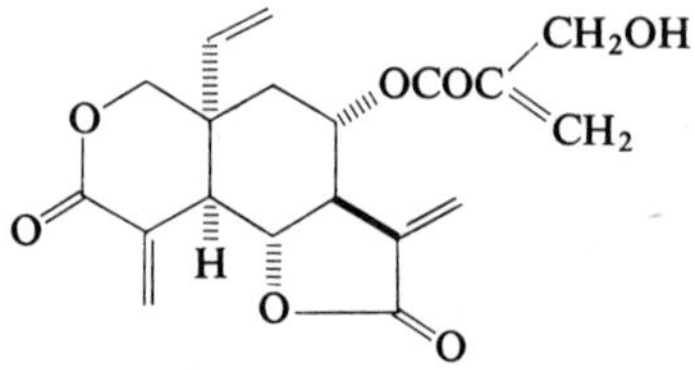

(82) Vernodalin

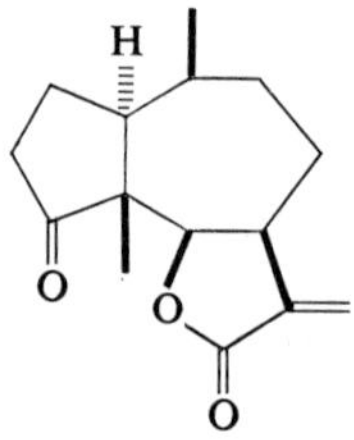

(83) Damsin

In connection with this remark it should be mentioned that the bitter principles, many of which are of the group of sesquiterpene lactones from the Compositae, were in popular and official medical use for centuries as tonics. At present they are quite commonly used, usually not *per se*, but in a more or less natural form. Lactucin and lactucopicrin (84a,b) (see Dolejš *et al.*, 1958) are bitter principles from common vegetables as for example lettuce and cichorium buds; absinthin (85) (Vokáč *et al.*, 1968) is the bitter principle of the drug *Herba absinthii* and is still used in the form of decoctions or in some alcoholic beverages (Vermouth), etc.

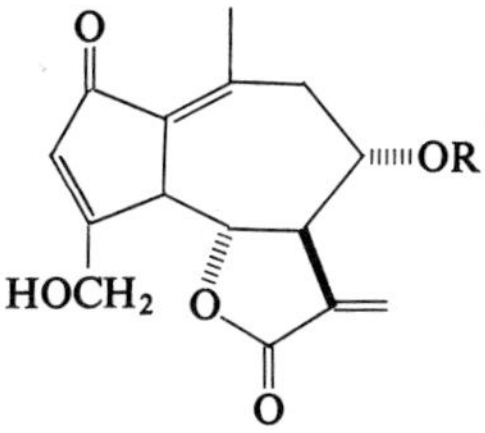

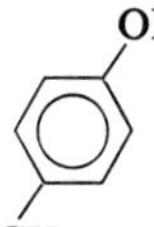

(84a) Lactucin, R = H
(84b) Lactucopicrin, R = —O—CO—CH_2

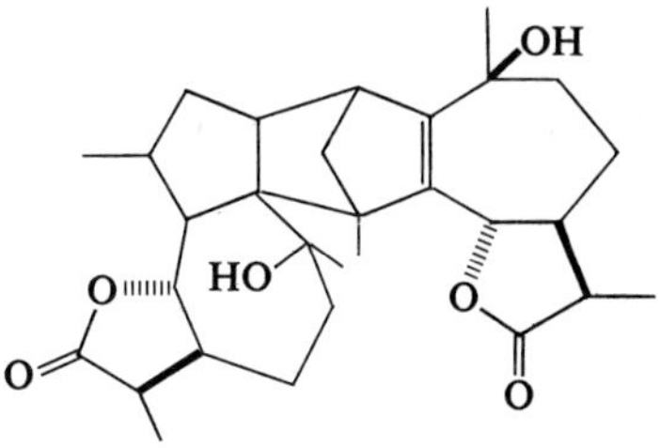

(85) Absinthin

8. Organoleptic Properties of Sesquiterpenoids

The bitter taste just mentioned is typical of a great number of sesquiterpenoids, being particularly marked in the majority of lactones present in the family of Compositae. It seems remarkable that unsubstituted or little substituted sesquiterpene lactones, as for example alantolactone and costunolide (86 and 87 respectively) give only a hot burning taste, while further substitution with hydroxy, keto, or ester groups leads to the formation of a bitter taste (see Herout, 1966). This may be in agreement with the observation of Kubota and Kubo (1969) that the presence of an electron donor and an electron acceptor in appropriate distance (1·5 Å) in the molecule is necessary for its formation.

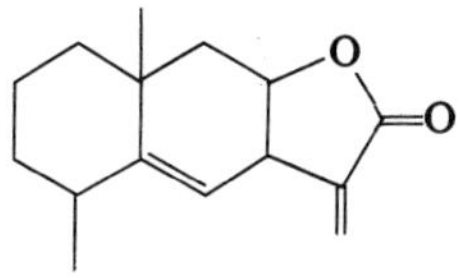

(86) Alantolactone

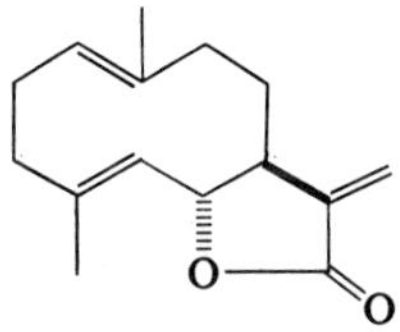

(87) Costunolide

A sharp peppery taste is typical of the dialdehyde polygodial (= tadeonal) (88), the hot component of some plants of *Polygonum* genus (*P. hydropiper*; Barnes and Loder, 1962).

The odour of sesquiterpenoids is limited in view of their low volatility. A pronounced odour is observed only in derivatives with simple substitution, as for example in alcohols or ketones, and to a lesser extent in oxides and esters. In comparison with monoterpenes sesquiterpenoids recede into the background, although some of them are undoubtedly appreciated in perfumery for their spicy odour. As examples farnesol or nerolidol (89, 90) should be mentioned; they have a flower-like aroma. The newly discovered sinensal α and β (91a,b) (Flath *et al.*, 1965) have the typical odour of mandarin peel

(*Citrus sinensis*). The classical aroma of the cedar tree and the heavy oriental-like aroma of some essential oils, e.g. sandal oil and vetiver oil, are also due to sesquiterpenoids.

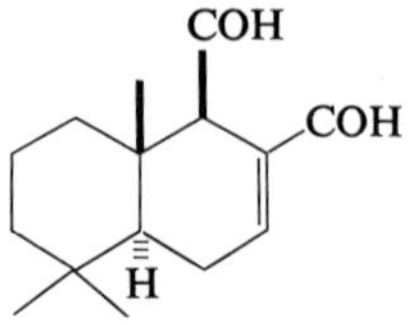

COH
COH
H

(88) Polygodial

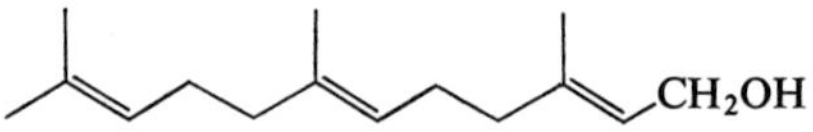

CH_2OH

(89) Farnesol

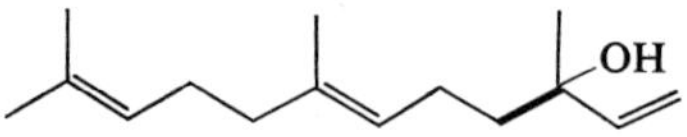

OH

(90) Nerolidol

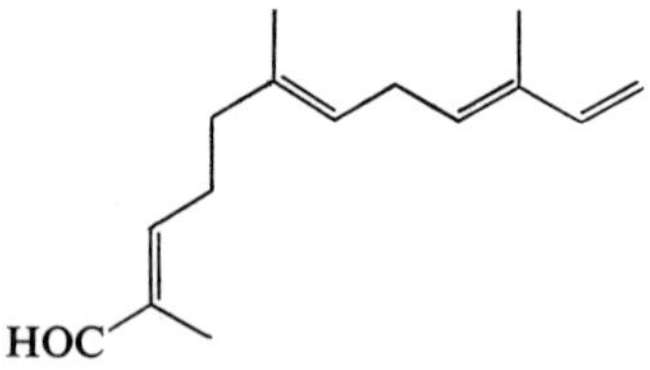

HOC

(91a) α-Sinensal

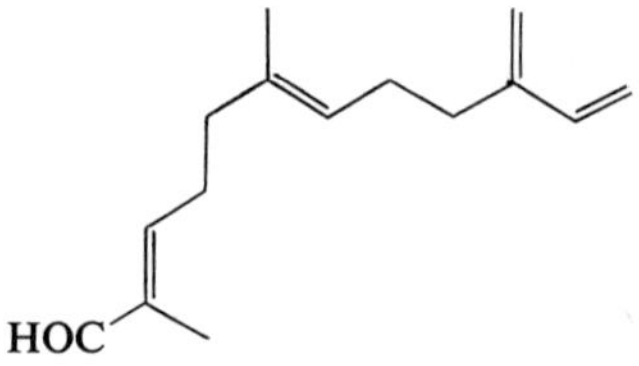

HOC

(91b) β-Sinensal

9. *Other Biological Activities of Sesquiterpenoids*

The functions of sesquiterpenoids in plants are far from all having been discovered. It can be expected that a deeper collaboration between organic chemists, plant physiologists, entomologists, and experts from other natural science disciplines will make new discoveries. In support of this statement it is possible to quote two examples: (1) Hartley and Fawcett (1969a,b) suggest that a greater or lesser resistance to powdery mildew of various cultures of hops (*Humulus lupulus* L.) could be connected with the presence of some sesquiterpenic hydrocarbons in them; (2) Schwär (1962) has described the influence of absinthin and its metabolites on the germination of the seeds of some plants.

B. FUNGAL KINGDOM

1. *Antibiotics of Sesquiterpenoid Structure*

In the last quarter of a century great attention has been devoted to the chemistry of fungi mainly because scientists were looking for new antibiotics. Among the great number of substances of this type discovered up to date,

(92a) Trichodermol, $R = R_2 = H$; $R_1 = OH$
(92b) Scirpendiol, $R = R_1 = OH$; $R_2 = H$
(92c) Scirpentriol, $R = R_1 = R_2 = OH$

(93) Verrucarin A
$R_1 = OH$
$R = H$

(94) Verrucarin B
$R_1 = \diagdown$ O
$R = \diagup$

TABLE VIII
The sesquiterpenoid antibiotics of trichothecane type

Substance	Occurrence	Formula	References
Verrucarin A (Mucomycin A)	*Myrothecium verrucaria*	93	Gutzwiller and Tamm, 1965a,b
Verrucarin B	*M. verrucaria*	94	
Verrucarins C, D, E, F, G, H	*M. verrucaria*	?	Böhmer *et al.*, 1965
Rorridin C (Trichodermol)	*Myrothecium rorridum*	95a	Gutzwiller *et al.*, 1964
Rorridins A, B, D, E	*Myrothecium rorridum*	?	Härri *et al.*, 1962
Mucomycin B	*Myrothecium verrucaria*	?	Wittimberga and Wittimberga, 1965
Trichothecin	*Trichothecium roseum*	95b	Godfredsen and Vangendahl, 1964
Trichodermin	A strain of *Trichoderma*	96	Godfredsen and Vangendahl, 1964
Nivalenol	*Fusarium nivale*	97	Tatsuno *et al.*, 1968
Fusarenone	*Fusarium nivale*	98	Grove, 1969
Diacetoxyscirpenol	*Fusarium* spp. *equiseti*, *diversisporum*, *sambucinum*, *scirpi*, *tricinctum*; *Gibberella intricans*	99	Flury *et al.*, 1965
8α(3-Methylbutyryloxy)-4β,15-diacetoxyscirp-9-en-8-one	*Fusarium tricinctum*	100	Bamburg *et al.*, 1968
4,15-Diacetoxy-3α,7α-dihydroxyscirp-9-en-8-one	*Fusarium equiseti*	101	

very often of most diverse and uncommon structures, a series of substances has been found which can be safely classified as sesquiterpenoids. Of antibiotics isolated from actinomycetes, the low organized mycelial bacteria belonging to the kingdom of Monera (see Whittaker, 1969), none has a sesquiterpenoid structure; however, numerous metabolites of higher fungi do. In this respect the discovery of geosmin (43) produced by a series of actinomycete strains (mainly Streptomycetes) and also by a species of an anuclear blue alga (Gerber and Lechevalier, 1965) is interesting. However, it is still questionable whether the biosynthetic pathway, i.e. the formation of a sesquiterpene by the loss of an isopropyl group, is correct.

Among known sesquiterpene antibiotics the most numerous are in the group of substances of the trichothecane type. A selection of the most typical ones is presented in Table VIII. They often exhibit a complex substitution of the basic sesquiterpenoid moiety of hydroxylated trichothecane as for example in trichodermol, or scirpenediol or triol (92a–c); for a newer proposal of nomenclature see Godfredsen *et al.* (1967). Verrucarin A and B (93, 94; Gutzwiller and Tamm, 1965a,b) can serve as examples.

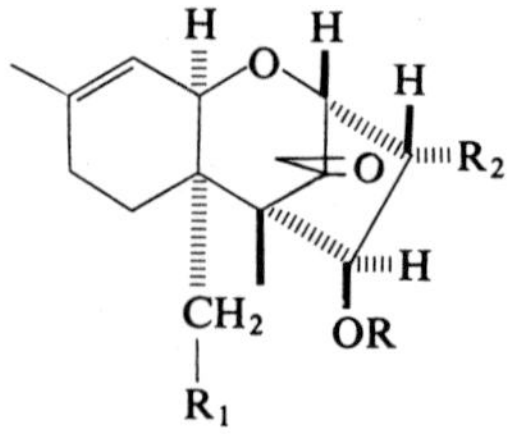

(95a) Rorridin C, $R = R_1 = R_2 = H$

(95b) Trichothecin, $R = H$, R = iso-Crotonyl, $R = H$

(96) Trichodermin, $R_1 = R_2 = H$, $R = Ac$

(99) Diacetoxyscirpenol, $R = R_1 = Ac$, $R_2 = OH$

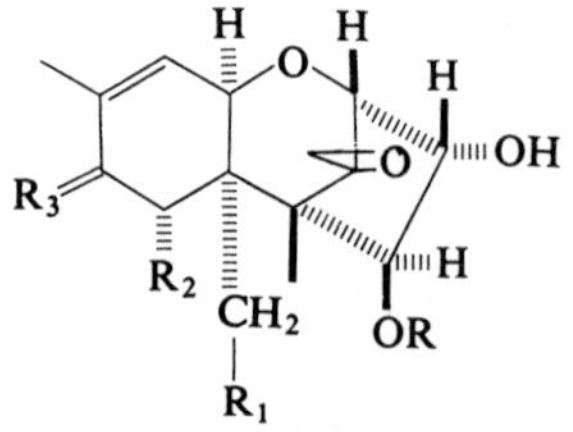

(97) Nivalenol, $R_1 = R_2 = OH$, $R = H$, $R_3 = O$

(98) Fusarenone, $R_1 = R_2 = OH$, $R = Ac$, $R_3 = O$

(100) $R = Ac$, $R_1 = OH$, $R_2 = H$, $R_3 = \langle{}^{H}_{OCOCH(CH_3)_2}$

(101) $R = Ac$, $R_1 = OAc$, $R_2 = H$, $R_3 = O$

All these substances were isolated from higher fungi, usually those parasitizing plants. They have as a rule a strong antifungal or antibacterial action; some also show marked cytotoxic activity. However, their utilization is prevented by their appreciable toxicity in mammals. It is known that feeding with spoiled fodder, for example that infected by various fusarioses (scabby grain), leads to losses among cattle.

Culmorin isolated originally from *Fusarium culmorum* is the first antibiotic with a longifolane skeleton. The structure (102) was proved by Barton and Werstiuk (1968); it is remarkable that the configuration of this longifolane skeleton is the opposite to that common in derivatives isolated from plants.

Substances listed in Table IX represent another common group of sesquiterpenoid antibiotics. They are related not only by their occurrence in higher fungi (Ascomycetes and Basidiomycetes) but also by the supposed identical biosynthetic pathway through a humulene-type precursor. These substances were also discovered and isolated for their antibacterial, antifungal

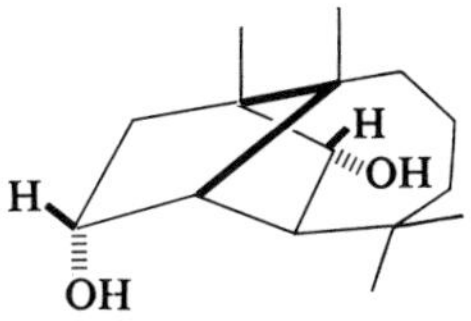

(102) Culmorin

and antitumor activity. Some of them (illudins, dihydroilludin) are supposed to be the bioluminescent principles of maternal toadstools (e.g. "Jack o'lantern" mushroom—*Clitocybe illudens*, or the Japanese mushroom *Lamptero-myces japonicus*) (Tada *et al.*, 1964; Ischihara *et al.*, 1969).

A group of substances of helminthosporal type is connected biosynthetically with sativene. This group occurs mainly in *Helminthosporium sativum* P. K. and B. (= *Bipolaris sorokiana* (Sacc. in Sorok.) Shoemaker). Sativene (109) was isolated only as a minor component in addition to helminthosporal (110) (de Mayo *et al.*, 1961, 1963; de Mayo and Williams, 1965). Japanese scientists

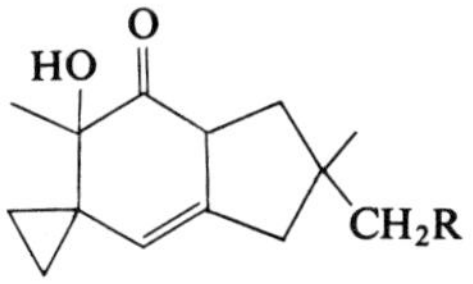

(103a) Illudin S, R = OH
(103b) Illudin M, R = H

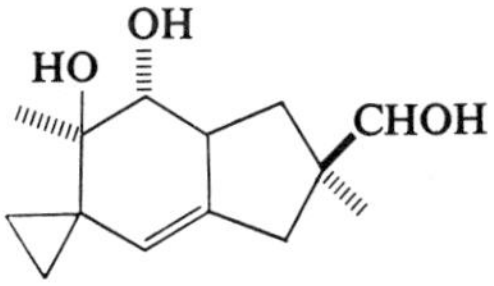

(104) Dihydroilludin

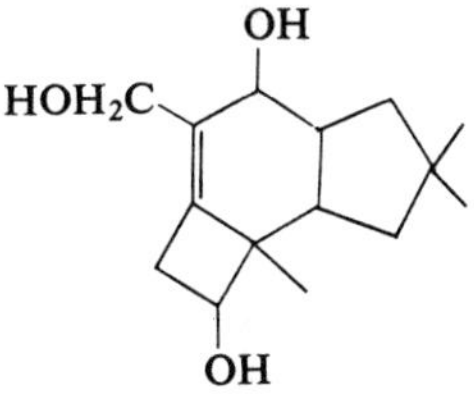

(105) Illudol

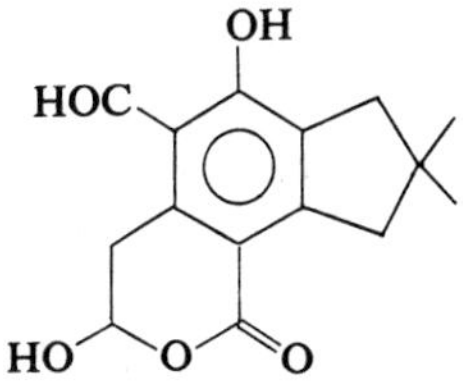

(106) Illudalic acid

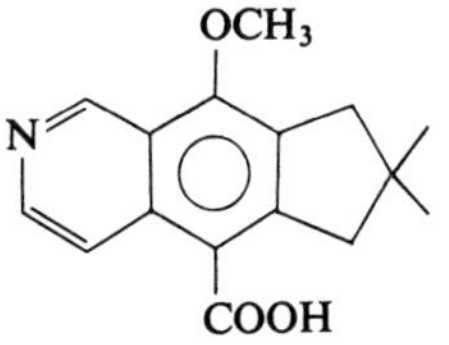

(107) Illudinine

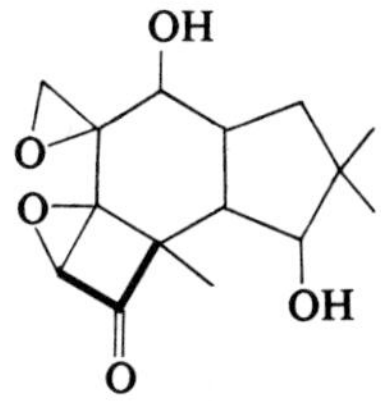

(108) Coriolin

TABLE IX
The sesquiterpene antibiotics derived from a humulane precursor

Substance	Origin	Formula	References
Illudin S (Lampterol)	*Clitocybe illudens, Lampteromyces japonicus*	103a	McMorris and Anchel, 1965
Illudin M	*Clitocybe illudens*	103b	McMorris and Anchel, 1965
Dihydroilludin	*C. illudens*	104	Ichihara *et al.*, 1969
Illudol	*C. illudens*	105	McMorris *et al.*, 1967
Illudalic acid	*C. illudens*	106	Nair *et al.*, 1969
Illudinine	*C. illudens*	107	Nair *et al.*, 1969
Coriolin	*Coriolus consors*	108	Takahashi *et al.*, 1969
Marasmic acid	*Marasmius conigenus*	15	Dugan *et al.*, 1965
Hirsutic acid	*Stereum hirsutum*	16	Corner *et al.*, 1965
Fomannosin	*Fomes annosus*	17	Kepler *et al.*, 1967

have isolated further metabolites such as helminthosporol (57), prehelminthosporal, and prehelminthosporol (111 and 112 resp.) (Tamura *et al.*, 1963). To a certain extent the difference of their properties is interesting. Helminthosporal was discovered as a toxin causing enormous losses in cereal production, while helminthosporol—at least at certain concentrations—is considered to be a growth-promoting factor.

Recently a whole series of substances was isolated from *H. siccans*, having a similar biosynthetic origin partly from a farnesane unit and partly from

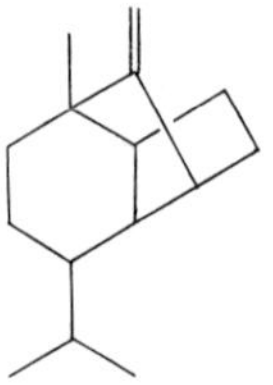

(109) Sativene

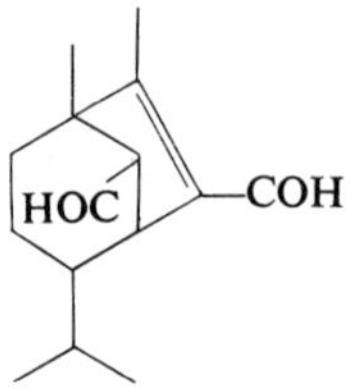

(110) Helminthosporal

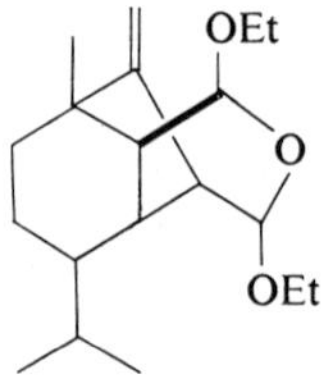

(111) Prehelminthosporal

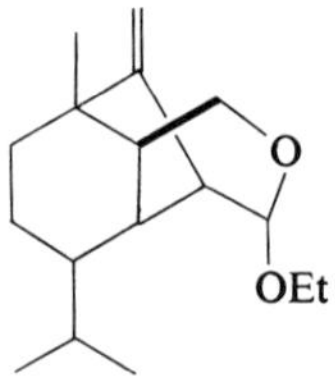

(112) Prehelminthosporal

an orcinol residue. Of the numerous isolated products some suggest a gradual cyclization of the terpenoid part, up to the drimane type. See for example the formulae of siccanochromene A (113) and E (114) and of the end product—siccanine (115). The formula of methyl presiccanochromenate (116) also suggests an intermediary in the biosynthesis of the orcinol moiety (see Hirai *et al.*, 1967; Nozoe and Suzuki, 1969). Siccanin possesses a distinct antifungal activity and its structure is not unique. Other antibiotics, as for example grifolin (117) from *Grifola confluens* (Gato *et al.*, 1963) or tauranin (118) from *Oospora aurantia* (Cooke) Sacc. et Vogl. (Kakisawa *et al.*, 1964) have a similar structure.

Helicobasidin (119) from *Helicobasidium mompa* Tanaka is a sesquiterpenoid antibiotic of quinonoid character (Natori *et al.*, 1964).

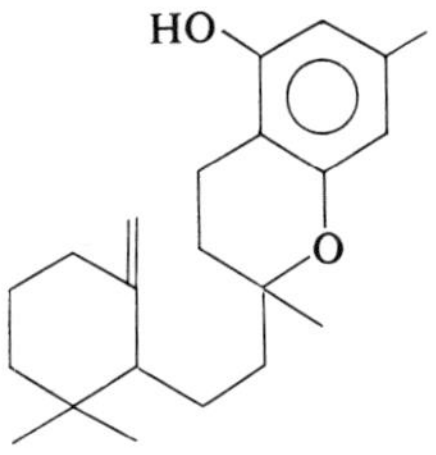

(113) Siccanochromene A

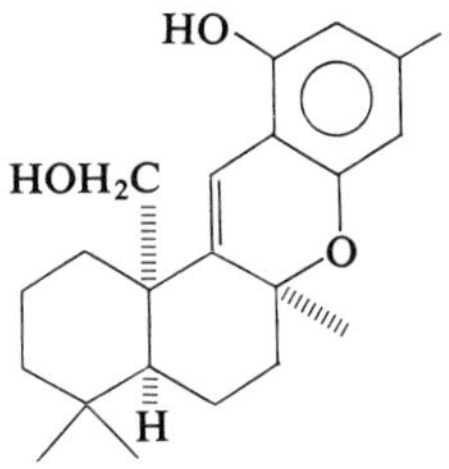

(114) Siccanochromene E

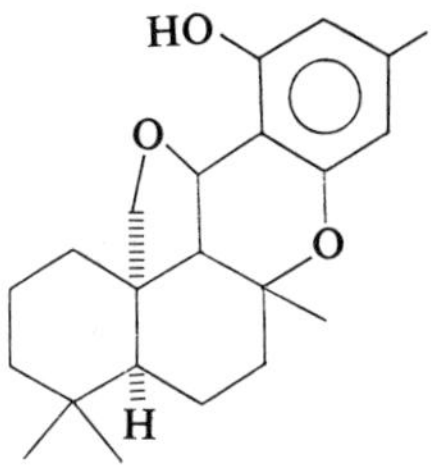

(115) Siccanine

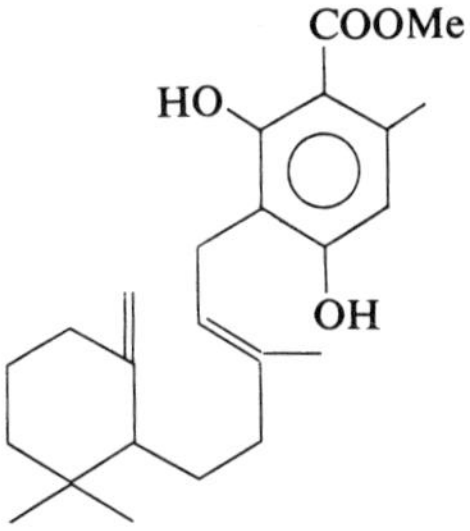

(116) Methyl presiccanochromenate

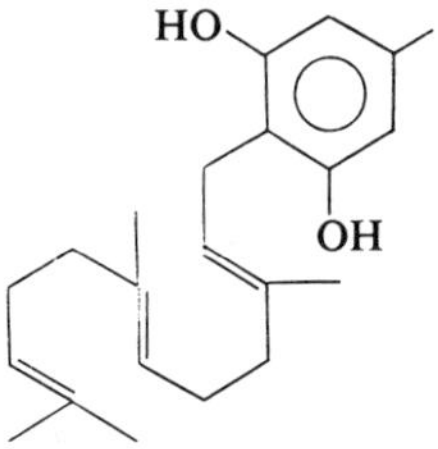

(117) Grifolin

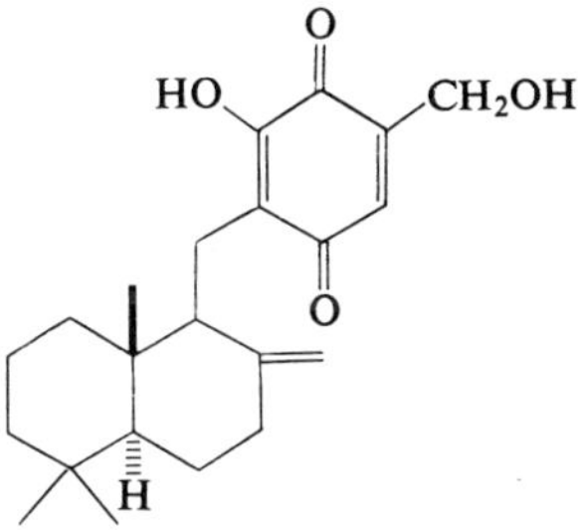

(118) Tauranin

The product of the metabolism of the mould *Phytophora infestans* parasitizing, for example, the tubers of *Solanum tuberosum* and *S. demisum* is called rishitin (28). In view of its activity it is called "phytoalexin" (Katsui *et al.*, 1968). The formation of ipomoeamarone (120) of the farnesane type is analogous, i.e. it arises on infection of sweet potatoes (*Ipomoea batatas*) by *Ceratostomella fimbriata*, causing the so-called "black disease". This sesquiterpenoid of bitter taste the structure of which was determined by Kubota (1958) has antibacterial activity.

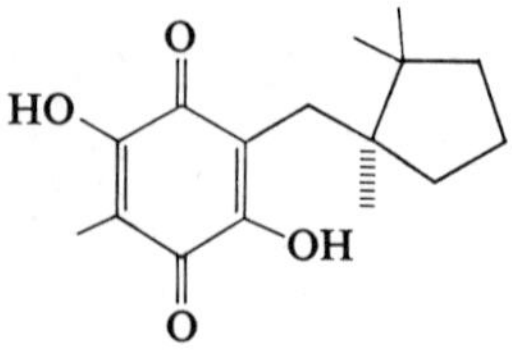

(119) Helicobasidin

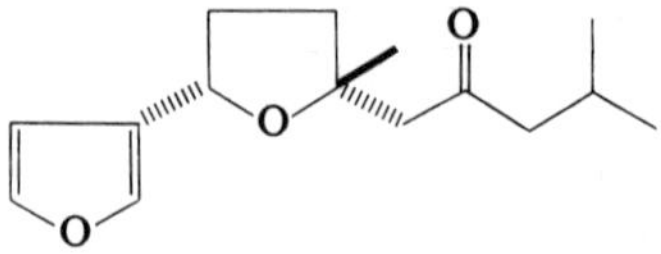

(120) Ipomeamarone

2. *Sex Attractants of Fungi*

Up to now three compounds of terpenoid nature are known to have a function in the sexual reproduction of fungi: a steroidal compound antheridiol, trisporic acids B and C possibly derived from carotenoids, and sesquiterpenoid derivative sirenin (121) (for a review see Barksdale, 1969). The structure of sirenin is derived from sesquicarene (Nutting *et al.*, 1966, 1968); it is the sperm attractant produced by female gametes of the water mould *Allomyces*.

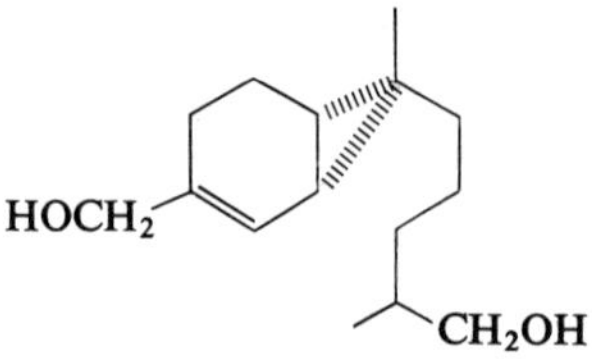

$HOCH_2$ CH_2OH

(121) Sirenin

3. *Coloured Components of* Lactarius deliciosus *L.*

The chemical nature of the mushroom dyes, at least those already studied, is of a very diverse nature. However, presently only one group of higher fungi is known which contains substances of the character of azulenoid precursors with a guaiane skeleton as the coloured components. Such are some species of the *Lactarius* genus; up to now only *L. deliciosus* has been thoroughly investigated. When its fruiting bodies were worked up in the conventional manner, without excluding the activity of oxidases or dehydrogenases, lactarazulene and lactaroviolin (Benešová *et al.*, 1954, 1955), structures (122) and (123), were isolated. More recently lactarofulvene was also isolated (124) (Bertelli and Crabtree, 1968). On very mild treatment, in which the influence of the above mentioned enzymes and of oxygen was excluded, native precursors of these substances have now been isolated which have a distinct orange colour. According to their structure determined by Vokáč *et al.* (1970) they are true natural fulvenes (125a and b, respectively).

HCO

(122) Lactarazulene

(123) Lactaroviolin

R—CH_2

(124) Lactarofulvene

(125a) R = OH
(125b) R = O—Stearoyl

4. *Other Fungal Metabolites of Sesquiterpenoid Nature*

According to some observations (Sprecher, 1965) the presence of monoterpenoids in fungi could not be proved. Neither could the ability of fungi to metabolize monoterpenes be experimentally confirmed. However, several examples are known where fungi during their life cycle intervene in the biosynthesis of sesquiterpenoids in plants. As examples the above mentioned ipomoeamarone can be mentioned, and the formation of agarospirol (126) which occurs only in fungus-infected agarwood (*Aquillaria agallocha* Roxb.) (Varma *et al.*, 1965). Of course, it was not possible to decide if this is fungal synthesis or a stimulation of higher plant synthesis by the fungus.

OH

(126) Agarospirol

C. ANIMAL KINGDOM

1. *Sesquiterpenoids of Marine Fauna*

During the last decade the presence of common and also less common sesquiterpenoids was demonstrated in sea animals of various classes. Probably the oldest papers are those of the Ciereszko group (Ciereszko *et al.*, 1960) concerning the presence of sesquiterpenes in gorgonians (class Anthozoa, subclass Alcyonaria, order Gorgonacea). For example in *Plexaura crassa* a hydrocarbon was discovered giving rise to (+)-cadinene dihydrochloride; in *Eunicea mamosa* eunicin probably also of terpenoid nature was found and more recently a whole group of sesquiterpenoid hydrocarbons was found in *Pseudopterogorgia americana* (Weinheimer *et al.*, 1968), as for example γ-maaliene (127), 9-aristolene (128), 1(10)-aristolene (129), and the previously unknown hydrocarbon β-gorgonene (39); for mention of its probable biosynthesis see p. 63.

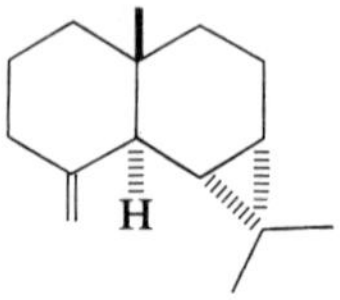

(127) (+)-γ-Maaliene 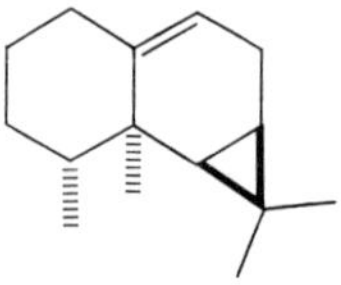 (128) 9-Aristolene 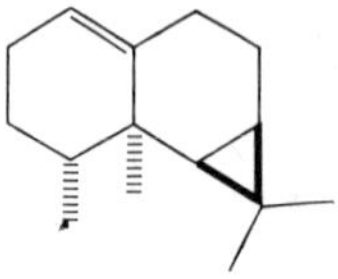 (129) 1(10)-Aristolene

Similarly in sea hares (Mollusca, class Gastropoda, family Aplysiidae) sesquiterpenes were discovered, as for example brominated derivatives as aplysin (5), aplysinol (130), and a related substance, debromoaplysin (131) from *Aplysia kurodai* (Yamamura and Hirata, 1963, 1964). As regards the origin of these substances, it can be said with great probability that they are metabolites of sesquiterpenoids present in some seaweeds. Thus, for example, in *Laurentia glandulifera* Kützing, or *L. nipponica* Yamada (Rhodomelaceae) (Irie *et al.*, 1969a,b), on which sea hares feed, laurene (132) or its bromine containing derivative laurenisol (4) is present. By the way, the biosynthesis of sesquiterpenoids in the above mentioned gorgonians is also not quite clear. Ciereszko *et al.* (1960) mentions that these animals live associated with zooxanthellae and one must consider the role of these algal cells in the formation of the terpenoid compounds.

A similar situation can also exist in the occurrence of sesquiterpenoids in some corals (*Gorgonia* sp.) which also live in symbiosis with algae.

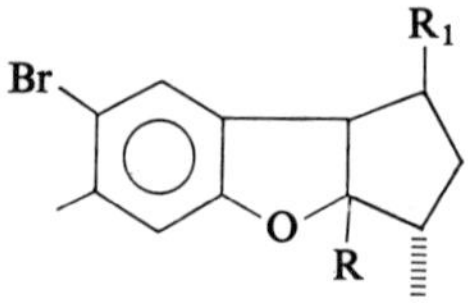

(130) Aplysinol
$R = CH_3$, $R_1 = CH_2OH$
or $R = CH_2OH$, $R_1 = CH_3$

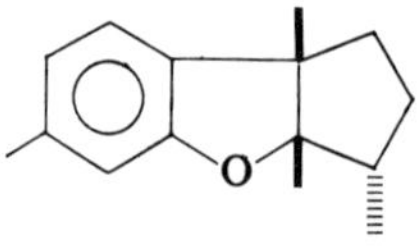

(131) Debromoaplysin

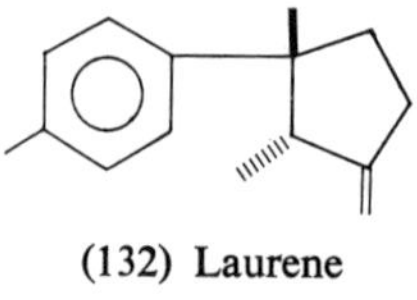

(132) Laurene

2. The Occurrence of Sesquiterpenoids in Arthropods

This section concerning the occurrence of sesquiterpenoids in arthropods can also be introduced by a doubt: is any insect species capable of full biosynthesis of a terpene derivative? However, the occurrence and the biological functions of these substances are especially numerous and also vitally important in insects. Let us mention here only the evident relation of the juvenile hormone to sesquiterpenoids. A whole series of substances serving insects for communication within a given species, or for deterring predators is of sesquiterpene nature. However, their structure is often not typically sesquiterpenoid. They are probably formed as insect metabolites of terpenoid substances of higher molecular weight, for example from carotenoids. As an example the allenic sesquiterpene of *Romalea* (already mentioned on p. 54) may be mentioned, or also the product (133) of the "hair pencil" of the male Monarch Butterfly (*Danaus plexipus*; see Meinwald *et al.*, 1968), i.e. *trans-trans*-10-hydroxy-3,7-dimethyl-2,6-decadienoic acid. Often we do not yet know fully the function of some of these sesquiterpenoids discovered in insects, as for example the presence of α-farnesene (134) in Dufour's gland of the myrmicine ant *Aphaenogaster longiceps* (Cavill *et al.*, 1967), of dendrolasin (135) in *Dendrolasius fuliginosus* (Pavan, 1956). A farnesol derivative, 2,3-dihydro-*trans*-farnesol (136) (Bergström *et al.*, 1967) (called newly terrestrol; Stöllberg-Stenhagen, 1970) is regarded as a territory-marking substance. It is excreted from the mandibular glands of male bumble-bees (*Bombus terrestris*). By depositing this substance on places regularly visited this insect evidently attracts the females of its species. Similarly nerolidol (90) is considered to be a sex-attractant of some other Hymenoptera (Kullenberg, 1956).

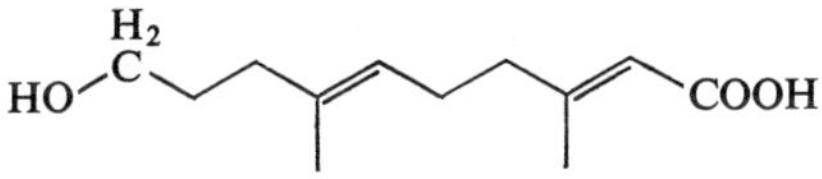

(133) "Hair pencil" acid

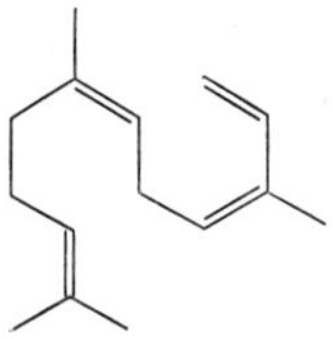

(134) α-Farnesene

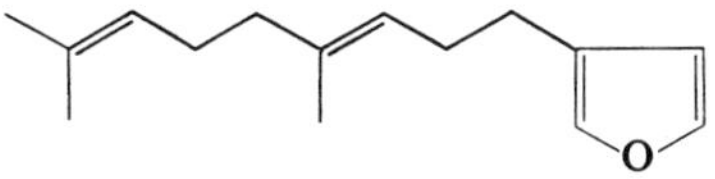

(135) Dendrolasin

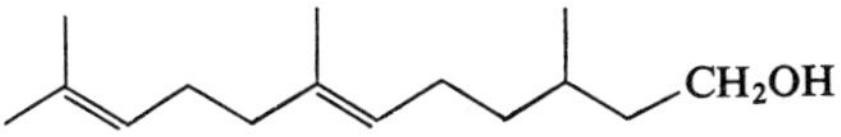

(136) Terrestrol

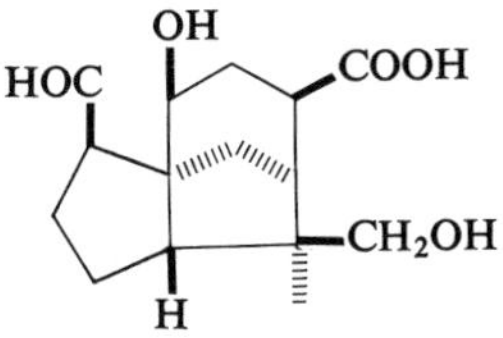

(137) Jallaric acid

The natural raw material for the preparation of shellac, so-called seedlack, represents a special case of the utilization of sesquiterpenoids by insect organism. This secretion of protective character with which the insect *Laccifer lacca* Kerr. encrusts itself is evidently a metabolite of vegetable origin. Its main component is jallaric acid (137), a cedrane type sesquiterpenoid (Wadia *et al.*, 1969).

REFERENCES

Aebi, A., Waaler, T. and Büchi, J. (1958). *Pharm. Weekbl. Med.* **93**, 397.

Bamburg, J. R., Riggs, N. V. and Strong, F. M. (1968). *Tetrahedron* **24**, 3329.

Barksdale, A. W. (1969). *Science, N.Y.* **166**, 831.

Barnes, C. S. and Loder, J. W. (1962). *Aust. J. Chem.* **15**, 322.

Barton, D. H. R. and Werstiuk, N. H. (1968). *J. chem. Soc.* (*C*) 148.

Benešová, V., Herout, V. and Šorm, F. (1954). *Coll. Czech. Chem. Commun.* **19**, 357.

Benešová, V., Herout, V. and Šorm, F. (1955). *Coll. Czech. Chem. Commun.* **20**, 510.

Benešová, V., Samek, Z., Herout, V. and Šorm, F. (1969). *Coll. Czech. Chem. Commun.* **34**, 582.

Bergström, G., Kullenberg, B., Stöllberg-Stenhagen, S. and Stenhagen, E. (1967). *Ark. kemi* **28**, 453.

Berkoff, Ch. E. (1969). *Q. Rev.* **XXIII**, No. 3, p. 372–391.

Bertelli, D. J. and Crabtree, J. H. (1968). *Tetrahedron* **24**, 2079.

Birch, A. J., Hochstein, F. A., Quartey, J. A. K. and Turnbull, J. P. (1964). *J. chem. Soc.* 2923.

Birnbaum, G. I. (1965). *Tetrahedron Letters* 4149.

Bohlmann, F. and Zdero, C. (1969). *Tetrahedron Letters* 1003.

Bohlmann, F., Zdero, C. and Grenz, M. (1969). *Tetrahedron Letters* 2417.

Böhmer, B., Fetz, E., Häri, E., Sigg, H. P., Stoll, Ch. and Tamm, Ch. (1965). *Helv. Chim. Acta* **48**, 1079.

Bowers, W. S. and Thompson, M. J. (1963). *Science, N.Y.* **142**, 1469.

Bowers, W. S., Fales, H. M., Thompson, M. J. and Uebel, E. C. (1966). *Science, N.Y.* **154**, 1020.

Bryant, R. (1969). *In* "Rodd's Chemistry of Carbon Compounds" (S. Coffey, ed.), 1st Ed., Vol. II, Part C, pp. 256–368, Elsevier, Amsterdam.

Büchi, G., Goldman, M. I. and Mayo, D. W. (1966). *J. Am. chem. Soc.* **88**, 3109.

Cagliotti, L., Naef, H., Arigoni, D. and Jeger, O. (1958). *Helv. Chim. Acta* **41**, 2278.

Cagliotti, L., Naef, H., Arigoni, D. and Jeger, O. (1959). *Helv. Chim. Acta* **42**, 2557.

Cavill, G. W. K., Williams, P. J. and Whitfield, F. B. (1967). *Tetrahedron Letters* 2201.

Černý, V., Dolejš, L., Lábler, L., Šorm, F. and Sláma, K. (1967). *Coll. Czech. Chem. Commun.* **32**, 3926.

Ciereszko, L. S., Sifford, D. H. and Weinheimer, A. J. (1960). *Am. N.Y. Acad. Sci.* **90**/3, 917.

Corner, T. W., McCapra, F., Qureshi, I. H., Tretter, J. and Scott, A. I. (1965). *Chem. Comm.* 310.

Cornforth, J. W., Milborrow, B. V. and Ryback, G. (1965). *Nature, Lond.* **206**, 715.

Cornforth, J. W., Draber, W., Milborrow, B. V. and Ryback, G. (1967). *Chem. Comm.* 114.

Coscia, C. J. (1969). *In* "Cyclopentanoid Terpene Derivatives" (W. I. Taylor and A. R. Battersby, eds), pp. 147–201, Dekker, New York.

Cruickshank, A. M. and Perrin, D. R. (1964). *In* "Biochemistry of Phenolic Compounds (J. B. Harborne, ed.), p. 511. Academic Press, London and New York.
Dolejš, L. and Herout, V. (1961). *Coll. Czech. Chem. Commun.* **26**, 2045.
Dolejš, L., Souček, M., Horák, M., Herout, V. and Šorm, F. (1958). *Chemy Ind.* (*Lond.*) 530.
Doskotch, R. W. and Hufford, C. D. (1969). *J. Pharm. Sci.* **58**, 186.
Dugan, J. J., de Mayo, P., Nisbet, M. and Anchel, M. (1965). *J. Am. chem. Soc.* **87**, 2768.
Eagles, C. F. and Wareing, P. F. (1963). *Nature, Lond.* **199**, 874.
Edwards, O. E. (1969). *In* "Cyclopentanoid Terpene Derivatives" (W. I. Taylor and A. R. Battersby, eds), pp. 357–408, Dekker, New York.
Edwards, O. E. and Elmore, N. F. (1962). *Can. J. Chem.* **40**, 256.
Endo, K. and de Mayo, P. (1967). *Chem. Comm.* 89.
Erdtman, H. (1939). *Ann. Chem. Liebig's* 539, 116.
Flath, R. A., Lundin, R. E. and Terranishi, R. (1965). *J. Org. Chem.* **30**, 1690.
Flury, E., Mauli, R. and Sigg, H. P. (1965). *Chem. Comm.* 26.
Frachebond, M., Rowe, J. W., Scott, R. W., Fanega, S. M., Bukl, A. J. and Toda, J. K. (1968). *Forest Products J.* **18**, 37.
Galston, A. W. and Davies, P. J. (1969). *Science, N.Y.* **163**, 1288–1297.
Gato, T., Kakisawa, H. and Hirata, Y. (1963). *Tetrahedron* **19**, 2079.
Gerber, N. N. and Lechevalier, H. A. (1965). *Appl. Microbiol.* **13**, 935.
Gilbert, L. I. and Goodfelow, R. D. (1965). *Zool. Jb. Physiol. Bd.* **11**, 718.
Godfredsen, W. O. and Vangendal, S. (1964). *Proc. Chem. Soc.* 188.
Godfredsen, W. O., Grove, J. F. and Tamm, Ch. (1967). *Helv. Chim. Acta* **50**, 1666.
Gough, J., Powell, A. and Sutherland, M. D. (1961). *Tetrahedron Letters* 763.
Grove, J. F. (1969). *Chem. Comm.* 1260.
Gutzwiller, J. and Tamm, Ch. (1965a). *Helv. Chim. Acta* **48**, 157.
Gutzwiller, J. and Tamm, Ch. (1965b). *Helv. Chim. Acta* **48**, 177.
Gutzwiller, J., Mauli, R., Sigg, H. P. and Tamm, Ch. (1964). *Helv. Chim. Acta* **47**, 2234.
Harri, F., Loeffler, W., Sigg, H. P., Stähelin, H., Stoll, Ch., Tamm, Ch. and Wiesinger, D. (1962). *Helv. Chim. Acta* **45**, 839.
Hartley, R. D. and Fawcett, R. W. (1969a). *Phytochemistry* **8**, 637.
Hartley, R. D. and Fawcett, R. W. (1969b). *Phytochemistry* **8**, 1793.
Hayashi, S., Matsuo, A. and Matsuura, T. (1969). *Experientia* **25**, 1139.
Hendrickson, J. B. (1959). *Tetrahedron* **7**, 82.
Herout, V. (1966). *Planta Medica* 14 Suppl., p. 97.
Herz, W., Sumi, Y., Sudarsanam, V. and Raulois, D. (1967). *J. Org. Chem.* **32**, 3658.
Hikino, H., Suzuki, N. and Takemoto, T. (1966). *Chem. Pharm. Bull.* **14**, 1441.
Hirai, K., Nozoe, S., Tsuda, K., Iitaka, Y., Ischibaschi, K. and Shirashaka, K. (1967). *Tetrahedron Letters* 2177.
Ina, K., Sakato, Y. and Fukami, H. (1968). *Tetrahedron Letters* 2777.
Inubushi, Y., Sasaki, Y., Tsuda, Y., Yasui, B., Konita, T., Matsumoto, J., Katarao, J. and Nakana, J. (1964). *Yakugaku Zasshi* **83**, 1184.
Irie, H., Kimura, H., Otani, N., Ueda, K. and Uyeo, S. (1967). *Chem. Comm.* 678.
Irie, T., Suzuki, T., Yasunari, Y., Kurasawa, E. and Masamune, T. (1969). *Tetrahedron* **25**, 459.
Irie, T., Fukuzawa, A., Izawa, M. and Kurasawa, E. (1969). *Tetrahedron Letters* 1343.
Ichihara, A., Shirahama, H. and Matsumoto, T. (1969). *Tetrahedron Letters* 3965.
Jommi, G., Manitto, P., Pelizzoni, F. and Scholastico, C. (1965). *Chim. Ind.* (*Milan*) **47**, 1328.

Jones, R. V. H. and Sutherland, M. D. (1968). *Chem. Comm.* 1229.
Jones, R. V. H. and Sutherland, M. D. (1968). *Aust. J. Chem.* **21**, 2255.
Kakisawa, K., Nakanishi, K. and Nishikawa, H. (1964). *Chem. Pharm. Bull. Japan* **12**, 796.
Kato, J., Shiotani, Y., Tamura, S. and Sakurai, A. (1964). *Naturwissenschaften* **51**, 341.
Katsui, N., Murai, A., Tagasuki, M., Imazumi, K. and Masamune, T. (1968). *Chem. Comm.* 43.
Kawasaki, I., Matsutani, S. and Kaneko, T. (1963). *Bull. Chem. Soc. Japan* **36**, 1474.
Kepler, J. A., Wall, M. E., Mason, J. E., Basset, C., McPhail, A. T. and Sim, G. A. (1967). *J. Am. chem. Soc.* **89**, 1260.
Kido, F., Uda, H. and Yoshikoshi, A. (1967). *Tetrahedron Letters* 2815.
Kolbe, M. and Westfelt, L. (1967). *Acta Chem. Scand.* **21**, 585.
Kotake, M., Kawasaki, I., Okamoto, T., Kusumoto, S. and Kaneko, T. (1960). *Annalen* **636**, 158.
Kotake, M., Kawasaki, I., Okamoto, T., Matsutani, S., Kusumoto, S. and Kaneko, T. (1962). *Bull. Chem. Soc. Japan* **35**, 1335.
Kubota, T. (1958). *Tetrahedron* **4**, 68.
Kubota, T. and Kubo, I. (1969). *Nature, Lond.* **223**, 97.
Kullenberg, B. (1956). *Zool. Bidrag, Uppsala* **31**, 254.
Kupchan, S. M., Aynehchi, A., Cassady, J. M., McPhail, A. T., Sim, G. A., Schnoes, H. K. and Burlingame, A. L. (1966). *J. Am. chem. Soc.* **88**, 3679.
Kupchan, S. M., Hemingway, R. J., Werner, D., Karim, A., McPhail, A. T. and Sim, G. A. (1968a). *J. Am. chem. Soc.* **90**, 3596.
Kupchan, S. M., Kelsey, J. E., Maruyama, M. and Cassady, J. M. (1968b). *Tetrahedron Letters* 3517.
Kupchan, S. M., Hemingway, R. J., Karim, A. and Werner, D. (1969). *J. Org. Chem.* **34**, 3908.
McMorris, T. C. and Anchel, M. (1965). *J. Am. chem. Soc.* **87**, 1594.
McMorris, T. C., Nair, M. S. R. and Anchel, M. (1967). *J. Am. chem. Soc.* **89**, 4562.
MacSweeney, D. F., Ramage, R. and Sattar, A. (1970). *Tetrahedron Letters* 557.
Marini-Bettolo, G. B., Casinovi, C. G. and Galeffi, C. (1965). *Tetrahedron Letters* 4857.
Marshall, J. A. and Johnson, P. C. (1967). *J. Am. chem. Soc.* **89**, 2750.
Martin-Smith, M. and Sneader, W. E. (1969). *In* "Biological Activity of the Terpenoids and Their Derivatives—Recent Advances", *in* "Progress in Drug Research" (E. Jucker, ed.), p. 11, Birkhauser, Basel.
Massy-Westropp, R. A., Reynolds, G. D. and Spotwood, T. M. (1966). *Tetrahedron Letters* 1939.
Mayo, P. de and Williams, R. E. (1965). *J. Am. chem. Soc.* **87**, 3275.
Mayo, P. de, Spencer, E. Y. and White, R. W. (1961). *Can. J. Chem.* **39**, 1608.
Mayo, P. de, Spencer, E. Y. and White, R. W. (1963). *Can. J. Chem.* **41**, 2996.
Menche, D. and Huneck, S. (1966). *Chem. Ber.* **99**, 2669.
Meinwald, J., Erickson, K., Hartshorn, M., Meinwald, Y. C. and Eisner, T. (1968). *Tetrahedron Letters* 2959.
Meinwald, J., Chalmers, A. M., Pliske, T. E. and Eisner, T. (1968). *Tetrahedron Letters* 4893.
Milborrow, B. V. (1968). *In* "Biochemistry and Physiology of Plant Growth Substances" (F. Wightman and G. Setterfield, eds), Runge Press, Ottawa.
Milborrow, B. V. (1969). *J. Chem. Soc.* (*D*) 966.
Morimoto, H., Sanno, Y. and Oshio, H. (1966). *Tetrahedron* **22**, 3173.

Nair, M. S. R., Takeshita, H., McMorris, T. C. and Anchel, M. (1969). *J. Org. Chem.* **34**, 240.
Natori, S., Nishikawa, H. and Ogawa, H. (1964). *Chem. Pharm. Bull.* (*Japan*) **12**, 236.
Nishimura, K., Shinoda, N. and Hirose, Y. (1969). *Tetrahedron Letters* 3097.
Novotný, L., Jizba, J., Herout, V. and Šorm, F. (1962). *Coll. Czech. Chem. Commun.* **27**, 1393.
Novotný, L., Herout, V. and Šorm, F. (1964). *Coll. Czech. Chem. Commun.* **29**, 2182.
Nozoe, S. and Suzuki, K. T. (1969). *Tetrahedron Letters* 2457.
Nozoe, T., Cheng, Y. S. and Toda, T. (1966). *Tetrahedron Letters* 3663.
Nutting, W. H., Williams, M. W. and Rapoport, H. (1966). *Biochemistry* **5**, 2147.
Nutting, W. H., Rapoport, H. and Machlis, L. (1968). *J. Am. chem. Soc.* **90**, 6434.
Ohashi, J. (1959). *J. Pharm. Soc. Japan* **79**, 729, 734.
Ohkuma, K., Lyon, J. L., Addicott, F. T. and Smith, O. E. (1963). *Science, N.Y.* 142.
Ohta, Y. and Hirose, Y. (1968). *Tetrahedron Letters* 1251.
Onaka, T., Kamata, S., Maeda, T., Kawazoe, Y., Natsume, M., Okamoto, T., Uchimaru, F. and Shimuru, M. (1964). *Chem. Pharm. Bull.* (*Tokyo*) **12**, 506.
Overbeek, J. van (1966). *Science, N.Y.* **152**, 721.
Parker, W., Roberts, J. S. and Ramage, R. (1967). *Q. Rev.* **XXI**, No. 3, 331–363.
Pavan, M. (1956). *Ric. Scien.* **26**, 144.
Porter, L. A. (1967). *Chem. Rev.* **67**, 441.
Richards, J. H. and Hendrickson, J. B. (1964). *In* "The Biosynthesis of Steroids, Terpenes and Acetogenins", W. A. Benjamin Inc., New York.
Röller, H. and Dahm, K. H. (1968). *In* "Recent Progress in Hormone Research", Vol. 24, pp. 651–671, Academic Press, New York and London.
Rowe, J. W. and Toda, J. K. (1969). *Chemy Ind.* (*Lond.*) 922.
Rücker, G. (1968). *Tetrahedron Letters* 3615.
Ruzicka, L. (1956). *In* "Perspectives in Organic Chemistry" (A. Todd, ed.), p. 297, Interscience, New York.
Ruzicka, L. (1959). *Proc. Chem. Soc.* 341.
Safferman, R. S., Rosen, A. A., Mashni, C. I. and Morris, M. E. (1967). *Environmental Sci. Technol.* **1**, 429.
Sakuma, R. and Yoshikoshi, A. (1968). *Chem. Comm.* 41.
Sakurai, A. and Tamura, S. (1965). *Agric. Biol. Chem.* **29**, 407.
Schwär, C. (1962). *Flora* 152, 509.
Sláma, K. and Williams, C. M. (1965). *Proc. Natn. Acad. Sci. U.S.A.* **54**, 411.
Smedman, L. Å. and Zavarin, E. (1968). *Tetrahedron Letters* 3833.
Šorm, F. and Dolejš, L. (1950). *Coll. Czech. Chem. Commun.* **15**, 96.
Sprecher, E. (1965). *Planta Medica* **13**, 418.
Stöllberg-Stenhagen, S. (1970). *Acta Chem. Scand.* **24**, 358.
Tada, M., Yamada, Y., Bhacca, N. S., Nakanishi, K. and Ohashi, M. (1964). *Chem. Pharm. Bull.* (*Tokyo*) **12**, 853.
Takahashi, S., Iimura, H., Takita, T., Maeda, K. and Umazawa, H. (1969). *Tetrahedron Letters* 4663.
Takeda, K., Minato, H., Horibe, I. and Myiawaki, M. (1967). *J. chem. Soc.* (*C*) 631.
Tamura, S., Sakurai, A., Kainuma, K. and Takai, M. (1963). *Agric. Biol. Chem.* (*Tokyo*) **27**, 738.
Tatsuno, T., Saito, M., Enomoto, M. and Tsunoda, H. (1968). *Chem. Pharm. Bull.* **16**, 2519.

Teisseire, P. (1969). *Recherches* 77.
Tidd, B. K. (1967). *J. chem. Soc.* (*C*) 218.
Tomita, B. and Hirose, Y. (1970). *Tetrahedron Letters* 143.
Treibs, W., Kirchhof, W. and Ziegenbein, W. (1955). *Fortschr. Chem. Forsch.* **3**, 334.
Varma, K. R., Maheswari, M. L. and Bhattacharyya, S. C. (1965). *Tetrahedron* **21**, 115.
Vlahov, R., Holub, M., Ognianov, I. and Herout, V. (1967). *Coll. Czech. Chem. Commun.* **32**, 808.
Vokáč, K., Samek, Z., Herout, V. and Šorm, F. (1968). *Tetrahedron Letters* 3855.
Vokáč, K., Samek, Z., Herout, V. and Šorm, F. (1970). *Coll. Czech. Chem. Commun.* **35**, 1296.
Vrkoč, J., Herout, V. and Šorm, F. (1961). *Coll. Czech. Chem. Commun.* **26**, 1343.
Wada, K. and Munakata, K. (1968). *Tetrahedron Letters* 4677.
Wada, K., Enomoto, Y., Matsui, K. and Munakata, K. (1968). *Tetrahedron Letters* 4673.
Wadia, M. S., Mhasakar, V. V. and Dev, S. (1969). *Tetrahedron* **25**, 3841.
Watt, J. M. and Breyer-Branwijk, M. G. (1962). *In* "The Medicinal and Paissonons Plants of Southern and Eastern Africa", 2nd Ed., Livingstone, Edinburgh.
Weinheimer, A. J. and Washecheck, P. H. (1969). *Tetrahedron Letters* 3315.
Weinheimer, A. J., Washecheck, P. H., Helm, D. and Hossain, M. B. (1968). *Chem. Comm.* 1070.
Whittaker, R. H. (1969). *Science, N.Y.* **163**, 150–160.
Winters, T. E., Geissman, T. A. and Safir, D. (1969). *J. Org. Chem.* **34**, 153.
Wittimberga, J. S. and Wittimberga, B. M. (1965). *J. Org. Chem.* **30**, 746.
Wolff, G. and Ourisson, G. (1968). *Tetrahedron Letters* 3849.
Wolff, G. and Ourisson, G. (1969). *Tetrahedron* **25**, 4903.
Yamada, K., Takada, S., Nakamura, S. and Hirata, Y. (1965). *Tetrahedron Letters* 4797.
Yamamura, S. and Hirata, Y. (1963). *Tetrahedron* **19**, 1485.
Yamamura, S. and Hirata, Y. (1964). *Tetrahedron Letters* 79.
Yoshihara, K., Ohta, Y., Sakai, T. and Hirose, Y. (1969). *Tetrahedron Letters* 2263.
Zalkow, L. H., Zalkow, V. B. and Brannon, D. R. (1963). *Chemy Ind.* (*Lond.*) 38.
Zalkow, V. B., Shaligram, A. M. and Zalkow, L. H. (1964). *Chemy Ind.* (*Lond.*) 194.

CHAPTER 4

Juvenile Hormones

A. PFIFFNER

Forschungsabteilung der F. Hoffmann-La Roche & Co. AG, Basel, Switzerland

I. Introduction

The juvenile hormone (JH) secreted by the corpora allata is one of three hormones that control the complicated postembryonic development of insects.

Since the pioneering studies of Sir Vincent Wigglesworth (1934), who established the existence of JH, and the successful preparation of the first extract with strong JH-activity from the abdomens of the adult male *cecropia* moth by Williams (1956), great efforts have been made to isolate the natural hormone and to characterize its structure (Williams and Law, 1965; Schmialek, 1963; Meyer, 1965; Meyer *et al.*, 1965). However, the very small amount present in the moth abdomen and the difficulties of rearing and obtaining sufficient moths to permit isolation allowed little but speculation about the structure of JH.

In 1967 the difficult goals of isolating and identifying the elusive and long-sought JH were at last attained by a team of chemists and biochemists headed by Röller of the University of Wisconsin in the United States. In a series of brilliant biological and chemical investigations Röller and Dahm and their colleagues (Röller and Bjerke, 1965; Röller *et al.*, 1965; Röller *et al.*, 1967a,b; Dahm *et al.*, 1967, 1968; Röller *et al.*, 1969) established that the JH isolated from the giant silk worm moth *Hyalophora cecropia*, is methyl 10,11-epoxy-7-ethyl-3,11-dimethyl-10,11-*cis*,2-*trans*,6-*trans*-tridecadienoate (1). Shortly afterwards Meyer and his team (Meyer *et al.*, 1968a,b) at Case Western

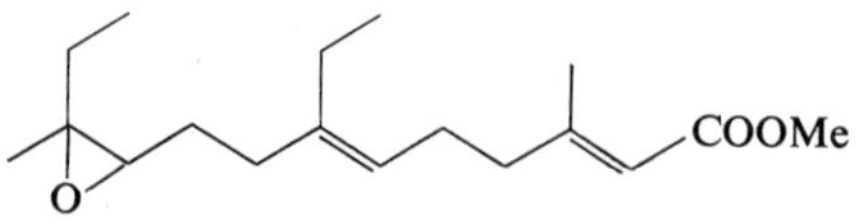

COOMe

cis *trans* *trans*

(1)

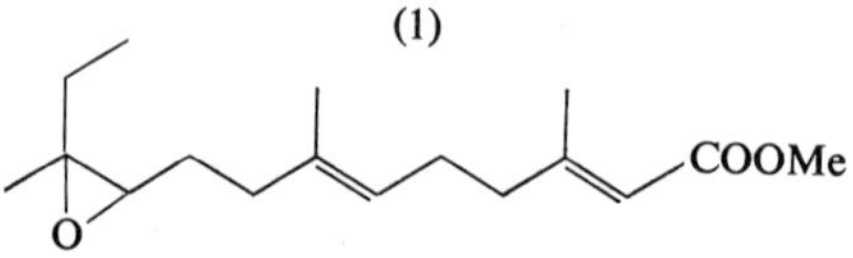

COOMe

cis *trans* *trans*

(2)

COOMe

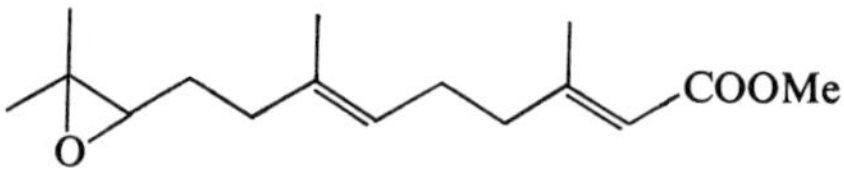

trans *trans*

(3)

Reserve University in Cleveland, Ohio, were successful in isolating a second JH (2) from the *cecropia* silk moth.

The molecule was shown to be identical in biological activity and structure to (1) except that a methyl group is present at C-7 instead of an ethyl group. A stereoselective total synthesis (Johnson *et al.*, 1969) of methyl 10,11-epoxy-3,7,11-trimethyl-10,11-*cis*,2-*trans*,6-*trans*-tridecadienoate (2) confirmed the structure. This second hormone is responsible for about 13–20% of the endocrine activity (Williams and Robbins, 1968) of the *cecropia* oil.

Structurally, JH (1), as well as the second natural hormone (2), are closely related to isoprenoid compounds. They can be considered as the di- and mono-homologues respectively of the known sesquiterpenoid 10,11-epoxy-ester (3), which was synthesized by Van Tamelen *et al.* (1965) in connection with fundamental cyclization studies. Compound (3) was found to have marked JH-activity in the *Tenebrio* assay by Bowers *et al.* (1965).

II. Function of Juvenile Hormone

JH is secreted by the corpora allata and acts directly on individual cells of the insect. It ensures the appearance and maintenance of those structural features which characterize the larval stage (Wigglesworth, 1936). Metamorphosis can be totally or partially prevented by additional application of JH. This results in the development of giant, but morphologically perfect, superlarvae or intermediate forms having both larval and adult characteristics. It has also been demonstrated (Röller and Bjerke, 1965) that the corpus allatum hormone can induce reversal of adult to larval integument. This morphogenetic activity is undoubtedly the most striking property of JH, but many other important functions, especially the gonadotropic and prothoracotropic effect, as well as changes in the protein and lipid metabolism, have been attributed to it.

JH is necessary for yolk formation in the female and for full activity of the accessory glands in the male (Wigglesworth, 1965). It controls the ovarian and egg development by control of protein metabolism (Thomas and Nation, 1966), and it may be indispensable for oocyte maturation because of its ability to enhance lipid synthesis in the ovary during oogenesis (Gilbert, 1967).

Moreover, JH activates the prothoracic gland in *Lepidoptera* (Wigglesworth, 1965). It has been suggested that the metabolic effects of JH are due to its action on prothoracic glands (Oberlander and Schneiderman, 1966), but Wigglesworth (1965) regards the so-called "metabolic action" of JH with scepticism.

JH is responsible for the stimulation of protein and protocatechuic acid glucoside synthesis in the accessory sex glands of the allatectomized adult female cockroach, *Periplaneta americana* (Shaaya, 1969). Allatectomy of the male desert locust, *Schistocerca gregaria*, leads to considerable accumulation of lipids, a condition which can be reversed by implantation of active corpora allata. Since allatectomy also results in loss of spontaneous motor activity, it is believed (Odhiambo, 1966) that JH regulates the intensity of motor activity by a direct effect on the central nervous system.

A female specific protein, essential for egg maturation in *Leucophaea maderae*, is synthesized in isolated abdomens of allatectomized females treated with the all-*trans*-isomer of authentic JH. The JH itself also stimulates to a similar degree the synthesis of other proteins which appear to be essential for egg maturation as well, but these are not specific to females (Engelmann, 1969; Wiss-Huber and Lüscher, 1969).

Though several studies concerning the biochemical aspects of JH-action have been reported (Wigglesworth, 1965; Minks, 1967), a great deal of research still needs to be done in order to fully understand the polyfunctionality of JH, especially with regard to the intricate neuroendocrine system.

III. Juvenile Hormone-like Substances

A few years before Röller and his colleagues (1965, 1967) worked out the structure of authentic JH, investigators at several laboratories had synthesized a number of substances with impressive JH-activity. Historically, the first compounds with demonstratable JH-activity, farnesol (4; $X = CH_2OH$) and farnesal (4; $X = CHO$), were isolated by Schmialek (1961) from the faeces of the beetle, *Tenebrio molitor*. Farnesol was therefore proposed by Schmialek (1961) and Wigglesworth (1961) as the JH, but other investigators (Chen *et al.*, 1962; Yamamoto and Jacobson, 1962; Schneiderman and Gilbert, 1964; Meyer, 1965; Williams, 1956) expressed serious doubts about this

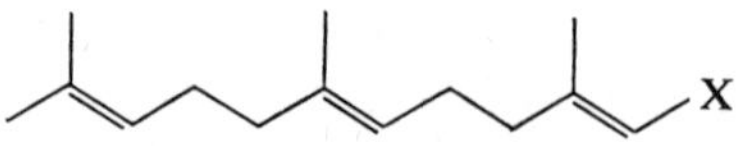

X

(4)

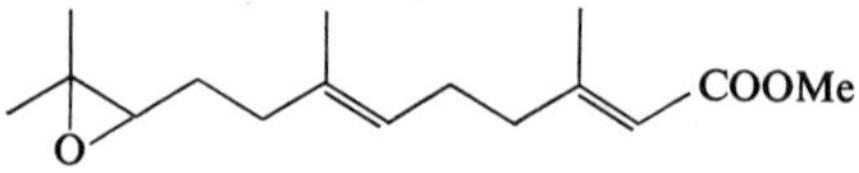

COOMe

O

(3)

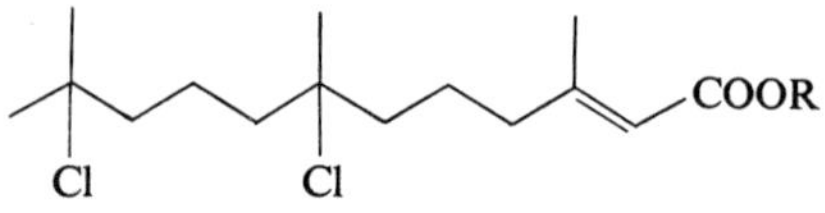

COOR

Cl Cl

(5)

because of its low degree of activity and because its chemical and chromatographic behaviour differs from that of the natural hormone. The evaluation of the JH-activity of a wide variety of natural and synthetic farnesane derivatives, analogues and homologues followed (Rüegg and Schmialek, 1962; Bowers and Thompson, 1963; Krishnakumaran and Schneiderman, 1965; Levinson, 1966). Among the more potent compounds were farnesyl methyl ether (4; $X = CH_2OMe$), N,N-diethylfarnesylamine (4; $X = CH_2NEt_2$). A very significant increase in the JH-activity of these sesquiterpenoid compounds was accomplished by Bowers *et al.* (1965) by the introduction of the epoxide-function at the 10,11 position. The synthetic methyl 10,11-epoxy-3,7,11-trimethyl-2,6-dodecadienoate (3) was 16 times more active than farnesyl methyl ether (4; $X = CH_2OMe$) and 1600 times as active as farnesol (4; $X = CH_2OH$).

Law *et al.* (1966) discovered in the course of their chemical studies on farnesane derivatives that the reaction between hydrogen chloride and an alcoholic solution of farnesenic acid (4; $X = COOH$) produced a mixture of at least six compounds of extremely high JH-activity. The authors observed that the activity of the reaction products was dependent on the alcohol used, ethanol giving maximum activity. Emergence of adult yellow-fever mosquitoes, *Aedes aegypti*, is blocked after the addition of 1 part of crude reaction products in 100,000 parts of water (Spielman and Williams, 1966). This so-called "Synthetic Juvenile Hormone" (SJH) proved also to be an effective insecticide and ovicide for human body lice, *Pediculus humanus* var. *corporis* (Vinson and Williams, 1967) and is able to block the embryonic development of silkworm eggs (Riddiford and Williams, 1967). It was later found by Romaňuk *et al.* (1967) that, for the hemipteran bug, *Pyrrhocoris apterus*, the most active compounds of this so-called SJH-mixture were esters of farnesenic acid (4; $X = COOH$) in which two double bonds were saturated by the addition of hydrogen chloride (5). They obtained, in almost quantitative yield, the methyl dichlorotetrahydrofarnesenoate (5; $R = Me$) by passing gaseous hydrogen chloride through a methanolic solution of methyl farnesenoate (4; X = COOMe). In their subsequent search for new JH-analogues, Sláma *et al.* (1969) have synthesized a number of dichlorotetrahydrofarnesenic ester derivatives (5; R = alkyl, C_1–C_6; cycloalkyl; benzyl; etc.). The most active compound was the ethyl ester of *trans*-dichlorotetrahydrofarnesenic acid (5; R = Et). Their data supported the earlier observations of Law *et al.* (1966), who found that the reaction between hydrogen chloride and farnesenic acid (4; X = COOH) in the presence of ethanol produced the most active JH-materials (SJH). As little as 0·1 ng of ethyl dichlorotetrahydrofarnesenoate (5; R = Et) applied topically to *Pyrrhocoris* larvae produced adultoids which were unable to survive and reproduce. Unlike the hemipteran larvae of *Pyrrhocoris* and *Dysdercus*, the coleopteran pupae of *Tenebrio* were quite insensitive to esters of dichlorotetrahydrofarnesenic acid (5).

In further investigations, Masner *et al.* (1968) obtained results with methyl dichlorotetrahydrofarnesenoate (5; R = Me) which could have profound insecticidal significance. Their initial studies showed that as little as 1 μg of this dichlorotetrahydroester per specimen is enough to induce permanent sterility in the adult female of *P. apterus*. These studies were extended to the treatment of males of *Pyrrhocoris*. When this dichlorotetrahydroester was applied to adult male linden bugs, it had a super-sterilizing effect on reproduction, without interfering with the bug's ability to mate. In mating, the treated males transmitted enough of the material to cause females to lay sterile eggs. And when these females later mated with untreated males, they passed on enough of the material to sterilize in turn another group of females. This is thus a possible way of controlling insect pests through the use of infectious sterilizing agents. Other studies with this ester suggest that it may have utility in preventing development of insect pests contaminating stored grain (Thomas and Bhatnagar-Thomas, 1968).

These surprising developments, especially the possibility of the use of JH and related compounds as insecticides (Williams, 1967), have induced a considerable amount of activity in this field. In their search for JH-active compounds, investigators in several laboratories (Siddall and Calame, 1967; Pfiffner and Schwieter, 1968, 1969; Braun *et al.*, 1968; Cavill *et al.*, 1969; Hoffmann *et al.*, 1969; Jarolim *et al.*, 1969; Mori *et al.*, 1969; Ratusky *et al.*, 1969; Schwarz *et al.*, 1969; Wakabayashi *et al.*, 1969; etc.) have synthesized an immense number of JH isomeric mixtures and analogues.

All these compounds have a carbon skeleton similar to JH and can be

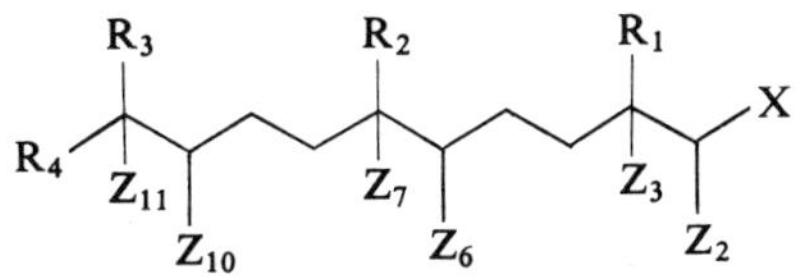

R_3 R_2 R_1 X R_4 Z_{11} Z_{10} Z_7 Z_6 Z_3 Z_2

(6)

described by the general formula (6), wherein the radicals R_1–R_4 may be hydrogen or alkyl groups (C_1–C_6), the radical X an alkoxycarbonyl-, alkoxymethylene-, aminocarbonyl-, aminomethylene-, phosphonomethylene- or a cyanogroup. Z_2 and Z_3, Z_6 and Z_7, Z_{10} and Z_{11} together may represent a carbon–carbon bond, an oxygen-, sulfur-, methylene-, dichloromethylene- or a difluoromethylenebridge and Z_2 hydrogen or hydroxy, Z_6 and Z_{10} hydrogen, hydroxy and halogen and Z_3, Z_7 and Z_{11} may represent hydrogen, hydroxy, alkoxy or halogen.

A particularly interesting development in JH-like compounds began when Sláma and Williams (1965) observed that when the European linden bug, *P. apterus*, was transported from Prague to Boston, it failed to undergo normal metamorphosis and without exception all the animals died without completing metamorphosis (Williams and Sláma, 1966). The source of JH-activity was eventually traced to exposure of the bugs to a specific brand of paper towel placed in the rearing jars. Following the evaluation of the JH-like activity of other samples of paper it was solemnly declared that the *New York Times*, *Wall Street Journal*, *Science* and the *Scientific American* were active, whereas the *London Times* and *Nature* were inactive. The active principle in American paper, the so-called "Paper Factor", originated in the balsam fir, *Abies balsamea*. From this wood, Bowers and his colleagues (1966) isolated, purified and characterized the active principle, (+)-juvabione (8), which is the methyl ester of the known todomatuic acid (7) (Tutihasi and Hanazawa, 1940;

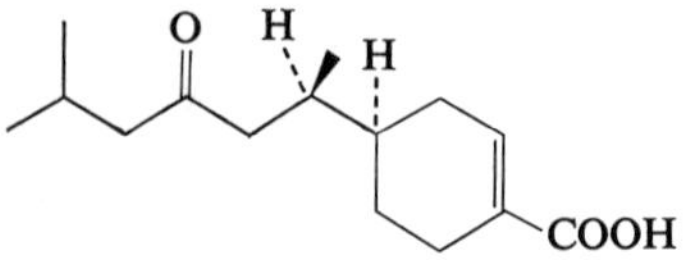

O H H COOH

(7)

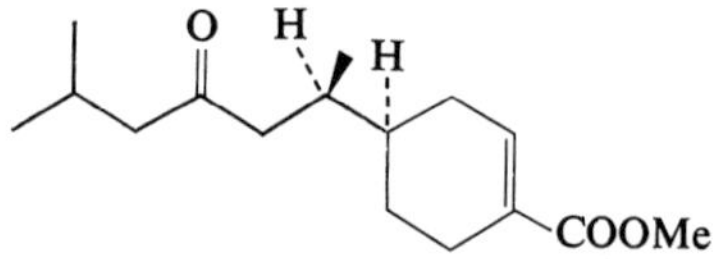

O H H COOMe

(8)

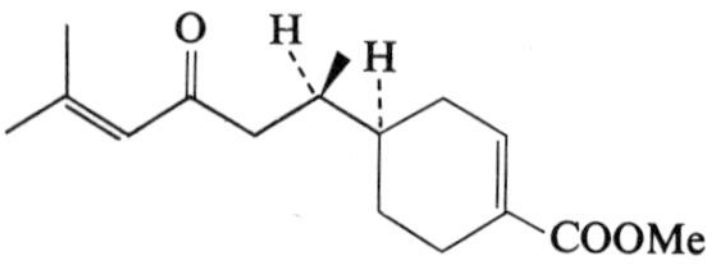

O H H COOMe

(9)

O H H COOMe

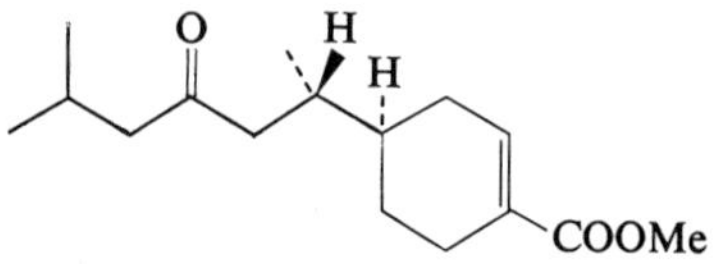

(10)

Momose, 1941; Nakazaki and Isoe, 1963). In the course of simultaneous work on the same problem, Černý and his co-workers (1967) found in the wood of a balsam fir growing in Czechoslovakia two active components, (+)-juvabione (8) and the slightly less active (+)-dehydrojuvabione (9).

Mori and Matsui (1968), followed by Subrahmania Ayyar and Krishna Rao (1968), were the first to describe syntheses of racemic juvabione (8).

Birch and co-workers (1969) were successful in carrying out a stereospecific synthesis of racemic juvabione (8) while Pawson and colleagues (1968) succeeded in the stereospecific synthesis of the natural (+)-juvabione (8).

They assigned the *R*-configuration to the two asymmetric centres, based upon the stereochemistry deduced for the parent (+)-todomatuic acid (7). A short time later, the same group (Blount *et al.*, 1969) was able to show by X-ray analysis of a key intermediate that their synthetic (+)-juvabione (8), as well as natural (+)-juvabione (8) and (+)-todomatuic acid (7), contrary to the assignment deduced previously, have the *R*,*S*-configuration (10). This so-called "Plant-JH", (+)-juvabione (10), not only selectively blocks the metamorphosis or sexual maturation of *Pyrrhocoris* larvae but also prevents the hatching of *Pyrrhocoris* eggs either before or immediately after oviposition. It does not, however, have similar effects on other species of laboratory insects (Sláma and Williams, 1966; Saxena and Williams, 1966). (+)-Juvabione (10) is therefore an effective and selective ovicide. Extending their interest in juvabione

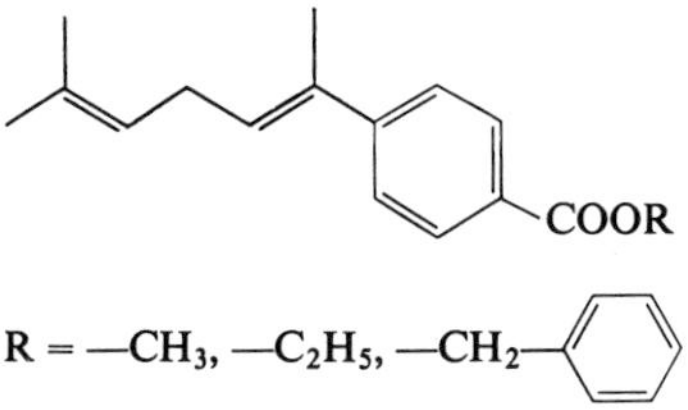

COOR

R = $—CH_3$, $—C_2H_5$, $—CH_2—$

(11)

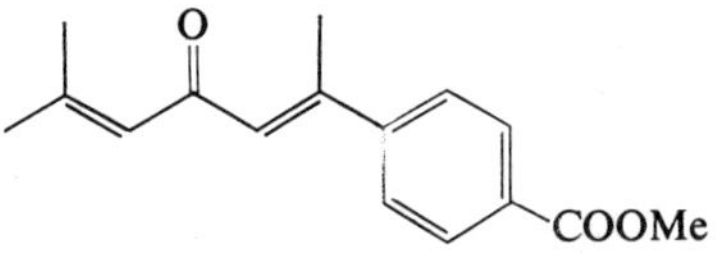

O

COOMe

(12)

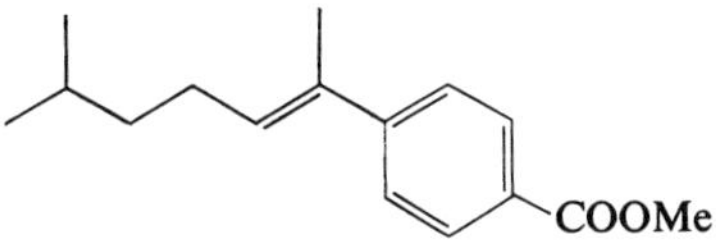

COOMe

(13)

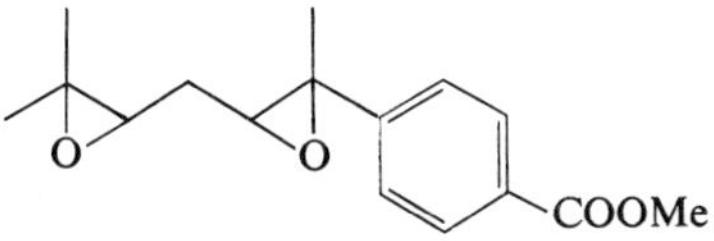

O O

COOMe

(14)

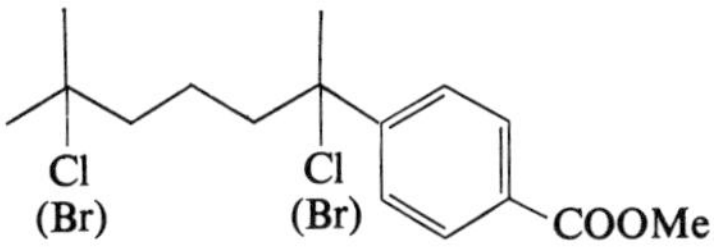

Cl (Br) Cl (Br)

COOMe

(15)

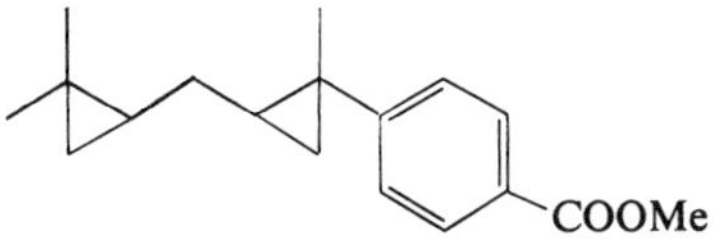

COOMe

(16)

chemistry, the Prague group (Sláma *et al.*, 1968; Suchý *et al.*, 1968) have prepared the synthetic analogues, (11), (12), (13), (14), (15), and (16) of juvabione and dehydrojuvabione, in which the alicyclic ring is replaced by an aromatic ring. Some of these synthetic derivatives are 100 times more active than natural (+)-juvabione (10), while retaining their specific action on the Pyrrhocorid bugs, *Pyrrhocoris* and *Neodysdercus*.

For *Pyrrhocoris*, the methyl ester (11) exhibited JH-activity which was about 10 times greater than that of (+)-juvabione (10). The introduction of a keto group (compound 12) did not lead to a further increase in activity. The derivative (13), which is a dihydro derivative of the methyl ester (11), has approximately 10 times lower activity than compound (12). The introduction of oxirane rings (compounds 14) was followed by a slight decrease of JH-activity. Addition of hydrogen chloride or hydrogen bromide (compound 15), or methylene bridges (compound 16), caused a slight increase in activity.

For *Neodysdercus*, most of the derivatives (11–16) especially compounds (13–16) were more active than for *Pyrrhocoris*. As in the case of the assays performed on *Pyrrhocoris*, all compounds were substantially more active than (+)-juvabione (10).

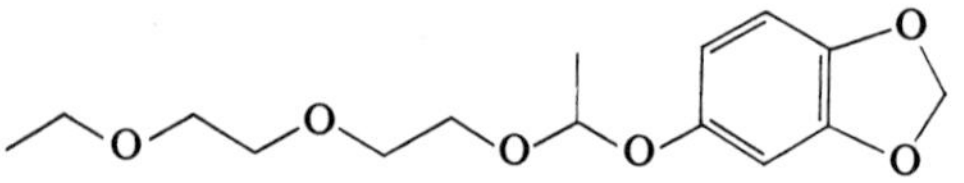

(17) Sesoxane

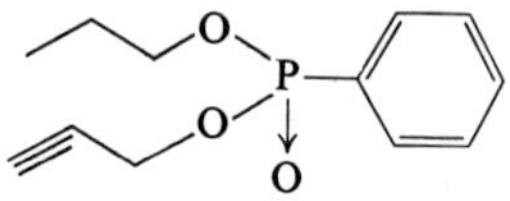

(18) Niagara 16388

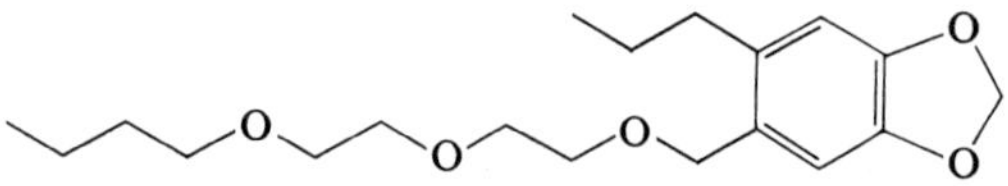

(19) Piperonyl butoxide

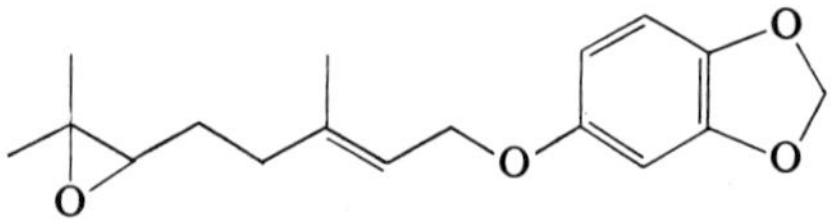

(20)

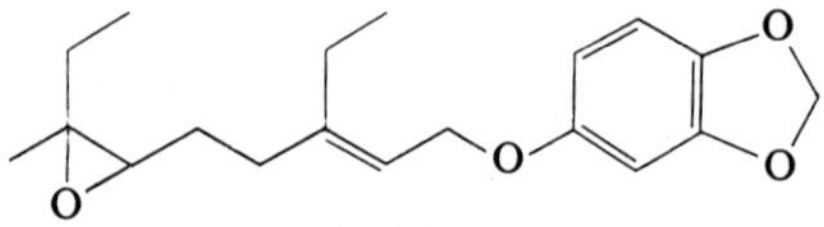

(21)

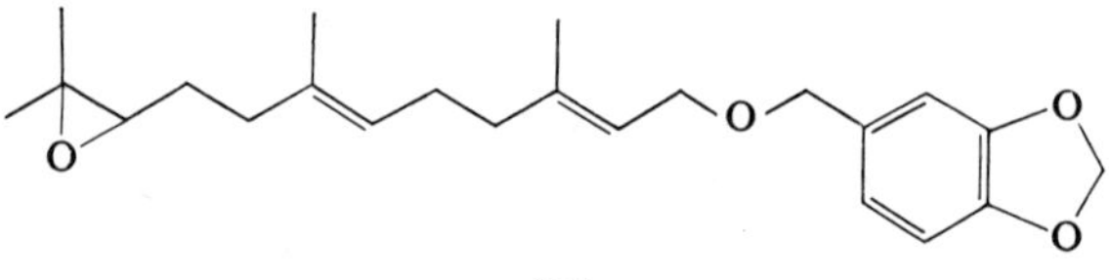

(22)

Interestingly, Bowers (1968, 1969) has found that a number of commercial non-sesquiterpenoidal synergists such as "Sesoxane" (17), propargyl propyl phenylphosphonate (18) and piperonyl butoxide (19) possess *per se* significant JH-activity.

In extending these studies, Bowers (1968, 1969) observed that the aromatic terpenoid ethers (20), (21) and (22) possess a high degree of morphogenetic activity when assayed on the yellow meal worm, *Tenebrio molitor*, and the milk weed bug, *Oncopeltus fasciatus*. The most active compound, the methylenedioxyphenyl ether of 6,7-dihydro-6,7-epoxydihomogeraniol (21), proved to be several hundred times as active as the insects' own juvenile hormone (1), whereas the phenyl or monosubstituted phenyl ethers of 6,7-dihydro-6,7-epoxygeraniol (23), (24) and (25) were significantly less active. The methylenedioxyphenyl group seems, therefore, to be very important for the activity of aromatic terpenoid ethers.

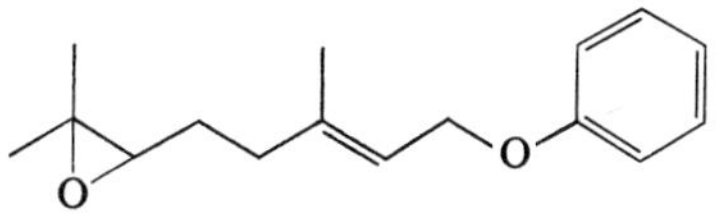

(23)

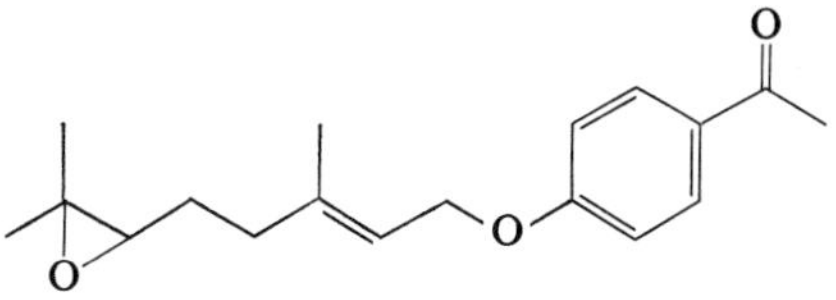

(24)

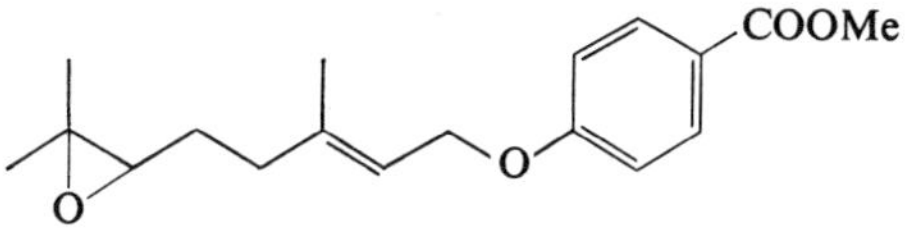

COOMe

(25)

IV. Chemistry of Juvenile Hormone

Since 1956, when Williams announced that the abdomen of the male silk moth, *Platysamia cecropia* L., is an exceptionally rich source of JH, several attempts to isolate JH from abdominal extracts of adult male *Hyalophora cecropia* have been reported. Williams and Law (1965), after a six-step process which effected a 50,000-fold purification, obtained an apparently homogeneous, active fraction by gas-liquid chromatography (GLC). 9,10-Epoxyhexadecanoic acid methyl ester was identified as a component of this active fraction, but upon synthesis of all isomers of this ester, neither the DL-*trans*-pair nor the DL-*cis*-pair possessed JH-activity. It was concluded that JH was but a minor component of the purified active fraction. Meyer and co-workers (1965) obtained, with GLC as the terminal procedure of a six-step process, two fractions represented by two peaks which accounted for the majority of recovered JH-activity. The authors proposed that these two peaks represented pyrolytic breakdown products of authentic JH. The brilliant teamwork of Röller and his colleagues (1965a,b, 1969) led first to a 100,000-fold purification of the *cecropia* extract, and thence to the elaboration (Röller *et al.*, 1967) of the structure of the JH-molecule as methyl 10,11-epoxy-7-ethyl-3,11-dimethyl-2,6-tridecadienoate (1). From their spectroscopic data, they were able to assign the *trans*-configuration to the 2,3 double bond, and a probable *trans*-configuration to the 6,7 double bond. About the stereochemistry of the oxirane ring, however, no assertion could be made. This formula allows for sixteen isomers—the DL-pairs of eight geometrical isomers.

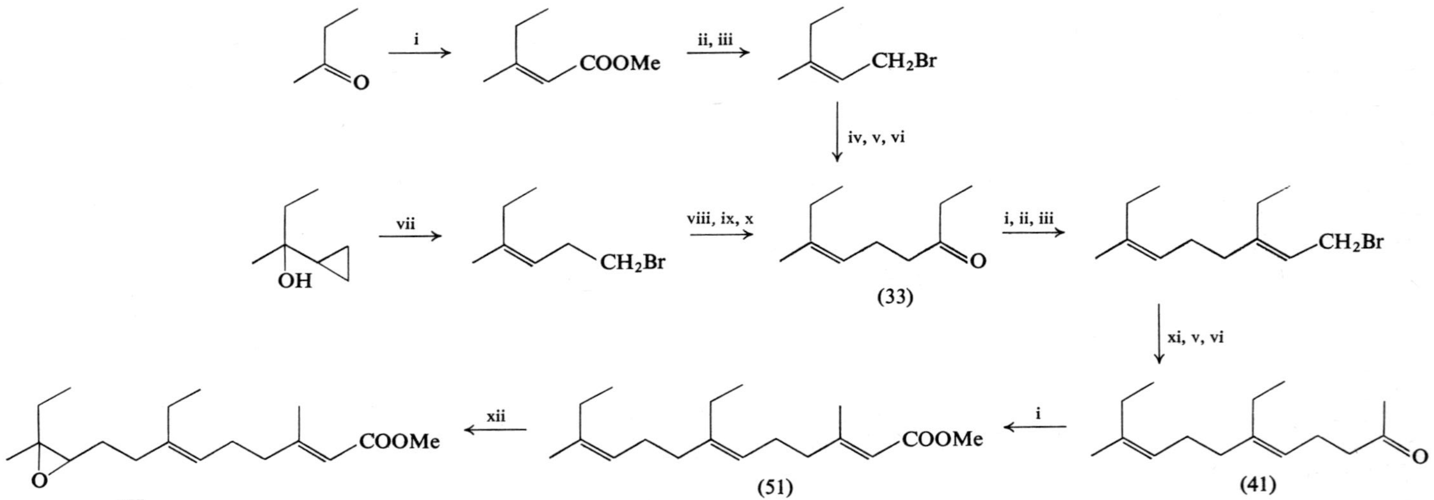

Scheme 1.

Reagents: i, $(MeO)_2POCH_2COOMe$; ii, $LiAlH_4$; iii, PBr_3;
iv, NaOEt—$CH_3CH_2COCH_2COOEt$; v, NaOH; vi, H^+;
vii, 48% HBr; viii, Mg; ix, CH_3CH_2CHO; x, CrO_3;
xi, NaOEt—CH_3COCH_2COOEt; xii, *m*-chloroperbenzoic acid.

From its NMR-spectrum and the subsequent synthesis (Dahm *et al.*, 1967, 1968) of four of the possible eight racemates, Röller and his team were able to refine their earlier structural conclusions and deduce that JH (1) is methyl 10,11-epoxy-7-ethyl-3,11-dimethyl-10,11-*cis*,2-*trans*,6-*trans*-tridecadienoate (1). The quantity of JH (1) isolated was insufficient for measuring its optical activity. Therefore, the absolute configuration of the oxirane ring could not be determined. Their synthesis of racemic JH (1) by successive application of the Wadsworth-Emmons reaction was accomplished as shown in Scheme 1.

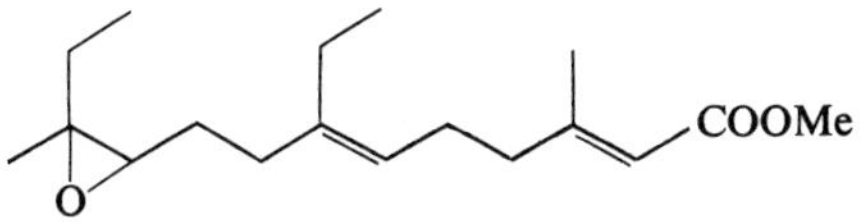

COOMe

O

D or L *cis* or *trans* *trans* or *cis* *trans* or *cis*

(1)

This synthetic racemic JH (1) was found to be identical with the natural JH (1) in many physical respects and in biological activity. Since this pioneer work by Röller and his colleagues, more than half a dozen synthetic routes for the preparation of JH (1) and its isomers have been described in the chemical literature.

Schulz and Sprung (Schering, Berlin; 1969) reported a convenient four-step procedure for racemic JH (1), using the Wittig reaction (Scheme 2). A short while later, Findlay and MacKay (1969) published a similar synthesis of racemic JH (1) (Scheme 3). Both synthetic routes use the same Röller intermediates (33) and (41).

Starting from 2-ethylfuran, Cavill *et al.* (1969) obtained by their synthetic route (Scheme 4) a racemic mixture of *erythro*- and *threo*-triols which they could separate on an alumina column impregnated with boric acid. These were then transformed into the corresponding *cis*- and *trans*-isopropylidene-dioxy-ketones. Using the *trans*-isopropylidenedioxy-ketone as a key intermediate, the elaboration of racemic JH (1) was accomplished by known methods.

The first of a series of stereospecific syntheses was a beautifully conceived route to the Röller intermediate (41) (Zurflüh *et al.*, 1968; Syntex, Palo Alto). The synthesis (Scheme 5) is based on sequential fragmentation of a bicyclic precursor. A second stereoselective synthesis of racemic JH (1) (Scheme 6), based on the stereospecific modification of the Julia method for introducing *trans*-trisubstituted double bonds, was carried out by Johnson *et al.* (1968). A third stereospecific total synthesis of racemic JH (1), starting with a benzenoid precursor, has been accomplished by Corey *et al.* (1968). Their route (Scheme 7) embodies an elegant application of no less than five novel synthetic processes.

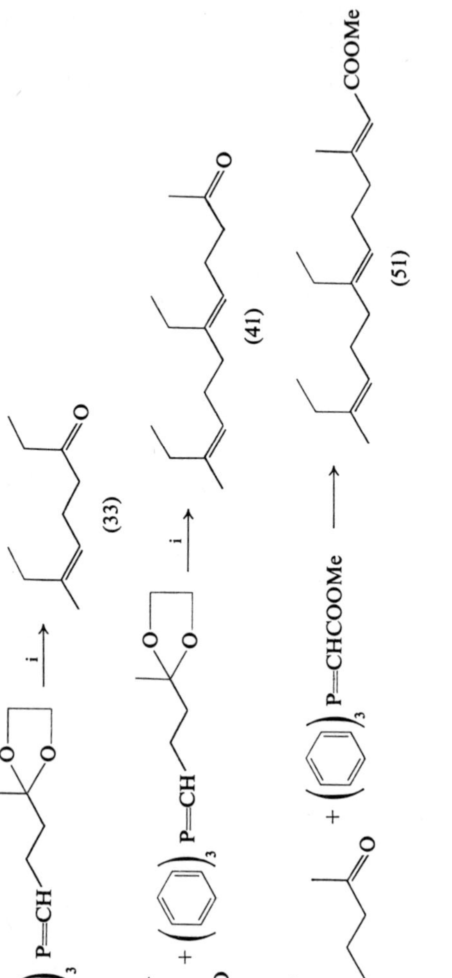

+ $(C_6H_5)_3P{=}CH$ →[i] (33)

(33) + $(C_6H_5)_3P{=}CH$ →[i] (41)

(41) + $(C_6H_5)_3P{=}CHCOOMe$ → COOMe (51)

→[ii] rac. JH (1)

Scheme 2.

Reagents: i, 3*N*-perchloric acid;
ii, *m*-chloroperbenzoic acid.

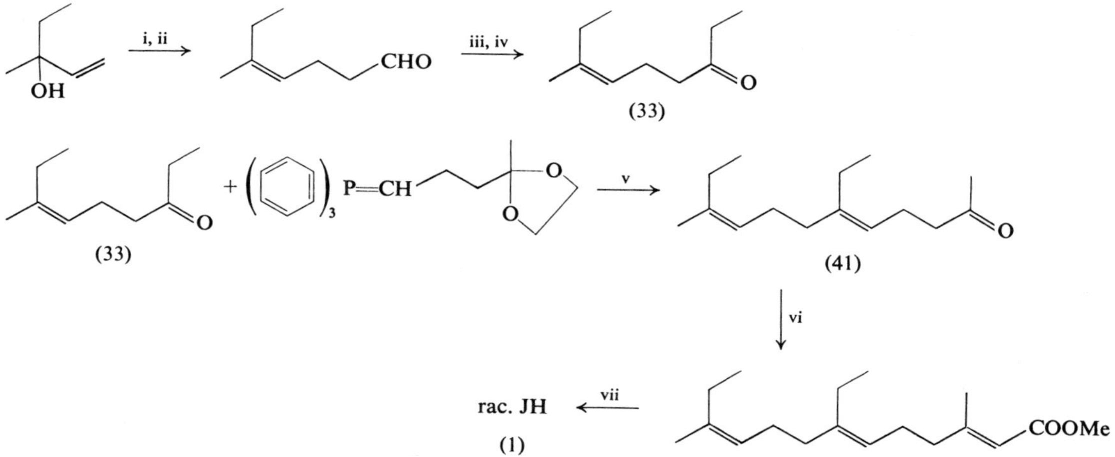

OH →(i, ii) CHO →(iii, iv) (33)

(33) + $(C_6H_5)_3P{=}CH$ →(v) (41)

(41) →(vi) (51) COOMe

(51) →(vii) rac. JH (1)

Scheme 3.

Reagents: i, Ethyl vinyl ether-H_3PO_4; ii, H_3PO_4-Δ; iii, EtMgBr; iv, CrO_3; v, H^+; vi, $(MeO)_2POCH_2COOMe$; vii, *m*-chloroperbenzoic acid.

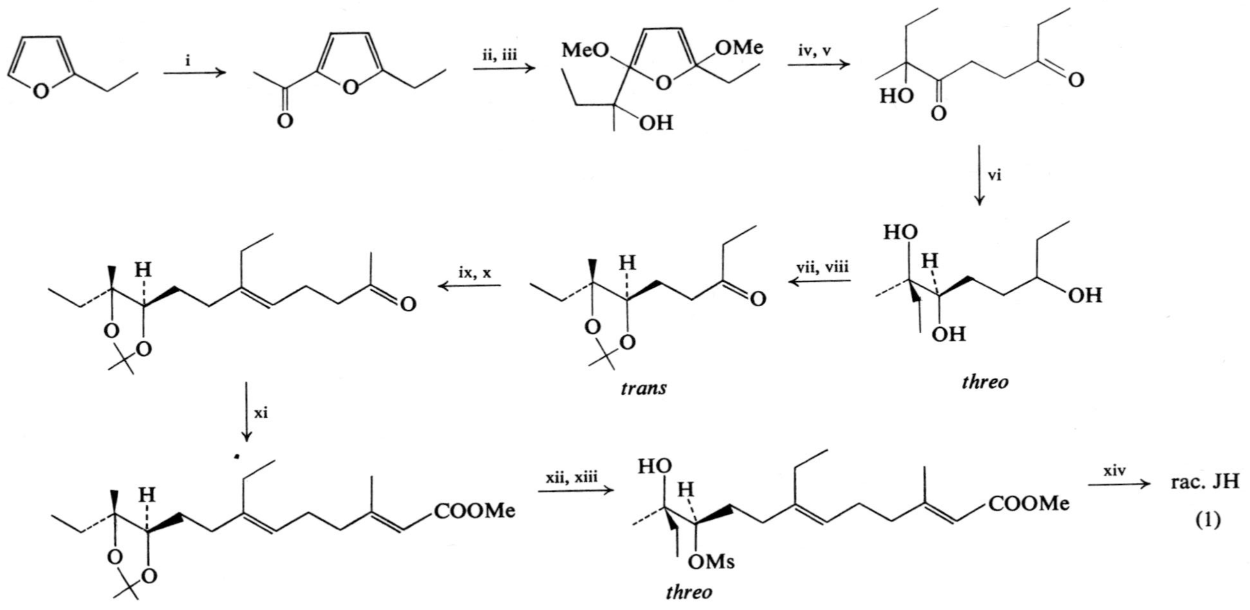

Scheme 4.

Reagents: i, Ac_2O—$SnCl_4$; ii, EtMgI; iii, Br_2 in MeOH; iv, 0·002*N* HCl; v, H_2/Pd—C; vi, $NaBH_4$; vii, CH_3COCH_3/*p*-TsOH; viii, $NaIO_4$;

ix, $(Ph)_3P{=}CHCH_2CH_2C(CH_3)$(—O—$CH_2CH_2$—O—); x, *p*-TsOH;

xi, $(MeO)_2POCH_2COOMe$; xii, 0·1*N* HCl—MeOH—H_2O; xiii, Py/CH_3SO_2Cl; xiv, 0·1*N* NaOMe.

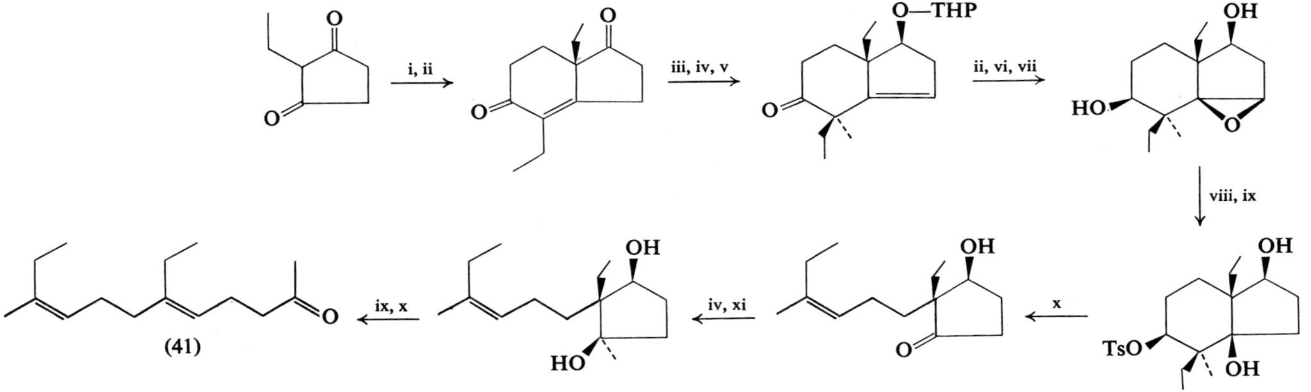

Scheme 5.

Reagents: i, Propyl vinyl ketone; ii, H^+; iii, $NaBH_4$;
iv, tetrahydropyranyl (THP) etherification; v, MeI;
vi, $LiAlH(O\text{-}t\text{-}Bu)_3$; vii, *m*-chloroperbenzoic acid; viii, $LiAlH_4$;
ix, tosylation; x, NaH; xi, MeLi.

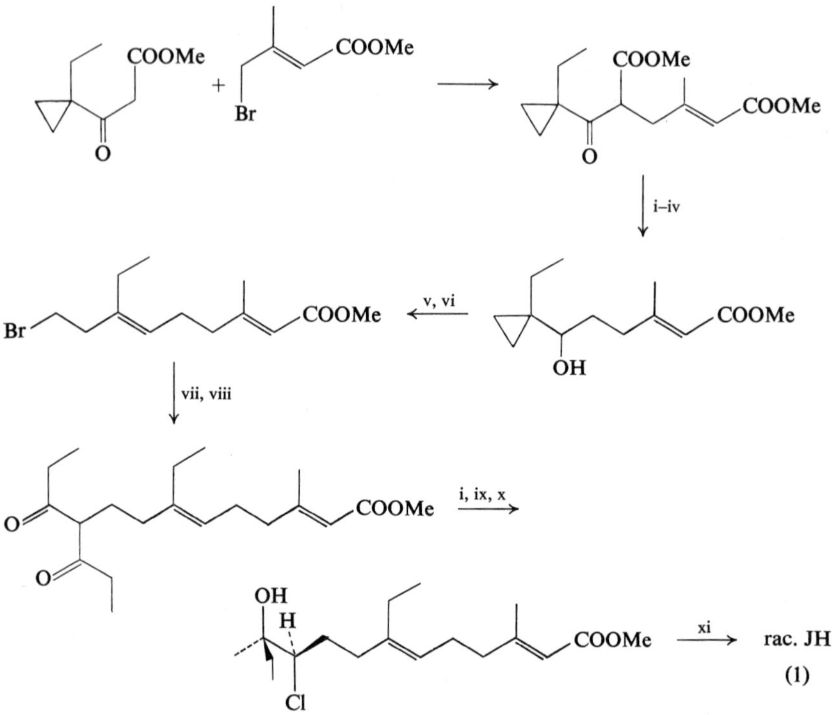

COOMe + COOMe Br → COOMe COOMe O

i–iv

Br COOMe ← v, vi COOMe OH

vii, viii

O O COOMe → i, ix, x

OH H Cl COOMe → xi rac. JH (1)

Scheme 6.

Reagents: i, $Ba(OH)_2$; ii, H^+; iii, CH_2N_2; iv, $NaBH_4$; v, PBr_3—LiBr; vi, $ZnBr_2$; vii, NaI; viii, Li enolate of 3,5-heptanedione; ix, $CuCl_2$/LiCl; x, MeMgCl; xi, K_2CO_3.

Most of the published syntheses of racemic JH (1) are academic in nature and too complex to allow the preparation of sufficient amounts of racemic JH (1) for extensive investigations. Moreover, the stereospecific syntheses provide only racemic JH (1) while it would be more desirable to have all the isomers for testing purposes. Most of the nonstereospecific syntheses of racemic JH (1) and its isomers suffer from the disadvantage that they involve separation methods such as preparative GLC, preparative thin-layer chromatography (TLC) or extensive column chromatographic separation procedures. These are feasible for the preparation of small amounts of racemic JH (1) and its isomers, but not for preparation of larger amounts.

We report now the Roche synthesis (Atherton and Pfiffner, 1967) of all eight possible racemates of JH (1), which was already carried out at the end of 1967. When we started our synthesis of racemic JH (1) and its isomers, no information was available about the stereochemistry of the oxirane ring or about the configuration of the 6,7 double bond (Röller *et al.*, 1967). Moreover, we did not possess any natural JH (1) for analytical, spectroscopic, or

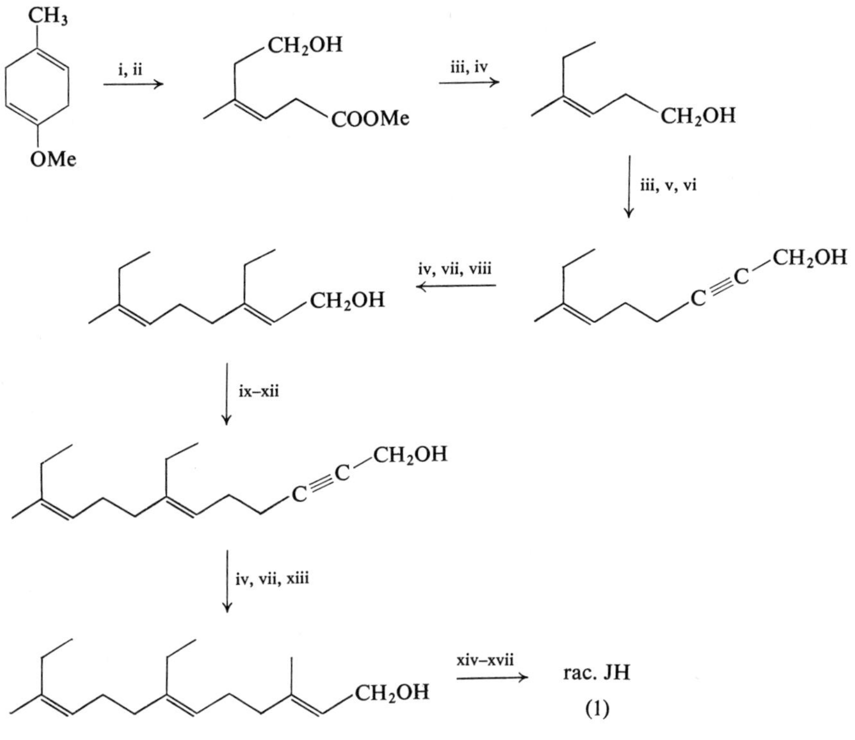

CH_3 OMe $\xrightarrow{\text{i, ii}}$ CH_2OH COOMe $\xrightarrow{\text{iii, iv}}$ CH_2OH

iii, v, vi

CH_2OH $\xleftarrow{\text{iv, vii, viii}}$ $C \equiv C$ CH_2OH

ix–xii

$C \equiv C$ CH_2OH

iv, vii, xiii

CH_2OH $\xrightarrow{\text{xiv–xvii}}$ rac. JH (1)

Scheme 7.

Reagents: i, $O_3/CH_3OH/(CH_3)_2S$; ii, $NaBH_4$; iii, TsCl; iv, $LiAlH_4$; v, $LiC{\equiv}CCH_2OTHP$; vi, H^+; vii, I_2; viii, $(Et)_2LiCu$; ix, PBr_3; x, $LiCH_2C{\equiv}CSi(Me)_3$; xi, Ag^+, then CN^-; xii, BuLi then CH_2O; xiii, $(Me)_2LiCu$; xiv, MnO_2; xv, MeOH/NaCN; xvi, N-bromosuccinimide (NBS); xvii, *i*-PrO^-.

biological comparison purposes. We therefore developed a synthetic procedure which made possible the preparation of relatively large amounts of racemic JH (1) and all its isomers. Ethynylation of ethyl methyl ketone (26) (cf. Scheme 8), followed by selective hydrogenation, allylic bromination, condensation with ethyl propionylacetate, saponification and decarboxylation (Isler *et al.*, 1958), gave the *cis*/*trans*-ketone mixture (29). Since this mixture (cf. Scheme 9) could not be separated by fractional distillation, we transformed it into the cyclic *cis*/*trans*-ketal mixture (30), which could be separated very smoothly by careful fractional distillation with a Podbielniak "Heli-Grid" column. The course of fractionation was controlled by GLC. The pure ketals (31) and (32), obtained in 16% and 45% yield, respectively, based on the isomeric ketal mixture (34), were then hydrolysed to the *cis*-(33) and *trans*-(34)-ketones which were shown to be gas-chromatographically pure (less than 5% of the other isomer).

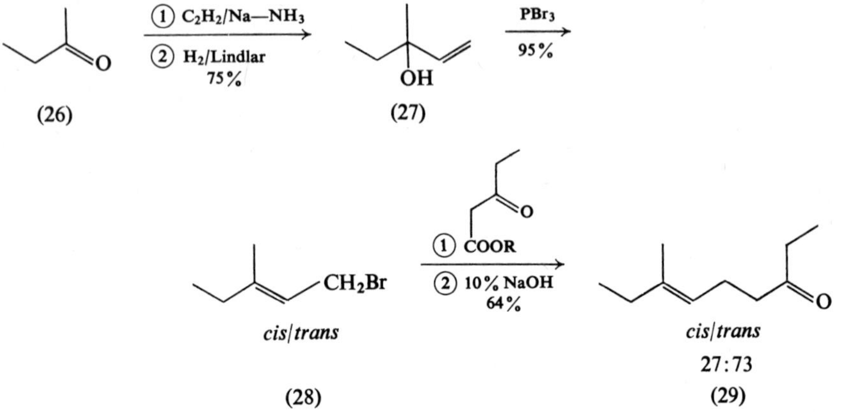

① $C_2H_2/Na—NH_3$
② H_2/Lindlar
75%

OH

PBr_3
95%

(26) (27)

O

① COOR
② 10% NaOH
64%

CH_2Br

cis/trans

cis/trans
27:73

(28) (29)

Scheme 8.

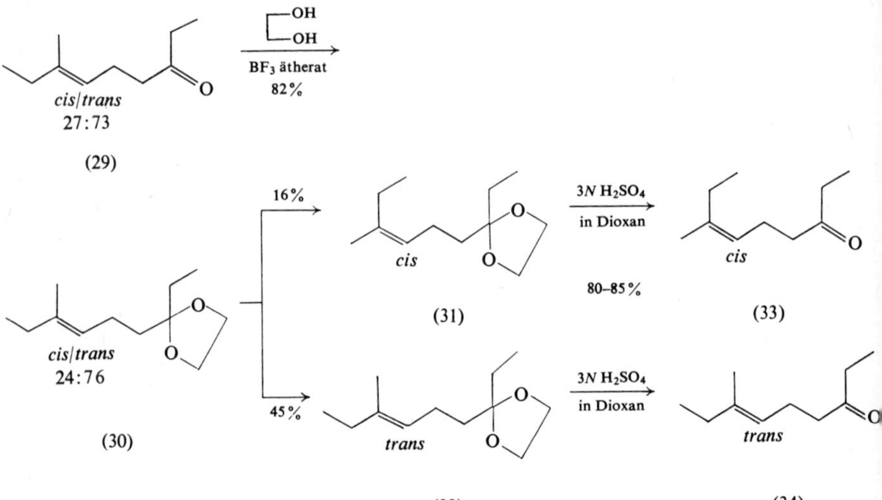

OH
OH

BF_3 ätherat
82%

cis/trans
27:73

(29)

16%

3*N* H_2SO_4
in Dioxan

80–85%

cis

cis

(31) (33)

cis/trans
24:76

(30)

45%

3*N* H_2SO_4
in Dioxan

trans

trans

(32) (34)

Scheme 9.

The *cis*-(33) and *trans*-(34)-ketones (cf. Scheme 10) were individually ethynylated and selectively hydrogenated. The tertiary vinylcarbinols (35) and (36) were transformed into the two isomeric 5-*cis*/*trans*-ketones (38) and (39) either by allylic bromination, condensation with acetoacetate, saponification and decarboxylation or by a one step reaction with isopropenyl methyl ether (Saucy and Marbet, 1967). The isomeric ketones (38) and (39) (cf. Scheme 11) could be separated by fractional distillation into the four gas-chromatographically pure ketones (40–43), which are key intermediates for the preparation of racemic JH (1) and its isomers. The refractive indices of the two 9-*trans*-ketones (42) and (43) are significantly higher than those of the corresponding 9-*cis*-ketones (40) and (41).

The absorption of the C-10 methyl group of the 9-*cis*-ketone pair (40) and (41) occurs in the NMR 0·08–0·09 ppm downfield from that of (42) and (43)

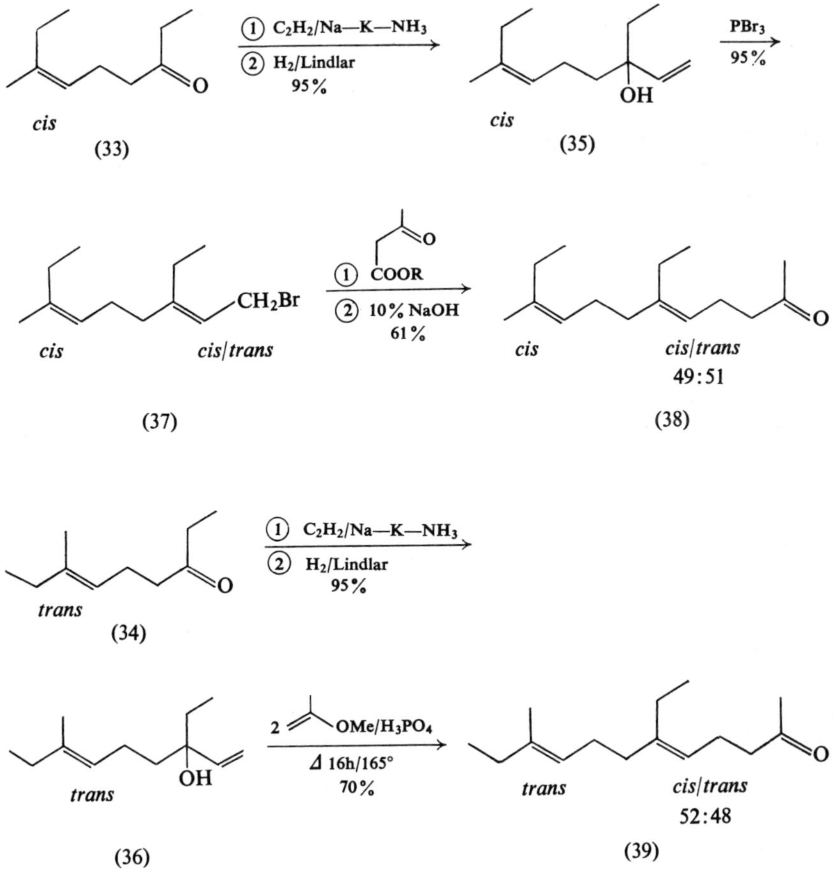

cis (33) → ① C_2H_2/Na—K—NH_3 ② H_2/Lindlar 95% → cis (35) OH → PBr_3 95% →

cis, cis/trans (37) CH_2Br → ① COOR ② 10% NaOH 61% → cis, cis/trans 49:51 (38)

trans (34) → ① C_2H_2/Na—K—NH_3 ② H_2/Lindlar 95% →

trans (36) OH → 2 OMe/H_3PO_4, Δ 16h/165°, 70% → trans, cis/trans 52:48 (39)

Scheme 10.

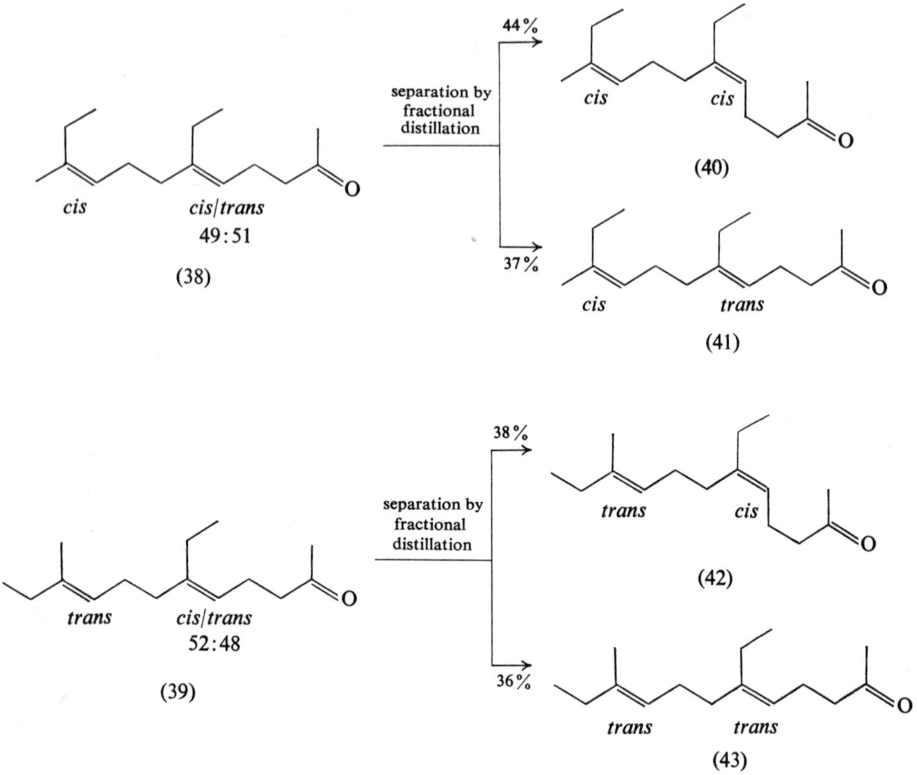

cis cis/trans 49:51 (38)

separation by fractional distillation

44% cis cis (40)

37% cis trans (41)

trans cis/trans 52:48 (39)

separation by fractional distillation

38% trans cis (42)

36% trans trans (43)

Scheme 11.

(see Table I). Reaction of the ketones (40), (41), (42) and (43) with the sodium salt of methyl (diphenoxyphosphinyl)acetate yielded the corresponding 2-*cis*/*trans*-ester mixtures (44), (45), (46) and (47) which were then separated by column chromatography to give the eight isomeric esters (48–55) (cf. Scheme 12). All of the isomeric esters (48–55) were obtained gas-chromatographically pure and the appropriate physical data are summarized in Table II. The conversion of these eight esters into the corresponding racemic epoxides was achieved in two ways. As an example, the conversion of the 2-*trans*,6-*trans*,10-*cis*-ester (51) to racemic JH (1) is shown in Scheme 13.

In path Ⓐ, the conversion was carried out with *m*-chloroperbenzoic acid. The racemic JH (1) thus formed had to be separated from the corresponding racemic 6,7-epoxy-ester, the racemic diepoxy-ester and unreacted starting ester by chromatographic methods. The chromatographic separation of the racemic 6,7-epoxy-ester from the corresponding racemic 10,11-epoxy-ester proved to be very difficult with some JH-isomers. The difference in the R_f-values is very small and a separation of larger amounts of these isomers could

TABLE I

Physical data of all four geometrical isomers (40–43)

$$\underset{(B)}{CH_3}-\overset{\overset{(C)\quad(A)}{CH_2-CH_3}}{C}=\underset{(E)}{CH}-\underset{(C)}{CH_2}-\underset{(C)}{CH_2}-\overset{\overset{(C)\quad(A)}{CH_2-CH_3}}{C}=\underset{(E)}{CH}-\underset{(C)}{CH_2}-\underset{(C)}{CH_2}-\overset{\overset{(D)}{CH_3}}{C}=O$$

Isomer	5-*cis*,9-*cis*	5-*trans*,9-*cis*	5-*cis*,9-*trans*	5-*trans*,9-*trans*
b.p.	80 ± 2°/0·4 ± 0·1 mm	80 ± 2°/0·4 ± 0·1 mm	78 ± 2°/0·3 ± 0·1 mm	78 ± 2°/0·3 ± 1 mm
n_D^{20}	1·4705	1·4708	1·4722	1·4722
NMR: (A)	0·97, t, J = 7	0·96, t, J = 7	0·98, t, J = 7	0·96, t, J = 7
(B)	1·69, d, J < 1	1·67, d, J < 1	1·60, d, J < 1	1·59, bs
(C)	1·78–2·60, m	1·78–2·60, m	1·75–2·60, m	1·75–2·60, m
(D)	2·14, s	2·13, s	2·13, s	2·12, s
(E)	5·09, m (bt)	5·05, m (bt)	5·06, m (bt)	5·04, m (bt)

In Tables I–IV the refractive indices were measured with a Zeiss refractometer at room temperature and corrected to 20°C. The NMR-spectra were recorded at 60 Mc in $CDCl_3$ as solvent and chemical shifts are given in ppm with internal TMS = 0. The coupling constants J are given in Hz. s = singlet; bs = broad singlet; d = doublet; t = triplet; bt = broad triplet; q = quadruplet; m = multiplet.

cis cis (40)

$(PhO)_2P(O)\cdot CH_2COOMe$

2 NaH/THF

78%

cis cis cis/trans 40:60 COOMe (44)

cis cis cis COOMe (48)

cis cis trans COOMe (49)

cis trans (41)

$(PhO)_2P(O)\cdot CH_2COOMe$

2 NaH/THF

82%

cis trans cis/trans 42:58 COOMe (45)

cis trans cis COOMe (50)

cis trans trans COOMe (51)

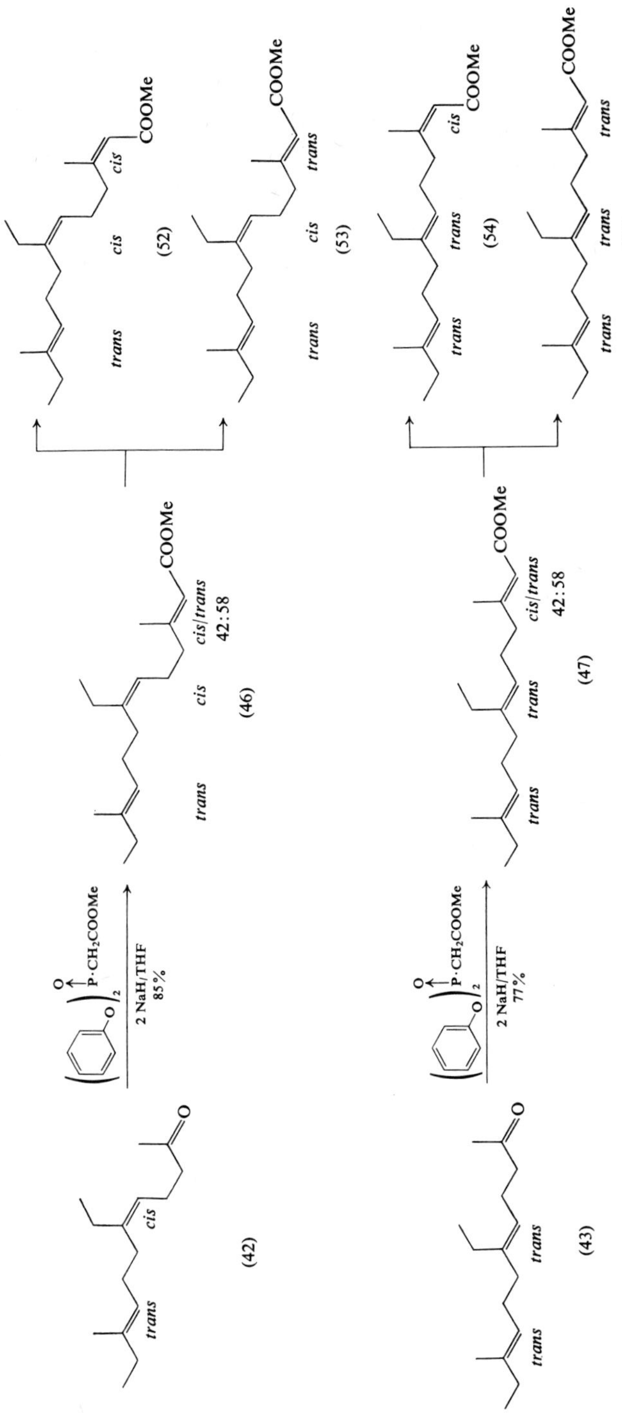

(42) *trans* *cis*

(43) *trans* *trans*

$(PhO)_2P(\to O)\cdot CH_2COOMe$, 2 NaH/THF, 85%

$(PhO)_2P(\to O)\cdot CH_2COOMe$, 2 NaH/THF, 77%

(46) *trans* *cis* *cis/trans* 42:58 COOMe

(47) *trans* *trans* *cis/trans* 42:58 COOMe

(52) *trans* *cis* *cis* COOMe

(53) *trans* *cis* *trans* COOMe

(54) *trans* *trans* *cis* COOMe

(55) *trans* *trans* *trans* COOMe

Scheme 12.

TABLE II
Physical data of all eight geometrical isomers (48–55)

```
 (C)  (A)             (C)  (A)          (D)
 CH2—CH3             CH2—CH3            CH3
 |                   |                  |
CH3—C=CH—CH2—CH2—C=CH—CH2—CH2C=CH—COOCH3
(B)    (G) (C)  (C)      (G) (C) (E)  (H)   (F)
```

Isomer	2-*cis*,6-*cis*,10-*cis*	2-*trans*,6-*cis*,10-*cis*	2-*cis*,6-*trans*,10-*cis*	2-*trans*,6-*trans*,10-*cis*
b.p. (°C)	93–95°/0·04 mm	97–99°/0·03 mm	97°/0·02 mm	98°/0·02 mm
n_D^{20}	1·4846	1·4859	1·4845	1·4860
IR (KBr)	1724 cm^{-1}, 1651 cm^{-1}	1725 cm^{-1}, 1652 cm^{-1}	1725 cm^{-1}, 1652 cm^{-1}	1724 cm^{-1}, 1652 cm^{-1}
NMR: (A)	0·99/0·975, 2t, J = 7	0·99/0·975, 2t, J = 7	0·965, t, J = 7	0·96, t, J = 7
(B)	1·69, d, J < 1	1·69, d, J < 1	1·69, bs	1·69, bs
(C)	1·77–2·47, m	1·77–2·40, m	1·77–2·47, m	1·77–2·40, m
(D)	1·86, d, J = 1·4 ± 0·25	2·19, d, J = 1·25 ± 0·25	1·89, d, J = 1·4 ± 0·25	2·18, d, J = 1·25 ± 0·25
(E)	2·47–2·85, m	1·77–2·40, m	2·47–2·85, m	1·77–2·40, m
(F)	3·69, s	3·70, s	3·67, s	3·68, s
(G)	5·18, bt, J = 6·5 ± 0·5	5·12 m	5·13, bt, J = 6·5 ± 0·5	5·07 m
(H)	5·69, bs	5·71, bs	5·67, bs	5·68, bs

Isomer	2-*cis*,6-*cis*,10-*trans*	2-*trans*,6-*cis*,10-*trans*	2-*cis*,6-*trans*,10-*trans*	2-*trans*,6-*trans*,10-*trans*
b.p. (°C)	80–81°/0·007 mm	87–88°/0·01 mm	93°/0·04 mm	98°/0·04 mm
n_D^{20}	1·4852	1·4866	1·4856	1·4869
IR (KBr)	1726 cm^{-1}, 1654 cm^{-1}	1725 cm^{-1}, 1652 cm^{-1}	1725 cm^{-1}, 1654 cm^{-1}	1724 cm^{-1}, 1651 cm^{-1}
NMR: Ⓐ	0·97, t, J = 7	0·98, t, J = 7	0·97, t, J = 7	0·96/0·97, 2t, J = 7
Ⓑ	1·60, bs	1·60, d, J < 1	1·60, bs	1·60, bs
Ⓒ	1·75–2·46, m	1·75–2·40, m	1·75–2·48, m	1·75–2·40, m
Ⓓ	1·88, d, J = 1·4 ± 0·25	2·17, d, J = 1·25 ± 0·25	1·88, d, J = 1·4 ± 0·25	2·17, d, J = 1·25 ± 0·25
Ⓔ	2·46–2·88, m	1·75–2·40, m	2·48–2·90, m	1·75–2·40, m
Ⓕ	3·66, s	3·67, s	3·65, s	3·67, s
Ⓖ	5·15, bt, J = 6·5 ± 0·5	5·10, m	5·11, bt, J = 6·5 ± 0·5	5·07, m
Ⓗ	5·67, bs	5·68, bs	5·65, bs	5·67, bs

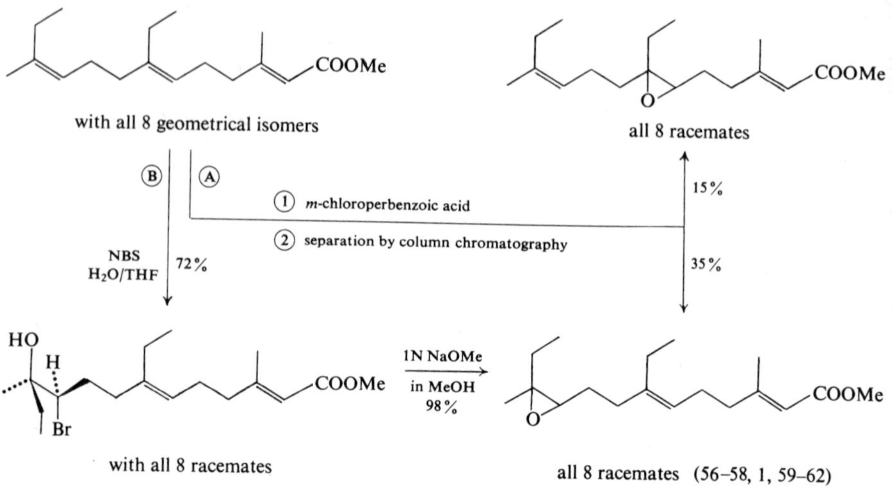

COOMe

with all 8 geometrical isomers

COOMe

all 8 racemates

Ⓑ Ⓐ

① *m*-chloroperbenzoic acid

② separation by column chromatography

15%

35%

NBS
H_2O/THF 72%

HO H

Br

COOMe

1N NaOMe in MeOH 98%

COOMe

with all 8 racemates

all 8 racemates (56–58, 1, 59–62)

Scheme 13.

only be attained by repeated chromatography, with a corresponding lowering of yield.

In path Ⓑ we prepared the racemic bromohydrins of all eight isomeric esters by a method elaborated by Van Tamelen and co-workers (Van Tamelen and Curphey, 1962; Van Tamelen *et al.*, 1965; Van Tamelen and Sharpless, 1967). Thus, N-bromosuccinimide (NBS) in aqueous tetrahydrofuran selectively oxidized the isomeric esters (48–55) at the C-10,11 double bonds. The conversion of these terminal bromohydrins to epoxides was readily accomplished by treatment with methanolic sodium methoxide solution. Starting with the 10-*cis*-ester isomers (48–51), the racemic *threo*-bromohydrins were obtained. These were converted to the corresponding racemic 10,11-*cis*-epoxy-ester isomers (1) (racemic JH) and (56–58). In an analogous manner, the 10-*trans*-ester isomers (52–55) gave the other four racemic 10,11-*trans*-JH isomers (59–62) in stereochemically pure form via the corresponding racemic *erythro*-bromohydrins (cf. Scheme 14). The identity of the racemates of the eight geometric racemic JH-isomers, (1, 56–62), was indicated by the known stereochemical course of the reactions employed in generating each double bond and was fully supported by spectroscopic data. A comparison of the physical data (see Tables III, IV and V) of these JH-isomers illustrated clearly that the most significant differences are caused by *cis*/*trans*-isomerism at position 2,3, whereas the configuration at positions 6,7 and 10,11 is of secondary importance.

In the NMR-spectra, the signal of the protons of the methyl group on the oxirane ring appears as a sharp singlet at 1·27–1·28 ppm in the 10,11-*cis*-isomers, (1, 56–58). In the corresponding 10,11-*trans*-isomers, (59–62), the same signal appears at 0·03–0·04 ppm higher field. The signal of the proton attached to the oxirane ring appears as triplet at 2·70–2·73 ppm in all eight

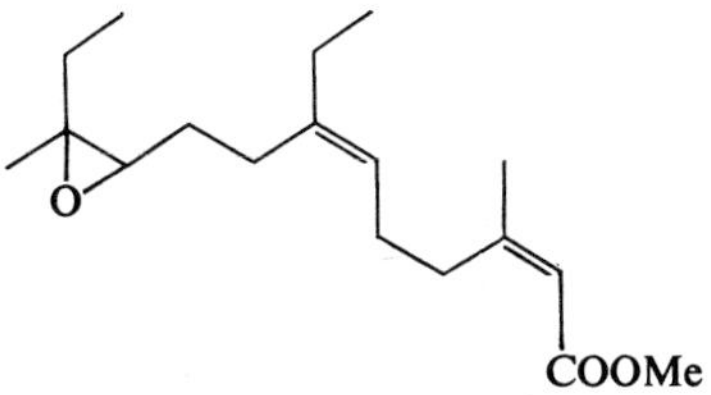

COOMe
cis *cis* *cis*
(56)

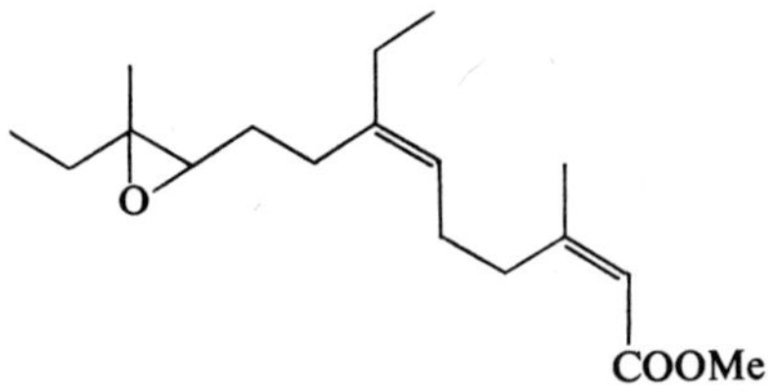

COOMe
trans *cis* *cis*
(59)

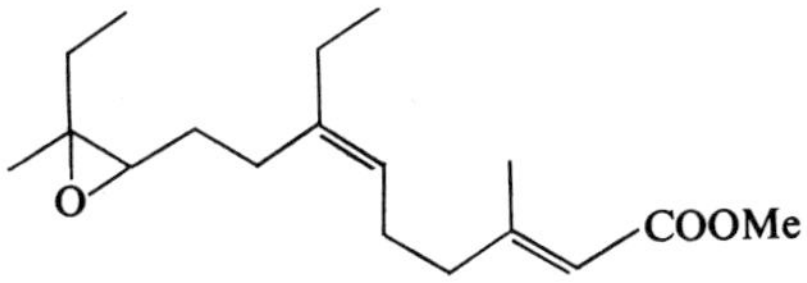

COOMe
cis *cis* *trans*
(57)

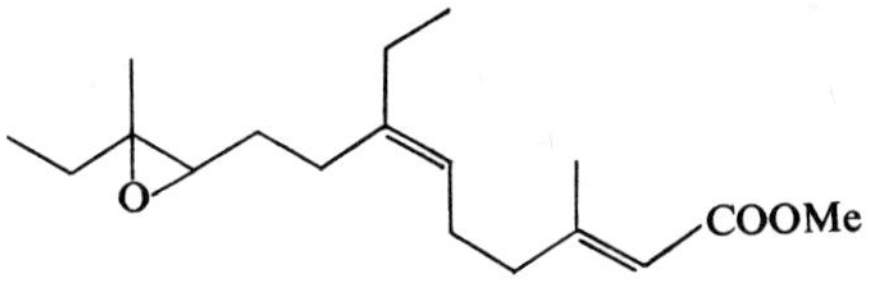

COOMe
trans *cis* *trans*
(60)

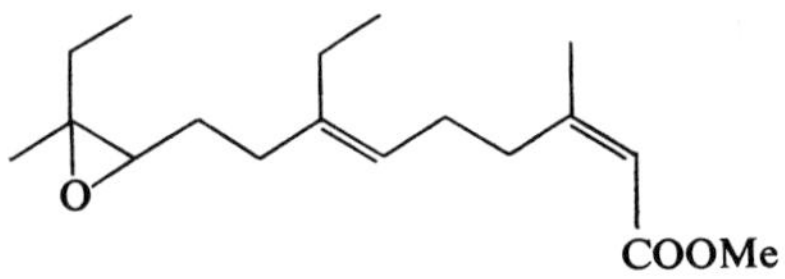

COOMe
cis *trans* *cis*
(58)

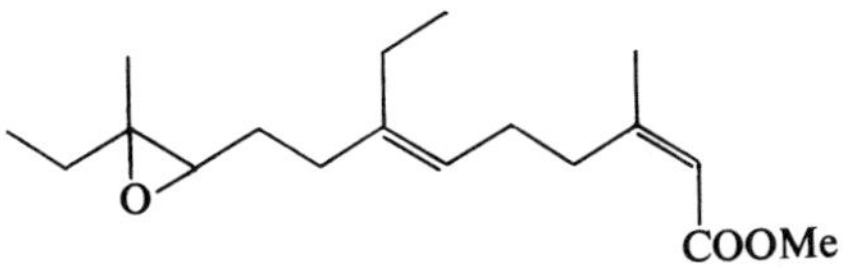

COOMe
trans *trans* *cis*
(61)

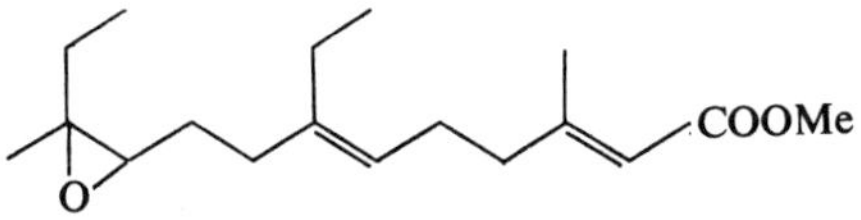

COOMe
cis *trans* *trans*
(1)

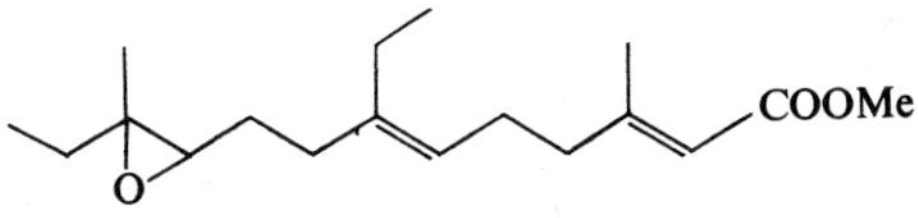

COOMe
trans *trans* *trans*
(62)

Scheme 14.

isomers. The β-methyl group and the methylene group at C-4 *cis* to the carbonyl group experience a larger deshielding than the *trans* ones. Thus, in the 2-*trans*-isomers, the signal of the methyl group at C-3 appears at 0·28 ppm lower field than in the corresponding 2-*cis*-isomers and the methylene multiplet of the C-4 methylene group in the 2-*cis*-isomers appears at approximately 0·55 ppm lower field than in the corresponding 2-*trans*-isomers (see Table IV).

Gas liquid chromatography of all intermediates revealed that the *cis*-isomers had shorter retention times than the corresponding *trans*-isomers. The 10,11-*cis*-isomers appear about 0·6 min before the corresponding 10,11-*trans*-isomers, whereas the conversion of the C-6,7 double bond from *trans* to *cis* in the 2-*cis*-isomers shortens the retention time by about 2·5–2·6 min and in the 2-*trans*-isomers by about 3·1–3·3 min. The 2-*cis*-isomers with

TABLE III

NMR spectral data of all eight racemic bromohydrins

```
 (A)  (C)                      (C)   (A)                (D)
 CH3—CH2   Br                  CH2—CH3                  CH3
      |   /                    |                        |
 CH3—C—CH—CH2—CH2—C=CH—CH2—CH2—C=CH—COOCH3
      |
      HO
 (B) (C) (G) (C) (C)        (H) (C)  (E)          (I)  (F)
```

Isomer	2-*cis*; 6-*cis*; 10,11-threo	2-*trans*; 6-*cis*; 10,11-threo	2-*cis*; 6-*trans*; 10,11-threo	2-*trans*; 6-*trans*; 10,11-threo
NMR: (A)	0·93/0·99, 2t, J = 7	0·93/0·99, 2t, J = 7	0·925/0·975, 2t, J = 7	0·925/0·97, 2t, J = 7
(B)	1·26, s	1·26, s	1·26, s	1·26, s
(C)	1·40–2·50, m	1·40–2·50, m	1·40–2·50, m	1·40–2·50, m
(D)	1·91, d, J = 1·4 ± 0·25	2·18, d, J = 1·25 ± 0·25	1·89, d, J = 1·40 ± 0·25	2·17, d, J = 1·25 ± 0·25
(E)	2·50–2·90, m	1·40–2·50, m	2·50–2·90, m	1·40–2·50, m
(F)	3·66, s	3·68, s	3·66, s	3·68, s
(G)	4·10, m	4·06, m	4·10, m	4·08, m
(H)	5·22, bt, J = 6·5 ± 0·5	5·15, m	5·19, bt, J = 6·5 ± 0·5	5·13, m
(I)	5·67, bs	5·68, bs	5·67, bs	5·68, bs

Isomer	2-*cis*; 6-*cis*; 10,11-erythro	2-*trans*; 6-*cis*; 10,11-erythro	2-*cis*; 6-*trans*; 10,11-erythro	2-*trans*; 6-*trans*; 10,11-erythro
NMR: Ⓐ	0·93/0·97, 2t, J = 7	0·93/0·98, 2t, J = 7	0·94/0·98, 2t, J = 7	0·94/0·975, 2t, J = 7
Ⓑ	1·27, s	1·27, s	1·28, s	1·28, s
Ⓒ	1·40–2·50, m	1·40–2·50, m	1·40–2·50, m	1·40–2·50, m
Ⓓ	1·91, d, J = 1·40 ± 0·25	2·18, d, J = 1·25 ± 0·25	1·91, d, J = 1·40 ± 0·25	2·18, d, J = 1·25 ± 0·25
Ⓔ	2·50–2·90, m	1·40–2·50, m	2·50–2·90, m	1·40–2·50, m
Ⓕ	3·66, s	3·68, s	3·67, s	3·68, s
Ⓖ	3·99, m	3·96, m	4·02, m	3·99, m
Ⓗ	5·21, bt, J = 6·5 ± 0·5	5·15, m	5·21, bt, J = 6·5 ± 0·5	5·14, m
Ⓘ	5·67, bs	5·69, bs	5·68, bs	5·69, bs

TABLE IV
Physical data of all eight racemates (56–58, 1, 59–62)

$$CH_3\text{—}\underset{\diagdown O \diagup}{C(CH_2\text{—}CH_3)\text{—}CH}\text{—}CH_2\text{—}CH_2\text{—}C(CH_2\text{—}CH_3)\text{=}CH\text{—}CH_2\text{—}CH_2\text{—}C(CH_3)\text{=}CH\text{—}COOCH_3$$

Labels on the formula (top row): Ⓒ Ⓐ over the first CH_2—CH_3; Ⓓ Ⓐ over the second CH_2—CH_3; Ⓔ over CH_3. Bottom row: Ⓑ Ⓖ Ⓒ Ⓓ Ⓘ Ⓓ Ⓕ Ⓚ Ⓗ

Isomer	2-*cis*; 6-*cis*; 10,11-*cis*	2-*trans*; 6-*cis*; 10,11-*cis*	2-*cis*; 6-*trans*; 10,11-*cis*	2-*trans*; 6-*trans*; 10,11-*cis*
b.p. (°C)	99–100°/0·002 mm	103–105°/0·002 mm	100–102°/0·002 mm	102–103°/0·002 mm
n_D^{20}	1·4802	1·4819	1·4807	1·4819
IR (KBr; cm^{-1})	1722, 1651, 1242, 1160	1723, 1652, 1227, 1156	1725, 1654, 1241, 1166	1724, 1652, 1227, 1154
NMR: Ⓐ	0·99, t, J = 7	1·0, t, J = 7	0·97/0·99, 2t, J = 7	0·97/0·99, 2t, J = 7
Ⓑ	1·27, s	1·28, s	1·27, s	1·27, s
Ⓒ	1·35–1·82, m	1·36–1·82, m	1·35–1·84, m	1·33–1·82, m
Ⓓ	1·82–2·50, m	1·82–2·40, m	1·84–2·40, m	1·82–2·40, m
Ⓔ	1·89, d, J = 1·4 ± 0·25	2·17, d, J = 1·25 ± 0·25	1·88, d, J = 1·4 ± 0·25	2·16, d, J = 1·25 ± 0·25
Ⓕ	2·50–2·85, m	1·82–2·40, m	2·50–2·87, m	1·82–2·40, m
Ⓖ	2·73, t, J = 6	2·72, t, J = 6	2·72, t, J = 6	2·71, t, J = 6
Ⓗ	3·67, s	3·68, s	3·67, s	3·67, s
Ⓘ	5·20, bt, J = 6·5 ± 0·5	5·13, m	5·17, bt, J = 6·5 ± 0·5	5·10, m
Ⓚ	5·67, bs	5·69, bs	5·67, bs	5·67, bs

Isomer	2-*cis*; 6-*cis*; 10,11-*trans*	2-*trans*; 6-*cis*; 10,11-*trans*	2-*cis*; 6-*trans*; 10,11-*trans*	2-*trans*; 6-*trans*; 10,11-*trans*
b.p. (°C)	100–101°/0·002 mm	103–105°/0·002 mm	103–104°/0·002 mm	108–109°/0·005 mm
n_D^{20}	1·4798	1·4812	1·4802	1·4816
IR (KBr; cm^{-1})	1724, 1652, 1242, 1161	1720, 1650, 1226, 1154	1725, 1653, 1241, 1166	1723, 1650, 1227, 1154
IR (KBr;				
NMR: Ⓐ	0·93/0·98, 2t, J = 7	0·94/0·99, 2t, J = 7	0·94/0·98, 2t, J = 7	0·94/0·97, 2t, J = 7
Ⓑ	1·24, s	1·24, s	1·23, s	1·24, s
Ⓒ	1·32–1·81, m	1·32–1·80, m	1·32–1·82, m	1·30–1·82, m
Ⓓ	1·81–2·45, m	1·80–2·40, m	1·82–2·45, m	1·82–2·40, m
Ⓔ	1·89, d, J = 1·4 ± 0·25	2·17, d, J = 1·25 ± 0·25	1·89, d, J = 1·4 ± 0·25	2·17, d, J = 1·25 ± 0·25
Ⓕ	2·50–2·86, m	1·80–2·40, m	2·50–2·87, m	1·82–2·40, m
Ⓖ	2·72, t, J = 6	2·70, t, J = 6	2·71, t, J = 6	2·71, t, J = 6
Ⓗ	3·66, s	3·68, s	3·66, s	3·67, s
Ⓘ	5·19, bt, J = 6·5 ± 0·5	5·12, m	5·17, bt, J = 6·5 ± 0·5	5·12, m
Ⓚ	5·67, bs	5·67, bs	5·67, bs	5·67, bs

TABLE V
Gas chromatographic data of all eight racemates

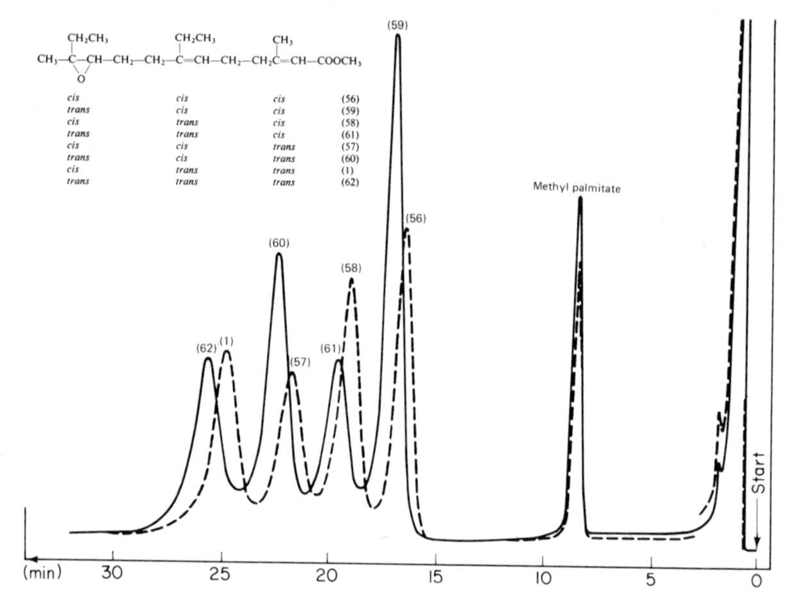

Isomer	R_t (min)	$R_{rel.}$
2-*cis*; 6-*cis*; 10,11-*cis* (56)	16·35	1·97
2-*cis*; 6-*cis*; 10,11-*trans* (59)	16·90	2·04
2-*cis*; 6-*trans*; 10,11-*cis* (58)	18·85	2·27
2-*cis*; 6-*trans*; 10,11-*trans* (61)	19·45	2·34
2-*trans*; 6-*cis*; 10,11-*cis* (57)	21·60	2·60
2-*trans*; 6-*cis*; 10,11-*trans* (60)	22·25	2·68
2-*trans*; 6-*trans*; 10,11-*cis* (1)	24·67	2·97
2-*trans*; 6-*trans*; 10,11-*trans* (62)	25·55	3·08

R_t, retention time; $R_{rel.}$, retention time relative to methyl palmitate. Column, glass, 3 m × 2·2 mm ID, with 5% OV 210 (Trifluoropropyl Methylsilicone) on 80–100 mesh Gas Chrom Q, isothermal at 180°C, vapour 190°C.

cis-configuration at the 6,7 double bond show a 5·2–5·4 min shorter retention time and those with *trans*-configuration at 6,7 about 5·8–6·1 min, when compared with the corresponding 2-*trans*-isomers (see Table V).

The IR-spectra show two characteristic absorption bands of the second ester absorption (–C–O–) at 1160–1166 cm^{-1} and 1241–1242 cm^{-1} in all 2-*cis*-isomers and at 1154–1156 cm^{-1} and 1226–1227 cm^{-1} in the corresponding 2-*trans*-isomers. The absorption band at higher frequency (1226–1227 cm^{-1})

in the 2-*trans*-isomers is considerably more intensive than that (1241–1242 cm^{-1}) of the corresponding 2-*cis*-isomers.

The mass spectra of the eight racemic JH-isomers are very similar, as expected. Only some variations in peak intensities are noted. The most significant one occurs in the peak at m/e 247 (M minus MeOH, minus CH_3) which is 3·0–3·5% of base peak intensity (peak m/e 57) in the 2-*cis*-isomers and only 0·7–1·2% of base peak intensity in the corresponding 2-*trans*-isomers.

The significant differences indicated by NMR, IR and mass spectroscopy as well as GLC are manifested also in the refractive indices. Whereas the difference of the measured refractive indices between the 10,11-*cis* and 10,11-*trans*-isomers is about 5.10^{-4}, the difference between the 2-*cis* and the corresponding 2-*trans*-isomers is about 14.10^{-4}.

A consideration of the physical and spectroscopic data for a given JH isomer can thus allow the assignment of its structure. The very important question of the biogenetic origin of the insect JH can not yet be answered with confidence. Four years before the authentic JH was isolated and its structure established, Schmialek (1963) was able to demonstrate that in *Samia cynthia* (2-^{14}C) mevalonate gives rise to labelled farnesol, farnesal and nerolidol, but incorporation of neither mevalonate nor methionine into JH can yet be demonstrated. The structural resemblance of the hormone to acyclic sesquiterpenoids, and the known presence of the latter in insects, makes it seem possible that an acyclic sesquiterpenoid, for example farnesol, could, by two C-methylations, account for the presence of JH. The isolation of JH-like compound (2) (Meyer *et al.*, 1968a,b), may have biogenetic relevance as an important stage in the biosynthesis of the authentic JH. The biogenetic origin of the so-called "Plant-JH", juvabione, and of dehydro-juvabione may be easier to explain. The monocyclic sesquiterpenoidal character of these two compounds would logically suggest their farnesol origin (Parker *et al.*, 1967), perhaps by cyclization of farnesyl pyrophosphate.

V. Chemical Structure and Biological Activity

Since the first successful isolation of an extract with strong JH-activity from the adult male *Hyalophora cecropia* by Williams in 1956, a rapid assay method has been required to facilitate the isolation of the active material. Wigglesworth (1958), while conducting extensive studies on the South American blood-sucking bug, *Rhodnius prolixus*, developed a simple, rapid test using pupae of the yellow meal worm, *Tenebrio molitor*. This *Tenebrio*-test, subsequently used by Schmialek (1961) and in a modified manner by Röller and Bjerke (1965) became a standard method for identifying, measuring and comparing JH-active substances. The effects of the test substances have been measured by a system of scoring for juvenile characteristics (cf. Williams, 1961; Bowers and Thompson, 1963; Romaňuk *et al.*, 1967; Rose *et al.*, 1968). One criterion for the comparison of JH-activities of derivatives from different sources is the degree of inhibition of metamorphosis per specimen, as indicated by the amount of substance in micrograms per specimen, which causes half-larval, half-adult intermediates in hemimetabolous insects or half-pupal, half-adult intermediates in holometabolous insects.

Extensive studies have been made on JH and related compounds in an attempt to establish meaningful structure-activity correlations (Jarolim *et al.*, 1969; Ratusky *et al.*, 1969; Röller *et al.*, 1969; Rose *et al.*, 1968; Schwarz *et al.*, 1969; Wakabayashi *et al.*, 1969; Westermann *et al.*, 1968). Despite the relatively comprehensive and consistent results obtained by Wigglesworth (1969a,b) on *Rhodnius*, no clear relationship has been established until now. The choice of the degree of metamorphosis inhibition as the only activity criterion, as well as the use of different application methods and different activity scoring systems has made the determination of structure-activity relationships extremely difficult. In addition, these test methods have no relationship to practical large scale application techniques, and thus cannot be used as a basis for prediction of activity under these conditions. Furthermore, the sterilizing and ovicidal effects must be taken into account when considering practical applications.

Besides the simple and rapid *Tenebrio* assay, carried out either by topical application or by injection of JH-active material, we have used a contact method for assaying JH-, as well as sterilizing and ovicidal, activity. In plastic boxes, freshly moulted last instar larvae are placed on filter papers of defined area which have been impregnated with a solution of the active material in acetone. The larvae are fed and are in permanent contact with the active material. The JH-activity is expressed in terms of the amount of active material per square centimeter of impregnated surface which is necessary to obtain a certain degree of metamorphosis inhibition. That inhibition dose (ID) which will give a median score or degree of metamorphosis inhibition is defined as metamorphosis ID_{50}. For the determination of the sterilizing effect, freshly moulted imagines are placed in a plastic cage which is lined with an impregnated filter paper in such a manner that the insect, but not the laid eggs, is in permanent contact with the active material. The eggs are collected daily and allowed to hatch in separate boxes.

To determine the ovicidal effect, freshly laid eggs are placed on filter papers or woollen discs which have been impregnated with the active substance. The mortality of the embryos or larvae soon after hatching is then evaluated. The sterilizing and ovicidal activity are indicated by the amount of active material per square centimeter of impregnated surface which causes a certain mortality of the embryos (blocking of the embryonic development) or larvae immediately after hatching. Those doses which will give rise to a 50% blocking of the embryonic development or mortality of freshly hatched larvae are defined as sterilizing ED_{50} and ovicidal ED_{50}, respectively.

The ID_{50} and ED_{50} values are convenient units for biological comparisons and can serve as a basis for consistent correlations between activity and chemical structure. We have tested a large number of substances on various insect species. Of those compounds tested, the eight racemic JH-isomers have given the most interesting and useful data for structure-activity correlation studies. A comparison of some test results for the eight racemic JH-isomers, (1, 56–58, 59–62) and of the Bowers compound (21) with the test results of Wigglesworth on last instar larvae of the South American blood-sucking bug, *Rhodnius prolixus*, is summarized in Table VI. Our experiments were performed by the contact method on larvae of the German cockroach, *Blattella germanica*, on larvae or freshly moulted imagines of the Indian cotton stainer, *Dysdercus cingulatus*, and on freshly moulted imagines or freshly laid eggs of the clothes moth, *Tineola bisselliella*. The results are expressed as ID_{50} and ED_{50}, indicating the amount of active substance in 10^{-x} g per larva (topical application) in the case of *Rhodnius*, and in 10^{-x} g of active substance per square centimeter filter paper in our experiments. As the results of Wigglesworth (1969b) on *Rhodnius* agree very closely with those reported by Röller and Dahm (1968) based on the *Tenebrio* test, we have not presented our *Tenebrio* results.

Consideration of the test results indicates the importance of the geometrical configuration of the JH-isomers on their activity. Significantly, this influence of geometrical isomerism on activity is species dependent. On *Tineola*, the *trans,trans,trans*-JH-isomer (62) and the natural racemic JH (1) (*trans,trans,cis*) are much more active than the other isomers, while on *Rhodnius*, *Blattella* and *Dysdercus* the four 2-*trans*-JH-isomers (1), (57), (60) and (62) were significantly more active than the corresponding four 2-*cis*-JH-isomers (56), (58), (59) and (61).

It has been known for a long time (Sláma *et al.*, 1969) that some compounds with JH-activity are more active on certain insects species, while they are less active or inactive on others. For example, the methylenedioxyphenyl ether of 6,7-dihydro-6,7-epoxydihomogeraniol (21) which Bowers (1968, 1969) reported to be several hundred times more active on *Tenebrio* than natural racemic JH (1), is far less active on *Blattella*, *Dysdercus* and *Tineola*. It was somewhat surprising, however, to find that there are large differences in insect sensitivity to geometrical isomers of the same molecule.

The highest JH-activity of all JH-isomers on *Rhodnius*, *Blattella* and *Dysdercus* was obtained with the natural racemic JH (1), as was the case with *Tenebrio*. However, the differences in level of JH-activity of the JH-isomers, especially the 2-*trans*-JH-isomers, becomes less in the order *Rhodnius* > *Blattella* > *Dysdercus*.

As seen in Table VI, the larvae of *Dysdercus* are about equally sensitive to all four 2-*trans*-JH-isomers (1, 57, 60 and 62). Incidentally, however, the *trans,cis,cis*-JH-isomer (57) has a significantly higher sterilizing effect than the natural racemic JH (1).

TABLE VI
Chemical structure and biological activity

	Topical appl. 10^{-x} g/larva	Inhibition dose$_{50}$ (ID_{50}) and Effective dose$_{50}$ (ED_{50}) in 10^{-x} g active subst./cm^2 filter paper				
	Metamorphosis ID_{50}			Sterilizing ED_{50}	Ovicidal ED_{50}	
	Rhodnius	*Blattella*	*Dysdercus*	*Dysdercus*	*Tineola*	
	X	X	X	X	X	X
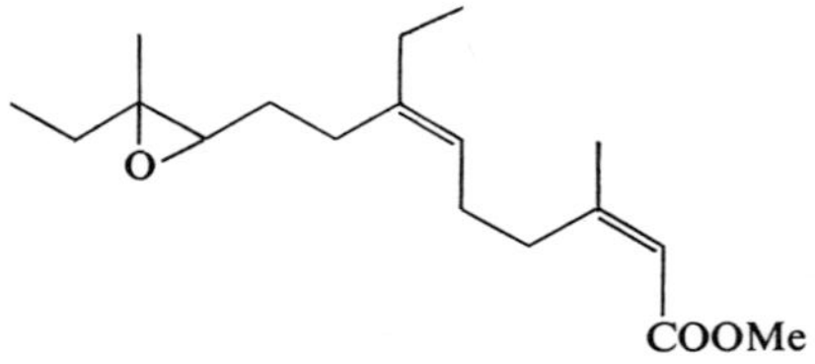 (59)	3·75	< 4·0	~ 3·5	< 4·0	< 3·0	~ 3·2
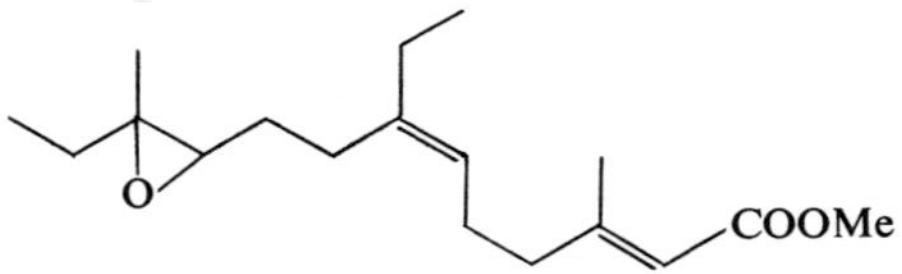 (60)	4·52	5·2–5·5	6·5–6·6	5·9–6·4	< 3·0	4·0–5·0
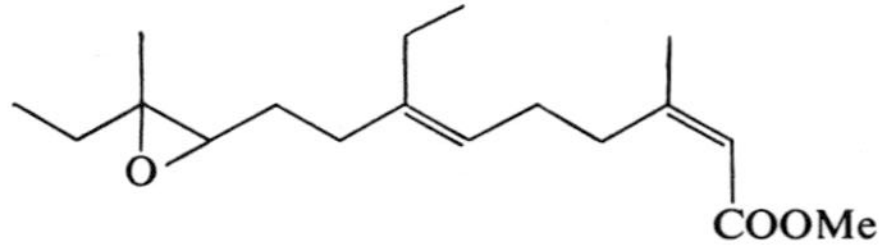 (61)	4·48	5·1–5·5	4·4–4·5	< 4·0	< 3·0	~ 4·3
 (62)	5·75	6·2–6·5	5·9–6·5	5·0–5·5	~4·9	~ 6·5

Structure	No.						
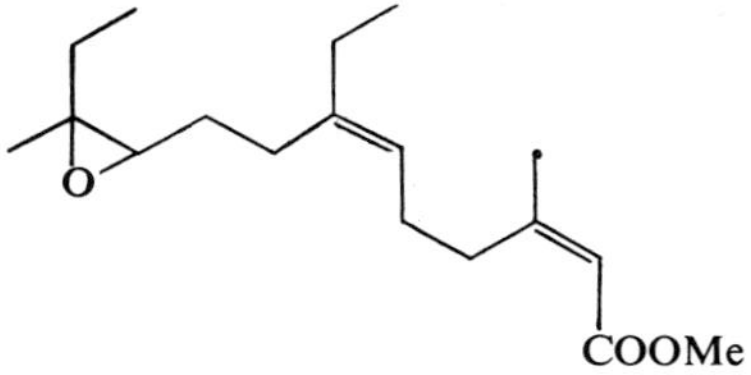 COOMe	(56)	3·64	3·0–4·0	5·0–5·6	< 4·0	< 3·0	~ 3·3
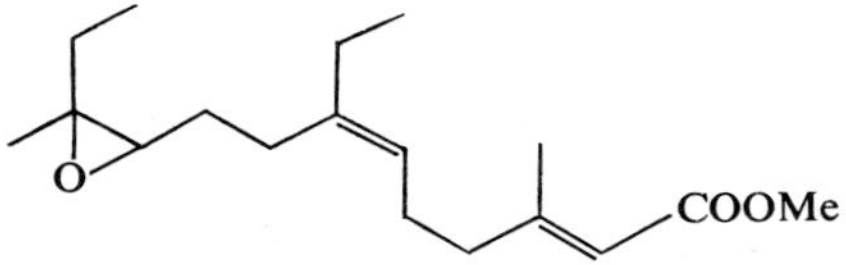 COOMe	(57)	5·17	5·0–5·5	6·4–6·8	7·2–7·5	< 3·0	4·0–5·0
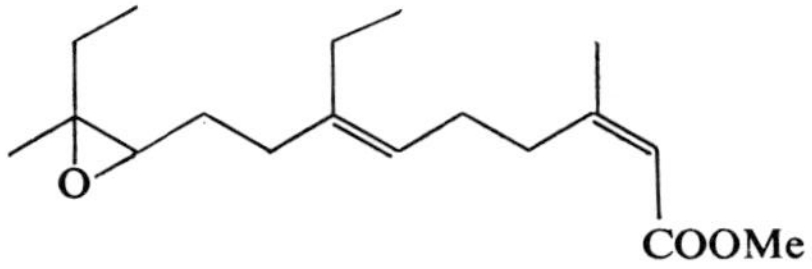 COOMe	(58)	4·82	4·8–5·3	~ 3·5	< 4·0	< 3·0	4·0–5·0
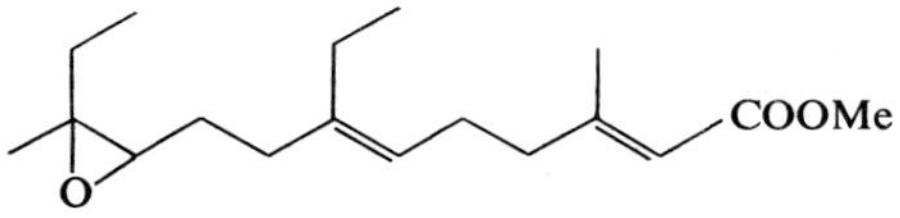 COOMe	(1)	6·30	6·0–6·5	6·8–6·9	6·0–6·7	~5·7	> 6·0
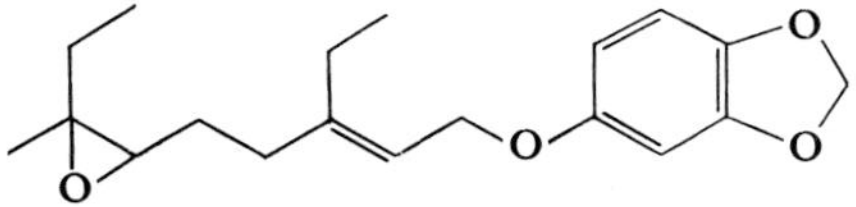 O O O	(21)	—	4·5–5·0	< 4·0	< 4·0	< 3·0	4·5

From the data summarized in Table VI it is possible to correlate stereochemical features with JH-activity. For the JH-isomers (1, 56–58, 59–62), the *trans*-configuration of the double bonds at C-2,3 and C-6,7 is obviously of great importance for biological activity, while the conformation of the oxirane ring seems to be not so important.

Wigglesworth (1969a,b) has shown that, for *Rhodnius*, not only proper double bond geometry but also a suitable balance between polar and apolar properties is important for JH-activity. Results from our testing of a large number of farnesene derivatives, analogues, and homologues on various insect species confirm this finding. Thus the presence of functional groups such as hydroxy, amino or carboxy reduces or abolishes the activity, while completely lipophilic derivatives are also virtually inactive. The epoxide at C-10,11 is important for activity in farnesene derivatives and homologues, but it is not always essential. A transfer of this function from C-10,11 to C-2,3 or to C-6,7 greatly reduces the activity. The presence of a second epoxide function reduces or increases the activity, depending upon the terminal function and the test insect.

At the present time, it is not known whether the site of action of JH and JH-mimics is the cytoplasm or the nucleus. Wigglesworth (1969a,b) has suggested that the JH and JH-mimicking compounds could be involved in the regulation of permeability relations within the cell, in such a way that the gene-controlled enzyme system responsible for larval characters is brought increasingly into action when JH-active compounds are present. This would indicate that a physico-chemical action is the actual basis of JH-activity (Wigglesworth, 1969a,b). A recent report (Baumann, 1968) that JH-active derivatives have a depolarizing effect on the plasma membrane of the salivary glands in *Galleria*, apparently by raising the conductivity of the membrane, seems to confirm this hypothesis. The striking differences in the relative activity of various JH-active compounds in different insect species would then be readily explained, since species differences in the properties of cell membranes have long been recognized.

We hope that our structure-activity correlation studies with the JH-isomers will encourage further studies in this area.

ACKNOWLEDGEMENT

The author thanks Drs R. Rüegg and U. Schwieter for stimulating discussions during the synthetic work carried out in our laboratory; Dr L. Chopard-dit-Jean for IR-spectra and distillative separations of isomeric intermediates; Drs G. Englert and W. Arnold for NMR-spectroscopy; Dr W. Vetter for mass spectroscopy; and Dr M. Vecchy for GLC-analyses. Further I am indebted to Miss B. Peyer and Mr E. Homberger of Dr R. Maag Ag., Chemische Fabrik, Dielsdorf, for the biological tests.

REFERENCES

Atherton, F. R. and Pfiffner, A. (1968). *Belg. Pat.* 725,576. Part of this work was presented by Pfiffner, A., on the occasion of the 148 Jahresversammlung der Schweizerischen Naturforschenden Gesellschaft in Einsiedeln (27–29 Sept., 1968).

Baumann, G. (1968). *J. Insect Physiol.* **14**, 1365.

Birch, A. J., Macdonald, P. L. and Powell, V. H. (1969). *Tetrahedron Letters* 351.

Blount, J. F., Pawson, B. A. and Saucy, G. (1969). *Chem. Commun., Lond.* 715.

Bowers, W. S. and Thompson, M. J. (1963). *Science, N.Y.* **142**, 1469.

Bowers, W. S., Thomson, M. J. and Uebel, E. C. (1965). *Life Sci.* **4**, 2323.

Bowers, W. S., Fales, H. M., Thompson, M. J. and Uebel, E. C. (1966). *Science, N.Y.* **154**, 1020.

Bowers, W. S. (1968). *Science, N.Y.* **161**, 895.

Bowers, W. S. (1969). *Science, N.Y.* **164**, 323.

Braun, B. H., Jacobson, M., Schwarz, M., Sonnet, P. E., Wakabayashi, N. and Waters, R. M. (1968). *J. Econ. Entomol.* **61**, 866.

Cavill, G. W. K. and Williams, P. J. (1969). *Aust. J. Chem.* **22**, 1737.

Cavill, G. W. K., Laing, D. G. and Williams, P. J. (1969). *Aust. J. Chem.* **22**, 2145.

Černý, V., Dolejš, L., Lábler, L., Šorm, F. and Sláma, K., (1967). *Coll. Czech. Chem. Commun.* **32**, 3926.

Chen, D. H., Robbins, W. E. and Monroe, R. E. (1962). *Experientia* **18**, 577.

Corey, E. J., Katzenellenbogen, J. A., Gilman, N. W., Roman, S. A. and Erickson, B. W. (1968). *J. Am. chem. Soc.* **90**, 5618.

Dahm, K. H., Trost, B. M. and Röller, H. (1967). *J. Am. chem. Soc.* **89**, 5292.

Dahm, K. H., Röller, H. and Trost, B. M. (1968). *Life Sci.* **7**, 129.

Engelmann, F. (1969). *Science, N.Y.* **165**, 407.

Findlay, J. A. and MacKay, W. D. (1969). *Chem. Commun., Lond.* 733.

Gilbert, L. I. (1967). *Comp. Biochem. Physiol.* **21**, 237.

Hoffmann, W., Pasedach, H. and Pommer, H. (1969). *Ann. Chem.* **729**, 52.

Isler, O., Rüegg, R., Chopard-dit-Jean, L., Winterstein, A. and Wiss, O. (1958). *Helv. Chim. Acta* **41**, 786.

Jarolim, V., Hejno, K., Sehnal, F. and Šorm, F. (1969). *Life Sci.* **8**, 831.

Johnson, W. S., Faulkner, T. Li. D. J. and Campbell, S. F. (1968). *J. Am. chem. Soc.* **90**, 6225.

Johnson, W. S., Campbell, S. F., Krishnakumaran, A. and Meyer, A. S. (1969). *Proc. Natn. Acad. Sci. U.S.A.* **62**, 1005.

Krishnakumaran, A. and Schneiderman, H. A. (1965). *J. Insect Physiol.* **11**, 1517.

Law, J. H., Yuan, C. and Williams, C. M. (1966). *Proc. Natn. Acad. Sci. U.S.A.* **55**, 576.

Levinson, H. Z. (1966). *Riv. Parassitol.* **27**, 47.

Masner, P., Slama, K. and Landa, V. (1968). *Nature, Lond.* **219**, 395.

Meyer, A. S. (1965). *Analyt. Biochem.* **11**, 290.

Meyer, A. S., Schneiderman, H. A. and Gilbert, L. I. (1965). *Nature, Lond.* **206**, 272.

Meyer, A. S., Schneiderman, H. A. and Hanzmann, E. (1968a). *Fedn. Proc.* **27**, 393.

Meyer, A. S., Schneiderman, H. A., Hanzmann, E. and Ko, J. H. (1968b). *Proc. Natn. Acad. Sci. U.S.A.* **60**, 853.

Minks, A. K. (1967). *Arch. Nerl. Zool.* **17**, 175.

Momose, T. (1941). *J. Pharm. Soc. Japan* **61**, 288; (1950). *Chem. Abstr.* **44**, 9383.

Mori, K. and Matsui, M. (1968). *Tetrahedron* **24**, 3127.

Mori, K., Stalla-Bourdillon, B., Ohki, M., Matsui, M. and Bowers, W. S. (1969). *Tetrahedron* **25**, 1677.
Nakazaki, M. and Isoe, S. (1963). *Bull. Chem. Soc. Japan* **36**, 1198.
Oberlander, H. and Schneiderman, H. A. (1966). *J. Insect Physiol.* **12**, 37.
Odhiambo, T. R. (1966). *J. exp. Biol.* **45**, 51.
Parker, W., Roberts, J. S. and Ramage, R. (1967). *Q. Rev., Lond.* **21**, 331.
Pawson, B. A., Cheung, H.-C., Gurbaxani, S. and Saucy, G. (1968). *Chem. Commun., Lond.* 1057.
Pfiffner, A. and Schwieter, U. (1968). *Belg. Pat.* 725,578 and 725,579.
Pfiffner, A. and Schwieter, U. (1969). *Belg. Pat.* 739,353.
Ratusky, J., Sláma, K. and Šorm, F. (1969). *J. Stored Prod. Res.* **5**, 111.
Riddiford, L. M. and Williams, C. M. (1969). *Proc. Natn. Acad. Sci. U.S.A.* **57**, 595.
Romaňuk, M., Sláma, K. and Šorm, F. (1967). *Proc. Natn. Acad. Sci. U.S.A.* **57**, 349.
Rose, M., Westermann, J., Trautmann, H. and Schmialek, P. (1968). *Z. Naturf.* **23b**, 1245.
Röller, H. and Bjerke, J. S. (1965). *Life Sci.* **4**, 1617.
Röller, H., Bjerke, J. S. and McShan, W. H. (1965). *J. Insect Physiol.* **11**, 1185.
Röller, H. and Dahm, K. H. (1968). *In* "Recent Progress in Hormone Research", Vol. 24, pp. 651–671, Academic Press, New York and London.
Röller, H., Dahm, K. H., Sweeley, C. C. and Trost, B. M. (1967a). *Angew. Chem.* **79**, 190; (1967a). *Angew. Chem. Intern. Ed.* **6**, 179.
Röller, H., Dahm, K. H., Trost, B. M. and Sweeley, C. C. (1967b). *Chem. Eng. News* **45**, 48.
Röller, H., Bjerke, J. S., Holthaus, L. M., Norgard, D. W. and McShan, W. H. (1969). *J. Insect Physiol.* **15**, 379.
Rüegg, R. and Schmialek, P. (1962). *Belg. Pat.* 617,175.
Saucy, G. and Marbet, R. (1967). *Helv. Chim. Acta* **50**, 2091.
Saxena, K. N. and Williams, C. M. (1966). *Nature, Lond.* **210**, 441.
Shaaya, E. (1969). Paper presented at the 5th Conference of European Comparative Endocrinologists, Utrecht/Neth., Aug. 24–29.
Siddall, J. B. and Calame, J. P. (1967). *Neth. Pat. Appl.* 12568.
Sláma, K. and Williams, C. M. (1965). *Proc. Natn. Acad. Sci. U.S.A.* **54**, 411.
Sláma, K. and Williams, C. M. (1966). *Biol. Bull.* **130**, 235.
Sláma, K. and Williams, C. M. (1966). *Nature, Lond.* **210**, 329.
Sláma, K., Suchý, M. and Šorm, F. (1968). *Biol. Bull.* **134**, 154.
Sláma, K., Romaňuk, M. and Šorm, F. (1969). *Biol. Bull.* **135**, 91.
Suchý, M., Sláma, K. and Šorm, F. (1968). *Science, N.Y.* **162**, 582.
Schmialek, P. (1961). *Z. Naturf.* **16b**, 461.
Schmialek, P. (1963). *Z. Naturf.* **18b**, 462.
Schneiderman, H. A. and Gilbert, L. I. (1964). *Science, N.Y.* **143**, 325.
Schulz, H. and Sprung, I. (1969). *Angew. Chem.* **81**, 258.
Schwarz, M., Braun, B. H., Law, M. W., Sonnet, P. E., Wakabayashi, N., Waters, R. M. and Jacobson, M. (1969). *Ann. Entomol. Soc. Am.* **62**, 668.
Spielman, A. and Williams, C. M. (1966). *Science, N.Y.* **154**, 1043.
Subrahmania Ayyar, K. and Krishna Rao, G. S. (1968). *Can. J. Chem.* **46**, 1467.
Thomas, K. K. and Nation, J. L. (1966). *Biol. Bull.* **130**, 442.
Thomas, J. P. and Bhatnagar-Thomas, P. L. (1968). *Nature, Lond.* **219**, 949.
Tutihasi, R. and Hanazawa, T. (1940). *J. Chem. Soc. Japan* **61**, 1041; (1943). *Chem. Abstr.* **37**, 258.
Van Tamelen, E. E. and Curphy, T. J. (1962). *Tetrahedron Letters* 121.

Van Tamelen, E. E., Storni, A., Hessler, E. J. and Schwarz, M. (1965). *J. Am. chem. Soc.* **85**, 3295.
Van Tamelen, E. E. and Sharpless, K. B. (1967). *Tetrahedron Letters* 2655.
Vinson, J. W. and Williams, C. M. (1967). *Proc. Natn. Acad. Sci. U.S.A.* **58**, 294.
Wakabayashi, N., Sonnet, P. E. and Law, M. W. (1969). *J. Med. Chem.* **12**, 911.
Westermann, J., Rose, M., Trautmann, H. and Schmialek, P. (1968). *Z. Naturf.* **24b**, 378.
Wigglesworth, V. B. (1934). *Q. J. Micros. Sci.* **77**, 191.
Wigglesworth, V. B. (1936). *Q. J. Micros. Sci.* **79**, 91.
Wigglesworth, V. B. (1958). *J. Insect Physiol.* **2**, 73.
Wigglesworth, V. B. (1961). *J. Insect Physiol.* **7**, 73.
Wigglesworth, V. B. (1965). *Nature, Lond.* **208**, 522.
Wigglesworth, V. B. (1969a). *Nature, Lond.* **221**, 190.
Wigglesworth, V. B. (1969b). *J. Insect Physiol.* **15**, 73.
Williams, C. M. (1956). *Nature, Lond.* **178**, 212.
Williams, C. M. (1961). *Biol. Bull.* **121**, 572.
Williams, C. M. and Law, J. H. (1965). *J. Insect Physiol.* **11**, 569.
Williams, C. M. and Sláma, K. (1966). *Biol. Bull.* **130**, 247.
Williams, C. M. (1967). *Sci. Am.* **217**, 13.
Williams, C. M. and Robbins, W. E. (1968). *BioScience* **18**, 791.
Wiss-Huber, M. and Lüscher, M. (1969). Paper presented at the 5th Conference of the European Comparative Endocrinologists, Utrecht/Neth., Aug. 24–29.
Yamamoto, R. T. and Jacobson, M. (1962). *Nature, Lond.* **196**, 908.
Zurflüh, R., Wall, E. N., Siddall, J. B. and Edwards, J. A. (1968). *J. Am. chem. Soc.* **90**, 6224.

CHAPTER 5

Abscisic Acid

B. V. MILBORROW

Shell Research Limited, Milstead Laboratory of Chemical Enzymology, Broad Oak Road, Sittingbourne, Kent

I. Absolute Configuration of Abscisic Acid

The history of the discovery and isolation of abscisic acid (1) has been well documented and this plant hormone has now been isolated and identified from so many higher plant species and different parts of these plants that it can now be considered as ubiquitous a constituent as auxins, gibberellins, cytokinins and the indigent hormone ethylene. The action of none of these classes of hormone is at present understood in biochemical terms and, because two of the four are non-terpenoid, discussion of their action can be left to another occasion. In a symposium on terpenoids it is appropriate to concentrate attention on those aspects of the biochemistry of abscisic acid that are related to its terpenoid nature, in particular its formation and destruction.

The absolute configuration of the single centre of asymmetry of abscisic acid (ABA) has been determined by the application of Mills' rule to the *cis*- and *trans*-diols (3 and 4) prepared from (+)- and (−)-ABA resolved from synthetic, racemic material (Cornforth *et al.*, 1967). Sodium borohydride reduces the 4′-oxo group to a mixture of equal amounts of *cis*- and *trans*-diols and these can be separated chromatographically. The IR spectra of the diols are not sufficiently distinct to give a satisfactory identification but the *cis*-diol has been synthesized (5, 6, 7) by an unequivocal route and can be used as a chromatographic standard during the separation. The methyl ester of the

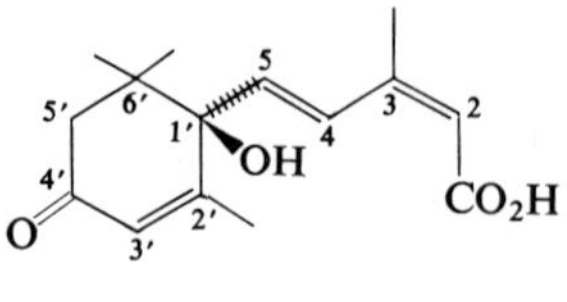

5′ 6′ 1′ 5 4 3 2 4′ OH CO_2H O 3′ 2′

(1) (+)-Abscisic acid

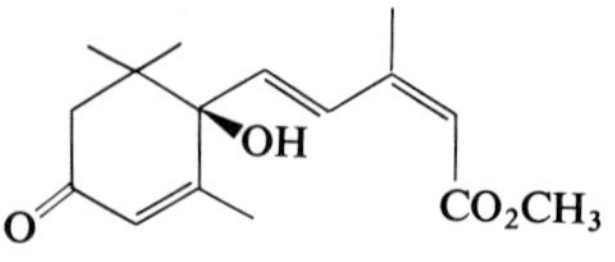

OH CO_2CH_3 O —NaBH$_4$→ H OH CO_2CH_3 HO

(2) Natural (+)-ABA methyl ester

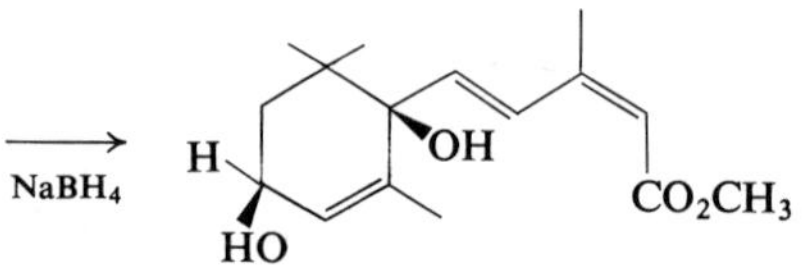

(3) *cis*-Diol of (+)-ABA methyl ester

NaBH$_4$

OH HO H CO_2CH_3

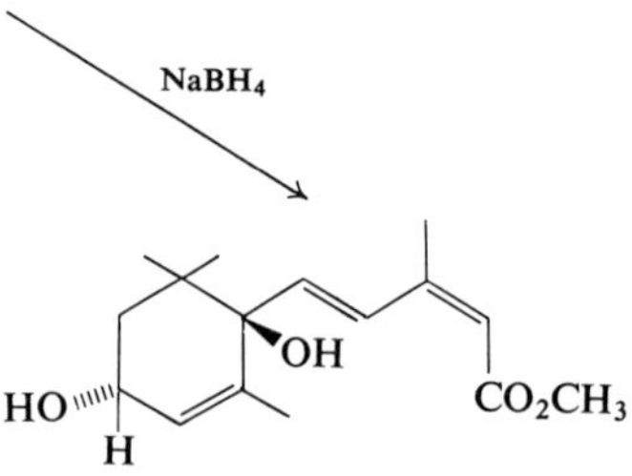

(4) *trans*-Diol of (+)-ABA methyl ester

CO_2CH_3 —O_2 light→ O O CO_2CH_3 O O CO_2CH_3 —H_2 Lindlar's catalyst→

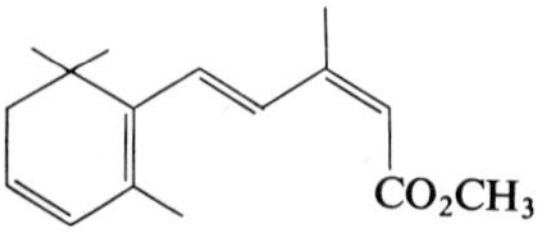

(5)

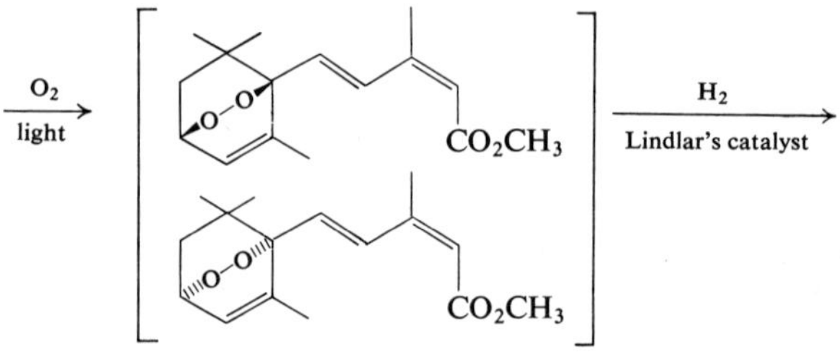

(6) Racemic epidioxide methyl ester

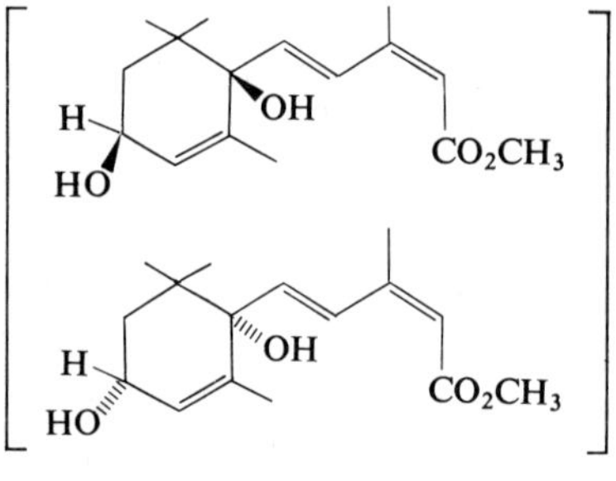

H OH HO CO_2CH_3 H OH HO CO_2CH_3

(7) Racemic *cis*-diol of ABA methyl ester

cis-diol runs closer to the origin than the ester of the *trans*-diol on silica-gel TLC plates developed in toluene, 50; ethyl acetate, 20; acetic acid, 3 (v/v).

At the D-line the *cis*-diol prepared from the naturally occurring (+)-enantiomer of ABA was more laevorotatory than the *trans*-diol, showing that the chirality of the 4′-hydroxyl in the former was as shown in (3). Therefore, because the tertiary hydroxyl is on the same side of the cyclohexenyl ring as the 4′-hydroxyl in the *cis*-diol, the absolute configuration of the original centre of asymmetry of (+)-ABA (1) can be deduced. It is (S). The ratio of the magnitudes, and of course the signs, of the specific rotations of the diols prepared from (–)-ABA were the converse of this.

II. Spectropolarimetric Methods of Measurements

Most of the identifications of abscisic acid (ABA) have made use of its strong optical rotatory dispersion (ORD) spectrum which has a highly characteristic shape, and all isolates of the natural material have been found to possess a positive dispersion spectrum. Where measurements of crystalline material have been reported, the specific rotations of the (+)-ABA have been so similar that it is probable that one enantiomer only is formed (Table I).

TABLE I
Specific rotations reported for natural (+)-abscisic acid isolated in crystalline form from different plant sources

λ nm	Rotation [α]	Molar amplitude	Source	ε	λ nm	Reference
589	+430	—	Sycamore	21,400	260	Milborrow, 1968
287	+24,000	—	Sycamore	21,400	262	Cornforth *et al.*, 1966
245	–69,000	—	Sycamore	—	—	Cornforth *et al.*, 1966
—	—	250,000°	Sycamore	—	—	Cornforth *et al.*, 1966
—	—	240,000°	Yellow lupin	—	—	Cornforth *et al.*, 1966
589	+424	—	Yellow lupin	21,800	258	Koshimizu *et al.*, 1966
589	+488	—	Pea	21,000	258	Isogai *et al.*, 1967
289	+30,700	—	Pea	—	—	Isogai *et al.*, 1967
245	–85,200	—	Pea	—	—	Isogai *et al.*, 1967

The optical rotatory dispersion spectrum of ABA has been used to determine its concentrations in plant extracts but the data are minimum values because of unknown losses during isolation.

The high specific rotation suggested a method of measuring the concentration of the hormone within plants, which is analogous to the inverse of the isotope dilution method but makes use of optically active and racemic material (Milborrow, 1967). The optically active (+)-ABA within a quantity of plant material is mixed with a known amount of synthetic, racemic material dissolved in methanol in which the plant material is homogenized. Provided that no preferential loss of one enantiomer occurs, by, for instance, enzyme action or crystallization, the ratio between the optically active material originally present within the plant and the added racemate is maintained, irrespective of losses during purification. Finally, when the extracted material is freed from all detectable UV absorbing impurities and the shape of the UV absorption spectrum is identical with that of authentic ABA, the amount of ABA can be determined photometrically (Table I). The amount of excess optically active ABA in the same solution can be measured spectropolarimetrically and the ratio between the excess (+)-ABA determined by this method and the racemic remainder can then be calculated. The amount of free ABA originally present in the plant can be calculated from this ratio and the amount of racemate added to the extract.

A. RACEMATE DILUTION METHOD

$$w = a \times \frac{b}{c - a}$$

where w = weight of free ABA present in the tissue
a = weight of (+)-ABA isolated, determined by ORD
b = weight of (±)-ABA added to the tissue homogenate
c = total weight of ABA isolated, by UV absorption.

This "Racemate Dilution" (RD) method has shown that during the usual isolation procedures between 50% and 70% of the (+)-ABA in a tissue is lost. Not surprisingly the data indicate that the lower the concentration of ABA within the tissues, the lower the percentage isolated. Later, when [2-^{14}C] labelled ABA was available, it was possible to compare the recoveries of natural (+)-ABA by the RD method with the parallel recovery of (+)-ABA and of labelled (±)-ABA.

It is also possible to use a known amount of hexadeuterio, racemic ABA as the diluent. The ether-soluble acid fraction is prepared and methylated with ethereal diazomethane. It is then analysed by combined gas-layer chromatography/mass spectrometry. The amount of ABA originally present is calculated from the peak heights of the normal, light parent ion, the parent ion of the deuterated material and the amount of deuterated material added to the extract (Table II).

III. Metabolism

When racemic [2-^{14}C] ABA is supplied to tomato shoots by placing their cut ends in solutions of 1 or 10 mg/ml, the labelled acid disappears rapidly so that some 10% only of the amount taken up remains as free acid after approximately 12–15 h. When the free ABA remaining was extracted its ORD spectrum showed that there was a preponderance of the unnatural (−) enantiomer. A majority of the radioactivity appeared in a water-soluble neutral material. This was purified and found to be abscisyl-β-D-glucopyranoside which had been isolated and identified from yellow lupin by Koshimizu,

TABLE II
Concentrations of free (+)-ABA within plants measured by four methods

Species	ABA, mg/kg fresh weight Standard method	RD method	Recovery of $[^{14}C]$ ABA	Recovery of hexa [D] ABA
Avocado, seed cv. Haff	0·034	0·11	—	—
Avocado, seed cv.	0·18	0·35	0·42	—
Cabbage, leaves of heart	0·056	0·24	—	—
Linden, fruit	0·046	0·14	—	—
Dog rose, stem	1·67	4·1	—	—
Field rose, stem	0·19	0·67	—	0·68

Inui, Fukui and Mitsui (1968). The material from tomato plants was identical with an authentic sample, provided by Professor Kosimizu, by NMR, IR, UV spectrometry and by chromatography and mass spectrometry of its tetraacetate.

The purified material from tomatoes was obtained as a colourless glass and yielded ABA and glucose on alkaline hydrolysis. The ORD of the ABA liberated was not positive as expected but showed that an excess of the (–) enantiomer was present.

We wanted to find out how the plants metabolized racemic ABA and were able to devise a method to do this (Milborrow, 1970) by making the reasonable assumptions that there is no inversion of configuration of ABA by the plant, that all the endogenous ABA is (+) and, as with the RD method, that the same percentage recovery obtains with the (+) as with the (–) enantiomer.

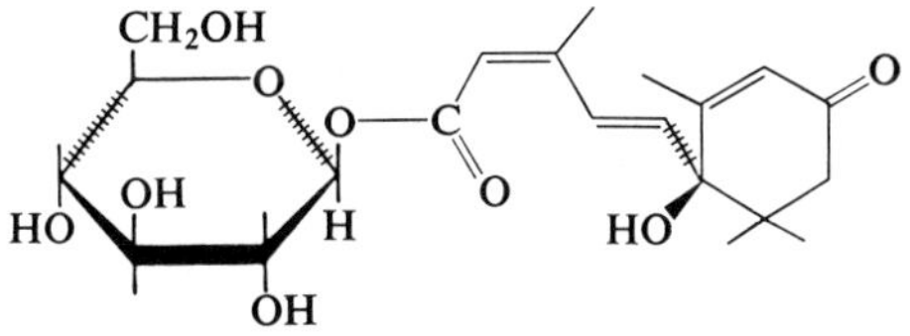

(8) (+)-Abscisyl-β-D-glucopyranoside

The total amount of ABA isolated in spectroscopically pure form from a tomato plant which has been supplied previously with racemic $[2\text{-}^{14}C]$ labelled ABA of known specific activity can be calculated from its UV absorption. The excess of optically active material, usually the (–) enantiomer, in the same solution is then measured spectropolarimetrically. This value can then be subtracted from the total ABA to give the amount of racemate present.

Total ABA by UV – (–)-ABA by ORD = Racemic ABA

The racemate is, by definition, half (+) and half (–) therefore all the (–)-ABA in the sample can be accounted for. The (–)-ABA is assumed to be entirely of synthetic origin and has the same specific radioactivity as the original racemate. The radioactivity of the sample attributable to the (–)-ABA can be calculated and this value subtracted from the total radioactivity of the sample to give the amount of (+)-[2-^{14}C] ABA remaining. The total amount of radioactive ABA subtracted from the total ABA measured photometrically gives the amount of unlabelled, endogenous (+)-ABA in the sample. The same reasoning can be applied to the measurements of ABA derived from the glucose ester by alkaline hydrolysis.

The data in Table III show that there is a preponderance of (–)-ABA and the missing (+)-ABA had been converted into another product. The only

TABLE III

Measurement of (+)-ABA, (+)-[2-^{14}C] ABA and (–)-[2-^{14}C] ABA as free acids and in the hydrolysate of abscisyl-β-D-glucopyranoside

	Free ABA, μg	ABA released by hydrolysis, μg
(a) Total ABA by UV absorption	199	305
(b) Excess (–)-ABA over racemate	56·2	203
Racemate = $(a - b)$	142·8	102
(c) Total (–)-ABA = $\frac{(a-b)}{2} + b$	127·6	254
(d) Radioactive ABA	195	269
(+)-ABA = $a - d$	4	36
(+)-[2-^{14}C] ABA = $d - c$	67·4	15
Metabolite C by measurement of its radioactivity	93	0

other radioactive product found in significant amounts after 10–15 h of treatment was an ether-soluble acidic material which moved at a lower R_f than ABA on TLC plates developed in toluene, 50; ethyl acetate, 30; acetic acid, 3 (v/v) and was referred to as metabolite C. It was isolated and showed a strong positive ORD curve similar to that of ABA but displaced slightly towards longer wavelengths ((+)-extremum = 298 nm, (–)-extremum = 254 nm). The material was methylated so that it could be dissolved in deuterio-chloroform for NMR analysis but was found to have rearranged to a slightly less polar compound (but still more polar than ABA) with a weak ORD spectrum. The properties of this rearrangement product were identical with those reported for phaseic acid which had been isolated from bean seeds (*Phaseolus multiflorus*) by MacMillan and Pryce (1968). They had tentatively assigned structure (9) to it but noted two properties, its resistance to alkali-induced rearrangement and a long-range coupling in the NMR spectrum between an epoxy-methylene proton and a methylene proton adjacent to the 4′-oxo group (ABA numbering, 1). It is difficult to reconcile this structure with the method of formation of phaseic acid and Cornforth proposed an alternative structure (11), which we found later had also been noted by MacMillan and Pryce. Structure (11) accounted for the physical properties and could also account for the rearrangement.

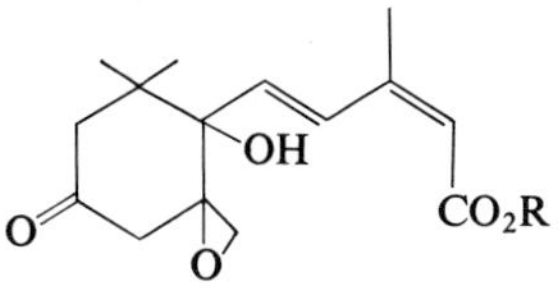

(9) R = H

(10) R = CH_3

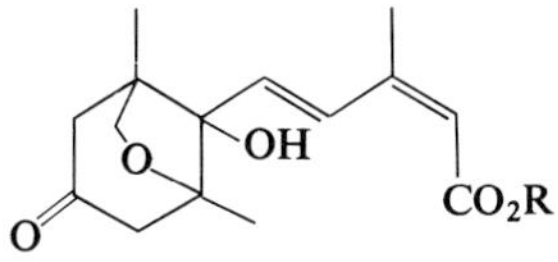

(11) R = H

(12) R = CH_3

A crucial difference between the two proposed structures is the origin of the two methyl signals at 1·01δ and 1·23δ in the deuteriochloroform NMR spectrum of methyl phaseate. In structure (10) they are attributed to *geminal* dimethyl groups whereas in structure (12) the signal at 1·01δ is attributed to the single remaining methyl group of the *geminal* pair of ABA and the one at 1·23δ to what had been the C-2′allylic methyl of ABA.

It is possible to exchange six of the skeletal hydrogen atoms of ABA in strong alkali and when this is done in *N* sodium deuterioxide, and then the deuterium "locked in" by neutralization with dideuterio oxalic acid, the octadeuterio ABA is obtained (13). This material immediately loses the deuterium atoms of the carboxyl and hydroxyl groups when dissolved in water but the other six are retained provided that the pH does not rise above 8.

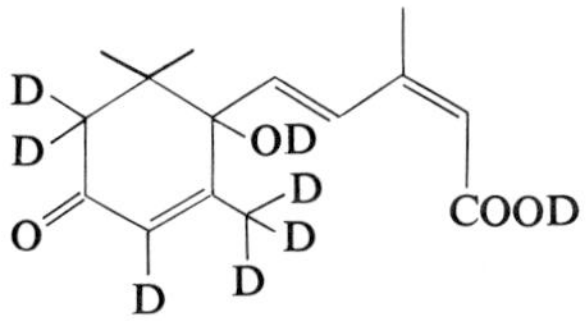

(13) Deuteriated abscisic acid

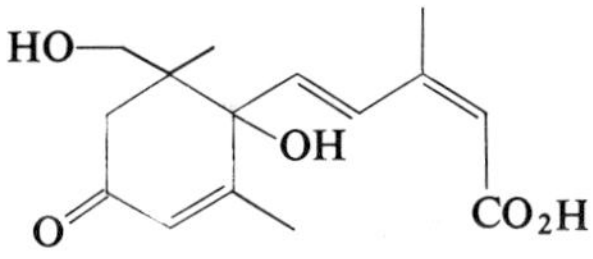

(14) Metabolite C

This deuteriated ABA was fed to tomato shoots and the phaseic acid isolated two days later. It was methylated and subjected to NMR analysis. The methyl signal at 1·23δ was absent (Fig. 1) showing that it had been derived from the 2′-trideuteriomethyl group of ABA. This confirmed structure (11) for phaseic acid (Milborrow, 1969).

The rearrangement of metabolite C can now be interpreted as a nucleophilic attack, by a hydroxyl group on one of the *geminal* dimethyl groups of ABA, on a double bond activated by an α-ketone. The structure for "metabolite C" is shown in (14).

The *geminal* methyl groups of ABA are not identical because one is *cis* to the tertiary hydroxyl group while the other is *trans*. We have tried to discover which one is hydroxylated to give metabolite C by looking at the region of the

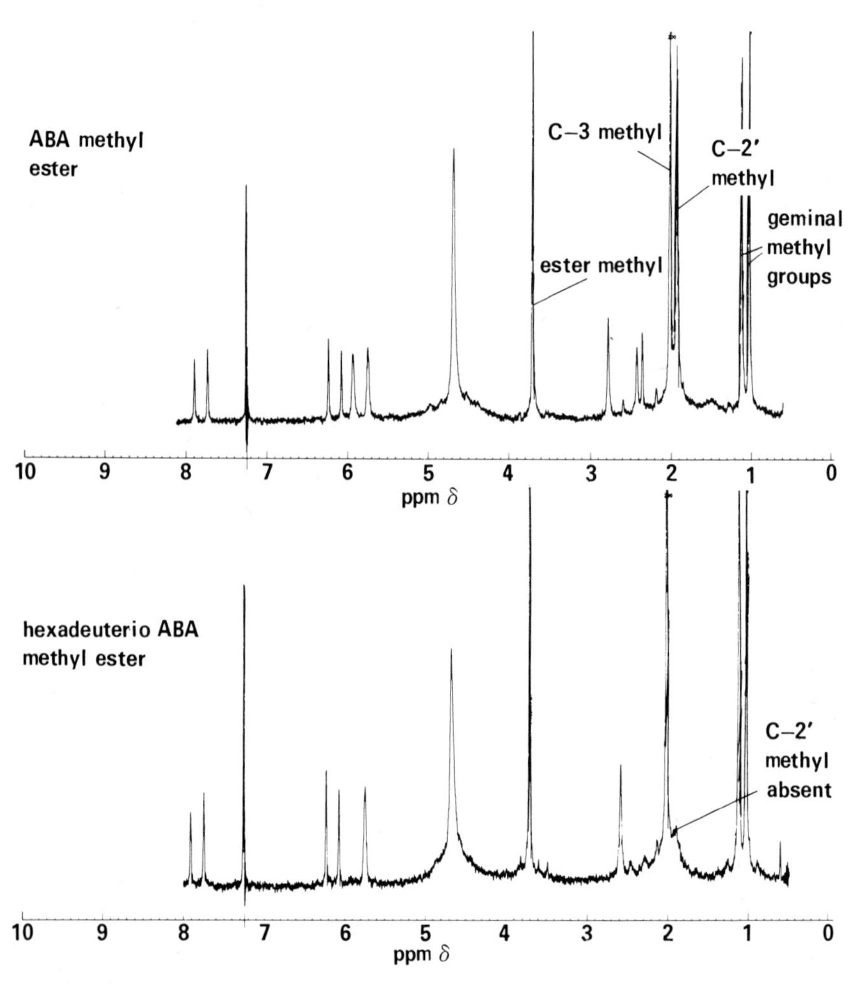

FIG. 1. NMR spectra ($CDCl_3$) of methyl esters of ABA, hexadeuterio ABA, phaseic acid and phaseic acid formed from hexadeuterio ABA. The absence of the signal at 1·23δ in the latter sample of phaseic acid identifies this methyl signal with the C-2′ methyl of ABA, thereby confirming structure (11) for phaseic acid.

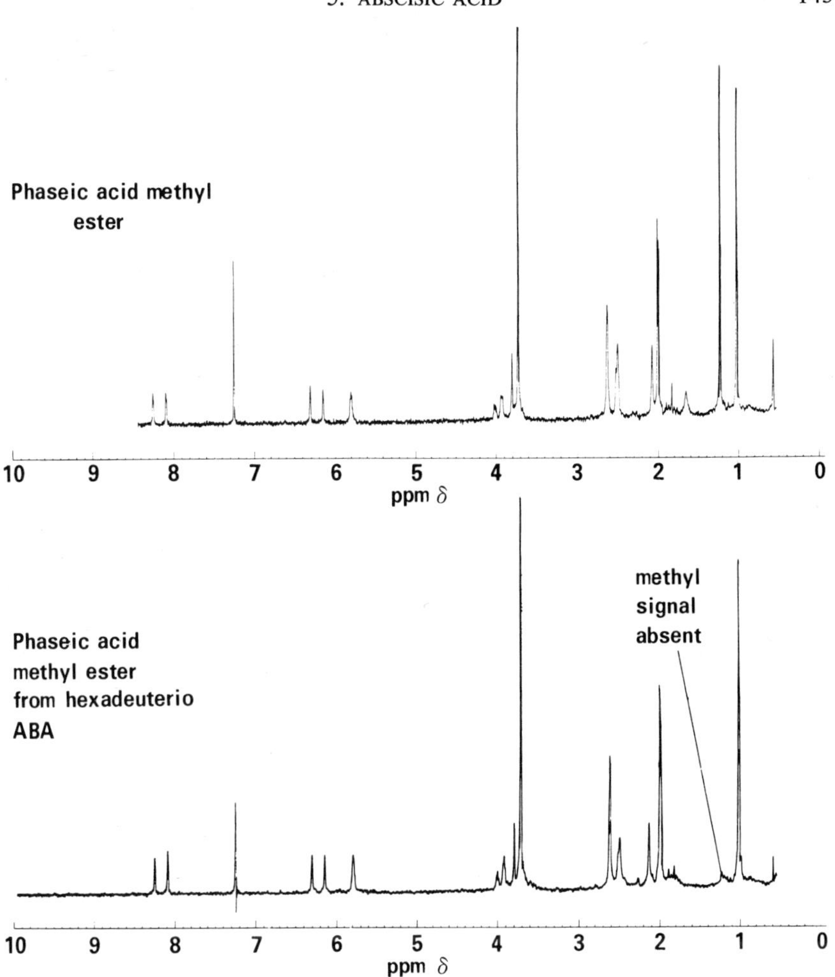

infra-red spectrum of methyl phaseate where hydrogen bonding of hydroxyl groups occurs. The position of the hydroxyl hydrogen in molecular models suggests that if the tetrahydrofuran bridge of phaseic acid were on the same side of the cyclohexyl ring as the tertiary hydroxyl group then it could bond to the ether oxygen. IR spectra of phaseic acid methyl ester at different dilutions in CS_2 failed to show signs of intramolecular hydrogen bonding; this suggests that the tetrahydrofuran bridge is *trans* to the tertiary hydroxyl group. It is recognized that this is negative and inconclusive evidence so the absolute configuration of phaseic acid (15) is tentative only. Further attempts to reisolate metabolite C have failed because phaseic acid was found each time. Whether phaseic acid is a natural metabolite formed *in vivo* or an artefact produced during the extraction is at present being investigated. The specific rotation of natural phaseic acid is, as near as we can measure, identical with

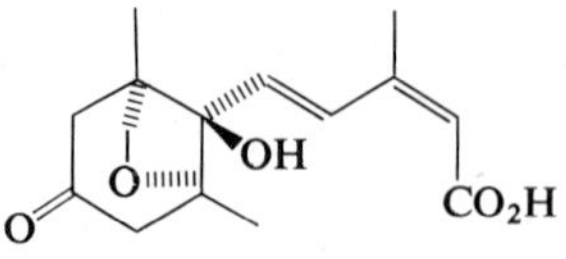

(15) Phaseic acid

that of phaseic acid derived from racemic ABA. This observation, together with the deficiency of (+)-[2-^{14}C] ABA as free acid and as the glucose ester in extracts of tomato shoots, suggests that the (+) is the major if not the sole enantiomer of ABA converted to metabolite C. However, after 48 h other oxygenated metabolites of [2-^{14}C] ABA can be isolated and two of these show a negative ORD spectrum closely similar to that of (–)-ABA. It is possible, therefore, that the hydroxylating enzyme is not completely specific for one enantiomer of ABA.

Another product from tomato shoots which contains [2-^{14}C] ABA was found to be an artefact of extraction when abscisyl-β-D-glucopyranoside was reacted with acetic anhydride in pyridine to make the tetraacetyl derivative. The product was contaminated with incompletely acetylated material but when the mixture was chromatographed it was found that the major product was ABA methyl ester. The residual acetic acid from the acetylation catalysed a *trans*-esterification between the glucose and the methanol in which the sample was dissolved to load onto the chromatogram. Subsequently, the supposed "metabolite A" (Milborrow, 1968) was identified as methyl ABA. It is formed by *trans*-esterification in the acidic methanol of the extraction medium.

IV. Biosynthesis

The structure of ABA suggests that it is assembled from three isoprene residues and Noddle and Robinson (1969) have recently shown that [^{14}C] mevalonic acid (MVA) is incorporated into ABA by avocado and tomato fruit. They supplied the labelled MVA to ripening fruit and extracted the ABA two days later. The radioactivity co-chromatographed with ABA in two solvent systems, with methyl ABA after methylation, and was divided between the two ABA-diols when these were formed by treatment with sodium borohydride in aqueous methanol. The ORD curves of all these derivatives matched those of authentic specimens.

This result showed that MVA is incorporated into ABA but there are two main, feasible, alternative routes by which this could take place. In the first of these the "direct synthesis pathway", it is postulated that a C-15 precursor is cyclized and then converted into ABA. In the second, the "carotenoid pathway", a cleavage of a carotenoid is postulated so that a terminal ring and six carbon atoms of the "backbone" are liberated as a C-15 moiety with the carbon skeleton of ABA.

Robinson and Ryback (1969) attempted to differentiate between the two pathways by investigating the stereochemistry of biosynthesis of the Δ^2 double-bond of ABA. When a *trans*-double-bond is formed in isoprene chains, between what had been the C-3 and C-4 atoms of MVA, then the (*S*) hydrogen on the C-4 of MVA is lost and the (*R*) is retained (Cornforth *et al.*, 1966). This is shown schematically in Fig. 2. With *cis*-double-bonds the (*R*) hydrogen is lost and the (*S*) is retained.

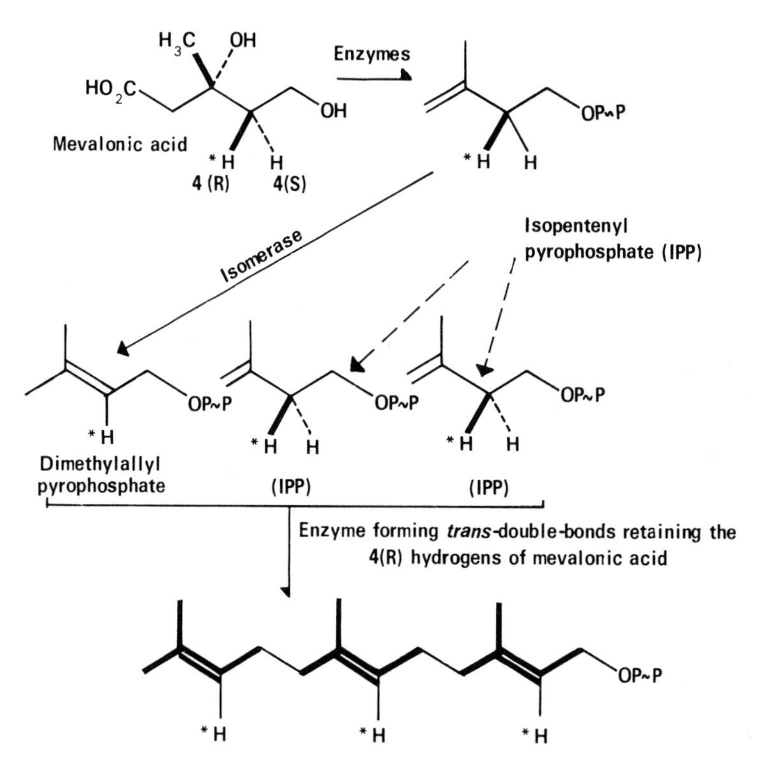

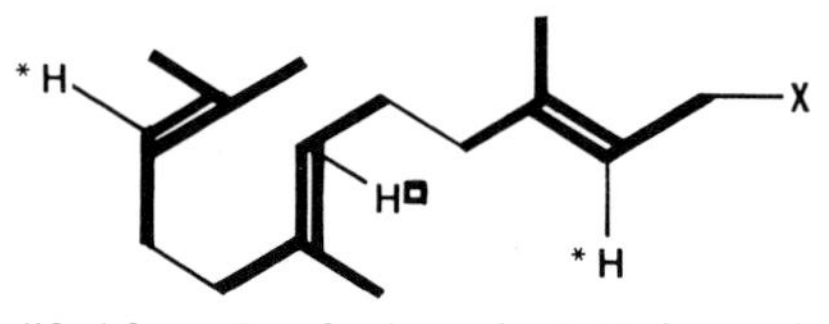

FIG. 2. Scheme (modified from Cornforth *et al.*, 1966) for the biosynthesis of *trans-,trans*-farnesyl pyrophosphate from mevalonic acid and showing the positions of the C-4 hydrogens of MVA. The expected positions of the C-4 hydrogens in a hypothetical C-15 precursor of ABA are also shown. A hypothetical C-15 precursor of ABA showing the positions where the 4(R) hydrogens of MVA would be expected to occur. The three MVA residues are drawn in heavy lines, the H□ would be expected to be lost during the biosynthesis to give a final ratio of two 4(R) hydrogen atoms to three MVA residues.

If the Δ^2 *cis*-double-bond of ABA were formed in the *cis* configuration then the (S) tritium would be retained, if the (R) tritium were retained it would show that the Δ^2 *cis*-double-bond had been formed originally in the *trans* configuration and isomerized at a later stage of biosynthesis.

Synthesis of a precursor with a Δ^2 *cis*-double-bond, that is the isomer in which the (S) tritium is retained on the C-2 of ABA, would exclude the carotenoid pathway because the double bonds of this part of the carotenoid "backbone" are *trans*. ABA biosynthesized by avocado fruit from (3 RS)-[2-^{14}C, (4R)-4-$^{3}H_1$] or (3 RS)-[2-^{14}C, (4S)-4-$^{3}H_1$] labelled MVA contained ratios of [^{14}C] to [^{3}H] of 3:1·93 and 3:0·05 respectively. 3:1·93 approximates to the expected ratio of three [^{14}C] to two (R) tritium atoms retained and one tritium atom was shown to be probably on C-2 of ABA. Therefore the Δ^2 *cis*-double-bond of ABA was probably formed in the *trans* configuration. Unfortunately this result could have been produced by biosynthesis via either pathway and so cannot give a definitive answer to the problem.

Taylor and Smith (1967) have noted the similarity of the structures of some xanthophylls with ABA, not only do these compounds contain the same carbon skeleton as ABA but the terminal rings of violaxanthin are oxygenated at the same positions as the ring of ABA. There is also some evidence that the action spectrum of some phototropic organs is similar to the absorption spectrum of carotenoids (Shropshire and Withrow, 1958) and Taylor (1968) has suggested that certain xanthophylls may act not only as photoreceptors but also as precursors of growth inhibitors. Taylor and Smith (1967) irradiated violaxanthin (16) with light from a tungsten lamp and found that a strongly growth-inhibitory, but neutral, material was produced.

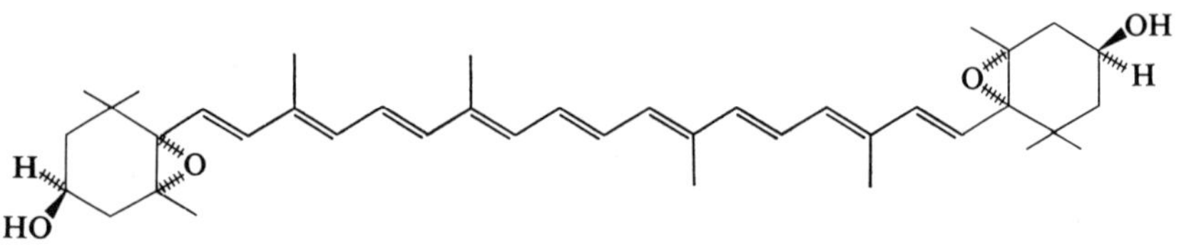

(16) Violaxanthin

It is impossible to prove a negative and show that in no tissue, under any conditions, is ABA formed by photolytic reactions. However, there is now good evidence that synthesis via a carotenoid stage does not occur in avocados.

These fruits form ABA rapidly from MVA in darkness showing that in this tissue the synthesis can occur without a photolytic step.

The second line of evidence comes from the unpublished work of Dr D. R. Robinson and I thank him for allowing me to quote his results here. He prepared [^{14}C] labelled phytoene, a compound considered to be an uncyclized precursor of carotenoids (Kushwaha *et al.*, 1969), and supplied it to avocado fruit together with [2-^{3}H] MVA. The ABA and carotenoids were isolated some hours later and although [^{14}C] labelling was found in carotenoids, showing that phytoene had entered the cells, there was no [^{14}C] in the ABA. On the other hand, a considerable amount of [^{3}H] label from MVA had been incorporated into the ABA. This result strongly indicates that MVA is incorporated directly into ABA and is not first assembled into a carotenoid.

The final line of evidence against the "carotenoid pathway" of ABA biosynthesis comes from work done by Bartlett *et al.*, (1969) on the absolute stereochemistry of violaxanthin. They tentatively allot the configuration shown in (16) to natural violaxanthin. If the absolute configurations of the epoxy-group of natural violaxanthin and the tertiary hydroxyl group of abscisic acid are confirmed then conversion of violaxanthin to ABA in plants, by a rearrangement of the epoxy group on the violaxanthin ring to give the tertiary hydroxyl and the Δ^2 double-bond of ABA, would require an inversion of configuration. If the neutral inhibitor produced from violaxanthin by *in vitro* photolytic reactions is converted to (–)-ABA by plants it would suppress growth because both (+) and (–) enantiomers of ABA are effective growth inhibitors (Milborrow, 1968).

Thus three distinct lines of evidence indicate that ABA is biosynthesized by a unique pathway and not by degradation of carotenoids.

V. Regulatory Mechanisms

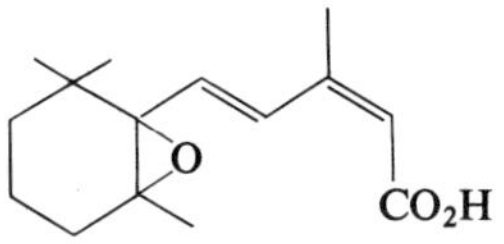

O

CO_2H

(17) 3-Methyl-5-(1,2-epoxy-2,6,6-trimethyl-1-cyclohexyl)-*cis*-, *trans*-2,4- pentadienoic acid

The amounts of hormones within plant tissues are of considerable interest to physiologists because correlations of physiological responses with changes in hormone concentrations provide a useful method for analysing the effects believed to be controlled by hormones *in vivo*. Abscisic acid, for instance, increases in concentration in the buds of trees in late summer when they become dormant. In spring, when the buds break, the concentrations fall and those of gibberellins and auxins rise (Phillips and Wareing, 1958). A knowledge of the mechanisms of biosynthesis and degradation of regulatory compounds is of considerable importance for understanding the way in which hormonal levels are maintained or adjusted. At present there is little knowledge of how the regulators are themselves regulated but a fascinating and important discovery by Wright and Hiron (1969) may provide an excellent test system for investigating this problem. They found that the concentration of ABA in wheat leaves rises 40-fold when the leaves wilt. We have found that if [^{3}H] mevalonic acid is supplied to a batch of leaves which is divided and half allowed to wilt, then these leaves incorporate more than ten times as much labelled MVA into ABA as do the unwilted controls. This treatment was later used to investigate the incorporation of (±)-[2-^{14}C] 3-methyl-5-(1,2-epoxy-2,6,6-trimethyl-1-cyclohexyl)-*cis-*,*trans*-2,4-pentadienoic acid (17) into ABA (Milborrow and Noddle, 1970). This compound has been synthesized by Anderson (1969) and Tamura and Nagao (1969) and reported to be a strong growth inhibitor, but no attribution of its inhibitory activity to its

TABLE IV
Incorporation of (±)-[2-^{3}H] MVA (93 mC/mM) and (±)-[2-^{14}C] 3-methyl-5-(1,2-epoxy-2,6,6-trimethyl-1-cyclohexyl)-*cis-*,*trans*-2,4-pentadienoic acid ("epoxide") (1·3 mC/mM) into ABA by wheat leaves during 6 h under wet and dry conditions in which the 60 g samples lost 29% of their original weights under dry conditions

Conditions	Compound supplied	[^{3}H] MVA incorporated into ABA expressed as ng ABA/kg fresh wt	[^{14}C] "epoxide" incorporated into ABA expressed as μg ABA/kg fresh wt
Wet	[^{3}H] MVA	0·181	—
Dry	[^{3}H] MVA	0·441	—
Wet	[^{3}H] MVA + [^{14}C] "epoxide"	0·202	0·148
Dry	[^{3}H] MVA + [^{14}C] "epoxide"	0·848	0·199
Wet	[^{14}C] "epoxide"	—	0·101
Dry	[^{14}C] "epoxide"	—	0·290

metabolism to ABA has been made. [2-^{14}C] Labelled material was synthesized for us by Ryback and Mallaby and was incorporated into ABA by tomato fruits. Two to three times as much of the "epoxide" was incorporated into ABA by wilted wheat leaves compared with unwilted controls (Table IV). These results indicate that the "epoxide", or a closely related compound derived from it, is on the biosynthetic pathway leading to ABA. The effects of wilting on the rate of its incorporation suggest that there is a site of regulation of biosynthesis between it and ABA. If a 1,2-epoxy-2-*cis*,4-*trans*-β-ionylidene acetic acid is found to be a naturally occurring intermediate, and is formed from a β-unsaturated ring, then ABA is an example of a route, albeit indirect, by which a β-unsaturated ring is converted to an α-unsaturated one (compare Williams *et al.*, 1967).

VI. SUMMARY

The concentrations of abscisic acid in plants have been determined by novel methods which make use of its special optical rotatory dispersion characteristics.

(±)-[2-^{14}C] Abscisic acid is converted into abscisyl-β-D-glucopyranoside, and to a more polar compound, metabolite C, which was identified as 6′-

hydroxymethyl abscisic acid. This compound rearranges to phaseic acid for which a new structure has been proposed.

The incorporation of MVA into abscisic acid is described and the evidence for the alternative routes of biosynthesis, the "direct synthesis pathway" and the "carotenoid pathway" is discussed.

REFERENCES

Anderson, M. (1969). British Patent 1,164,564.

Bartlett, L., Klyne, W., Mose, W. P., Scopes, P. M., Galasko, G., Mallams, A. K., Weedon, B. C. L., Szabolcs, J. and Tóth, G. (1969). *J. chem. Soc.* (C) 2527.

Cornforth, J. W., Cornforth, R. H., Donninger, C. and Popják, G. (1966a). *Proc. r. Soc.* B **163**, 492.

Cornforth, J. W., Milborrow, B. V. and Ryback, G. (1966b). *Nature, Lond.* **210**, 627.

Cornforth, J. W., Milborrow, B. V., Ryback, G., Rothwell, K. and Wain, R. L. (1966c). *Nature, Lond.* **211**, 742.

Cornforth, J. W., Draber, W., Milborrow, B. V. and Ryback, G. (1967). *Chem. Commun.* 114.

Isogai, Y., Okamoto, T., and Komoda, Y. (1967). *Chem. Pharm. Bull.* **15**, 1256.

Koshimizu, K., Fukui, H., Mitsui, T. and Ogawa, Y. (1966). *Agric. Biol. Chem.* **30**, 941.

Koshimizu, K., Inui, M., Fukui, H. and Mitsui, T. (1968). *Agric. biol. Chem.* **32**, 789.

Kushwaha, S. C., Subbarayan, C., Beeler, D. A. and Porter, J. W. (1969). *J. biol. Chem.* **244**, 3635.

MacMillan, J. and Pryce, R. J. (1968). *Chem. Commun.* 124.

Milborrow, B. V. (1967). *Planta* **76**, 93.

Milborrow, B. V. (1968). *In* "Biochemistry and Physiology of Plant Growth Substances" (F. Wightman and G. Setterfield, eds.), p. 1531, Runge Press, Ottawa.

Milborrow, B. V. (1969). *Chem. Commun.* 966.

Milborrow, B. V. (1970). *J. exp. Bot.* **21**, 117.

Milborrow, B. U. and Noddle, R. C. (1970). *Biochem. J.* **119**, 727.

Noddle, R. C. and Robinson, D. R. (1969). *Biochem. J.* **112**, 547.

Phillips, I. D. J. and Wareing, P. F. (1958). *J. exp. Bot.* **9**, 350.

Robinson, D. R. and Ryback, G. (1969). *Biochem. J.* **113**, 895.

Shropshire, W. and Withrow, R. B. (1958). *Pl. Physiol., Lancaster* **33**, 360.

Tamura, S. and Nagao, M. (1969). *Planta* **85**, 209.

Taylor, H. F. (1968). *In* "Plant Growth Regulators", S.C.I. Monograph No. 31, p. 22.

Taylor, H. F. and Smith, T. A. (1967). *Nature, Lond.* **215**, 1513.

Williams, R. J. H., Britton, G. and Goodwin, T. W. (1969). *Biochem. J.* **105**, 99.

Wright, S. T. C. and Hiron, R. W. P. (1969). *Nature, Lond.* **224**, 719.

CHAPTER 6

Diterpenes—The Gibberellins

J. MACMILLAN

Department of Chemistry, The University of Bristol, Bristol, England

I. Introduction

The gibberellins are diterpenoid acids. They were originally isolated from the fungus *Fusarium moniliforme* (*Gibberella fujikuroi*) as secondary metabolites which reproduced the overgrowth symptoms in rice seedlings, infected by this fungus. The gibberellins were subsequently found to affect many phases of plant growth and development and to occur in most, if not all, higher plants. Gibberellins or gibberellin-like substances have been detected in over 100 species of dicotyledons, in over 30 species of monocotyledons and in several conifers. There seems little doubt that the gibberellins are endogenous plant hormones.

The gibberellins can be subdivided into two groups—the C_{20}- and the C_{19}-gibberellins—with the basic structures shown in Fig. 1. Grove and Mulholland (1960) proposed a semi-systematic nomenclature based upon the trivial name, gibbane, for the basic ring system shown in Fig. 1. While the "gibbane" nomenclature system has proved very useful particularly for degradation products, the numbering system is confusing anent the generally accepted numbering of other diterpenes. Now that the stereochemistry and biogenesis of the gibberellins are known there are advantages in the recent

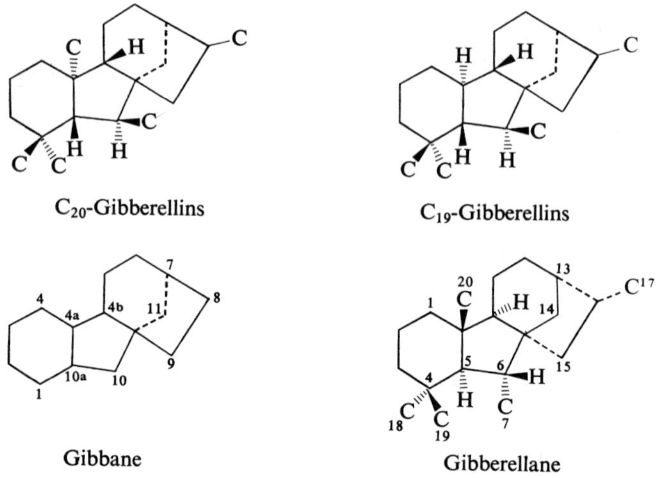

C_{20}-Gibberellins

C_{19}-Gibberellins

Gibbane

Gibberellane

FIG. 1. Nomenclature.

suggestion contained in proposals before the I.U.P.A.C. Commission on Organic Nomenclature (Rowe, 1968). These proposals contain the suggestion that the name, gibberellane, be used for the ring system, shown in Fig. 1, and numbered as shown and consistent with the numbering of other tetracyclic diterpenes. The gibberellins would all be enantiomers of gibberellanes. Thus the C_{20}-gibberellanes would be *ent*-gibberellanes and the C_{19}-gibberellins would be *ent*-20-norgibberellanes. The gibberellane system will be used in this chapter.

There is also a trivial nomenclature for the gibberellins in which they are accorded A-numbers (MacMillan and Takahashi, 1968). At the time of writing (April 1970) there are twenty-nine fully characterized gibberellins* which are commonly referred to as gibberellins A_1 to A_{29}. Their structures and sources of isolation are shown in Table I. Six conjugates of some of these gibberellins have been isolated (Table II).

II. BIOSYNTHESIS

The biosynthesis of the gibberellins has been comprehensively reviewed by Cross (1968) up to October 1966. This account deals mainly with work which has appeared since then.

The main biosynthetic pathway to the gibberellins from mevalonic acid, via geranylgeranyl pyrophosphate, *ent*-kaur-16-ene, and *ent*-gibberellan-7-al-19-oic acid, is shown in Fig. 2. While it has been demonstrated that C_{20}-gibberellins are converted *in vivo* into C_{19}-gibberellins, the precise level of oxidation at which these conversions occur is not known (see later). The small arrows in Fig. 2 indicate the known positions of oxidative decoration of

* Added in Proof: 36 Gibberellins are now known (March 1971).

TABLE I
Structures of the gibberellins and sources of isolation

Gibberellin	Structure	Source	Reference
A_1	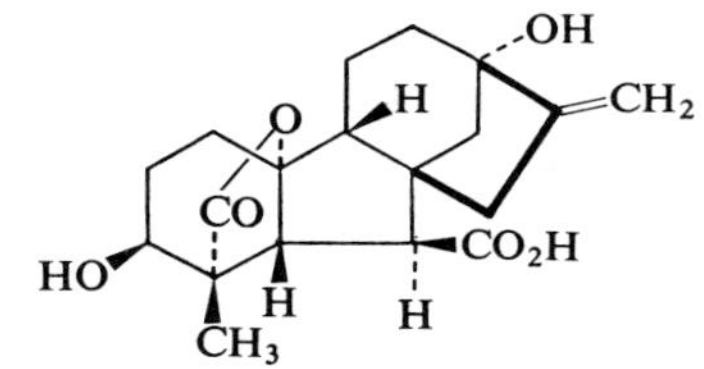	*G. fujikuroi* *Phaseolus multiflorus* *Phaseolus vulgaris* *Citrus reticulata* *Althaea rosea*	Stodola *et al.* (1955); Grove *et al.* (1958) MacMillan *et al.* (1960) West and Phinney (1959) Kawarada and Sumiki (1959) Harada and Nitsch (1968)
A_2	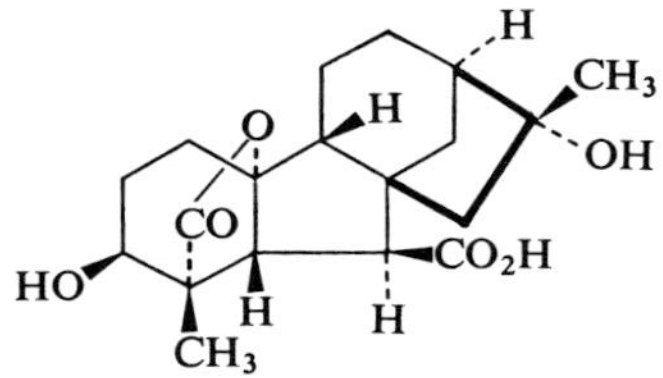	*G. fujikuroi*	Takahashi *et al.* (1955) Grove (1961)
A_3	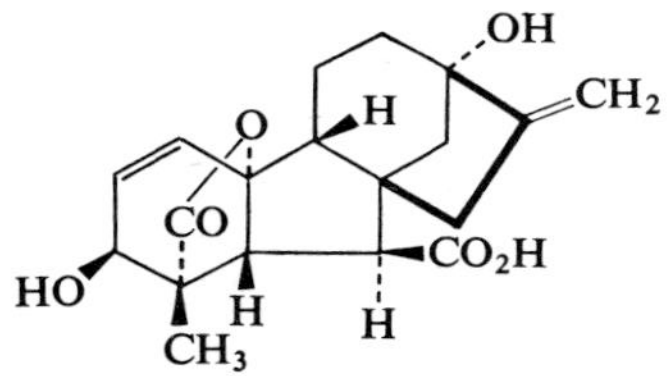	*G. fujikuroi* *Althaea rosea*	Stodola *et al.* (1955); Cross (1954) Harada and Nitsch (1967)

TABLE I—*continued*

Gibberellin	Structure	Source	Reference
A_4	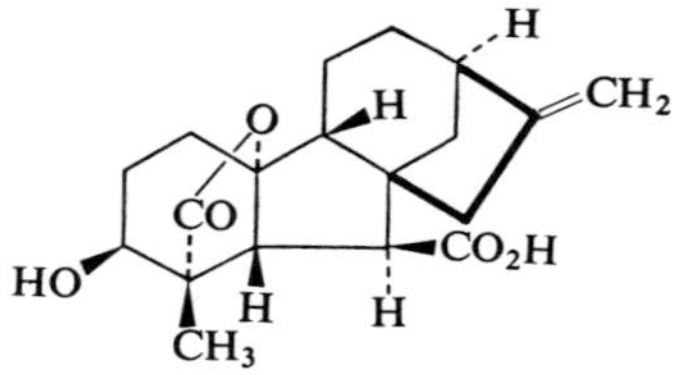	*G. fujikuroi*	Takahashi *et al.* (1959); Grove *et al.* (1960)
A_5	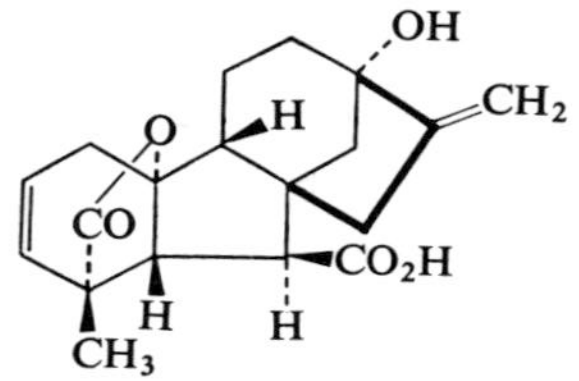	*P. multiflorus* *P. vulgaris* *Pharbitis nil*	MacMillan *et al.* (1960) West and Phinney (1959) Murofushi *et al.* (1968)
A_6	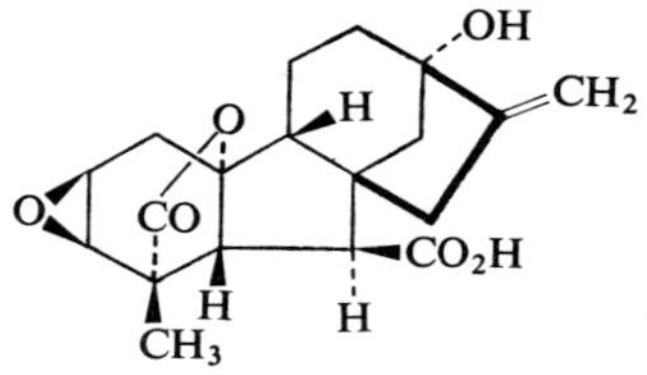	*P. multiflorus*	MacMillan *et al.* (1962)
A_7	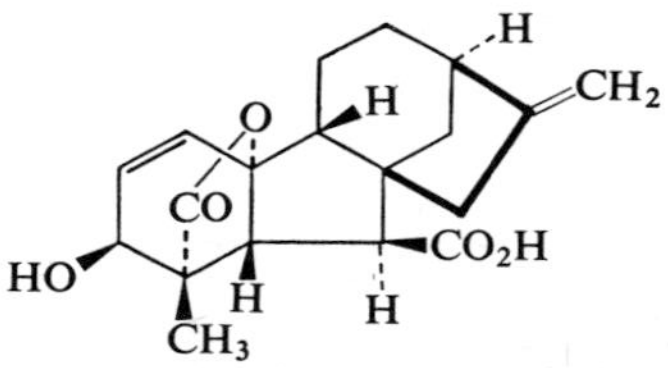	*G. fujikuroi*	Cross *et al.* (1962)

A_8	*P. multiflorus*	MacMillan *et al.* (1962) Sembdner *et al.* (1968)
A_9	*G. fujikuroi* *Althaea rosea*	Cross *et al.* (1962) Harada and Nitsch (1967)
A_{10}	*G. fujikuroi*	Hanson (1966)
A_{11}	*G. fujikuroi*	Brown *et al.* (1967)

TABLE I—*continued*

Gibberellin	Structure	Source	Reference
A_{12}	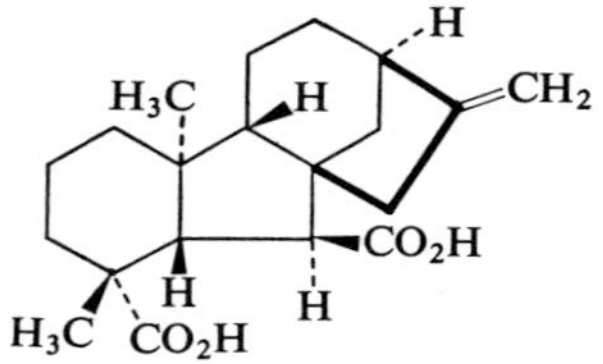 H; H_3C; H; CH_2; CO_2H; H; H; H_3C; CO_2H	*G. fujikuroi*	Cross and Norton (1965)
A_{13}	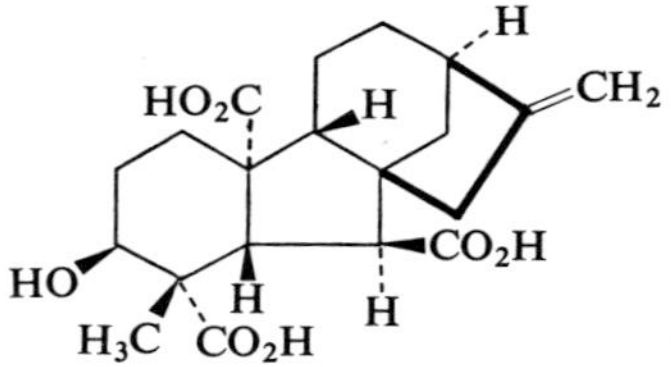 H; HO_2C; H; CH_2; CO_2H; HO; H; H; H_3C; CO_2H	*G. fujikuroi*	Galt (1965)
A_{14}	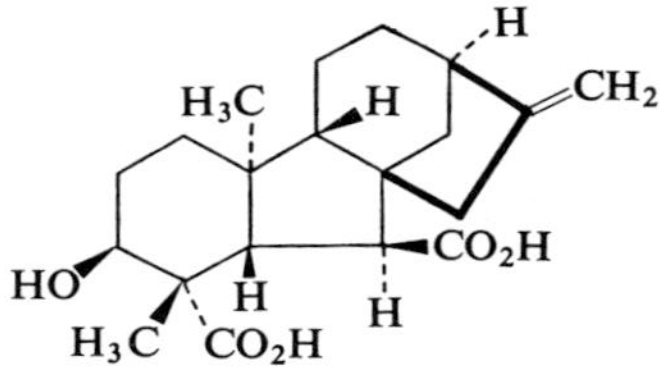 H; H_3C; H; CH_2; CO_2H; HO; H; H; H_3C; CO_2H	*G. fujikuroi*	Cross (1966)
A_{15}	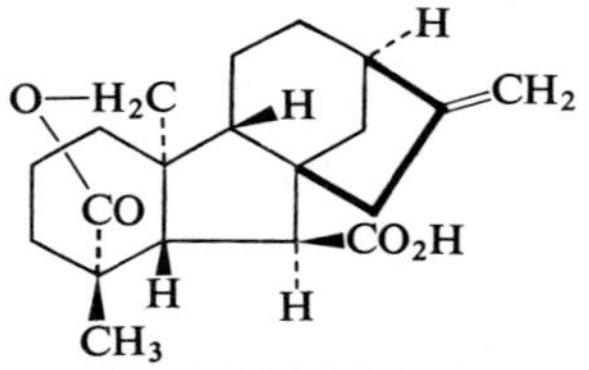 H; O—H_2C; H; CH_2; CO; CO_2H; H; H; CH_3	*G. fujikuroi*	Hanson (1967)

	Source	Reference
A_{16}	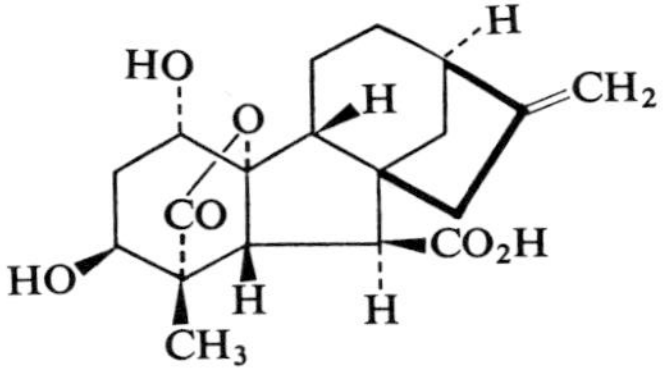 *G. fujikuroi*	Galt (1968)
A_{17}	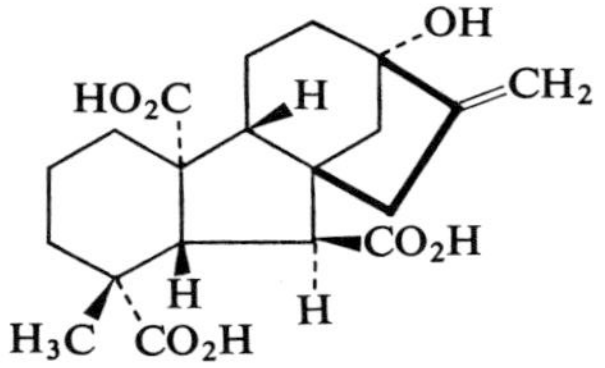 *P. multiflorus* *Calonyction aculeatum*	Pryce and MacMillan (1967) Takahashi *et al.* (unpublished work)
A_{18}	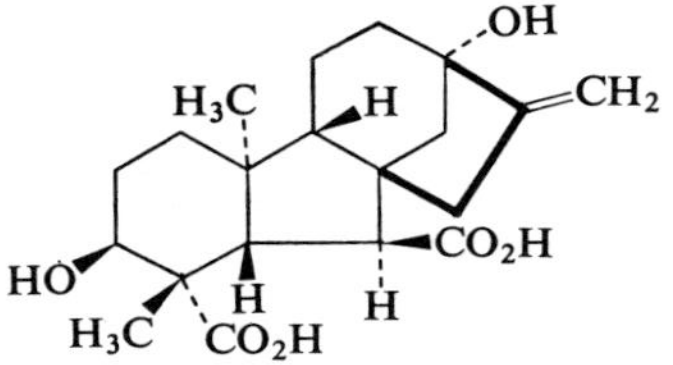 *Lupinus luteus*	Koshimizu *et al.* (1968b)
A_{19}	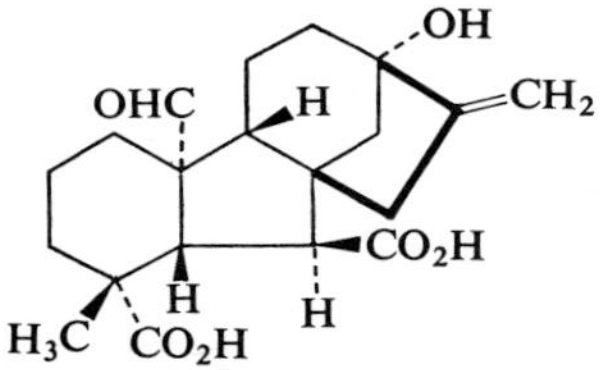 *Phyllostachys edulis*	Murofushi *et al.* (1966)

TABLE I—*continued*

Gibberellin	Structure	Source	Reference
A_{20}	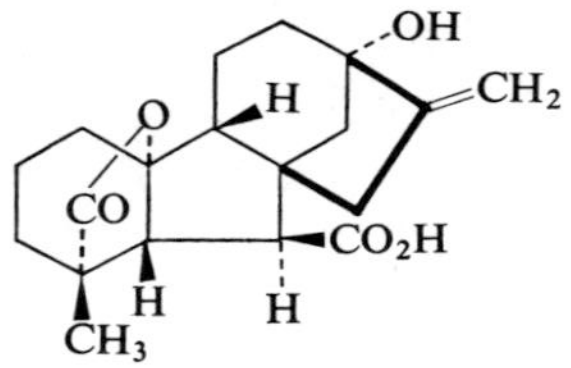 OH, CH_2, O, H, CO, CO_2H, H, H, CH_3	*Pharbitis nil*	Murofushi *et al.* (1968)
A_{21}	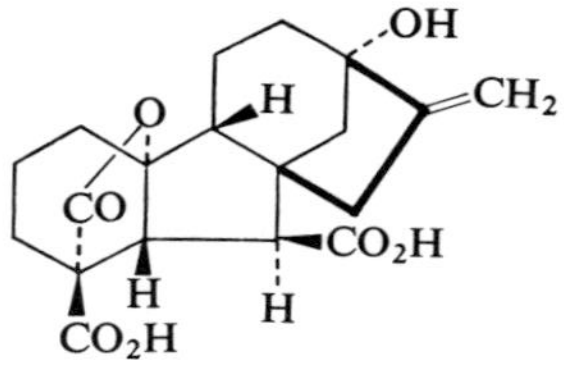 OH, CH_2, O, H, CO, CO_2H, H, H, CO_2H	*Canavalia gladiata*	Murofushi *et al.* (1969)
A_{22}	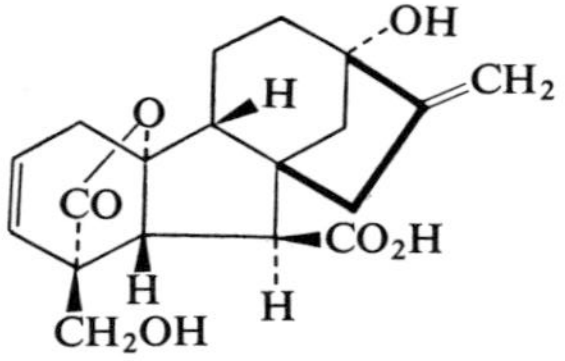 OH, CH_2, O, H, CO, CO_2H, H, H, CH_2OH	*Canavalia gladiata*	Murofushi *et al.* (1969)
A_{23}	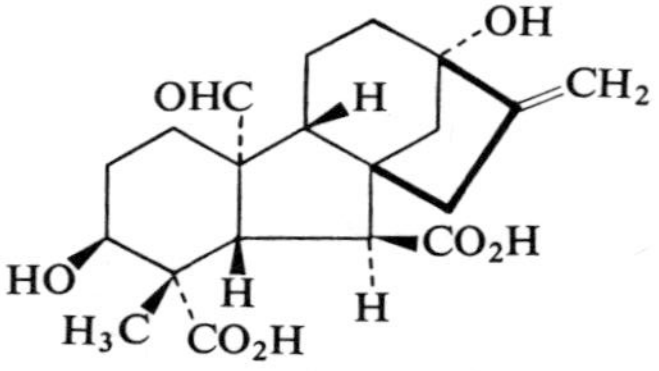 OH, OHC, H, CH_2, CO_2H, HO, H, H, H_3C, CO_2H	*Lupinus luteus*	Koshimizu *et al.* (1968a)

A_{24}	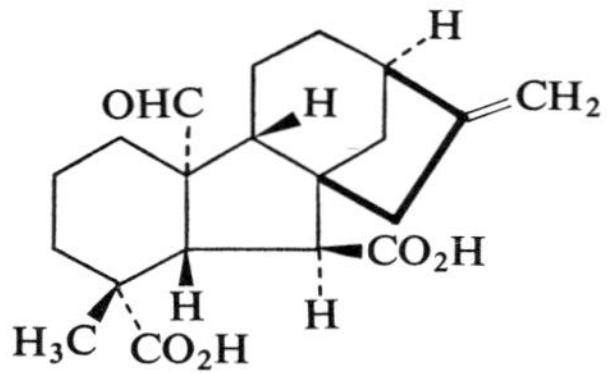 *G. fujikuroi*	Harrison *et al.* (1968)
A_{25}	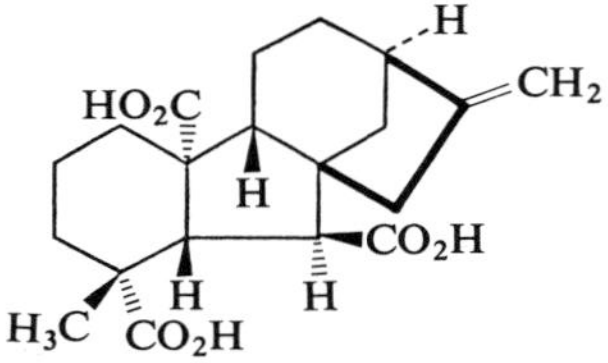 *G. fujikuroi*	Harrison and MacMillan (unpublished work)
A_{26}	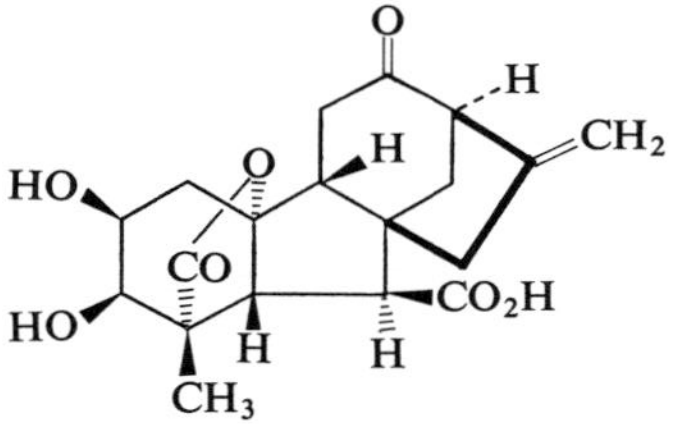 *Pharbitis nil*	Takahashi *et al.* (1969)
A_{27}	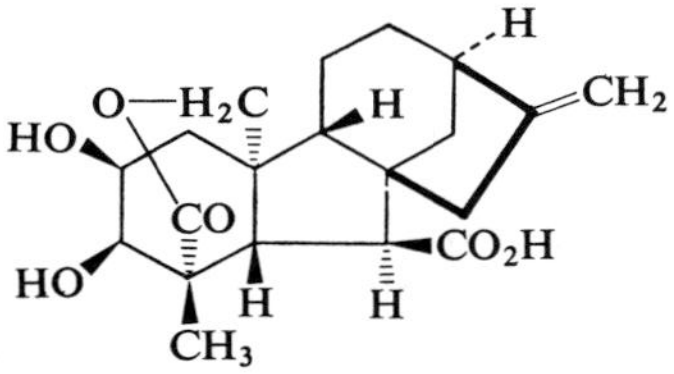 *Pharbitis nil*	Takahashi *et al.* (1969)

TABLE I—*continued*

Gibberellin	Structure	Source	Reference
A_{28}	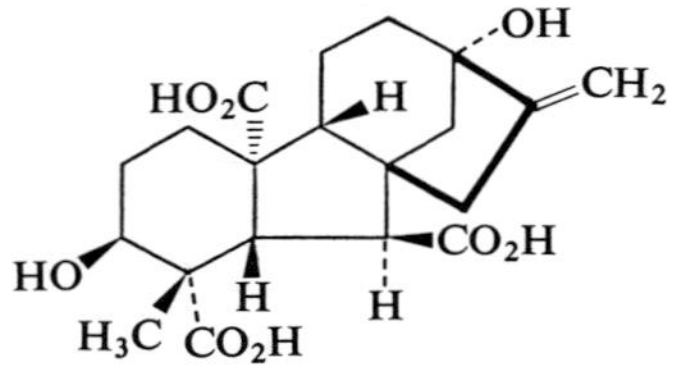	*Lupinus luteus*	Koshimizu (unpublished work)
A_{29}	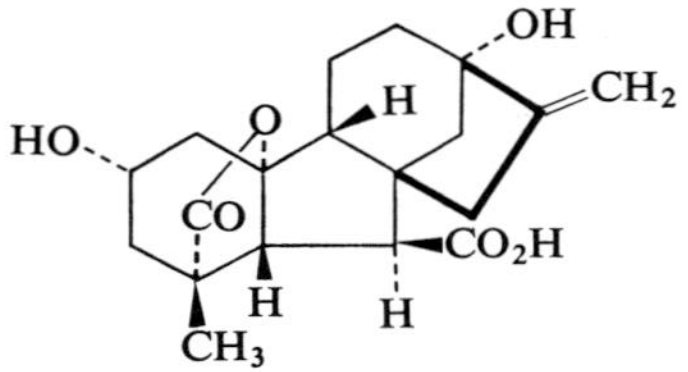	*Calonyction aculeatum*	Takahashi (unpublished work)

TABLE II
Conjugate gibberellins

3-*O*-β-Acetyl A_3	*Fusarium moniliforme*	Schreiber *et al.* (1966)
3-*O*-β-D-Glucopyranosyl A_3	*Pharbitis nil*	Yokota *et al.* (1969)
2-*O*-β-D-Glucopyranosyl A_8	*Phaseolus multiflorus*	Schreiber *et al.* (1967)
	Pharbitis nil	Yokota *et al.* (1969)
2-*O*-β-D-Glucopyranosyl A_{26}	*Pharbitis nil*	Yokota *et al.* (1969)
2-*O*-β-D-Glucopyranosyl A_{27}	*Pharbitis nil*	Yokota *et al.* (1969)
2-*O*-β-D-Glucopyranosyl A_{29}	*Pharbitis nil*	Yokota *et al.* (1970)

the basic *ent*-gibberellane and *ent*-20-norgibberellane structures. Gibberellin A_3 (gibberellic acid) is produced commercially by fermentation of *Fusarium moniliforme*. It is the most accessible gibberellin and consequently most of the biosynthetic studies on the gibberellins have been directed towards the production of gibberellin A_3 in cultures of this fungus.

The diterpenoid nature of gibberellin A_3 was first established by Birch *et al.* (1959). Degradation of gibberellin A_3, isolated from the fungus fed with [2-^{14}C]-mevalonic acid lactone or with [1-^{14}C]-acetic acid, was consistent with a biogenesis via geranylgeraniol and a tetracyclic diterpene. This early work showed that carbon-19 of gibberellin A_3, and therefore of the tetracyclic intermediate, was not labelled from [2-^{14}C]-mevalonic acid; thus ring A cyclization of the then presumed geranylgeraniol was stereospecific. Also carbon-7 of gibberellin A_3 was shown to be derived from carbon-2 of mevalonic

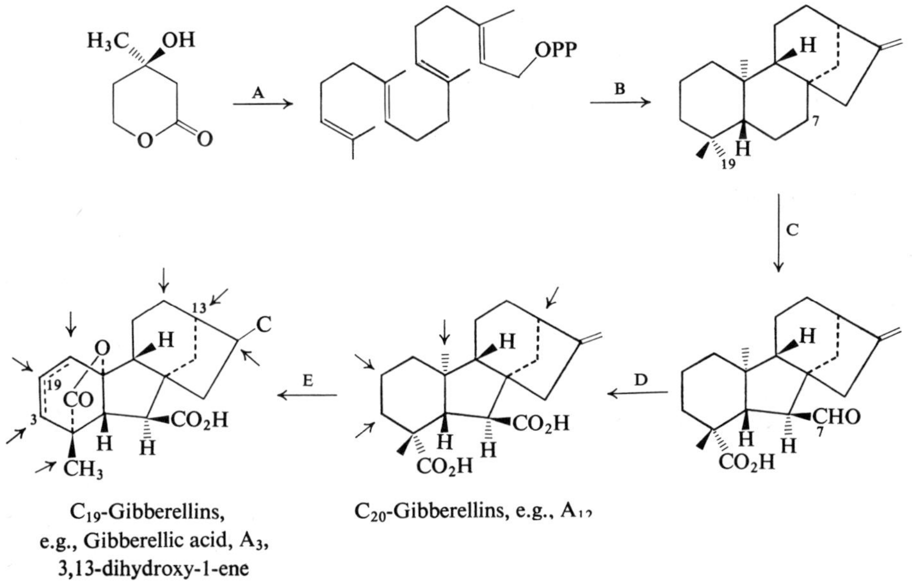

C_{19}-Gibberellins,
e.g., Gibberellic acid, A_3,
3,13-dihydroxy-1-ene

C_{20}-Gibberellins, e.g., A_{12}

FIG. 2. Main biosynthetic pathway.

acid and therefore from carbon-7 of the then presumed tetracyclic diterpenoid intermediate (see Fig. 2). Subsequently Cross *et al.* (1964) found that *ent*-kaur-16-ene both occurred in the fungus and was specifically incorporated into gibberellin A_3 in fungal cultures.

Detailed information on the various steps A–E, depicted in Fig. 2, is now available from tracer work with cultures of *F. moniliforme* and with cell-free enzyme preparations from both the fungus and plants.

A. STAGE A

Notable contributions on the early stages of the biosynthesis of gibberellins and related diterpenes have been made by West and his colleagues using cell-free enzyme systems. These workers prepared supernatant and particulate fractions by centrifugation at 105,000 × ***g*** of the filtered homogenate from the liquid endosperm of unripe seed of *Echinocystis macrocarpa* (wild cucumber). This supernatant fraction, with added adenosine triphosphate and Mg^{2+}, catalyses the conversion of [2-^{14}C]-mevalonic acid into *ent*-kaurene

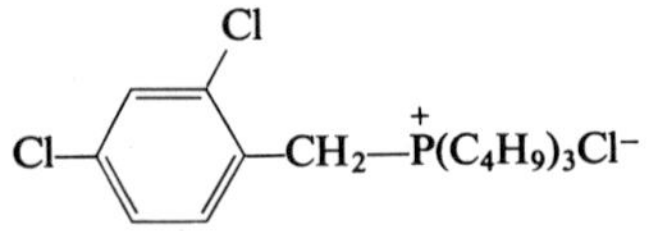

$Cl-C_6H_3(Cl)-CH_2-\overset{+}{P}(C_4H_9)_3Cl^-$

Phosfon D

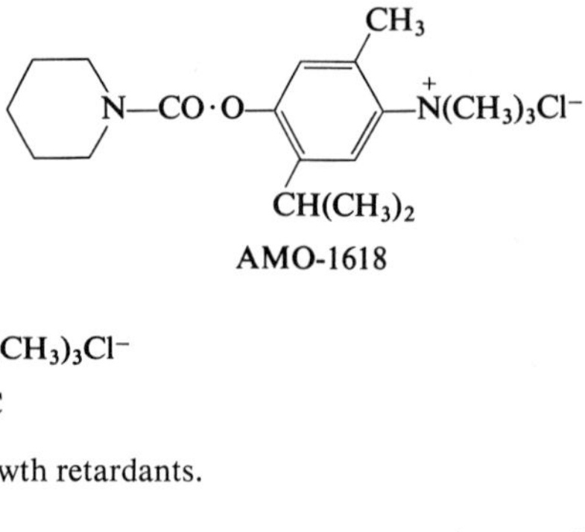

$C_5H_{10}N-CO\cdot O-C_6H_2(CH_3)(CH(CH_3)_2)-\overset{+}{N}(CH_3)_3Cl^-$

AMO-1618

$ClCH_2CH_2\overset{+}{N}(CH_3)_3Cl^-$

CCC

FIG. 3. Synthetic growth retardants.

(Graebe *et al.*, 1965). It is possible to stop this sequence at geranylgeranyl pyrophosphate by the addition of the synthetic plant growth retardant AMO-1618 (Fig. 3). Oster and West (1968) were able to isolate *trans*-[^{14}C]-geranylgeranyl pyrophosphate from such incubations which also produced small quantities of geranyl and farnesyl pyrophosphates. Evidence was also obtained for the formation of the expected intermediates mevalonic acid-5-phosphate, mevalonic acid-5-pyrophosphate, and isopentenyl pyrophosphate from homogenates incubated with [2-^{14}C]-mevalonic acid, ATP and $MgCl_2$ in the presence of 5 mM iodoacetamide.

B. STAGE B

In the absence of the growth retardant AMO 1618, the *Echinocystis* supernatant fraction converts [^{14}C]-geranylgeranyl pyrophosphate into [^{14}C]-*ent*-kaurene in 90% yield. This conversion provided the first direct experimental evidence for the long-standing postulate (see Ruzicka, 1959) that geranylgeraniol, or a derivative, was the acyclic precursor of cyclic diterpenes. The free geranylgeraniol which can accumulate as a result of phosphatase activity is not converted into *ent*-kaurene by the soluble enzyme preparation (Upper and West, 1967).

Since the early studies by West and his colleagues with the soluble enzyme preparation from *Echinocystis* endosperm, the conversion of mevalonic acid or geranylgeranyl pyrophosphate into *ent*-kaurene has been achieved using cell-free enzyme preparations from pea seed (Anderson and Moore, 1967), from pea fruit (Graebe, 1968) and from seed of *Cucurbito pepo* (Graebe, 1969). In the latter case the yield of [^{14}C]-kaurene is almost 40%, based upon 3*R*-[2-^{14}C]-mevalonic acid. While the enzyme systems from seed appear to produce *ent*-kaurene and no other diterpene hydrocarbons, Robinson and West (1970) have recently reported that cell-free enzyme preparations from

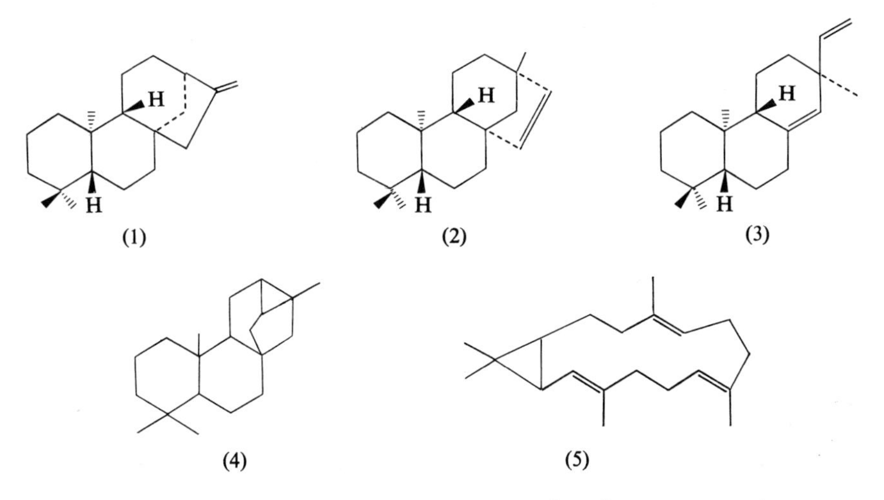

FIG. 4. Diterpenes from cell-free enzyme preparation of *Ricinus communis*.

two- or three-day old seedlings of *Ricinus communis* (castor bean) catalysed the conversion of mevalonic acid and geranylgeranyl pyrophosphate into several cyclic diterpenes (Fig. 4). *ent*-Kaurene (1), *ent*-beyerene (2), and *ent*-isopimaradiene [(+)-sandaracopimaradiene] (3) were fully characterized as products. *ent*-13,16-Cycloatisane (trachylobane) (4) was not completely characterized and insufficient was obtained for the determination of optical rotation. A fifth hydrocarbon, named casbene, was tentatively assigned the structure (5). Robinson and West (1970b) effected a partial purification of the enzymes responsible for the cyclization of geranylgeranyl pyrophosphate to these diterpenes. The results suggested that separate enzymic components participate in the cyclization to each of the diterpenes except for *ent*-kaurene (1) and 13,16-cycloatisane (4) where no evidence for the separation of activities was obtained. It is interesting to note that the stereochemistry at carbon-13 in the *ent*-isopimaradiene (3) is the opposite to that required for a tricyclic intermediate in the bio-conversion of geranylgeranyl pyrophosphate to *ent*-kaurene (see below).

In their studies with the soluble enzyme preparation from the endosperm of *Echinocystis* seed, West and his co-workers were unable to detect any intermediates between geranylgeranyl pyrophosphate and *ent*-kaurene.

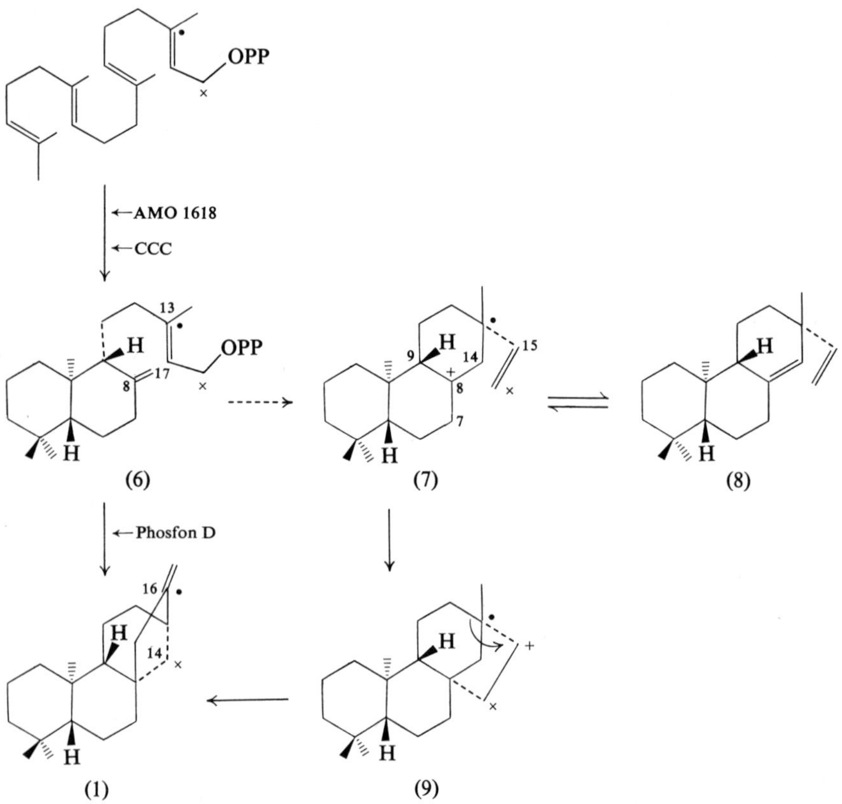

FIG. 5. Cyclization of geranylgeranyl pyrophosphate.

However, from a cell-free extract of mycelia of the fungus, *Fusarium moniliforme*, Shechter and West (1969) obtained a soluble enzyme system which formed the bicyclic diterpene, *ent*-labda-8(17),13-dien-15-ol (copalol) (see Fig. 5) and its pyrophosphate (6) when incubated with *trans*-[^{14}C]-geranylgeranyl pyrophosphate. The pyrophosphate (6) was converted into *ent*-kaurene by the fungal enzyme preparation and by the soluble enzyme system from *Echinocystis* seed. The pyrophosphate (6) was also transformed into the five diterpenoid hydrocarbons, shown in Fig. 4, by the enzyme preparation from *Ricinus communis*. *ent*-Labda-8(17),13-dien-15-yl pyrophosphate (6) is thus clearly established as an intermediate in the biosynthesis of these cyclic diterpenoid hydrocarbons and therefore of gibberellin A_3 via *ent*-kaurene. Indeed Hanson and White (1969a) have demonstrated that the pyrophosphate (6) is specifically incorporated into gibberellic acid by cultures of *F. moniliforme*.

No tricyclic intermediates between *ent*-labda-8(17),13-dien-15-yl pyrophosphate (6) and *ent*-kaurene have been isolated yet from cell-free enzyme systems or from cultures of *F. moniliforme*. The presumed pathway (6 → 7 → 9 → 1), shown in Fig. 5, is probably correct in essentials. Certainly the rearrangement of the tetracyclic carbonium ion is supported by the early work of Birch *et al.* (1959) who showed that the label from 1-[^{14}C]-acetate, and

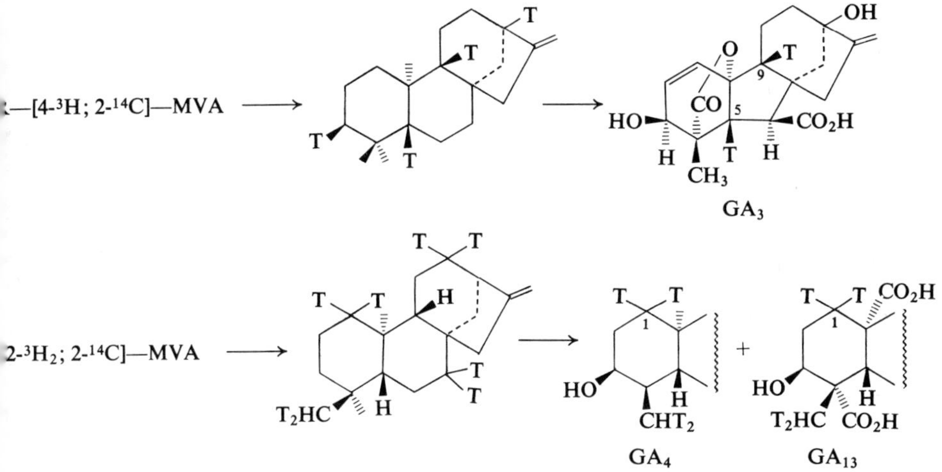

—[4-^{3}H; 2-^{14}C]—MVA ⟶ ⟶ GA$_3$

2-$^3H_2$; 2-^{14}C]—MVA ⟶ ⟶ GA$_4$ + GA$_{13}$

FIG. 6. Incorporation of [4R-^{3}H]- and [2-$^3H_2$]-mevalonic acid.

therefore from carbon-13 in the bicyclic intermediate (6), appears at position-16 in gibberellin A_3 and hence at position-16 in *ent*-kaurene (1) (see Fig. 5). Also Hanson and White (1969a) have demonstrated that a 15-[^{3}H]-atom in the bicyclic pyrophosphate (6) appears exclusively at carbon-14 in gibberellin A_3, and hence at the same position in *ent*-kaurene (1). The possible intermediacy of *ent*-pimara-7,15-, -8,15-, and 8(14),15-dienes (see 7 and 8, Fig. 5) has been investigated by Hanson and White (1969a) by feeding doubly-labelled mevalonic acid lactone to cultures of *F. moniliforme*. A more detailed discussion of this work is presented later. Suffice it to say here that the retention of [^{3}H]-label at position-9 in gibberellin A_3 derived from 4*R*-[4-^{3}H; 2-^{14}C] mevalonic acid (see Fig. 6) clearly precluded the intermediate formation of a Δ-8-double bond in a tricyclic intermediate (e.g. from 7, Fig. 5) between the bicyclic pyrophosphate (6) and *ent*-kaurene (1). Also the retention of all eight [^{3}H]-atoms from [2-$^3H_2$; 2-^{14}C]-mevalonic acid lactone in *ent*-kaurene (Fig. 6) clearly precluded an *ent*-pimara-7,15-diene as an intermediate. *ent*-[16-^{3}H]-Pimara-8(14),15-diene (8, Fig. 5) was specifically incorporated by fungal cultures into [14-^{3}H]-gibberellin A_3 but with low percentage incorporation (0·024%). As Hanson and White point out this low incorporation may be a measure of an unfavourable equilibrium between the added [16-^{3}H]-pimara-8(14),15-diene (8) and the corresponding 8-carbonium ion (7), bound perhaps to an enzyme.

The role of the synthetic plant growth retardants on the cyclization of geranylgeranyl pyrophosphate is of some interest since it has been suggested that these growth retardants may act by preventing the biosynthesis of plant gibberellins. The effect of a wide selection of these retardants on the overall conversion of geranylgeranyl pyrophosphate to *ent*-kaurene was investigated by Dennis *et al.* (1965). More recently Shechter and West (1969) have re-investigated the three growth retardants shown in Fig. 3 and found that Phosfon at 1 mM concentration inhibited the conversion of *ent*-labda-8,(17),13-dien-15-yl pyrophosphate (6) into *ent*-kaurene (1) (see Fig. 5) whereas the other two did not. Since CCC and AMO-1618 inhibit the overall conversion they must inhibit the cyclization of geranylgeranyl pyrophosphate to the bicyclic intermediate (6) (see Fig. 5).

C. STAGE C

The supernatant fraction obtained by centrifugation of the homogenate from *Echinocystis* endosperm takes mevalonic acid only as far as *ent*-kaurene. However, the particulate fraction catalyses the stepwise oxidation (Fig. 7) of *ent*-kaurene (1) to the 19-ol (10), the 19-al (11), the 19-oic acid (12) and finally the 7β-hydroxy acid (13) (Graebe *et al.*, 1965; West *et al.*, 1968). These reactions are probably of the mixed function oxidase type analogous to those found in mammalian systems (Dennis and West, 1967; Murphy and West, 1969). On

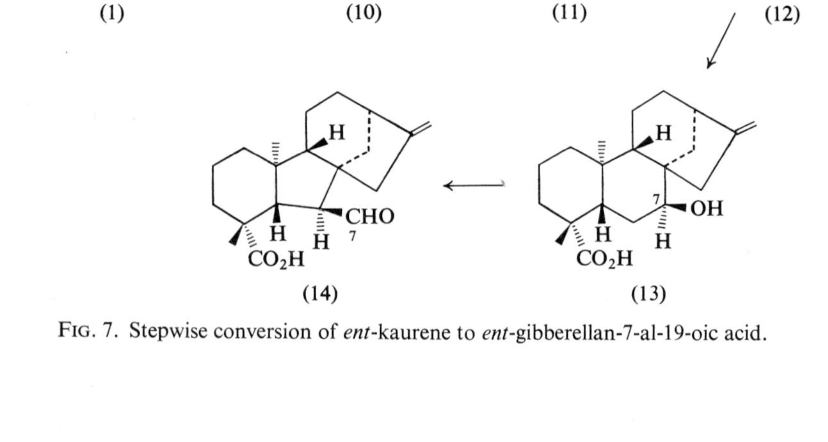

FIG. 7. Stepwise conversion of *ent*-kaurene to *ent*-gibberellan-7-al-19-oic acid.

present knowledge, the enzyme preparations from *Echinocystis* seed stop at the 7β-hydroxy acid (13) although *Echinocystis* seed appear to be a rich source of gibberellins (Phinney *et al.*, 1957). However, the oxidation products of *ent*-kaurene do appear to be intermediates in gibberellin biosynthesis. This is clearly so in *Fusarium moniliforme* where *ent*-kaurenoic acid (12) (Cavell and MacMillan, 1967) and the 7β-hydroxy acid (13) (Hanson and White, 1969b) occur and where all the oxidation products (10–13) are incorporated into gibberellin A_3 (Geissman *et al.*, 1966; Galt, 1965; Graebe *et al.*, 1965; Dennis and West, 1967; West *et al.*, 1968). The evidence for the intermediacy of *ent*-kaurene (1) and its oxidation products (10–13) in the biosynthesis of gibberellins in higher plants is circumstantial. *ent*-Kaurene (1) and its oxidation products (10), (12) and (13) show weak but significant biological activity in some bio-assay systems highly specific for gibberellins (Katsumi *et al.*, 1964). The inference is that these compounds are converted into gibberellins by the plants used in the bio-assay. Also *ent*-[17-^{14}C]-kauren-19-oic acid is incorporated into gibberellin-like substances by leaves of *Trifolium pratense* (Stoddart and Lang, 1968) and also by chloroplast preparations from *Brassica oleracea* (Stoddart, 1969). However, more precise information on the steps subsequent to the 7β-hydroxy-acid (13) comes entirely from tracer studies with cultures of *F. moniliforme*.

The next recognized intermediate is the aldehyde (14) shown in Fig. 7. This aldehyde was first shown by Cross *et al.* (1968) to be converted into gibberellin A_3 with a high percentage incorporation by fungal cultures. This aldehyde was subsequently shown to be present in fungal cultures together with *ent*-7α-hydroxykaurenoic acid (13) by isotope dilution analysis (Hanson and White, 1969b). Details of the ring contraction of *ent*-7α-hydroxykaurenoic acid (13) to the *ent*-gibberellan-7-al-19-oic acid (14) have been studied by Hanson and White (1969c) using doubly-labelled mevalonic acid in fungal cultures.

The [^{3}H]-labelling of *ent*-kaurene (15) which one would expect (Popják and Cornforth, 1966) from 2*R*-[2-^{3}H]-mevalonic acid is shown in Fig. 8. Hanson and White found that the two kaurenolides (17) and (19), shown in Fig. 8 and which co-occur with gibberellins in *F. moniliforme*, had the same ^{14}C:^{3}H ratio as the mevalonic acid substrate and therefore contained four [^{3}H]-atoms. One of these [^{3}H]-atoms was located at position-7 by oxidation of 7β-hydroxykaurenolide (*ent*-6β,7α-dihydroxykaur-16-en-19-oic acid 19,6-lactone) (17) to the 7-keto-derivative. This fact established that 7β-hydroxylation had occurred with retention of configuration (see 16). Gibberellin A_3 (see 18), derived from 2*R*-[2-^{3}H]-mevalonic acid lactone was found to possess a [^{3}H]:[^{14}C] ratio indicating the presence of three [^{3}H]-atoms, having lost as expected the [^{3}H]-label at position-7 in the intermediate (16) during ring contraction.

From 5*R*-[5-$^3H_1$]-mevalonic acid, the expected [^{3}H]-labelling of *ent*-7α-hydroxykaurenoic acid (see 20) is shown in Fig. 9. The 7β-hydroxykaurenolide,

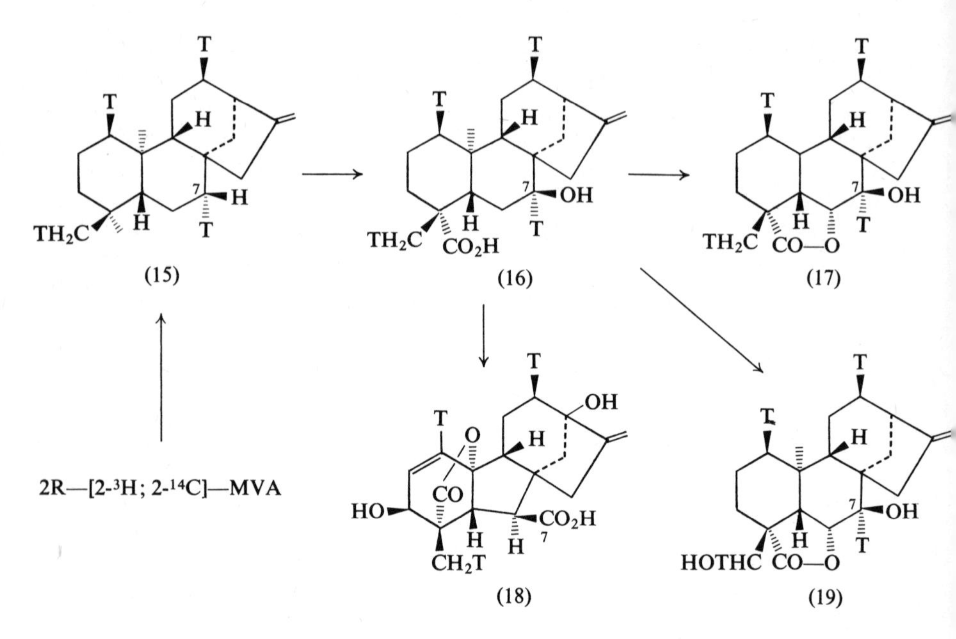

FIG. 8. Incorporation of [2R-³H]-mevalonic acid.

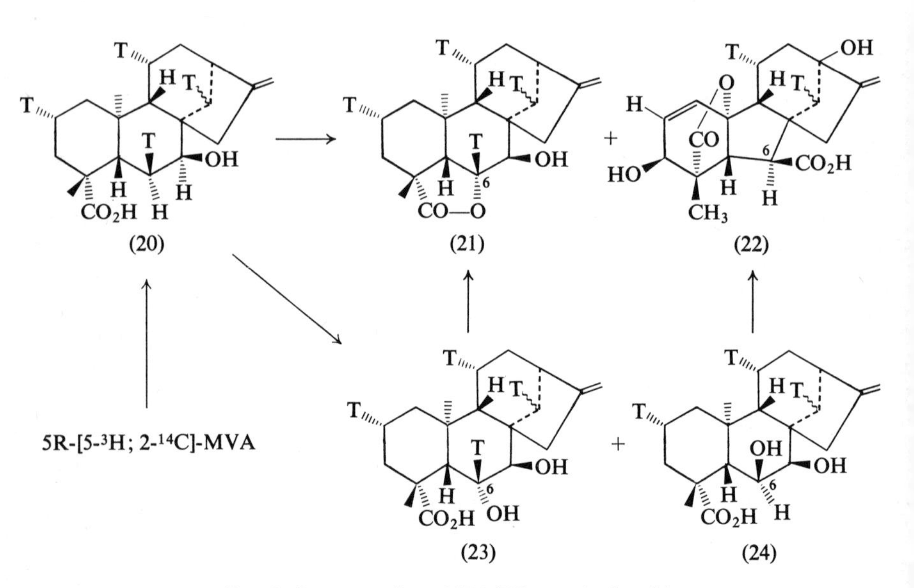

FIG. 9. Incorporation of [5*R*-³H]-mevalonic acid.

isolated from fungal cultures fed 5*R*-[5-$^3H_1$; 2-^{14}C]-mevalonic acid lactone, possessed a [^3H]:[^{14}C] ratio which indicated the retention of all four [^3H]-atoms and in particular one at position-6 (see 21). From the same experiment gibberellin A_3 was isolated with only two [^3H]-atoms and none at position-6 (see 22). From these results Hanson and White (1969c) deduced that formation of the kaurenolides involved hydroxylation at position-6 in the 7β-hydroxy acid (20) with retention of configuration and therefore of the 6-[^3H]-label (see 23). In contrast ring contraction to the gibberellins appears to require hydroxylation at the 6-position from the β-face with loss of the [^3H]-label (see 24). The latter conclusion was originally based upon the plausible assumption that a 6-hydroxylation also occurred with retention of configuration and that a 6α-[^3H]-atom would be retained during ring contraction. Retention of one of the two [^3H]-atoms at position-6 in gibberellin A_3, biosynthesized from [1-^3H]-geraniol, has now been demonstrated by J. R. Hanson (unpublished work).

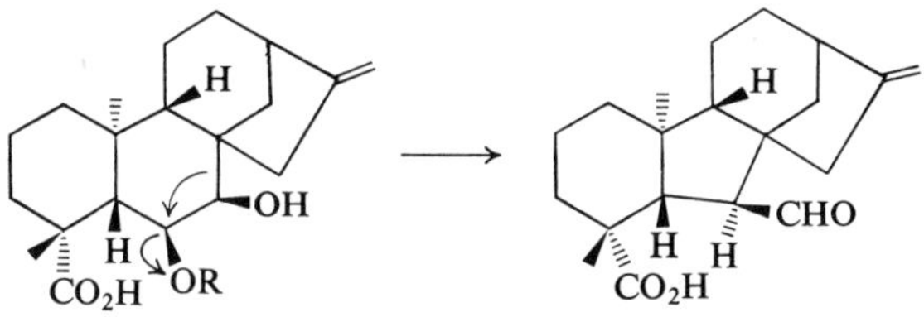

FIG. 10. Ring B contraction.

The ring contraction of *ent*-7α-hydroxykaurenoic acid to *ent*-gibberellan-7-al-19-oic acid would therefore appear to be initiated by β-oxygenation at position-6. The most likely process (Fig. 10) is the loss of a 6β-hydroxyl group, as perhaps a phosphate ester, with concerted rearrangement. *ent*-7β-Hydroxy kauren-19-oic acid, *ent*-kaur-6,16-dien-19-oic acid, and *ent*-6β,7α-dihydroxy-kaur-16-en-19-oic acid 19,6-lactone are not incorporated into gibberellin A_3 in fungal cultures (Cross *et al.*, 1968a).

D. STAGES D–E

There is little direct evidence on the further conversion of *ent*-gibberellan-7-al-19-oic acid to the C_{20}- and C_{19}-gibberellins. On the basis of comparative percentage incorporations of *ent*-gibberellan-7-al-19-oic acid, gibberellin A_{12} and the corresponding diol, and gibberellin A_{14} and the corresponding triol, Cross *et al.* (1968b) concluded that 3-hydroxylation of *ent*-gibberellan-7-al-19-oic acid occurs before oxidation of the aldehyde group. Indeed all further conversions may occur at the aldehyde oxidation level of carbon-7 and the gibberellins, such as gibberellins A_{12} and A_{14}, may be oxidation products of the true intermediates of gibberellin A_3 biosynthesis.

3β-Hydroxylation of *ent*-gibberellan-7-al-19-oic acid which has been shown to occur with retention of configuration (Hanson and White, 1969a), would thus appear to be a branching point (Fig. 11), the 3β-hydroxylated series leading to gibberellin A_4 via C_{20}-gibberellins and the non-hydroxylated branch leading to gibberellins A_9 and A_{10} perhaps via known C_{20}-gibberellins such as A_{15} (R = CH_2OH) and A_{24} (R = CHO). Certainly gibberellin A_9 does not appear to be a precursor of gibberellin A_3 in fungal cultures (Cross *et al.*, 1968a).

It is not known at which level of oxidation carbon-20 is lost in going from the C_{20}- to the C_{19}-gibberellins. From the double-labelling experiments, outlined in Fig. 6, Hanson and White (1969a) found that [^{3}H]-labels from

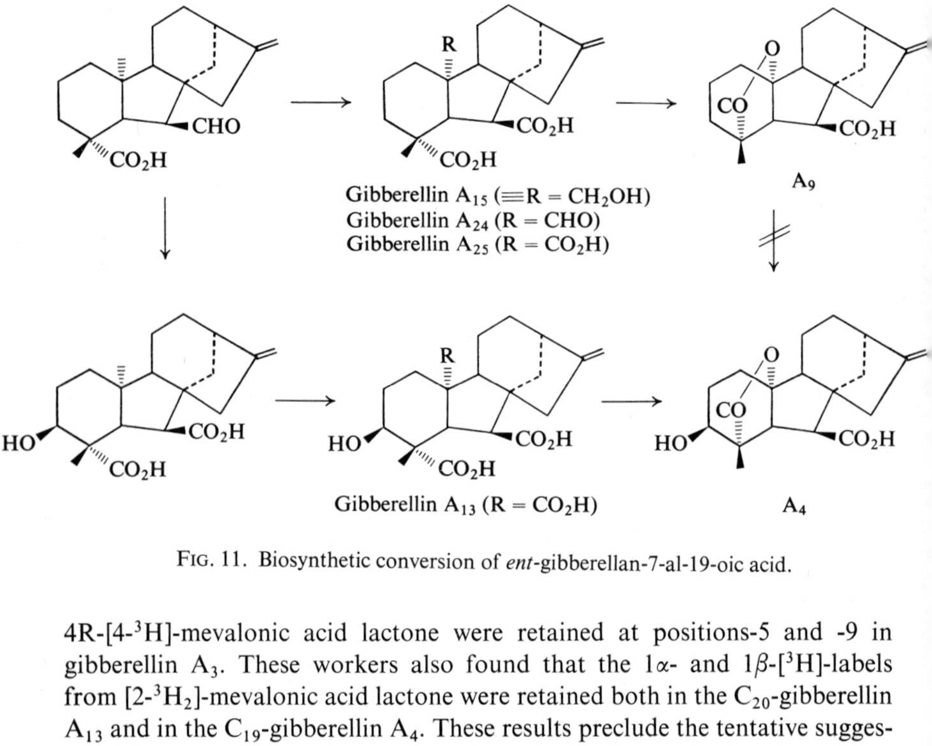

Gibberellin A_{15} (≡R = CH_2OH)
Gibberellin A_{24} (R = CHO)
Gibberellin A_{25} (R = CO_2H)

A_9

Gibberellin A_{13} (R = CO_2H)

A_4

FIG. 11. Biosynthetic conversion of *ent*-gibberellan-7-al-19-oic acid.

4R-[4-^{3}H]-mevalonic acid lactone were retained at positions-5 and -9 in gibberellin A_3. These workers also found that the 1α- and 1β-[^{3}H]-labels from [2-$^3H_2$]-mevalonic acid lactone were retained both in the C_{20}-gibberellin A_{13} and in the C_{19}-gibberellin A_4. These results preclude the tentative suggestion (Cross and Norton, 1966) that carbon-20 might be lost by decarboxylation of a β,γ-unsaturated acid. Gibberellin A_{13} (Fig. 11) is not converted into gibberellin A_3 by fungal cultures (Cross *et al.*, 1968b) and appears to be a biosynthetic dead-end from the way in which it accumulates in the culture medium. It may be, as Hanson and White (1969a) suggest, that loss of carbon-20 occurs at the aldehyde level by the biochemical equivalent of a Baeyer-Villiger oxidation. Thus gibberellin A_{24}, and not A_{25} (see Fig. 11), would be the precursor of gibberellin A_9.

13-Hydroxylation appears to be the final stage in the fungal biosynthesis of gibberellin A_3. In time-course studies Geissman *et al.* (1966) and Verbiscar *et al.* (1967) have shown (Fig. 12) that gibberellins A_4 and A_7 are produced earlier than gibberellins A_1 and A_3 and are converted into them. Spector and Phinney (1968) have identified two genes in *Fusarium moniliforme*. One (g_1) controls an early step in biosynthesis while the other (g_2) controls the 13-hydroxylation step. The latter step is an interesting one. The fungal enzyme appears to be influenced by temperature and pH since the proportions of gibberellin A_7 to gibberellin A_3 can be increased by raising the temperature (Kagawa *et al.*, 1965) and the pH (Cross and Hanson, 1964) of the culture medium.

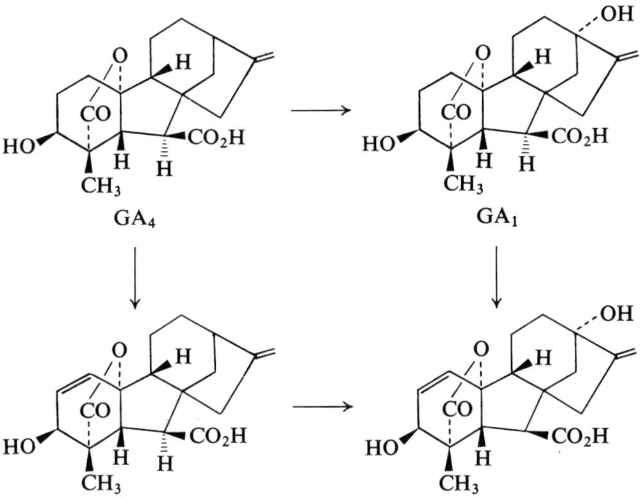

FIG. 12. 13-Hydroxylation.

The introduction of the Δ-1 double bond in gibberellin A_3 involves the loss of the 1α- and 2α-hydrogen atoms (Hanson and White, 1969c). This conclusion follows from the retention of a [^{3}H]-label at position-1 in gibberellin A_3 (18, Fig. 9) derived from 2*R*-[2-^{3}H]-mevalonic acid lactone (Fig. 8) and from the nonretention of a [^{3}H]-atom at position-2 in gibberellin A_3 (22, Fig. 9) derived from 5*R*-[5-^{3}H]-mevalonic acid lactone.

Apart from the studies with cell-free enzyme systems on the early biosynthetic steps, little is known of the biosynthesis of gibberellins in green plants. The indications are, however, that the main details are similar to those in *F. moniliforme*. One difference between the fungus and higher plants is that all fungal C_{20}-gibberellins have no 13-hydroxyl group whereas many, but not all, C_{20}-gibberellins from plants are 13-hydroxylated. Clearly 13-hydroxylation in plants can occur at an earlier stage in some plants. Of

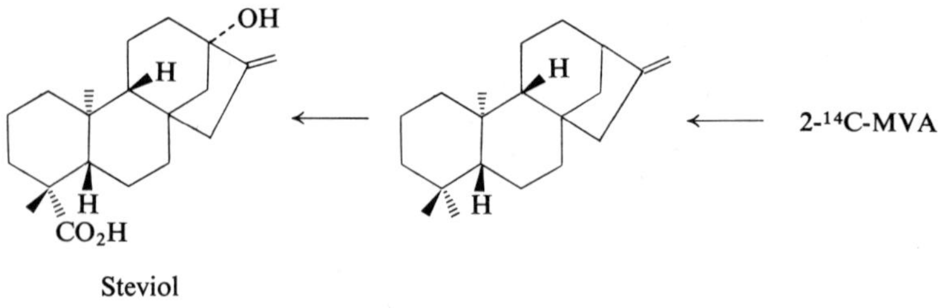

FIG. 13. Biosynthesis of Steviol.

interest in this connection is steviol (Fig. 13) which occurs in leaves of *Stevia rebandiana* as the glycoside, stevioside. Steviol shows low but significant gibberellin-like biological properties and may be a precursor of 13-hydroxylated C_{20}-gibberellins in plants. It has been shown to be produced from mevalonic acid lactone via *ent*-kaurene in *S. rebandiana* (Ruddat *et al.*, 1965; Bennett *et al.*, 1967; Hanson and White, 1968).

III. SITES OF BIOSYNTHESIS IN HIGHER PLANTS

There is indirect but convincing evidence for three distinct sites of biosynthesis of gibberellins in higher plants: (a) young apical leaves; (b) root tips; and (c) developing seed. Jones and Philips (1966) have shown that the sites of synthesis in young sunflower seedlings were in the young leaves in the apical bud, but not in the apical dome, and in the root tips. They compared by bioassay the amounts of gibberellin-like substances, which diffused into agar and which were extracted by aqueous methanol. The amounts which diffused were greater than the amounts obtained by solvent extraction, indicating synthesis at these sites during the period of diffusion. Supporting evidence comes from the use of the growth retardant CCC (Fig. 3) which greatly reduced the amounts of gibberellin-like substances which diffused from these sites (Jones and Philips, 1967). The older internodes and the subapical regions of the roots of young sunflower plants did not yield more gibberellin-like substances by diffusion than by extraction (Jones and Philips, 1966). Further indications of the biosynthesis of gibberellins in the root are that substantial amounts of gibberellin-like substances can be detected in the bleeding sap of plants over periods of several days (Sitton *et al.*, 1967). However, these levels decrease rapidly if the roots are treated with a synthetic growth retardant, such as those shown in Fig. 3 (Jones and Philips, 1966; Reid and Carr, 1967), or are subjected to water-logging (Reid *et al.*, 1969). Furthermore, incubation of root apices with 2-[^{14}C] mevalonic acid lactone yielded [^{14}C]-*ent*-kauren-19-ol, a known intermediate in gibberellin biosynthesis (Sitton *et al.*, 1967).

Developing seed contain relatively large amounts of gibberellins. Baldev *et al.* (1965) showed that the level of gibberellin-like substances in pea seed developing in excised fruit increased by a factor of 300; this increase and the inhibition of this increase by AMO 1618 (Fig. 3) clearly demonstrate that these pea seeds are actively biosynthesizing gibberellin-like substances.

IV. Identification and Characterization

Studies on the biosynthesis, sites of biosynthesis, metabolism and the biochemical function of gibberellins in higher plants require sensitive and definitive methods of analysis. The problem is made especially difficult by the large number of known gibberellins, by the potentially large number of minor structural variants yet to be discovered, and by the very low endogenous levels. Considerable use has been made of thin-layer chromatography (Sembdner *et al.*, 1962; MacMillan and Suter, 1963; Kagawa *et al.*, 1963; Cavell *et al.*, 1967) and, more recently, of gas chromatography (Ikekawa *et al.*, 1963; Cavell *et al.*, 1967). However, these methods provide, at best, only circumstantial evidence of identity even when authentic samples are available for direct comparison. Lately, combined gas chromatography-mass spectrometry (GC-MS) has provided a powerful method for the conclusive identification of known gibberellins (MacMillan *et al.*, 1967; Pryce *et al.*, 1967; MacMillan and Pryce, 1968) and for the characterization of new ones (Pryce and MacMillan, 1967). Reference spectra have been published (Binks *et al.*, 1969) for the methyl esters of gibberellins A_1 to A_{24} and of their methyl ester trimethylsilyl ethers where applicable. Reference spectra of these derivatives of the later gibberellins A_{25} to A_{29} are also available (P. Gaskin and J. MacMillan, unpublished work). A further recent refinement is the development of a system by Binks *et al.* (1970) for on-line processing of GC-MS spectra by a small computer which can also locate and compare the data, processed as a line diagram, with reference spectra stored in a reference library on magnetic tape. The processed line diagrams can be photographed from an oscilloscope display, presented as a line-diagram drawn by a slave flat-bed pen-recorder and illustrated in Fig. 14, or printed out in tabular form by a tele-type accessory.

V. Biochemical Function

The biochemical function can be considered at three levels—the whole plant, the cellular, and the molecular levels. The effects on the whole plant are well-documented (Brian *et al.*, 1960; Phinney and West, 1960; Paleg, 1965). These effects are frequently quite spectacular, for example in the restoration of the normal growth of the wild type in dwarf cultivars or in a single gene dwarf mutants and in the induction of flowering stalks in certain biennial plants. At the cellular level it has been established that internode lengthening induced by gibberellins is due to cell extension or cell division or both.

At the molecular level much work has been published on the changes in various cell constituents, notably carbohydrates and enzymes. Gibberellin A_3 apparently has the ability to enhance the activity of a large number of

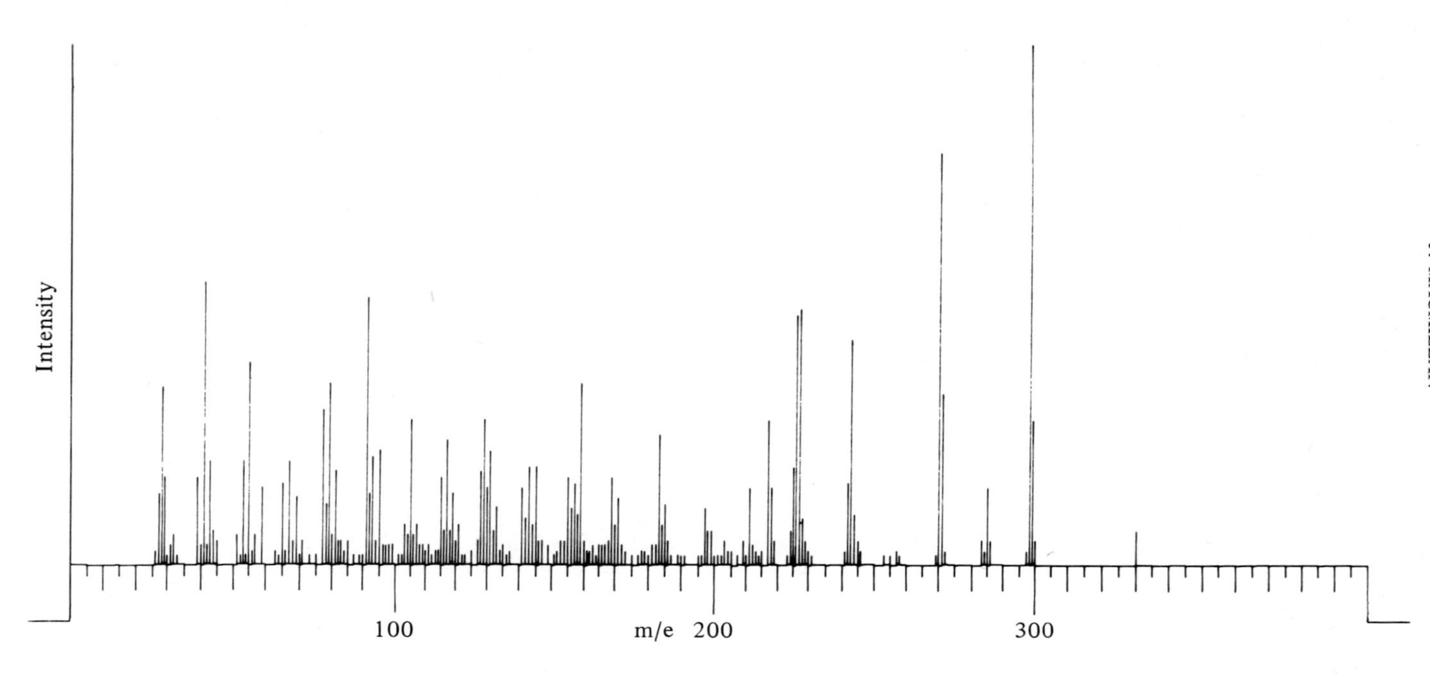

FIG. 14. Mass spectrum of gibberellin A_{13} methyl ester: computer display.

enzymes. For example, the specific activity of the primary photosynthetic enzyme, ribulose-1,5-diphosphate carboxylase, can be increased by raising the endogenous gibberellin level in leaves of *Brassica oleracea* (Treharne and Stoddart, 1968). Also, *de novo* synthesis of a fructose-1,6-diphosphatase is induced in the endosperm of germinating seed of *Ricinus communis* by gibberellin A_3 (Scala *et al.*, 1969). Gibberellin A_3 has been reported to stimulate α-amylase activity in wheat, in wild oats and in isolated aleurone layers from barley grain. In the latter case, this stimulation has been shown to be due to *de novo* synthesis of the enzyme (Filner and Varner, 1967). Other enzyme activities produced by the aleurone layers in response to gibberellin A_3 are proteases, ribonucleases, β-glucanases and pentanases (see Varner and Johri, 1968 for leading references).

It would seem, however, that these effects on enzymes are not direct effects. There is a lag period of about 8 h between application of the hormone and the onset of enzyme synthesis. During this period there are quite substantial changes occurring in the cell, both structurally as shown by electron microscopy (Jones, 1969) and biochemically. Pollard (1969) found, for example, that the first effect of gibberellin A_3 on isolated barley aleurone layers was the secretion of soluble carbohydrates some of which appeared to be a glucan with some β-1,3-linkages. Then after increased oxygen uptake, an increased secretion was observed of ATPase, GTPase, phytase, phosphomonoesterase, phosphodiesterase, inorganic phosphate, carbohydrases other than amylase, peroxidase, and finally amylase. The effect on some enzymes was on secretion only and on others on both synthesis and secretion.

It has been suggested that the regulatory action of gibberellin A_3 involves formation of RNA molecules specific for the synthesis of amylase and other hydrolases. Consistent with this idea is the report that RNA from germinating seed promoted α-amylase production in isolated aleurone layers of barley (Zolotov and Lesham, 1968). Also in sugar-cane slices evidence has been obtained that gibberellin A_3 affects the transcriptive, and not the translative systems involved in protein synthesis (Glasziou *et al.*, 1968). In isolated nuclei from pea seedlings, gibberellin A_3 has no effect on RNA synthesis if purified nuclei were pretreated with the hormone immediately before or during incubation. There was, however, a 60–100% increase in the incorporation of all four nucleotides when gibberellin A_3 at a concentration of 10^{-8} M was included in the medium used in the isolation of the nuclei (Johri and Varner, 1968). The synthesized RNA had a higher average molecular weight than the RNA from controls and also had a different nearest neighbour frequency. Thus gibberellin A_3 can clearly modify RNA synthesis in these isolated pea nuclei.

In experiments designed to identify the target tissue of gibberellins Musgrave *et al.* (1969) found that gibberellins A_1 and A_5 which elicit a growth response in pea seedlings accumulate in, and were retained by, the apical part of the stem of young pea seedlings but not by the basal part. Gibberellins with low biological activity were not taken up preferentially by the apical region of the stem. This accumulation and retention of gibberellins A_1 and A_5 in the hormone-responsive tissue and not in the non-responsive region may be due to binding of the hormone to specific gibberellin receptors. However, attempts to isolate the gibberellin-protein complex have so far failed (H. Kende, unpublished results).

REFERENCES

Anderson, J. D. and Moore, T. C. (1967). *Pl. Physiol.* **42**, 1527–1534.

Baldev, B., Lang, A. and Agatep, A. O. (1965). *Science, N.Y.* **147**, 155–157.

Bennett, R. D., Lieber, E. R. and Heftmann, E. (1967). *Phytochemistry* **6**, 1107–1110.

Binks, R., MacMillan, J. and Pryce, R. J. (1969). *Phytochemistry* **8**, 271–284.

Binks, R., Cleaver, R., Littler, J. S. and MacMillan, J. (1971). *Chemistry in Britain* **7**, 8–12.

Birch, A. J., Richards, R. W., Smith, H., Harris, A. and Whalley, W. B. (1959). *Tetrahedron* **7**, 241–251.

Brian, P. W., Grove, J. F. and MacMillan, J. (1960). *Fortschr. Chem. Org. Naturstoffe* **18**, 350–433.

Brown, J. C., Cross, B. E. and Hanson, J. R. (1967). *Tetrahedron* **23**, 4095–4103.

Cavell, B. D. and MacMillan, J. (1967). *Phytochemistry* **6**, 1151–1154.

Cavell, B. D., MacMillan, J., Pryce, R. J. and Sheppard, A. C. (1967). *Phytochemistry* **6**, 867–874.

Cross, B. E. (1954). *J. chem. Soc.* 4670–4676.

Cross, B. E. (1966). *J. chem. Soc.* (C) 501–504.

Cross, B. E. (1968). *In* "Progress in Phytochemistry" (L. Reinhold and Y. Liwschitz, eds.), Vol. 1, pp. 195–222, Wiley and Sons, New York.

Cross, B. E. and Hanson, J. R. (1964). *Chem. Abstr.* **74**, 5742.

Cross, B. E. and Norton, K. (1965). *J. chem. Soc.* 1570–1572.

Cross, B. E. and Norton, K. (1966). *Tetrahedron Letters* 6003–6007.

Cross, B. E., Galt, R. H. B. and Hanson, J. R. (1962). *Tetrahedron* **18**, 451–459.

Cross, B. E., Galt, R. H. B. and Hanson, J. R. (1964). *J. chem. Soc.* 295–300.

Cross, B. E., Galt, R. H. B. and Norton, K. (1968a). *Tetrahedron* **24**, 231–237.

Cross, B. E., Norton, K. and Stewart, J. C. (1968b). *J. chem. Soc.* (C) 1054–1063.

Dennis, D. T. and West, C. A. (1967). *J. biol. Chem.* **242**, 3293–3300.

Dennis, D. T., Upper, C. D. and West, C. A. (1965). *Pl. Physiol.* **40**, 948–952.

Filner, P. F. and Varner, J. E. (1967). *Proc. Natn. Acad. Sci. U.S.A.* **58**, 1520–1526.

Galt, R. H. B. (1965). *J. chem. Soc.* 3143–3151.

Galt, R. H. B. (1968). *Tetrahedron* **24**, 1337–1339.

Geissman, T. A., Verbiscar, A. J., Phinney, B. O. and Cragg, G. (1966). *Phytochemistry* **5**, 933–947.

Glasziou, K. T., Gaylor, K. R. and Waldron, J. C. (1968). *In* "Biochemistry and Physiology of Plant Growth Substances" (F. Wightman and G. Setterfield, eds.), pp. 433–442, Runge Press, Toronto.

Graebe, J. E. (1968). *Phytochemistry* **7**, 2003–2020.

Graebe, J. E. (1969). *Planta* **85**, 171–174.

Graebe, J. E., Dennis, D. T., Upper, C. D. and West, C. A. (1965). *J. biol. Chem.* **240**, 1847–1854.

Grove, J. F. (1961). *J. chem. Soc.* 3545–3547.

Grove, J. F. and Mulholland, T. P. C. (1960). *J. chem. Soc.* 3007–3022.

Grove, J. F., Jeffs, P. W. and Mulholland, T. P. C. (1958). *J. chem. Soc.* 1236–1240.

Grove, J. F., MacMillan, J., Mulholland, T. P. C. and Turner, R. B. (1960). *J. chem. Soc.* 3049-3057.
Hanson, J. R. (1966). *Tetrahedron* **22**, 701–703.
Hanson, J. R. (1967). *Tetrahedron* **23**, 733–735.
Hanson, J. R. and White, A. F. (1968). *Phytochemistry* **7**, 595–597.
Hanson, J. R. and White, A. F. (1969a). *J. chem. Soc.* (C), 981–985.
Hanson, J. R. and White, A. F. (1969b). *Chem. Commun.* 410–411.
Hanson, J. R. and White, A. F. (1969c). *Chem. Commun.* 1071–1072.
Harada, H. and Nitsch, J. P. (1967). *Phytochemistry* **6**, 1695–1703.
Harrison, D. M., MacMillan, J. and Galt, R. H. B. (1968). *Tetrahedron Letters* 3137–3139.
Ikekawa, N., Kagawa, T. and Sumiki, Y. (1963). *Proc. Japan Acad.* **39**, 507–512.
Johri, M. M. and Varner, J. E. (1968). *Proc. Natn. Acad. Sci. U.S.A.* **59**, 269–276.
Jones, R. L. (1969). *Planta* **88**, 73–86; and earlier papers.
Jones, R. L. and Philips, I. D. L. (1966). *Pl. Physiol.* **41**, 1381–1386.
Jones, R. L. and Philips, I. D. L. (1967). *Planta* **72**, 53–59.
Kagawa, T., Fukinbaba, T. and Sumiki, Y. (1963). *Agric. Biol. Chem.* (*Tokyo*) **27**, 598–599.
Kagawa, T., Fukinbaba, T. and Sumiki, Y. (1965). *Agric. Biol. Chem.* (*Tokyo*) **29**, 285–291.
Katsumi, M., Phinney, B. O., Jefferies, P. R. and Henrick, C. A. (1964). *Science, N.Y.* **144**, 849–850.
Kawarada, A. and Sumiki, Y. (1959). *Bull. Agric. Chem. Soc. Japan* **23**, 343.
Koshimizu, K., Fukui, H., Inui, M., Ogawa, Y. and Mitsui, T. (1968a). *Tetrahedron Letters* 1143–1147.
Koshimizu, K., Fukui, H., Kusaki, T., Mitsui, T. and Ogawa, Y. (1968b). *Agric. Biol. Chem.* (*Tokyo*) **32**, 1135–1140.
MacMillan, J. and Pryce, R. J. (1968). *Tetrahedron Letters* 1537–1542.
MacMillan, J. and Suter, P. J. (1963). *Nature, Lond.* **197**, 790.
MacMillan, J. and Takahashi, N. (1968). *Nature, Lond.* **217**, 170–171.
MacMillan, J., Seaton, J. C. and Suter, P. J. (1960). *Tetrahedron* **11**, 60–66.
MacMillan, J., Seaton, J. C. and Suter, P. J. (1962). *Tetrahedron* **18**, 349–355.
MacMillan, J., Pryce, R. J., Eglinton, G. and McCormick, A. (1967). *Tetrahedron Letters* 2241–2243.
Murofushi, N., Iriuchijima, S., Takahashi, N., Tamura, S., Kato, J., Wada, Y., Watanabe, E. and Aoyama, T. (1966). *Agric. Biol. Chem.* (*Tokyo*) **30**, 917–924.
Murofushi, N., Takahashi, N., Yokota, T. and Tamura, S. (1968). *Agric. Biol. Chem.* (*Tokyo*) **32**, 1239–1245.
Murofushi, N., Takahashi, N., Yokota, T. and Tamura, S. (1969). *Agric. Biol. Chem.* (*Tokyo*) **33**, 598–609.
Murphy, P. J. and West, C. A. (1969). *Archs Biochem. Biophys.* **133**, 395–407.
Musgrave, A., Kays, S. E. and Kende, H. (1969). *Planta* **89**, 165–177.
Oster, M. and West, C. A. (1968). *Archs Biochem. Biophys.* **127**, 112–123.
Paleg, L. G. (1965). *Ann. Rev. Pl. Physiol.* **16**, 291–322.
Phinney, B. O. and West, C. A. (1960). *Ann. Rev. Pl. Physiol.* **11**, 411–436.
Phinney, B. O., West, C. A., Ritzel, M. and Neely, P. M. (1957). *Proc. Natn. Acad. Sci. U.S.A.* **43**, 398–404.
Pollard, C. J. (1969). *Pl. Physiol.* **44**, 1227–1232.
Popják, G. and Cornforth, J. W. (1966). *Biochem. J.* **101**, 553–568.
Pryce, R. J. and MacMillan, J. (1967). *Tetrahedron Letters* 4173–4175.
Pryce, R. J., MacMillan, J. and McCormick, A. (1967). *Tetrahedron Letters* 5009–5011.

Reid, D. M. and Carr, D. J. (1967). *Planta* **73**, 1–11.
Reid, D. M., Crozier, A. and Harvey, B. M. R. (1969). *Planta* **89**, 376–379.
Rowe, J. R. (ed.) (1968). "The Common and Systematic Nomenclature of Cyclic Diterpenes", 3rd Rev.
Robinson, D. R. and West, C. A. (1970a). *Biochemistry* **9**, 70–79.
Robinson, D. R. and West, C. A. (1970b). *Biochemistry* **9**, 80–89.
Ruddat, M., Heftmann, E. and Lang, A. (1965). *Archs Biochem. Biophys.* **110**, 496–499.
Ruzicka, L. (1959). *Proc. chem. Soc.* 341–360.
Scala, J., Patrick, C. and MacBeth, G. (1969). *Phytochemistry* **8**, 37–44.
Schreiber, K., Schneider, K., Sembdner, G. and Focke, I. (1966). *Phytochemistry.* **5**, 1221–1225.
Schreiber, K., Weiland, J. and Sembdner, G. (1967). *Tetrahedron Letters* 4285–4288.
Sembdner, G., Cross, R. and Schreiber, K. (1962). *Experientia* **18**, 584–585.
Sembdner, G., Weiland, J., Aurich, O. and Schreiber, K. (1968). *In* "Plant Growth Regulators", S.C.I. Monograph No. 31, pp. 70–86.
Shechter, I. and West, C. A. (1969). *J. biol. Chem.* **244**, 3200–3209.
Sitton, D., Richmond, A. and Vaadia, Y. (1967). *Phytochemistry* **6**, 1101–1105.
Spector, C. and Phinney, B. O. (1968). *Physiol. Pl.* **21**, 127–136.
Stoddart, J. L. and Lang, A. (1968). *In* "Biochemistry and Physiology of Plant Growth Substances" (F. Wightman and G. Setterfield, eds.), pp. 1371–1387, Runge Press, Toronto.
Stoddart, J. L. (1969). *Phytochemistry* **8**, 831–837.
Stodola, F. H., Raper, K. B., Fennel, D. I., Conway, H. F., Sohns, V. E., Langford, C. T. and Jackson, R. W. (1955). *Archs Biochem. Biophys.* **54**, 240–245.
Takahashi, N., Kitamura, H., Kawarada, A., Seta, M., Takai, M., Tamura, S. and Sumiki, Y. (1955). *Bull. Agric. Chem. Soc. Japan* **19**, 267–277.
Takahashi, N., Seta, Y., Kitamura, H. and Sumiki, Y. (1959). *Bull. Agric. Chem. Soc. Japan* **23**, 405–407.
Takahashi, N., Yokota, T., Murofushi, N. and Tamura, S. (1969). *Tetrahedron Letters* 2077.
Treharne, K. J. and Stoddart, J. L. (1968). *Nature, Lond.* **220**, 457–458.
Upper, C. D. and West, C. A. (1967). *J. biol. Chem.* **242**, 3285–3292.
Varner, J. E. and Johri, M. M. (1968). *In* "Biochemistry and Physiology of Plant Growth Substances" (F. Wightman and G. Setterfield, eds.), pp. 793–814, Runge Press, Toronto.
Verbiscar, A. J., Cragg, G., Geissman, T. A. and Phinney, B. O. (1967). *Phytochemistry* **6**, 807–814.
West, C. A. and Phinney, B. O. (1959). *J. Am. chem. Soc.* **81**, 2424–2427.
West, C. A., Oster, M., Robinson, D., Lew, F. and Murphy, P. (1968). *In* "Biochemistry and Physiology of Plant Growth Substances" (F. Wightman and G. Setterfield, eds.), pp. 313–332, Runge Press, Toronto.
Yokota, T., Takahashi, N., Murofushi, N. and Tamura, S. (1969). *Tetrahedron Letters* 2081–2084.
Yokota, T., Murofushi, N. and Takahashi, N. (1970). *Tetrahedron Letters* 1489–1491.
Zolotov, Z. and Lesham, Y. (1968). *Plant Cell Physiol.* **9**, 831–832.

CHAPTER 7

Ecdysones

H. H. REES

Department of Biochemistry, The University of Liverpool, Liverpool, England

I. Introduction

The name "ecdysone" was originally given to the first pure crystalline moulting hormone isolated from insects. However, the term "ecdysones" is now collectively applied to a whole range of closely related compounds, showing moulting hormone activity. Since the original isolation and structural elucidation of the moulting and juvenile hormones, advances in the chemistry and biochemistry of insect hormones have been rapid and exciting. It is with these recent advances in the case of the moulting hormone that this review will be particularly concerned. For accounts of earlier work on insect hormones the reader is referred to a number of reviews (Karlson 1956a, 1963b, 1966, 1967a,b; Gilbert and Schneiderman, 1961; Gabe *et al.*, 1964; Gilbert, 1964; Karlson and Sekeris, 1964, 1966a; Etkin and Gilbert, 1968; Berkoff, 1969).

II. Insect Endocrinology

In order to reach their final, mature stage (the imago or adult), insects pass through a number of larval stages and in many groups (holometabolic insects) through a pupal stage also. Between each instar and the next, the insects moult, i.e. the old chitinous exoskeleton is thrown off and a new one formed. This process of moulting and metamorphosis is controlled by three hormonal glands: the neurosecretory cells of the brain, the prothoracic glands and the corpora allata. Moulting is initiated by the neurosecretory cells of the brain, which secrete a prothoracotropic hormone (the brain hormone). The latter stimulates the prothoracic glands to release moulting hormone (ecdysone), which acts mainly on the epidermis, causing it to moult. For preservation of the larval stage, the juvenile hormone of the corpora allata is necessary. The pupal and imaginal moults occur when the moulting hormone is more or less secreted alone. It has been suggested (Kobayashi, 1963) that the brain hormone, in addition to acting on the prothoracic gland, acts synergistically with ecdysone directly on the tissues during imaginal development. The hormonal control of post-embryonic insect development is represented schematically in Fig. 1.

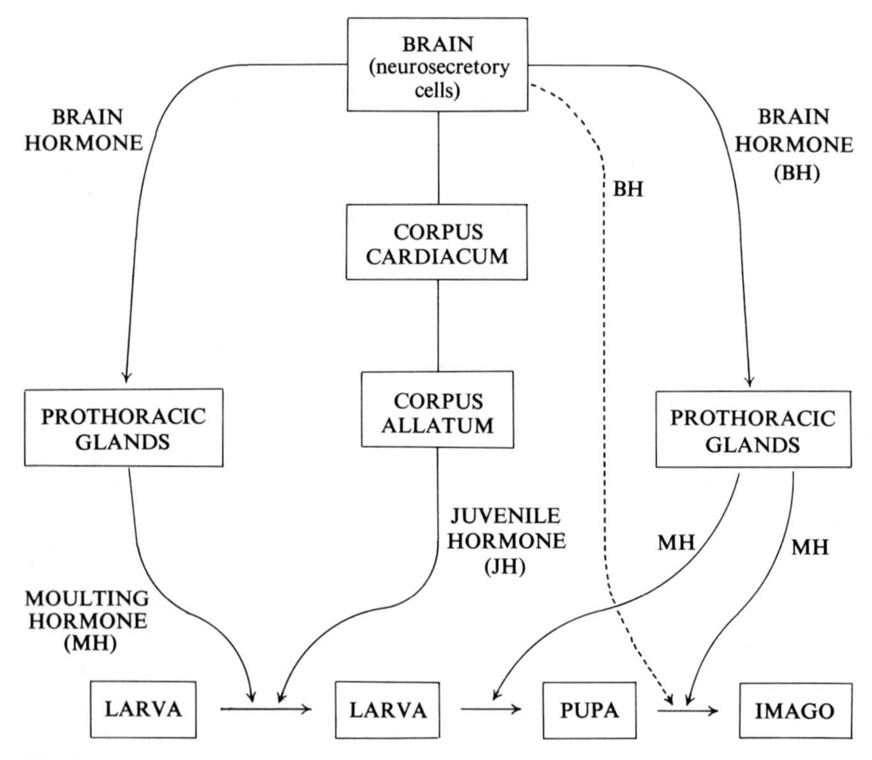

Fig. 1. Schematic representation of hormonal control of post-embryonic insect development.

III. Ecdysones in Insects and Other Arthropods

A. Isolation of Ecdysone

In 1954 Butenandt and Karlson carried out a mass extraction of 500 kg of silkworm pupae (*Bombyx mori*), yielding after a 20×10^6 fold purification, 25 mg of pure crystalline ecdysone (previously α-ecdysone). The purification procedure (Karlson *et al.*, 1963) was followed using the "*Calliphora* test" (Karlson, 1956a). This involves ligaturing mature larvae of *Calliphora erythrocephala*, thus preventing the hormone from the anterior portion from diffusing into the posterior part. After 24 h this usually yields 10–15% of animals which have pupated only in the anterior part, and are suitable for bioassay. Injection of hormone into the posterior part induces formation of a puparium, which is noticeable as a darkening and hardening of the cuticle. The *Calliphora* unit of biological activity was originally defined as the amount of material necessary to produce 50–70% pupation in ligated abdomens of the blowfly, *Calliphora*. Since this effect is usually obtained by 0·01 μg of pure ecdysone, the *Calliphora* unit has been standardized and redefined as the equivalent of 0·01 μg of pure ecdysone.

B. Structure

Following the elucidation of the main structural features of ecdysone by Karlson and co-workers (Karlson *et al.*, 1963; 1965 and references cited therein; for review see Karlson and Sekeris, 1966a), the complete structure was finally revealed by X-ray crystallographic analysis (Huber and Hoppe, 1965). Ecdysone was thus shown to be 2β,3β,14α, 22*R*, 25-pentahydroxy-5β-cholest-7-en-6-one (Table I, structure 1). This structure has been confirmed by several syntheses (Kerb *et al.*, 1966; Siddall *et al.*, 1966; Furlenmeier *et al.*, 1967; Harrison *et al.*, 1966; Mori *et al.*, 1968).

C. Other Ecdysones in Arthropods

Following the isolation of ecdysone from *Bombyx mori* pupae, Karlson and co-workers (Karlson 1956a) isolated smaller quantities of a second more polar substance (provisionally called β-ecdysone). The same compound having slightly higher biological activity than ecdysone was subsequently isolated from silkworm pupae by several workers and shown to be 20-hydroxyecdysone (2) (ecdysterone) (Hoffmeister, 1966; Hoffmeister and Grützmacher, 1966; Hocks and Wiechert, 1966; Hoffmeister *et al.*, 1967b; Hocks *et al.*, 1967). On the basis of bioassay of counter-current distribution fractions of *Bombyx* extracts, Burdette and Bullock (1963) have reported five separate active fractions.

Following the demonstration by Karlson and co-workers (Karlson 1956a,b; Karlson and Skinner, 1960) that partially purified extracts of crustaceans are active in the *Calliphora* bioassay, 2 mg of the pure active principle (crustecdysone) was finally isolated from 1 ton of the marine crayfish, *Jasus lalandei*, by

Hampshire and Horn (1966). Since crude extracts of crayfish at an intermoult stage, have less than $\frac{1}{100}$th the activity of comparable extracts of silkworm pupae (Hampshire and Horn, 1966), a new purification procedure was developed (Horn *et al.*, 1968). Crustecdysone was shown to be represented by structure (2) (Table I) (Hampshire and Horn, 1966; Horn *et al.*, 1966).

Horn and co-workers (1966) also demonstrated that crustecdysone and β-ecdysone are the same substance. It has been subsequently shown (Hocks *et al.*, 1967; Galbraith *et al.*, 1967) that structure (2) represents ecdysterone, β-ecdysone, crustecdysone, 20-hydroxyecdysone and the two moulting hormones isolated from plants, iso-inokosterone and polypodine A (see Section IV,A). This structure has been confirmed by syntheses (Kerb *et al.*, 1968; Huppi and Siddall, 1967; Mori and Shibata, 1969). In this review, compound (2) will be mainly called ecdysterone, or crustecdysone, for simplicity and clarity. Table I lists the compounds possessing insect-moulting hormone activity, which have been isolated from insects and crustacea. Since many of the ecdysterone isolations were carried out prior to its chemical synthesis, the stereochemistry at C-20 and C-22 was mainly assumed on the basis that ecdysone was the probable precursor and that the biological hydroxylation at C-20 occurred with retention of configuration, as in the case of cholesterol (Shimizu *et al.*, 1962).

In *B. mori* pupae α-ecdysone (ecdysone) was found to be the main hormone with smaller quantities of β-ecdysone (ecdysterone) (Karlson, 1956a). On the other hand, in the adult Moroccan locust, *Dociostaurus maroccanus* (Stamm, 1959) and in *Antherea pernyi* pupae (Horn *et al.*, 1966), ecdysterone (200 μg from 31 kg of *Antherea* pupae) predominates over ecdysone. Similarly in tobacco hornworm pupae (*Manduca sexta*), ecdysterone is accompanied by smaller amounts of ecdysone and 20,26-dihydroxy ecdysone (3) (Thompson *et al.*, 1967).

Karlson (1956a,b) reported that *Calliphora erythrocephala* pupae contained both ecdysone and ecdysterone (β-ecdysone), and considered the former as the principle moulting hormone. Later Shaaya and Karlson (1965a) followed the amount of moulting hormone in the body of *Calliphora* throughout development, using the *Calliphora* bioassay. It was found that there is a sharp rise in hormone level at the time of puparium formation, followed by a rapid drop to a minimal value approximately 24 h after pupation began. This was later followed by a broad maximum with a drop to a low level prior to adult emergence. Therefore high hormone levels were found at the beginning of puparium formation and during adult development. Since it was considered that a different hormone or different combination of hormones might be involved at each stage, Galbraith and co-workers (1969e) have recently determined the nature of the hormones produced in *Calliphora* at these two developmental stages. It was found that crustecdysone is the only hormone present in significant amounts at puparium formation in *C. stygia* and *C. vicina* (= *erythrocephala*), and also during adult development in *C. stygia*, whereas

TABLE I
Structures of ecdysones isolated from Arthropods

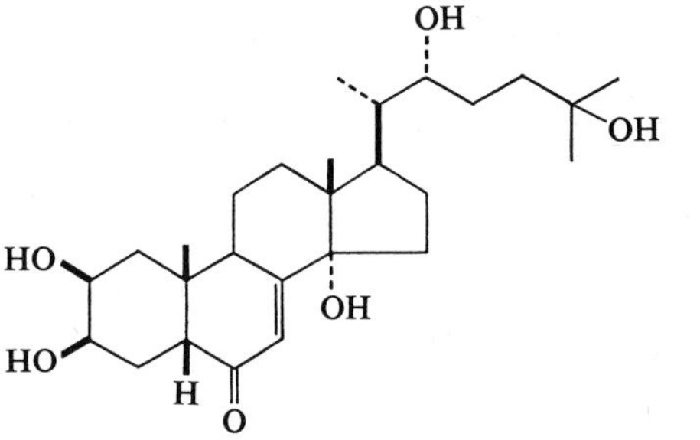

(1) Ecdysone (α-ecdysone)

Bombyx mori (silkworm) (Butenandt and Karlson, 1954). *Dociostaurus maroccanus* Thunberg (adult Moroccan locust) (Stamm, 1959). *Manduca sexta* (Johannson) (tobacco hornworm) (Kaplanis *et al.*, 1966b). *Antherea pernyi* (oak silkmoth) (Horn *et al.*, 1966).

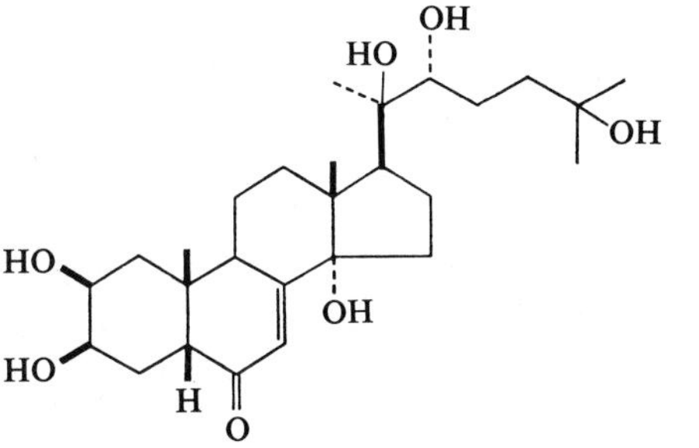

(2) Ecdysterone (crustecdysone, 20-hydroxyecdysone, β-ecdysone)

Insects
B. mori (Karlson, 1956a; Hocks and Wiechert, 1966; Hoffmeister and Grutzmacher, 1966). *D. maroccanus* (Stamm, 1959). *M. sexta* (Kaplanis *et al.*, 1966b). *A. pernyi* (Horn *et al.*, 1966). *Calliphora stygia* (Galbraith *et al.*, 1969e). *C. vicina* (Galbraith *et al.*, 1969e).

Crustacea
Jasus lalandei (sea-water crayfish) (Horn *et al.*, 1966). *Callinectes sapidus* (crab) (Faux *et al.*, 1969).

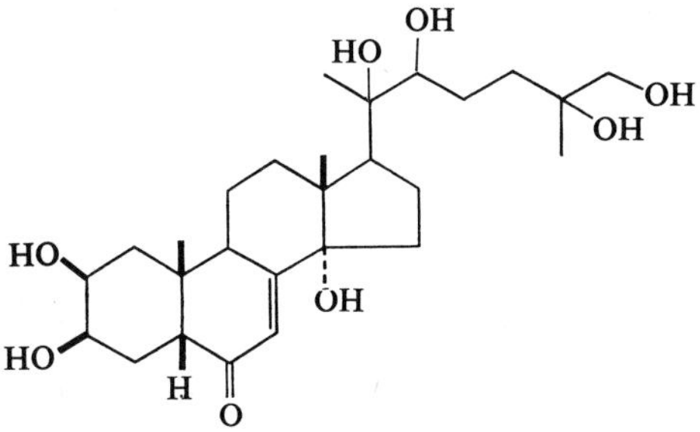

(3) 20,26-Dihydroxyecdysone

M. sexta (Johannson) (Thompson *et al.*, 1967).

TABLE I—*continued*

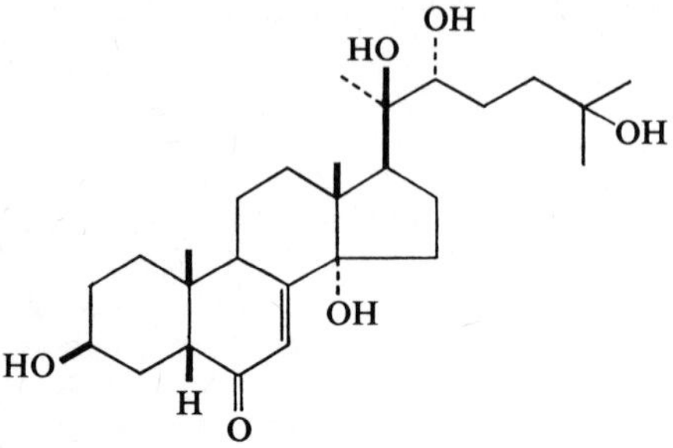

(4) 2-Deoxycrustecdysone	*J. lalandei* (crayfish) (Galbraith *et al.*, 1968a).
(5) Callinecdysone A (≡ inokosterone ?)	*C. sapidus* (Faux *et al.*, 1969).
(6) Callinecdysone B (≡ makisterone A ?)	(Faux *et al.*, 1969).

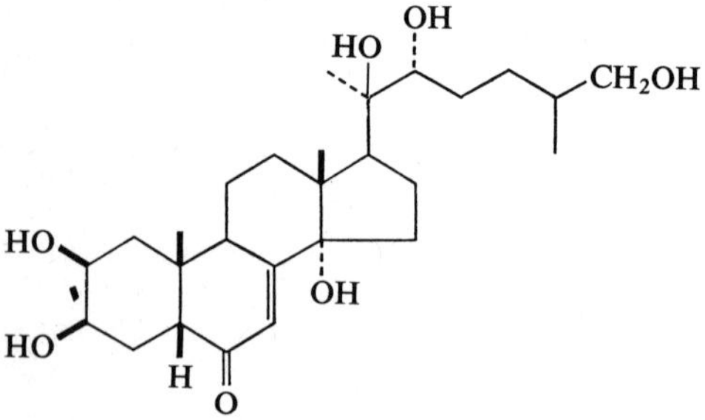

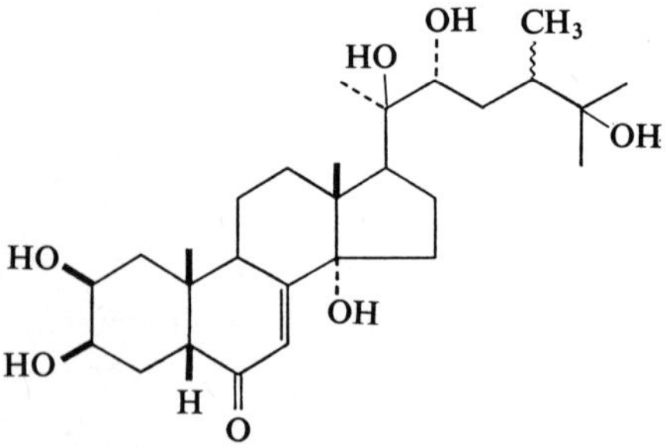

if ecdysone is present, the amount is quite small. They conclude that crustecdysone is probably the principle ecdysone-group hormone involved in the control of both these processes in *Calliphora*.

In contrast to reports (Burdette, 1962; Shaaya and Karlson, 1965a,b) of the occurrence in *Bombyx* of two distinct maxima in moulting hormone titre during pupal development, only a single peak was detected in *Manduca sexta* after pupation (Kaplanis *et al.*, 1966b). Another peak was found in the prepupa of the hornworm (*M. sexta*) as in the case of *Calliphora* and *Bombyx*. It is noteworthy that in *Lucilia cuprina*, there is a rise in the level of prothoracic gland hormone before emergence of the adult (Barritt and Birt, 1970). Thompson and co-workers (1967) have isolated and identified three different ecdysones [α-ecdysone (1), 20-hydroxyecdysone (2) and 20,26-dihydroxyecdysone (3)] from tobacco hornworm pupae during maximum titre of moulting hormone activity. On the basis that 20,26-dihydroxyecdysone is less active than α-ecdysone or 20-hydroxyecdysone during bioassay, these authors have suggested that increased hydroxylation is a mechanism in insects for the deactivation of the ecdysones. Thompson and co-workers (1967) further suggest that although both the structures and biological activity indicate that α-ecdysone, 20-hydroxyecdysone, and 20,26-dihydroxyecdysone are intermediates in a biosynthetic scheme, each could have a specific physiological function.

Reports that crustecdysone (ecdysterone) is the predominant moulting hormone in two crustacea, the marine crayfish *J. lalandei* (Horn *et al.*, 1966) and the crab *Callinectes sapidus* (Faux *et al.*, 1969) are in agreement with the finding that this hormone also predominates in most (not all) of the insect species so far investigated. Crustecdysone in *J. lalandei* was originally thought to be accompanied by smaller amounts of ecdysone (Horn *et al.*, 1966). However, on fractionation of a larger amount (3 tons) of crayfish waste, Galbraith and co-workers (1968a) isolated 200 μg of the less polar moulting hormone which they originally believed to be ecdysone (Horn *et al.*, 1966) and demonstrated that it was in fact 2-deoxycrustecdysone (4). This compound is about as active as crustecdysone in the *Calliphora* test.

Evidence has now been obtained that ecdysones induce moulting in crustaceans. The X-organ of the crustacean eyestalk produces a moult-inhibiting hormone (MIH) during the intermoult stages, which is believed to act by inhibiting the synthesis of moulting hormones in the Y-gland (an endocrine gland homologous to the prothoracic gland). MIH may also inhibit the action of the moulting hormone (Lowe *et al.*, 1968). A similar MIH is probably produced by the brain of locusts (Carlisle and Ellis, 1968b). It has been demonstrated (Lowe *et al.*, 1968) that injection of very small amounts of crustecdysone into the fresh-water crayfish, *Procambarus simulans*, from which the eyestalks had been removed, reduces the time taken to moult. No effect was obtained on injection of crustecdysone into intact animals at an intermoult stage. However, Krishnakaumaran and Schneiderman (1968, 1969) have shown that ecdysones control moulting in members of diverse classes of arthropods in addition to insects. Ecdysterone stimulated moulting in three members of the Crustacea: the terrestial isopod *Armadillium vulgare*, the fresh-water crayfish *Procambarus* and the fiddler crab *Uca pugilator*, together with three members of the Chelicerate subphylum: the spider *Araneus cornutus*, the tarantula *Dugesiella hentzi* and the horseshoe crab, *Limulus polyphemus*. They also found (Krishnakumaran and Schneiderman, 1969) that a number of other ecdysone analogues, namely ecdysone (1), inokosterone (Table II, 10), ponasterone A (7), cyasterone (24) and 2β,3β,14α-trihydroxy-5β-cholest-7-en-6-one, were approximately equally effective in inducing moulting in *Procambarus*. Other sterols were ineffective, e.g. β-sitosterol, 2β,3β,14α-trihydroxy-5α-cholest-7-en-6-one, 2β,3β-dihydroxy-5β-cholest-7-en-6-one and 2β,3β,14α-trihydroxy-5β-stigmast-7-en-6-one. Induction of moulting by ecdysones has also been demonstrated for two other crustaceans, the crab *Carcinus maenas* (Carlisle, 1965) and the prawn, *Penaeus japonicus* (Kurata, 1968). It is also interesting that ecdysone terminates diapause in larvae of the winter tick, *Dermacentor albipictus* (Wright, 1969).

Faux and co-workers (1969) have examined females of the marine crab *Callinectes sapidus* just before and after moulting, and have interestingly found a moulting hormone bearing a 24-methyl group (Table I, 6). In the early premoult stage, callinecdysone A (5) was the only hormone that could be detected, whereas at the later premoult stage, the amount of callinecdysone A had increased but was accompanied by a smaller amount of crustecdysone (2). After moulting, crustecdysone was the major hormone present together with a smaller amount of callinecdysone B (6). The authors suggest that the rising titre of ecdysones during moulting in *Callinectes* provides additional evidence that ecdysones function as moulting hormones in crustaceans. On the basis of the high level of crustecdysone present after moulting, they suggest that a high level of hormone is not only associated with cuticle shedding but also with cuticle hardening. Faux and co-workers conclude that events prior to moulting may be sequentially triggered partly by different ecdysones and partly by a rising hormone level, with final cuticle hardening occurring only at the highest hormone level.

IV. Ecdysones in Plants

A. Structure and Distribution

Widespread interest in compounds possessing moulting hormone activity was created by Nakanishi's report (Nakanishi *et al.*, 1966) of the isolation from leaves of the conifer, *Podocarpus nakaii*, of four active substances which were termed "ponasterones". The structure of Ponasterone A was elucidated as (7). This plant was reported to be a far richer source of insect hormones than insects. Almost simultaneously with this report, compounds possessing moulting hormone activity were isolated from an Australian species, *P. elatus* (Galbraith and Horn, 1966), rhizomes of the fern *Polypodium vulgare* (Jizba

TABLE II

Structures and some sources of plant ecdysones. (For other sources see Herout, 1970)

C_{27} Skeleton

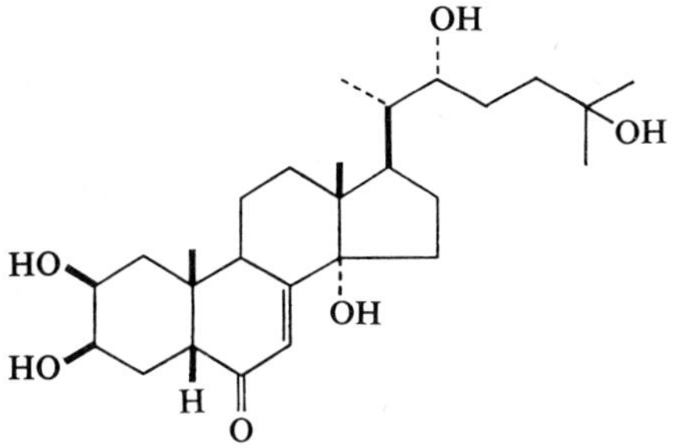

(1) Ecdysone

Pteridium aquilinum (Kaplanis *et al.*, 1967). *Polypodium vulgare* (Heinrich and Hoffmeister, 1967). *Lemmaphyllum microphyllum* (as quoted in Takemoto *et al.*, 1968f). *Osmunda japonica* and *O. asiatica* (Takemoto *et al.*, 1968g).

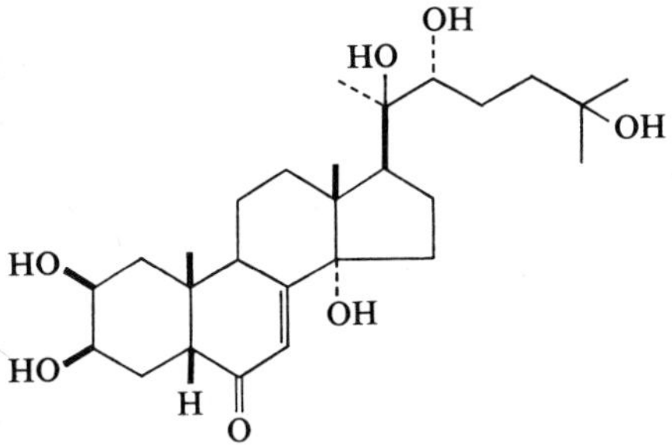

(2) Ecdysterone (crustecdysone, 20-hydroxyecdysone)

Podocarpus elatus (Galbraith and Horn, 1966, 1969). *P. macrophyllus* (Imai *et al.*, 1967). *Polypodium vulgare* (Jizba *et al.*, 1967a). *Pteridium aquilinum* (Kaplanis *et al.*, 1967). *Achyranthes fauriei* (Takemoto *et al.*, 1967c). *Taxus baccata* (Takemoto *et al.*, 1967f; Hoffmeister *et al.*, 1967a). *Lemmaphyllum microphyllum*, *Pleopeltis thunbergiana* and *Neocheiropteris ensata* (Takemoto *et al.*, 1968d). *Osmunda japonica* and *O. asiatica* (Takemoto *et al.*, 1968g). *Ajuga japonica* (Imai *et al.*, 1969b). *Blechnum amabile* and *B. niponicum* (Takemoto *et al.*, 1969b).

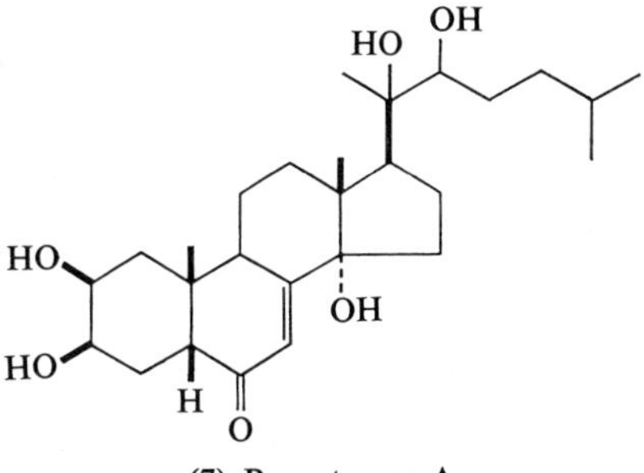

(7) Ponasterone A

Podocarpus nakaii (Nakanishi *et al.*, 1966; Nakanishi, 1969). *P. macrophyllus*, *P. chinensis* and *Taxus cuspidata* (Imai *et al.*, 1967). *T. baccata* (de Souza *et al.*, 1969). *Pteridium aquilinium* (Takemoto *et al.*, 1968b). *Osmunda japonica* and *O. asiatica* (Takemoto *et al.*, 1968g). *Blechnum amabile* and *B. niponicum* (Takemoto *et al.*, 1969b).

TABLE II—*continued*

Compound	Source
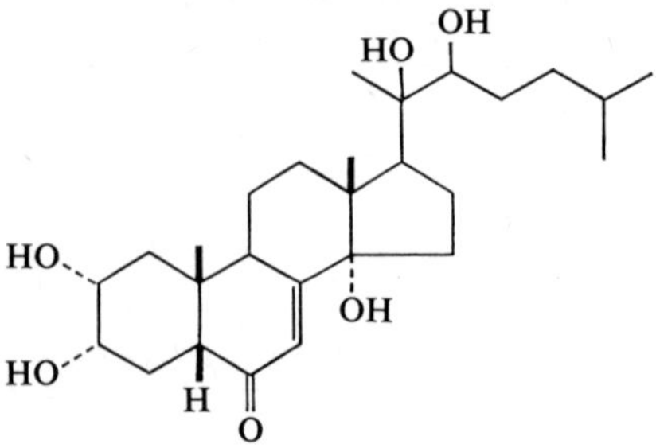 (8) Ponasterone B	*Podocarpus nakaii* (Nakanishi *et al.*, 1968; Nakanishi, 1969; Koreeda and Nakanishi, 1970).
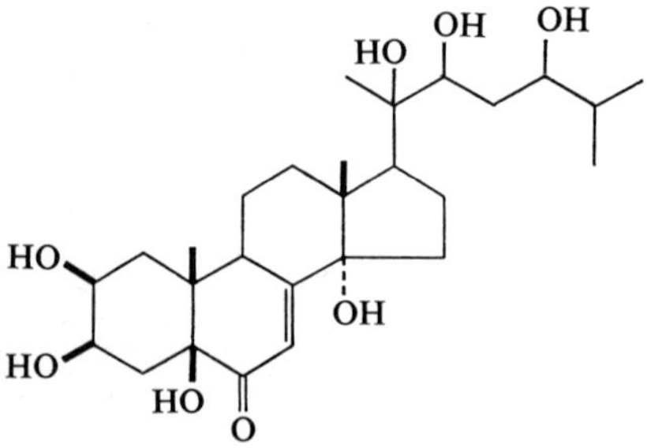 (9) Ponasterone C	*Podocarpus nakaii* (Nakanishi *et al.*, 1968; Nakanishi, 1969; Koreeda and Nakanishi, 1970).
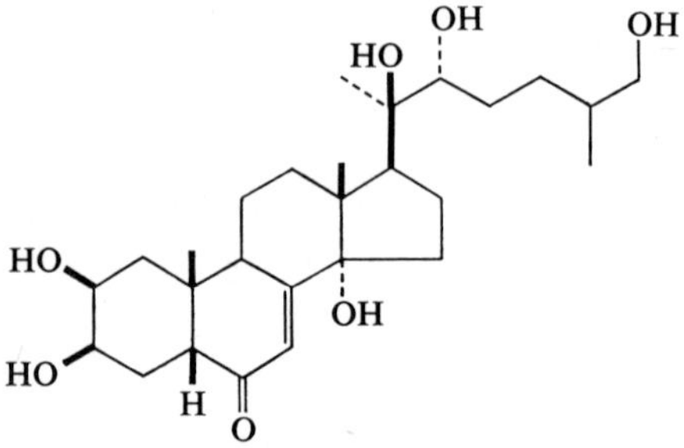 (10) Inokosterone	*Achyranthes fauriei* (Takemoto *et al.*, 1967d). *A. rubrofusca* (Takemoto *et al.*, 1967g). *A. longifolia*, *A. japonica* and *A. longifolia* (Takemoto *et al.*, 1968k).
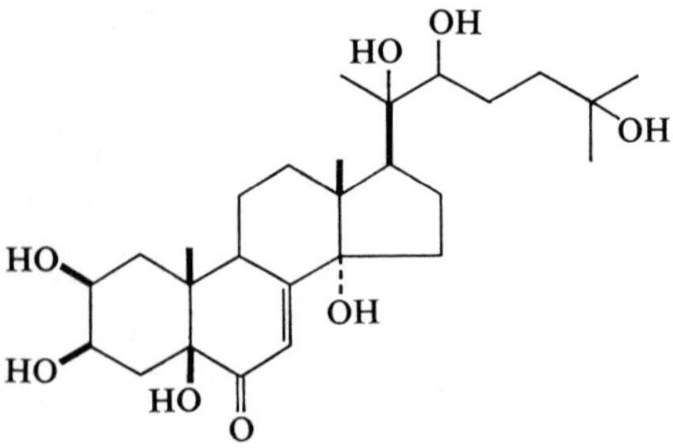 (11) Polypodine B (5β-hydroxy-ecdysterone, ajugasterone A)	*Polypodium vulgare* (Jizba *et al.*, 1967b; Heinrich and Hoffmeister, 1968). *Ajuga incisa* (Imai *et al.*, 1969a,b). *Vitex megapotamica* (Rimpler, 1969).

TABLE II—*continued*

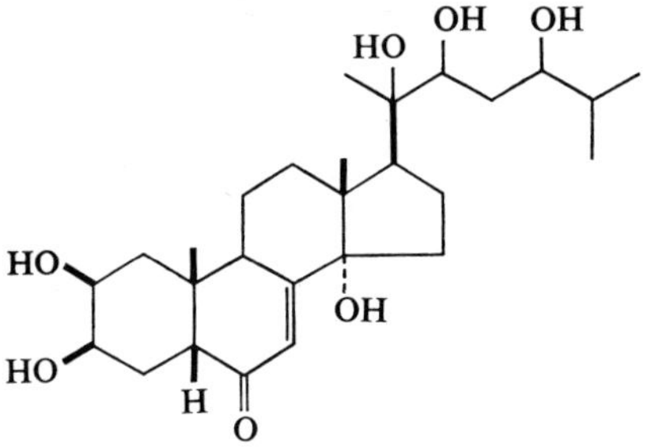

(12) Pterosterone

Onoclea sensibilis and *Lastrea thelypteris* (Takemoto *et al.*, 1967a). *Pteridium aquilinum* (Takemoto *et al.*, 1968b). *Vitex megapotamica* (Rimpler, 1969).

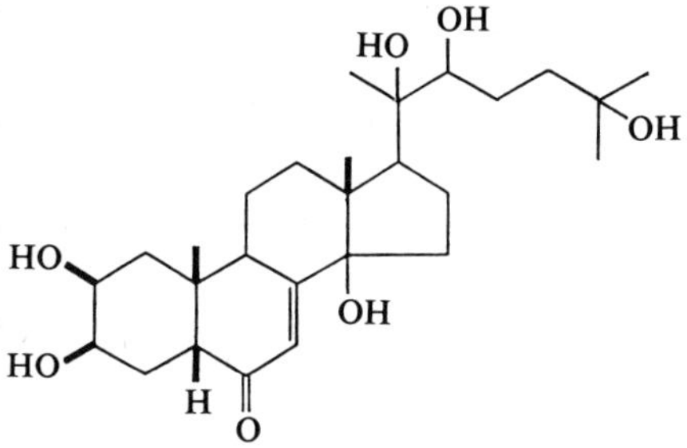

(13) Shidasterone (Stereoisomer of ecdysterone)

Blechnum niponicum (Takemoto *et al.*, 1968h).

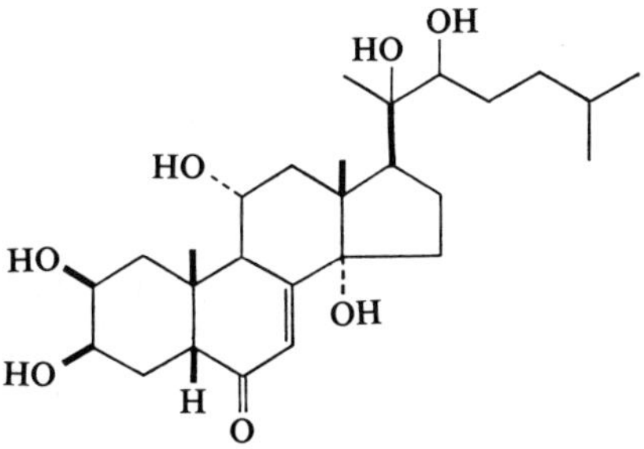

(14) Ajugasterone C (11α-hydroxy-ponasterone A)

Ajuga japonica and *A. decumbens* (Imai *et al.*, 1969b).

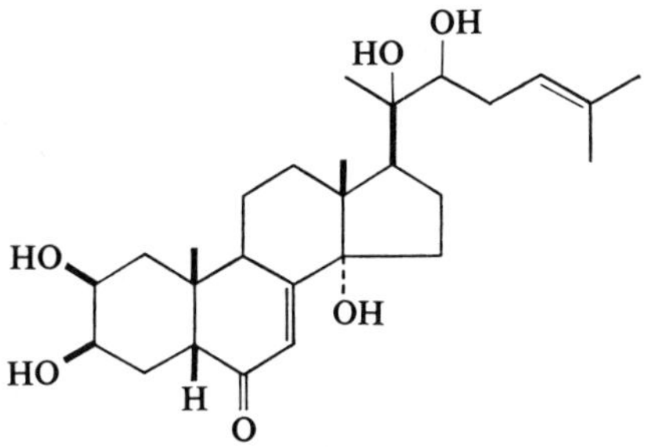

(15) Stachysterone C

Stachyurus praecox (Imai *et al.*, 1970).

TABLE II—*continued*

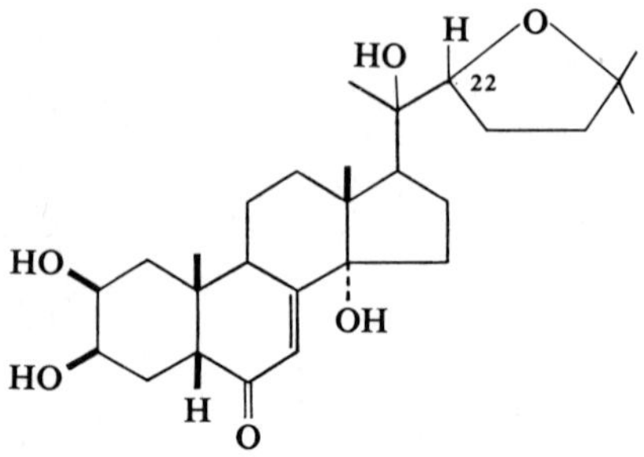

Compound	Source
(16) Stachysterone D (inactive)	*Stachyurus praecox* (Imai *et al.*, 1970).
(17) Viticosterone E	*Vitex megapotamica* (Rimpler, 1969).
(18) Ponasteroside A	*Pteridium aquilinum* (Takemoto *et al.*, 1968a).

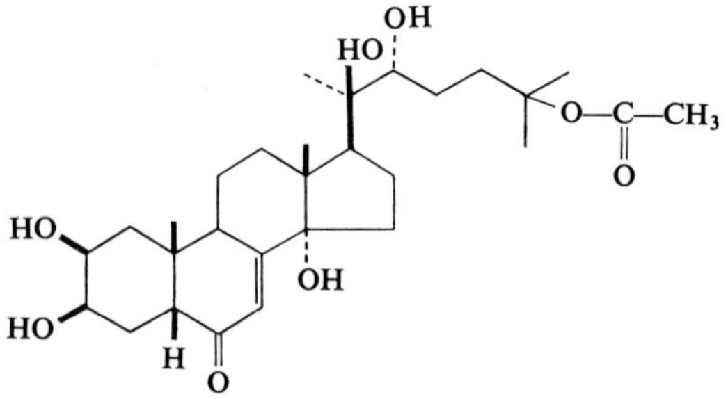

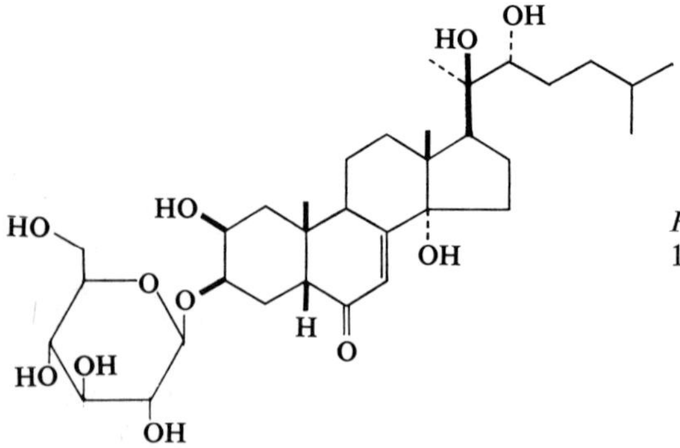

TABLE II—*continued*

HO, HO, H, O, HO, OH, OH

(19) Podecdysone B

Podocarpus elatus (Galbraith *et al.*, 1969c).

C_{28} *Skeleton*

HO, HO, H, O, OH, HO, OH, OH

(20) Makisterone A (podecdysone D)

Podocarpus macrophyllus (Imai *et al.*, 1968b). *P. elatus* (Galbraith *et al.*, 1969c).

HO, HO, H, O, OH, HO, OH, OH

(21) Makisterone B

Podocarpus macrophyllus (Imai *et al.*, 1968a).

TABLE II—*continued*

C_{29} Skeleton

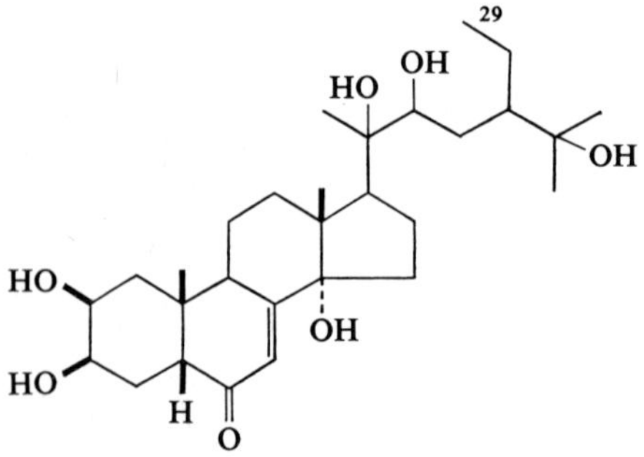

Compound	Source
(22) Makisterone C (lemmasterone, podecdysone A?)	*Podocarpus macrophyllus* (Imai *et al.*, 1968a). *Lemmaphyllum microphyllum* (Takemoto *et al.*, 1968c). Probably the same as podecdysone A from *P. elatus* (Galbraith *et al.*, 1968b).
(23) Makisterone D	*Podocarpus macrophyllus* (Imai *et al.*, 1968a).
(24) Cyasterone	*Cyathula capitata* (Takemoto *et al.*, 1967b). *Ajuga japonica* (Imai *et al.*, 1969b). *A. decumbens*, *A. incisa* and *A. nipponensis* (Imai *et al.*, 1969c).

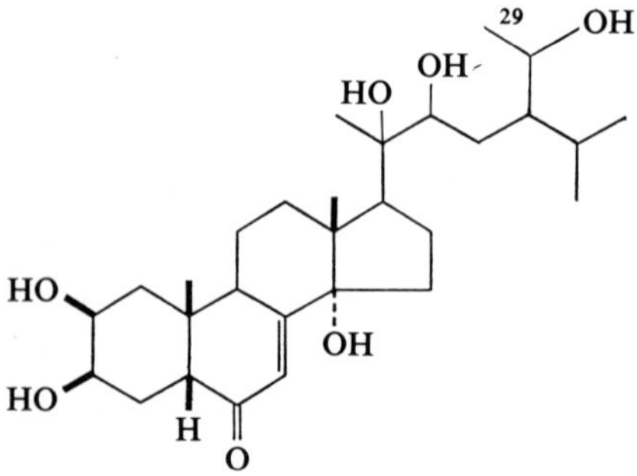

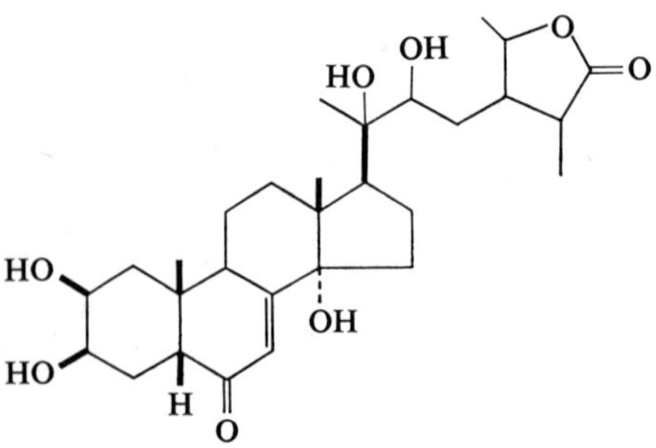

TABLE II—*continued*

Compound	Source
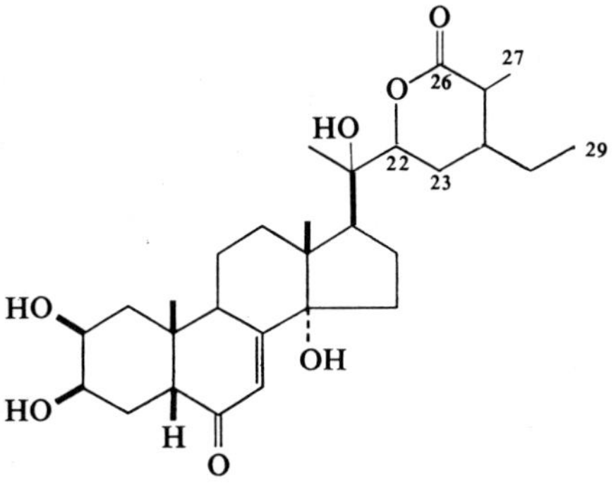 (25) Capitasterone	*Cyathula capitata* (Takemoto *et al.*, 1968j).
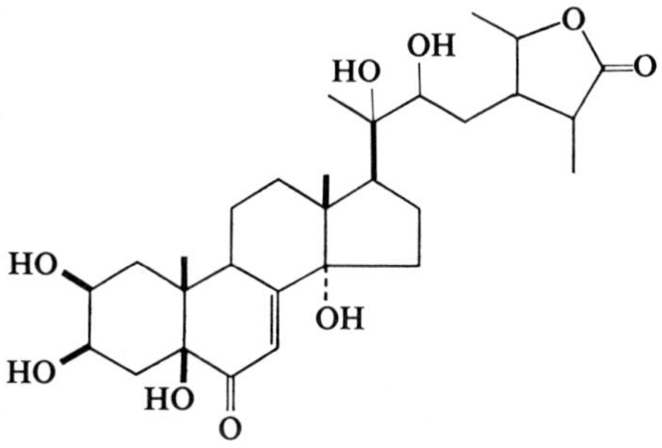 (26) Sengosterone	*Cyathula capitata* (Hikino *et al.*, 1969c).
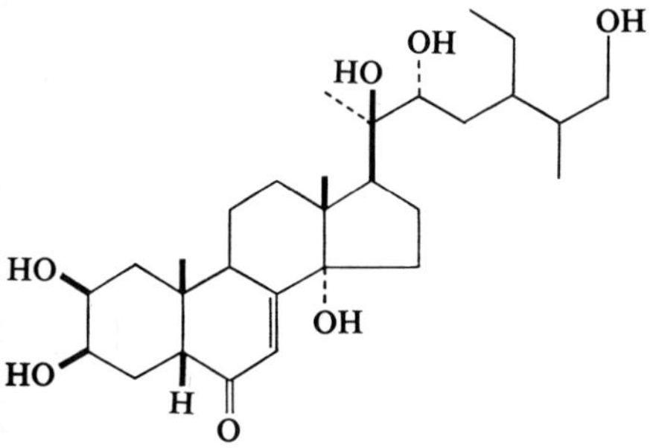 (27) Amarasterone A	*Cyathula capitata* (Takemoto *et al.*, 1968i).

TABLE II—*continued*

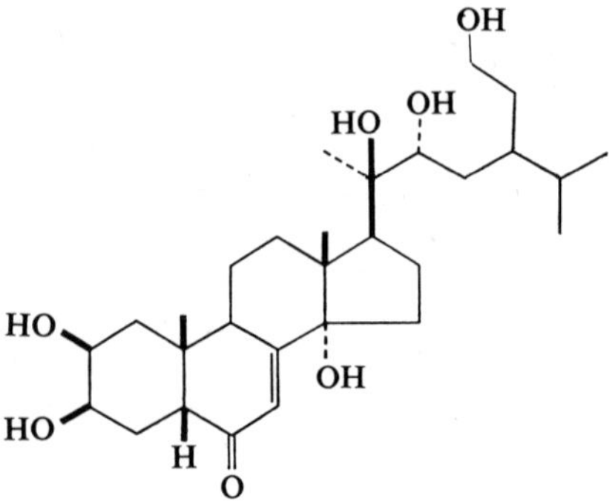

Compound	Source
(28) Amarasterone B	*Cyathula capitata* (Takemoto *et al.*, 1968i).
(29) Ajugasterone B	*Ajuga incisa* (Imai *et al.*, 1969a).
	C_{19} *Skeleton*
(30) Rubrosterone (inactive)	*Achyranthes rubrofusca* (Takemoto *et al.*, 1968f). *A. fauriei* (Takemoto *et al.*, 1969a). *A. obtusifolia* (Hikino *et al.*, 1969a).

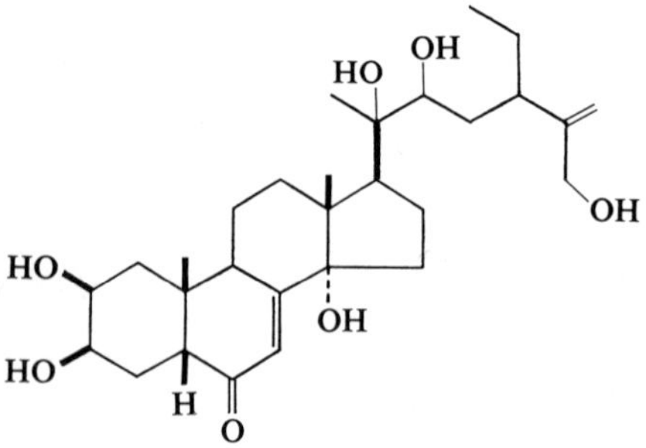

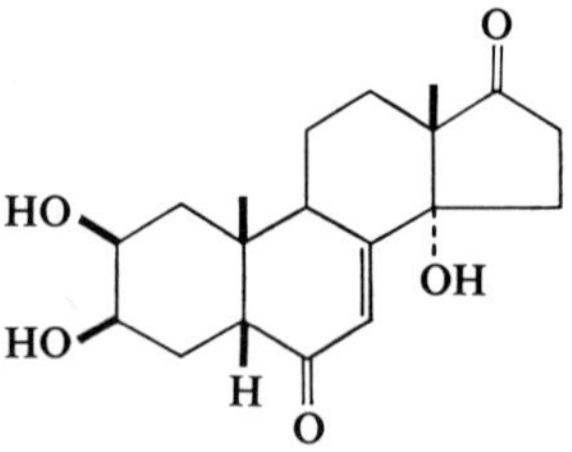

et al., 1967a,b) and roots of the Japanese plant *Achyranthes fauriei* (Takemoto *et al.*, 1967c). The rhizomes of *P. vulgare* were reported to contain as much as 2% of ecdysterone (2).

As a result of widespread screening of plants for insect-moulting activity, using various insect assays, at least 28 different active compounds have been isolated from the plant kingdom. These are listed in Table II, together with some of the sources from which they have been obtained. For other sources of ecdysones in plants see Herout (1970). It is interesting to note that C_{27}, C_{28} and C_{29} compounds have been isolated. The common structural features include a sterol nucleus possessing a Δ^7-6-oxo-grouping and an A/B *cis* ring junction, with hydroxyl substituents at C-2, C-3 (both β except in one case) and C-14 (α). The side chains are substituted with various oxygen functions.

From the results so far obtained (Table II and Herout, 1970), it appears that ecdysterone is probably one of the most widespread compounds possessing insect-moulting hormone activity. It has been demonstrated that the tetracycles of crustecdysone (ecdysterone) and ecdysone are identical (Siddall *et al.*, 1967). It seems from present data that ecdysone (1) might not be as widely distributed as ecdysterone (2) in the plant kingdom, although it must be remembered that only a comparatively small number of plants have been analysed. It is of interest that $2\beta,3\beta$- and $2\alpha,3\alpha$-hydroxy steroids (ponasterones A (7), B (8) and C (9) co-occur in the conifer *Podocarpus nakaii.* The skeletal structure of ponasterone A has been confirmed (Moriyama and Nakanishi, 1968) and its stereochemical identity with crustecdysone (2) established by synthesis (Huppi and Siddall, 1968). The high biological activity of the $2\alpha,3\alpha$-hydroxy steroid, ponasterone B, indicates that ring A hydroxyl configurations can be varied in addition to the side-chain structure for manifestation of moulting hormone activity (Kobayashi *et al.*, 1967; Hoffmeister *et al.*, 1968). Ponasterone C was also originally assigned a $2\alpha,3\alpha$-hydroxy grouping, but this has now been corrected (Koreeda and Nakanishi, 1970) to a $2\beta,3\beta,5\beta$-trihydroxy structure (9). Takemoto and co-workers (1968e) have investigated the absolute configuration of inokosterone (10) which they originally isolated from *A. fauriei* roots. They suggest that since inokosterone is most likely biosynthesized from ponasterone A, the absolute configurations at C-20 and C-22 are indicated to be both *R*, and also conclude that it is an epimeric mixture of $2\beta,3\beta,14\alpha,20(R),22(R),22(R)$, 26-hexahydroxy-25(*R* and *S*)-5β-cholest-7-en-6-ones.

Polypodine B (11) originally isolated from *Polypodium vulgare* (Jizba and Herout, 1967; Jizba *et al.*, 1967a) along with ecdysone and ecdysterone was assigned the structure $5\beta,20\xi$-dihydroxyecdysone, whereas Heinrich and Hoffmeister (1968) designated the compound isolated by them from the same source, 5β-hydroxyecdysterone. Recently, however, these two compounds have been shown to be identical and the $2\beta,3\beta,5\beta$-trihydroxy structure has been established (Nakanishi, K. as quoted in Hikino *et al.*, 1969c). Polypodine B is probably the most active moulting hormone-like substance so far known, being approximately four times more active in the test on *Calliphora* pupae than ecdysone (Jizba *et al.*, 1967a). A 5β-hydroxylation is also known in sengosterone (26) (Hikino *et al.*, 1969c) and ponasterone C (9) (Koreeda and Nakanishi, 1970).

Shidasterone (13) originally isolated from the fern, *Blechnum niponicum* is a stereoisomer of ecdysterone (Takemoto *et al.*, 1968h). However, direct comparison with synthetic 22-*epi*-ecdysterone (Hocks, 1968) revealed that the two substances were different. Viticosterone E (17) seems to be unique amongst the insect moulting-active compounds so far isolated in that the C-25 hydroxyl group is acetylated. Ponasteroside A (18) (originally warabisterone) isolated from *Pteridium aquilinum* var. *latiusculum*, was revealed on analysis to be ponasterone A 3-β-glycoside (Takemoto *et al.*, 1968a). It is interesting that ponasteroside A also exhibits high insect moulting hormone activity, although the authors remark that the possibility that this compound reveals this activity only after enzymic hydrolysis, cannot be excluded.

The structure assigned to podecdysone B (19) isolated from the bark of *Podocarpus elatus* (Galbraith *et al.*, 1969c) along with podecdysone A (22) and podecdysone D (20) denotes a substantial departure from the normal insect moulting hormone active compounds bearing a Δ^7-6-keto function. Since an analogous 8,14-diene structure has been assigned to one of the products obtained by acid treatment of ecdysone, Galbraith and co-workers suggest that podecdysone B may be biosynthesized from crustecdysone (also present in *P. elatus*) by dehydration. The authors present arguments suggesting that it is unlikely that podecdysone B is an artifact formed by dehydration of crustecdysone, but repeated isolation under milder conditions resulted in isolation of a different compound, podecdysone E. A third sample of bark collected from young trees yielded no phytoecdysones.

The leaves of the conifer, *P. macrophyllus* is the only known source to contain C_{27}, C_{28} and C_{29} ecdysones. The C_{27} compounds, ecdysterone (2) and ponasterone A (7), predominate (Imai *et al.*, 1967) and are accompanied by smaller amounts of the C_{28} compounds, makisterone A (20) and B (21), together with the C_{29} compounds, makisterone C (22) and D (23) (Imai *et al.*, 1968a,b). Makisterone A is identical (Galbraith *et al.*, 1969c) with podecdysone D isolated from *P. elatus*, whereas makisterone C is identical with lemmasterone isolated from *Lemmaphyllum microphyllum* (Takemoto *et al.*, 1968c) and also probably with podecdysone A isolated from *P. elatus* (Galbraith *et al.*, 1968b). The exact side-chain stereochemistry of these C_{28} and C_{29} compounds remains undefined.

The plant, *Cyathula capitata* contains C_{29} ecdysones possessing lactone rings [cyasterone (24), capitasterone (25) and sengosterone (26) which co-occur with other C_{29} ecdysones [amarasterone A (27) and amarasterone B (28)] (Takemoto *et al.*, 1967b, 1968i,j; Hikino *et al.*, 1969c). The side-chain configurations of these compounds remains to be determined. Cyasterone also occurs in several *Ajuga* species (Imai *et al.*, 1969b,c). *Ajuga japonica* and

A. decumbens also furnished 11α-hydroxy ponasterone A (ajugasterone C, 14) (Imai *et al.*, 1969b), whereas ajugasterone B (29) from *A. incisa* (Imai *et al.*, 1969a) is a rare example of an ecdysone possessing a double bond besides the ubiquitous Δ^7 one. Stachysterone C (15) possessing a Δ^{24} bond is another example of such a compound and was isolated from *Stachyurus praecox* together with stachysterone D (16), which has a side-chain lactone ring structure and is practically inactive in the insect moulting hormone assay (Imai *et al.*, 1970).

It is of interest that ecdysterone and inokosterone are accompanied in *Achyranthes rubrofusca*, *A. fauriei* and *A. obtusifolia* by small amounts of a C_{19} steroid (rubrosterone, 30) possessing the basic tetracyclic structure characteristic of ecdysones (Takemoto *et al.*, 1968f, 1969a; Hikino *et al.*, 1969a). Rubrosterone has only weak activity in the insect moulting assay. The chemical synthesis of rubrosterone has also been reported (Hocks *et al.*, 1968; Shibata and Mori, 1968).

The occurrence of these compounds possessing moulting hormone activity in plants raises the question as to how widely distributed are they in the plant kingdom. Extensive testing of conifer leaf extracts (Staal, 1967a) revealed that many species belonging to the Podocarpaceae and Taxaceae families showed high activity. Takemoto and co-workers (1967e) have screened a large number of crude drugs and plants for insect-moulting hormone activity, using various insect assays. Activity was found amongst crude drugs, ferns and higher plants, but no marked activity was found in mushrooms and seaweeds. A far more extensive examination of plants and crude drugs was carried out by other Japanese workers (Imai *et al.*, 1969d) who screened 1056 species of plants representative of a wide taxonomical distribution, for phyto-ecdysones. They employed ligated final instar larvae of the rice stem borer, *Chilo suppressalis* which did not pupate after 24 h. Insects were dipped for 5–10 sec in the methanol extract of 3 g of each plant and observed for sclerotization and tanning. Activity was observed in 24 species of Pteridophytes and in 30 species of Gymnosperms and Angiosperms, including both Monocotyledoneae and Dicotyledoneae. Therefore, these active polyhydroxy steroids are fairly widely distributed in the plant kingdom, but have only been discovered so far in perennial herbaceous and woody plants. No annual herbaceous plants have been found to be active. The probability of finding an active principle from plants belonging to the same genus is high, although there are exceptions.

B. POSSIBLE FUNCTION OF ECDYSONES IN PLANTS

The occurrence of insect-moulting active steroids in plants, raises the question as to why do certain plants synthesize such exotic sterols. As pointed out by Robbins and co-workers (Robbins *et al.*, 1968), it is not known for certain whether these ecdysones are end products of sterol metabolism, analogous to the bile alcohols or acids of vertebrates, as suggested by the A/B ring *cis* configuration, or have they a physiological or biochemical function in plant growth and development, as in insects. In this connection, it is interesting that Carlisle and co-workers (1963) have found that fractions of locust extracts containing large amounts of what they call "ecdysone-λ" exert a significant stimulation of growth of the internodes of dwarf pea plants in the standard bioassay for gibberellin activity. This stimulation corresponds to approximately 10% of that exerted by gibberellin-A_3. No other work seems to have been done on this. It is also interesting that gibberellic acid shortens the time between successive moults, in the same manner as ecdysone, when injected at the appropriate time into locust larvae. Carlisle and co-workers have also reported that ecdysone from *Bombyx*, fresh homogenates of active prothoracic glands of locusts, solutions of gibberellic acid and locust extracts containing "ecdysone-λ" all have specific effects on the nervous system of locusts that results in reduced locomotory activity. It therefore seems that the insect growth-substance "ecdysone-λ" and the plant growth-substance gibberellic acid have similar effects on both plants and locusts.

It is possible that the ecdysones have been elaborated by the plant to interfere with the growth processes of insect predators. So the first question that arises is whether or not ingested ecdysones can be absorbed from the gut into the blood. In this connection, it is known that the bracken fern *Pteridium aquilinum* contains active ecdysones (Kaplanis *et al.*, 1967; Takemoto *et al.*, 1968a,b), but Carlisle and Ellis (1968a) found that when bracken was fed to the desert locust (*Schistocerca gregaria*) as its sole or chief diet, it did not affect moulting, growth or development. They also found large amounts of ecdysone in the faeces and suggest that ecdysones probably cannot be absorbed from the gut. In support of this conclusion, is the observation by Staal (1967a; Staal and van der Burg, 1967b) that a very much larger dose of ecdysone (at least tenfold) has to be administered in the food than by injection, to upset development in larvae. The Prague group (Sláma, 1968) have also been unable to demonstrate the uptake of ecdysones from the insect gut. In addition, many insects live on plants known to be rich in insect-moulting substances, e.g. the silkworm, *Bombyx mori*, lives on mulberry leaves, a rich source of ecdysones (as quoted in Takemoto *et al.*, 1968b). On the basis of finding a very powerful enzymic system in *Calliphora*, which regulates the moulting hormone activity by forming a hormone-glycoside, Hoffmeister (1970) has suggested that possibly some insects protect themselves against the comparatively large quantities of hormones in their food by this mechanism. However, this observation is difficult to reconcile with the finding that ponasterone A glycoside (Ponasteroside A, 18) is active in the moulting hormone assay (Takemoto *et al.*, 1968a).

It is also known that the ecdysones do not pass through the insect cuticle (Ohtaki *et al.*, 1967; Herout, 1970). However, Williams (1968b) has shown that the ecdysones can penetrate the insect cuticle when topically applied in non-volatile solvents such as undecylenic acid, α-tocopherol, or caprylic acid, and has therefore suggested that it is just possible that in their native dissolved state within the plant, ecdysones can penetrate the insect cuticle on contact. It is also possible that surface leaf oils could aid passage of ecdysones through the insect cuticle. An excessive uptake of ecdysone would cause lethal derangement of development.

In contrast to other studies, Robbins and co-workers (Robbins *et al.*, 1968) have obtained evidence to suggest that certain ecdysones and analogues can be absorbed from the intestinal tract of insects when fed in the diet. The three synthetic ecdysone analogues, 2β,3β,14α-trihydroxy-5β-cholest-7-en-6-one (31), 2β,3β-dihydroxy-5β-cholest-7-en-6-one (32) and 2β,3β-dihydroxy-5β-cholestan-6-one (33) inhibited larval growth and development in several species of insects when fed in the diet, whereas ecdysterone was inactive. These three

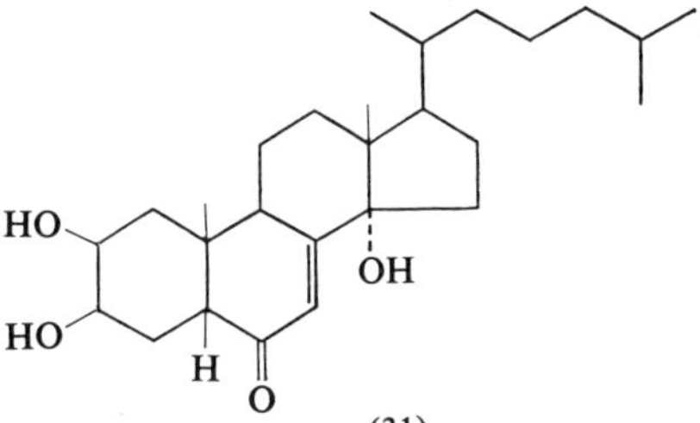

(31)

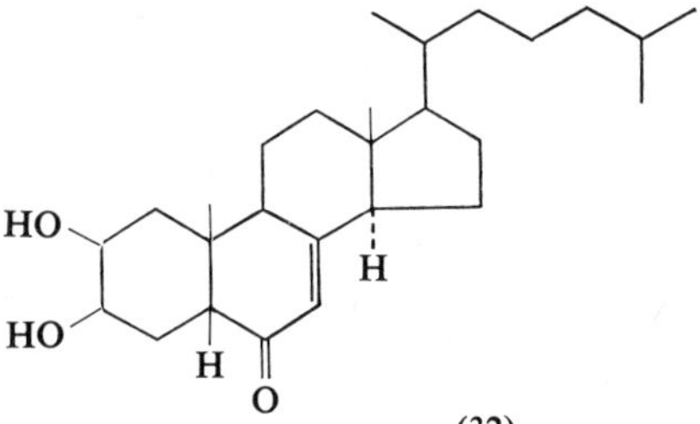

(32)

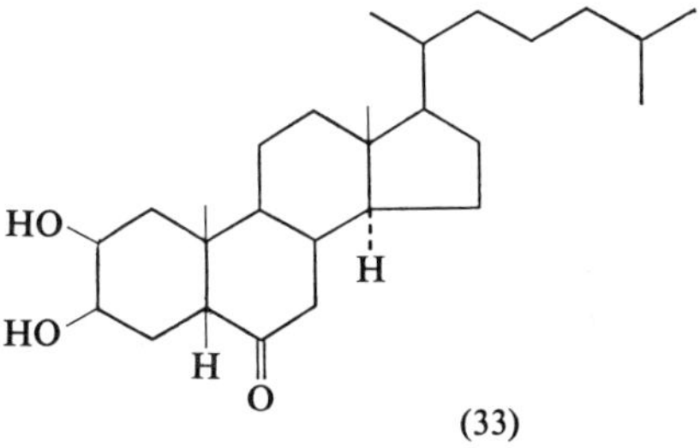

(33)

steroids (31, 32, 33) showed low activity in the house-fly moulting-hormone assay (Kaplanis *et al.*, 1966a). The corresponding 5α-analogues are hormonally inactive in the house-fly assay and are not inhibitors of growth and development. It is also interesting that ecdysterone (2), ponasterone A (7) and 2β,3β,14α-trihydroxy-5β-cholest-7-en-6-one (31) inhibit maturation of the ovaries and egg production in the adult house-fly. Some of these synthetic steroids showed inhibition activity at concentrations well within the concentrations of phytoecdysones in certain plants. It has been suggested (Robbins *et al.*, 1968) that these analogues, especially compounds (31) and (32), may well be similar to or even identical with intermediates in the biosynthesis of ecdysones. Therefore, structurally similar steroids to the ecdysones may be more likely to serve as plant protectants against immature insects than the insect ecdysones themselves. In conclusion, therefore, we can say that the real function(s) of ecdysones in plants is still uncertain.

V. Biosynthesis and Metabolism

A. Insects

Insects do not possess the complete sterol biosynthetic pathway and require a dietary supply (Clayton, 1964; Ritter and Wientjens, 1967). Karlson and Hoffmeister (1963) first demonstrated the conversion of [^{3}H]-cholesterol into ecdysone (1) in *Calliphora erythrocephala*, but only in very low yield (approx. 0·0001%). However, this is not surprising in view of recent findings by Australian workers (Galbraith *et al.*, 1969e) that ecdysterone is the only hormone present in detectable amounts in this insect. A more efficient conversion of cholesterol into ecdysterone (2) has now been observed in both *C. stygia* (Galbraith *et al.*, 1970) and C. *vicina* (= *C. erythrocephala*) (A. Willig, H. H. Rees and T. W. Goodwin, unpublished results). Nakanishi and co-workers (personal communication) have demonstrated that the conversion of cholesterol to ecdysone (1) and ecdysterone (2) occurs in aseptically reared silkworms.

The role of cholesterol as a precursor of ecdysones is not surprising in view of the observation that only two species of insects [*Drosophila pachea* (Heed and Kircher, 1965) and *Xyleborus ferrugineus* (Kok *et al.*, 1970)] have been reported, which cannot be maintained on a diet containing cholesterol as the sole sterol. The sterol requirements of the Mexican cactus fly, *D. pachea* can only be covered by stigmast-7-en-3β-ol (a cactus sterol) or other Δ^7 or $\Delta^{5,7}$ sterols such as cholest-7-en-3β-ol and cholest-5,7-dien-3β-ol (Heed and Kircher, 1965). It has been shown (Kok *et al.*, 1970) that larvae of the beetle, *X. ferrugineus* are incapable of pupating when reared under aseptic conditions on a defined medium containing cholesterol as the sole sterol. Ergosterol or 7-dehydrocholesterol was adequate as the sole sterol source for continued growth, development and reproduction of the aseptic beetle, whereas lanosterol, lumisterol, vitamin D_2 or D_3 and squalene were ineffective. The beetle normally obtains its sterol supply in nature from its gut fungal symbiont, *Fusarium solani*, which produces ergosterol as the only major sterol. It seems that both these insects cannot introduce the Δ^7 bond, which is essential for ecdysone formation. Phytophagous and omnivorous insects can obtain C_{27} sterols, such as cholesterol, not only from the food, but also by dealkylation of dietary C_{28} and C_{29} sterols (Clayton, 1964; Ritter and Wientjens, 1967; Svoboda *et al.*, 1969 and references cited therein). The beetle, *X. ferrugineus* must also presumably be capable of carrying out such a dealkylation of its ergosterol supply, prior to ecdysone synthesis.

Very little is known concerning the intermediate steps between cholesterol and the ecdysones. Cholesta-5,7-dien-3β-ol can maintain growth in several insects (Clayton, 1964; Ritter and Wientjens, 1967) and can be considered as a possible intermediate in the transformation of cholesterol to ecdysones. The conversion of cholesterol into cholesta-5,7-dien-3β-ol has been demonstrated under aseptic conditions in the cockroach, *Blatella germanica* (Robbins *et al.*,

1964) and in the housefly, *Musca domestica* (Monroe *et al.*, 1967). This conversion has also been demonstrated under non-aseptic conditions in other studies (Kaplanis *et al.*, 1960; Robbins *et al.*, 1961; Ishii *et al.*, 1963). The formation of cholesta-5,7-dien-3β-ol from cholesta-7-en-3β-ol has also been demonstrated under aseptic conditions in *D. pachea* (Kircher and Goodnight, 1970). The conversion of cholesta-5,7-dien-3β-ol into ecdysterone in *C. stygia* larvae has recently been demonstrated (Galbraith *et al.*, 1970). Since the conversion of radioactive cholesterol into ecdysterone in *C. stygia* was not reduced by simultaneously injecting non-radioactive cholesta-5,7-dien-3β-ol, these authors suggest that it appears unlikely that this compound in a free, unconjugated form, is an intermediate in ecdysterone biosynthesis in *C. stygia*. However, this point warrants further investigation, since it is based on negative evidence. *Calliphora stygia* (Galbraith *et al.*, 1970) in common with many other insects (see for example Beck and Kapadia, 1957; Schaefer *et al.*, 1965; Martin and Carls, 1968) contains small amounts of cholesta-5,7-dien-3β-ol. It is also interesting that during the last nymphal moult of *Periplaneta americana*, 25% of the sterols of the prothoracic glands, the source of ecdysones, consists of cholesta-5,7-dien-3β-ol (Robbins *et al.*, 1964).

There are a number of possible mechanisms for the formation of the A/B *cis* ring junction and the α,β unsaturated keto group. The A/B *cis* ring junction could be formed by reduction of a Δ^4-3-ketone, as in bile alcohol and acid formation in animals (Dorfman and Ungar, 1965). Alternatively, the formation of this ring junction could be coupled with the introduction of the oxygen function at C-6 by epoxidation of a Δ^5 bond, followed by opening of the epoxide ring. This could presumably occur in a Δ^5 or $\Delta^{5,7}$ sterol. The C-6 oxygen function could also be introduced by direct hydroxylation of 5β-cholestan-3β-ol or 5β-cholest-7-en-3β-ol, or hydroxylated derivatives thereof. Oxidation to give the C-6 ketone followed by enolization could give the correct configuration at C-5. An analogous oxidation of a C-6 hydroxyl group in Δ^7 sterols to a 6-keto function is known to occur in rat liver (Slaytor and Bloch, 1965). These are merely a few of the possibilities.

A number of publications have recently appeared concerning some of the later ecdysone biosynthetic steps. King and Siddall (1969) have demonstrated that both crustacea and insects can effect the conversion of ecdysone (1) into ecdysterone (2), using three species of animals. The groups which were presumably producing moulting hormones at rapid rates, actively moulting shrimp (*Crangon nigricauda*), premoult crab (*Uca pugilator*) and pupating blowfly (*Calliphora vicina*), showed the most efficient conversion (approx. 75%) of ecdysone to ecdysterone, whereas inter-moult crab (*U. pugilator*) only showed approximately 25% conversion. Since the presence of ecdysone has not yet been demonstrated in these species, whereas ecdysterone does occur, the authors suggested that ecdysone may possibly be a very short-lived intermediate in ecdysterone biosynthesis, if indeed it is a normal precursor. The conversion of ecdysone into ecdysterone has also been observed (Thomson *et al.*, 1969) in third instar larvae of *C. stygia*.

Kaplanis and co-workers (1969) have demonstrated that [1α-$^3H_1$]-2β,3β,14α-trihydroxy-5β-cholest-7-en-6-one (31) terminated diapause when injected into diapausing tobacco hornworm (*Manduca sexta*) pupae, and that it was efficiently converted into ecdysone and ecdysterone. They also had evidence of the presence of [^3H]-20,26-dihydroxyecdysone (3), the third moulting hormone isolated from the tobacco hornworm. The authors present evidence in support of their claim that in diapausing hornworm pupae, the triol (31) (or its metabolites) trigger the biosynthetic mechanism for the synthesis of the ecdysones from endogenous steroid precursors, in addition to serving as an efficient precursor of the moulting hormones. On the basis that the conversion of ecdysone into ecdysterone has been demonstrated in *Calliphora* and since the three hornworm ecdysones (1, 2 and 3) are biosynthesized from a common precursor, Kaplanis and co-workers suggest that the three ecdysones are metabolites in a biosynthetic degradative scheme of the ecdysones.

It has been demonstrated (Thomson *et al.*, 1969) that [23,24,24,25,25-$^3H_5$]-25-deoxyecdysone (34) injected into *C. stygia* at the time of puparium formation, is converted into radioactive ponasterone A (7), inokosterone (10) and ecdysterone (2), with no detectable activity in ecdysone (1). The authors suggest that ponasterone A is a precursor of inokosterone and ecdysterone, and that as ponasterone A and inokosterone were not detected on bulk extraction

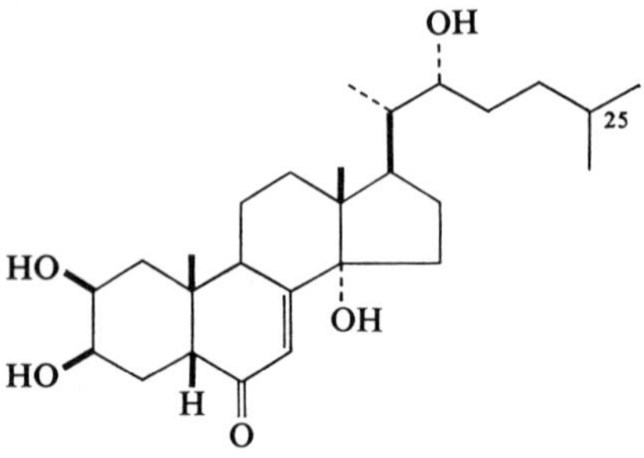

(34) 25-Deoxyecdysone

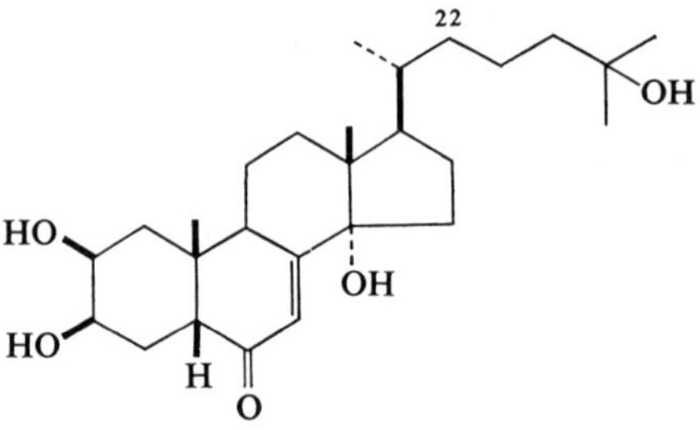

(35) 22-Deoxyecdysone

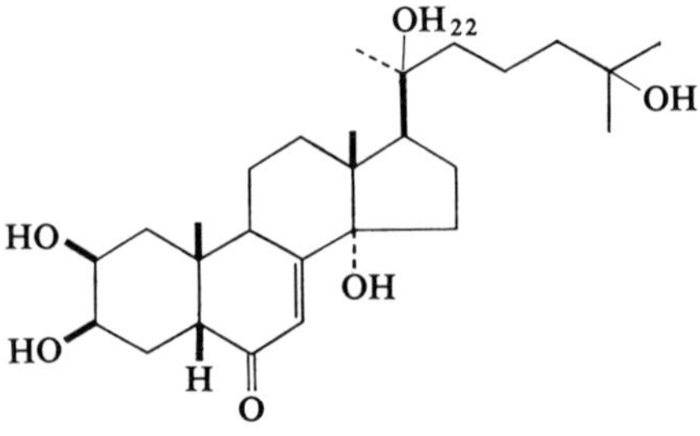

(36) 22-Deoxycrustecdysone

of *C. stygia*, 25-deoxyecdysone is not a normal major precursor of ecdysterone in *Calliphora*. Thomson and co-workers also suggest that it is unlikely that 22-deoxyecdysone (35) or 22-deoxycrustecdysone (36) can be a precursor of ecdysterone, but there is no direct evidence for this. This suggestion was partly based upon the very low biological activity of 22-deoxycrustecdysone in isolated abdomens of *Calliphora*, and hence the inability of this tissue to metabolize this compound to ecdysterone. It was also suggested that ecdysterone biosynthesis in *Calliphora* probably involves side-chain hydroxylation of precursor sterols at C-22 and C-25 prior to elaboration of the ring structure. However, this seems to be contradictory to the observed conversion of 2β,3β,14α-trihydroxy-5β-cholest-7-en-6-one (31) into ecdysone and ecdysterone in *M. sexta* (Kaplanis *et al.*, 1969). Thomson and co-workers (1969) suggest that in the crayfish, *Jasus lalandei*, 2-deoxyecdysterone (4), with a fully hydroxylated side chain may be expected to be the precursor of ecdysterone, since both have been isolated from this source (Galbraith *et al.*, 1968a). Clearly, there are a number of alternative possible biosynthetic pathways for ecdysone formation from cholesterol, involving different hydroxylation sequences, but a full consideration of these is beyond the scope of this review.

The further metabolism or deactivation of ecdysones has also been investigated in a few insect species. Karlson's group (Bode *et al.*, 1968; Karlson and Bode, 1969; Karlson, 1970) found that ecdysone injected into *Calliphora* larvae is rapidly inactivated, with a half-life of approximately 2 h, but this varies slightly with the age of the larvae. This is in agreement with other similar observations in *Calliphora* (Shaaya, 1969), *Sarcophaga peregrina* (Ohtaki *et al.*, 1968) and *Pyrrhocoris apterus* (Emmerich, 1970). In view of these findings, it is surprising that King and Siddall (1969) could account for a minimum of 25% of the injected ^3H from [23,24-^3H]-ecdysone as ecdysone and ecdysterone after 12–24 h in *Calliphora*. Karlson and Bode (1969) found that high inactivating activity was present in the fat body and that the inactivation is mediated by a soluble enzyme system, which has been partly purified by ammonium sulphate precipitation. These authors also demonstrated in *Calliphora* that the activity of the inactivating enzyme is nearly a mirror image of the ecdysone titre at different stages of development. By thin-layer chromatography of the products of radioactive ecdysone inactivation, Karlson (1970) demonstrated the presence of two radioactive metabolites, less polar than ecdysone, which means that they should contain at least one hydroxyl group less. Therefore, it seems that ecdysterone is not an obligatory product of ecdysone metabolism, as indicated by King and Siddall (1969). This discrepancy may possibly merely reflect the precise developmental stage of the animals at the time of experimentation.

The 20,22-diol structure of ecdysterone, like postulated intermediates (Shimizu *et al.*, 1962) in the biological side-chain cleavage of cholesterol prompted Horn and co-workers (1966) to suggest that ecdysterone could

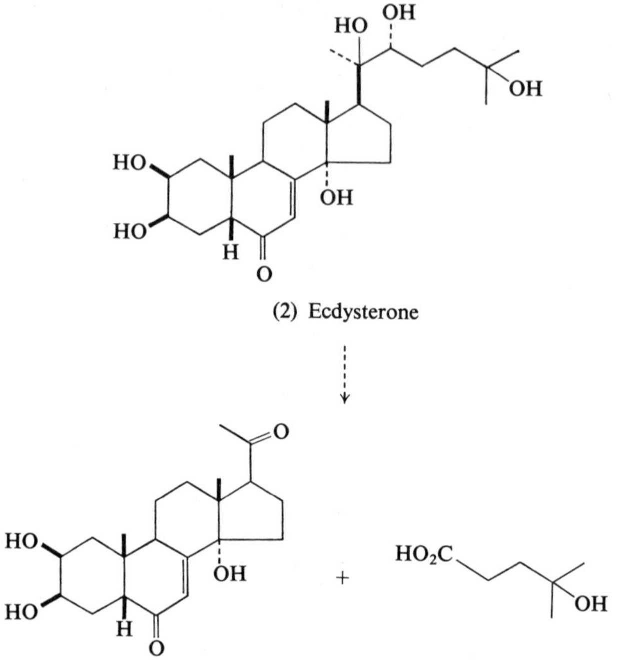

(2) Ecdysterone

(37) 2β,3β,14α-Trihydroxy-5β-pregn-7-en-6,20-dione

(38) 4-Hydroxy-4-methylpentanoic acid

FIG. 2. Suggested catabolic pathway of ecdysterone in *Calliphora* (Galbraith *et al.*, 1969d).

possibly be catabolized in an analogous manner to 2β,3β,14α-trihydroxy-5β-pregn-7-en-6,20-dione (Fig. 2, 37). This keto compound was synthesized but could not be detected in one ton of crayfish (Siddall *et al.*, 1967). When [23,23,24,24-$^3H_4$]-ecdysterone was injected into *Calliphora* prepupae 20 h after puparium formation and the insects extracted 6 h later, less than 10% of the radioactivity of the extract could be isolated as unchanged ecdysterone (Galbraith *et al.*, 1969d). Most of the radioactivity was present in the form of water-soluble metabolites which were not readily extracted by butanol. From these water-soluble metabolites, 4-hydroxy-4-methylpentanoic acid (38) was isolated as the lactone, after addition of carrier material, but accounted for only 0·3% of the total activity of the catabolic products. It was therefore suggested (Galbraith *et al.*, 1969d) that ecdysterone catabolism in *Calliphora* by direct C-20–C-22 bond scission may not be a major metabolic pathway. However, the low yield of side-chain fragment may possibly reflect to some extent the poor recoveries often obtained during isolation of such minor fragments from biological material. The 20-ketone (37) expected to be produced in the cleavage reaction has not been isolated to date from insects, but may undergo further rapid oxidation to a compound such as rubrosterone (30), as suggested in the case of plants (Takemoto *et al.*, 1969a).

Carlisle and Ellis (1968a) have suggested that ecdysone appears to have little effect on physiological processes in locusts and that it may possibly be an excretory metabolite of the true moulting hormone. Locusts would therefore excrete ecdysones by dehydroxylation. This is in contrast to the views of Thompson and co-workers (1967), who suggest that since 20,26-dihydroxy-ecdysone (3) is less active than either ecdysone or ecdysterone, increased hydroxylation is a mechanism in tobacco hornworm for the deactivation of the ecdysones.

B. PLANTS

It is well known that plants, in contrast to insects, possess the complete sterol biosynthetic pathway from mevalonic acid (see Goad, 1967). The biosynthesis from mevalonate of ponasterone A (7) and ecdysterone (2) in *Taxus baccata* seedlings (de Souza *et al.*, 1969) and of ecdysterone in *Polypodium vulgare* (de Souza *et al.*, 1970) has been demonstrated, by administration of the substrate to the leaves over a period of approximately 3 weeks.

Since cholesterol had been converted into ecdysones in insects, it seemed possible that the C_{27} ecdysones in plants could also possibly be formed from the same substance. Cholesterol has been isolated from a large number of plant families and its biosynthesis from mevalonate has been demonstrated (for references see de Souza *et al.*, 1970). The conversion of cholesterol into ecdysterone has been demonstrated in *Podocarpus elata* (Sauer *et al.*, 1968), into ecdysterone and ponasterone A in *P. macrophyllus* (Hikino *et al.*, 1970) and *T. baccata* (N. J. de Souza, E. L. Ghisalberti, H. H. Rees and T. W. Goodwin, unpublished results) and into ecdysone, ecdysterone and 5β-hydroxyecdysterone in *Polypodium vulgare* (de Souza *et al.*, 1970). In all these studies the cholesterol was applied to the leaves for periods of 4–8 weeks, and the maximum conversion into an ecdysone was in the region of 2%, but appreciably lower in many cases. That the comparatively low levels of incorporation of mevalonic acid and cholesterol into ecdysones is probably due to a slow biosynthetic rate rather than a high rate of turnover of the ecdysones, is indicated by the fact that short-term administration of [2-^{14}C]-mevalonic acid over periods of 1–7 days to *T. baccata* or *P. vulgare* results in virtually no incorporation of radioactivity into the ecdysones (N. J. de Souza, E. L. Ghisalberti, H. H. Rees and T. W. Goodwin unpublished results; de Souza *et al.*, 1970). Whether cholesterol is an obligatory intermediate in the biosynthesis of ecdysones in plants remains an open question. It is possible that the ecdysone pathway branches off the normal phytosterol pathway before the 4-desmethyl sterol stage.

It seems possible that in different plants, the sequence of side-chain hydroxylation could be different. For example, it has been suggested that in *T. baccata* (de Souza *et al.*, 1969), ecdysterone could be formed via ponasterone A, whereas in *P. vulgare*, ecdysone could be the precursor of ecdysterone (de

Souza *et al.*, 1970), which is probably further metabolized to 5β-hydroxy-ecdysterone (Fig. 3). However, the participation of ecdysone as a rapidly metabolized precursor of ecdysterone in *T. baccata* cannot be ruled out.

There are a number of possible mechanisms for the formation of the A/B *cis* ring structure in plants, as in insects. However, in the case of plants, it is possible that the ecdysone biosynthetic pathway branches off the usual phytosterol pathway before introduction of the Δ^5 bond. It is therefore just conceivable that the hydrogen originally at C-5 in cycloartenol, the first stable cyclization product of squalene-2,3-oxide during sterol biosynthesis in plants (see Goad, 1970) could be retained probably at another position. Preliminary results (N. J. de Souza, E. L. Ghisalberti, H. H. Rees and T. W. Goodwin, unpublished results) using 3*R*-[2-^{14}C-(4*R*)-4-^{3}H$_1$]-MVA, indicate that this hydrogen originally at C-5 is lost during ecdysterone biosynthesis in *T. baccata*, but the mechanism is uncertain. In view of the long time of radioactive incubations with labelled mevalonic acid (de Souza *et al.*, 1969, 1970), there was a possibility that scrambling of label might occur in the ecdysones, but these latter results also indicate that this is not so. Heftmann's group have considered the change from Δ^5 to 5β-H. In animals, this transformation involves a Δ^4-3-ketone intermediate, as previously indicated. Caspi and Hornby (1968) also demonstrated the same requirement in the plant, *Digitalis lanata*, where progesterone was found to be an obligatory intermediate between the Δ^5 precursor, pregnenolone and the 5β-cardenolides. An analogous mechanism has also been suggested by Heftmann's group (Sauer *et al.*, 1968) for biosynthesis of the ecdysones, but they failed to demonstrate any significant conversion of [4-^{14}C]-cholest-4-en-3-one into ecdysterone in *Podocarpus elata* plants. Another possibility in the formation of the A/B *cis* ring junction (Joly *et al.*, 1969) is the involvement of sterol-5, 6-epoxides followed by rearrangement to the 6-ketone; both 5α,6α- and 5β,6β-epoxides are possibilities since a 5α-H,6-ketone could be isomerized to the 5β-isomer by enolization. When labelled cholesterol 5α,6α-epoxide and cholesterol 5β,6β-epoxide were administered to *P. elata* seedlings, no incorporation into ecdysterone was obtained (Joly *et al.*, 1969). However, it is possible that these types of compounds could be intermediates in ecdysone biosynthesis at some other stage, e.g. after introduction of the Δ^7 bond or after partial hydroxylation. 25-Hydroxycholesterol-[26-^{14}C] did not serve as a precursor of ecdysterone in *Podocarpus*, indicating that hydroxylation at C-25 follows some other transformations and that the enzymes involved in the initial steps are sufficiently specific that they cannot use 25-hydroxycholesterol as substrate in place of cholesterol (Joly *et al.*, 1969). These authors also demonstrated that the cholesterol side-chain is incorporated intact into ecdysterone by feeding [26-^{14}C]-cholesterol.

It is interesting that Herout's group (Herout *et al.*, 1969) have isolated from *Polypodium vulgare* rhizomes two compounds having certain structural similarities to ecdysones, which might be considered (Herout, 1970) as possible

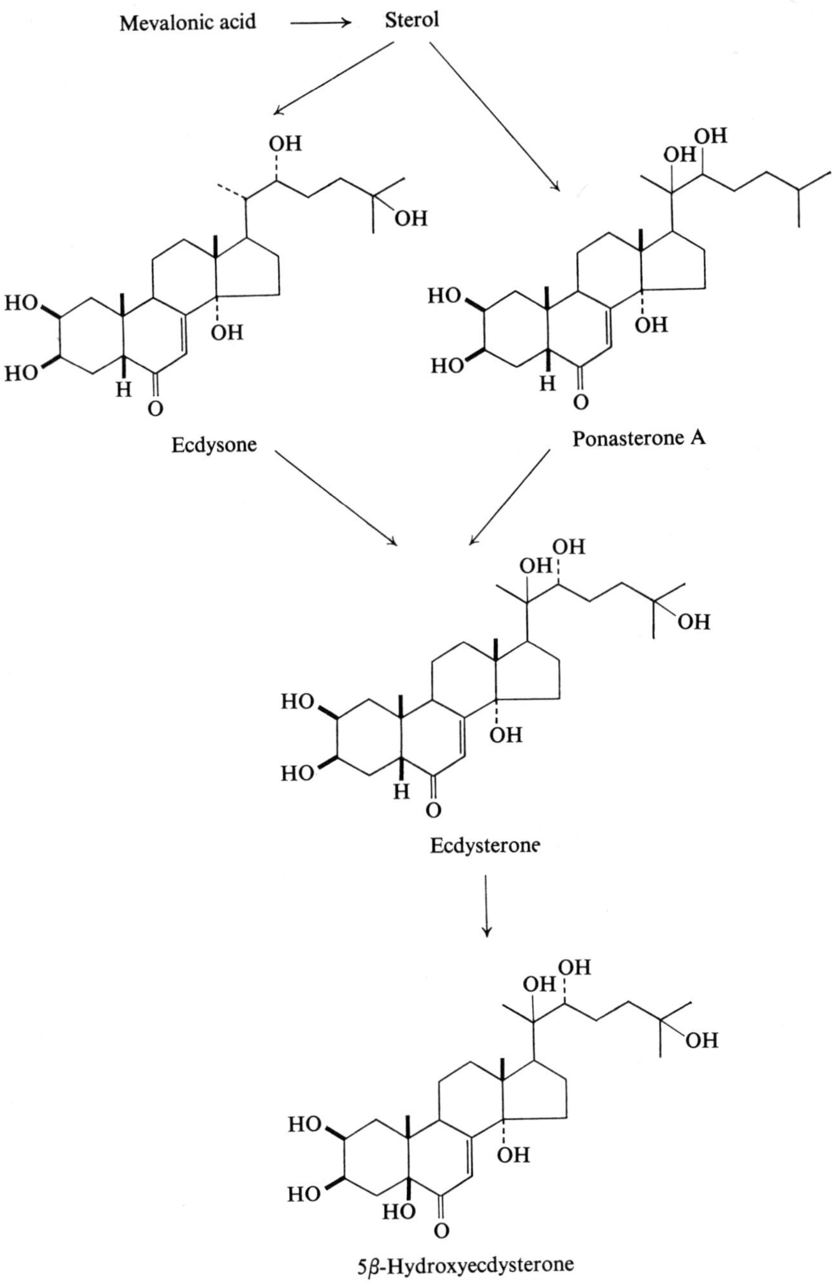

Mevalonic acid → Sterol

Ecdysone

Ponasterone A

Ecdysterone

5β-Hydroxyecdysterone

FIG. 3. Possible alternative biosynthetic pathways of ecdysones in plants (de Souza *et al.*, 1970).

intermediates in the biosynthesis of ecdysone-type compounds. These compounds are osladin (probable structure 39) and polypodosaponin A (40), which occur as their glycosides. It is interesting that these compounds have

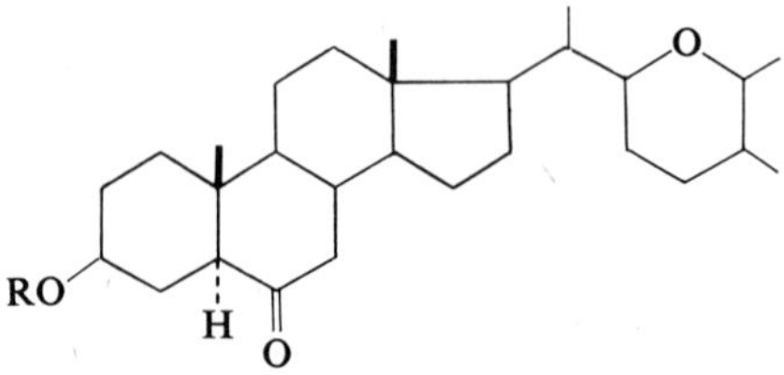

R = Dirhamnoglucosyl

(39) Osladin

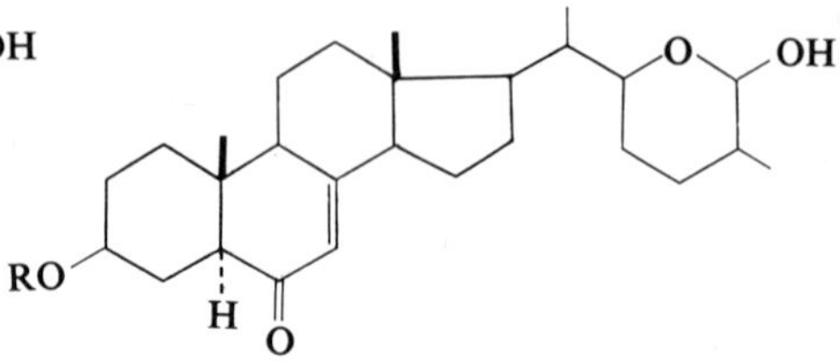

R = Rhamnoglucosyl

(40) Polypodosaponin A

the A/B ring *trans* configuration, but could easily change to 5β-H by enolization. Examination of molecular models of a sterol indicates that a hydroxyl at C-2 and the C-19 methyl are in close proximity when the A/B rings are *trans* fused, so that it is possible that the change to *cis* fusion in ecdysones may occur at roughly the same time as C-2 hydroxylation. Two other ecdysone analogues, cheilanthones A (41) and B (42) have been isolated together with

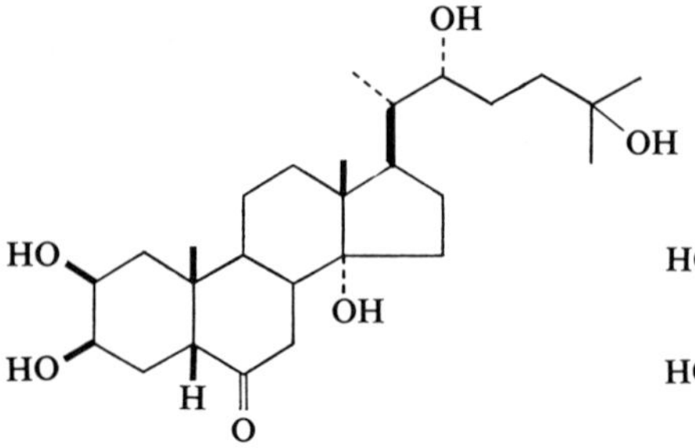

(41) Cheilanthone A

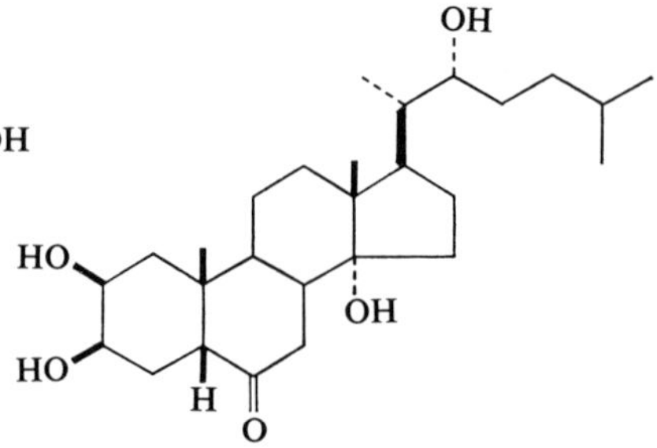

(42) Cheilanthone B

ecdysone from the fern, *Cheilanthes tenuifolia* (Faux *et al.*, 1970). It seems possible merely on structural grounds that the introduction of the Δ^7 bond and the 25-hydroxyl group may be the final stages in ecdysone biosynthesis in this plant. However, it is conceivable that the Δ^7 bond is reduced to give these structures (41 and 42). Cheilanthones A and B were biologically inactive in the *Calliphora* assay. On the basis that unlike ecdysone, cheilanthone A did not accelerate puparium formation in intact larvae, it was suggested (Faux *et al.*, 1970) that *Calliphora* is unable to introduce the Δ^7 bond into cheilanthone A and that this bond is probably introduced during ecdysone biosynthesis in this insect before the introduction of all the hydroxyl groups. Other sterols having certain structural features in common with the ecdysones have been isolated from plant material, e.g. peniocerol (43) (Djerassi *et al.*, 1961, 1964), macdougallin (44) (Djerassi *et al.*, 1963), desoxyviperidone (45) (Knight and Pettit, 1969), viperidone (46) and viperidinone (47) (Djerassi

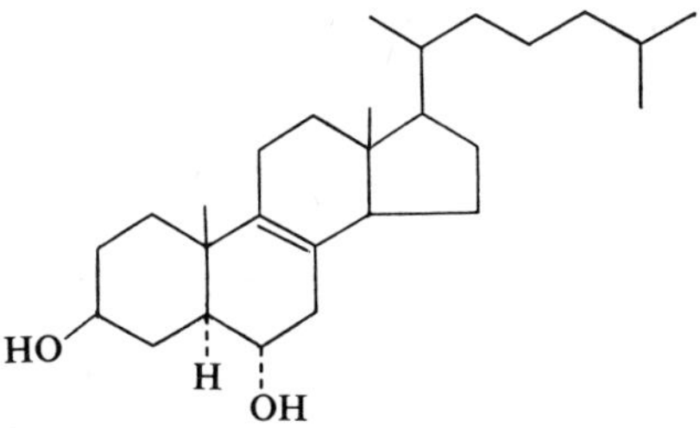

(43) Peniocerol

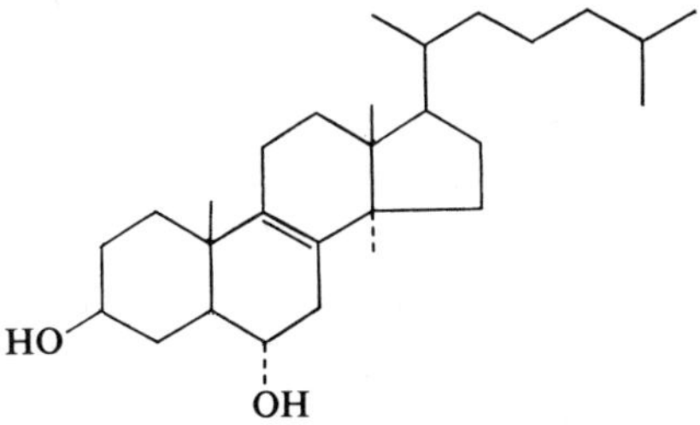

(44) Macdougallin

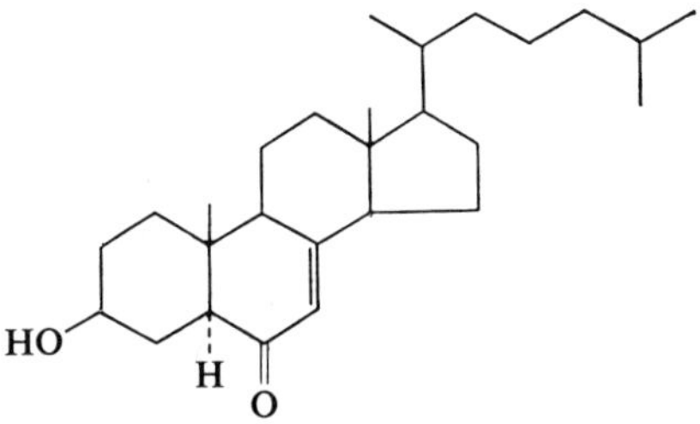

(45) Desoxyviperidone

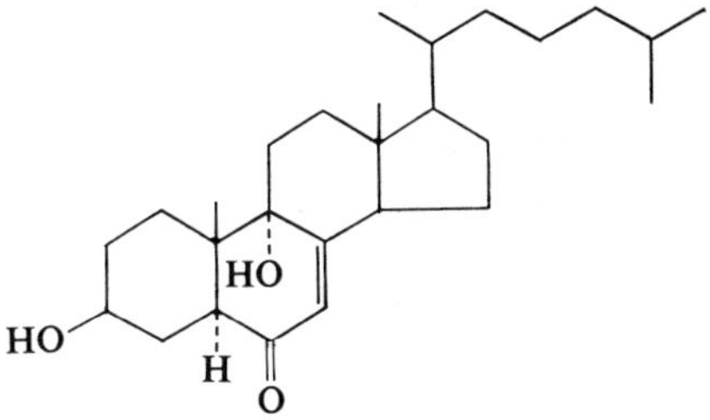

(46) Viperidone

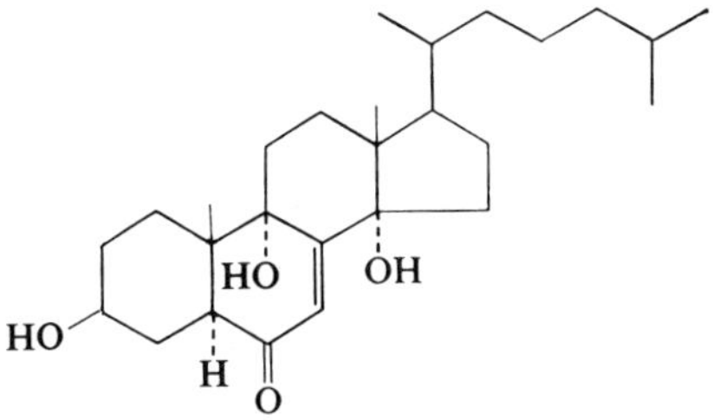

(47) Viperidinone

et al., 1964) have been isolated from various cacti and co-occur except for macdougallin in the cactus *Peniocereus greggii* (Knight and Pettit, 1969). It is possible that peniocerol (43) is formed by direct hydroxylation at C-6, and that this hydroxyl group undergoes oxidation to a ketone together with migration of the Δ^8 bond to the Δ^7 position to give desoxyviperidone (45). Again enolization could invert configuration at C-5. However, there is no information on whether or not the A/B ring structure in ecdysones could be formed from a Δ^8 compound by this type of mechanism.

In summary, it is not known at which stage in the normal phytosterol pathway, the biosynthetic pathway to the C_{27}, C_{28} and C_{29} ecdysones branches off. We do, however, know that it can occur at a late stage, i.e. a Δ^5 4-desmethyl sterol stage.

VI. Physiological and Biochemical Effects

A full discussion of the physiological and biochemical effects of ecdysones is well beyond the scope of this review, but a brief outline might be useful. Most of the physiological effects are concerned with moulting and metamorphosis, or with processes preparatory to moulting [for a summary of some of these effects see Karlson and Sekeris (1966a)]. It is still not certain whether different ecdysones or different combinations of ecdysones control moulting in different insects.

It has been demonstrated that regeneration of imaginal discs in *Galleria mellonella* requires ecdysone (1) (Madhavan and Schneiderman, 1969), whereas both ecdysterone (2) and cyasterone (24) induce metamorphosis of *Drosophila* imaginal discs cultured *in vitro* (Postethwait and Schneiderman, 1970). Observations that ecdysone, ecdysterone and inokosterone (10) all initiated metamorphosis of *G. mellonella* wing discs *in vitro*, whereas only ecdysone stimulated DNA synthesis, led to the suggestion that ecdysone and ecdysterone may have distinct roles *in vivo* (Oberlander, 1969a,b).

Berreur and Fraenkel (1969) demonstrated that ecdysone controls the puparial contraction as well as tanning in *Calliphora*. A neurohormone in *Sarcophaga bullata* has recently been reported (Zdarek and Fraenkel, 1969) to accelerate the formation and tanning of the fly puparium in the presence of ecdysone. On the basis of experiments with *S. peregrina* and the short half-life of ecdysone, it has been suggested (Ohtaki *et al.*, 1968) that "evidence points to the accumulation, not of the hormone itself, but the covert biochemical and biophysical effects of the hormone. The covert effects undergo spatial and temporal summation within the target organs and finally discharge the covert developmental response."

Whereas DOPA decarboxylase is induced by ecdysone (see below), it has been suggested that 5-hydroxytryptophan decarboxylase, which decarboxylates 5-hydroxytryptophan to serotonin, may possibly be repressed by ecdysone (Marmaras *et al.*, 1966). It has been demonstrated (Kobayashi and Kimura, 1967) that ecdysone injection into brainless pupae of *Bombyx mori* results in preferential conversion of labelled glucose to trehalose, whereas in control experiments glycogen was formed. It was therefore suggested that ecdysone affects the relative activities of enzymes at the branching point of the pathways of trehalose and glycogen synthesis in the fat body, so that the pathway of trehalose synthesis is activated.

Several ecdysones have been shown to stimulate RNA (Otaka and Uchiyama, 1969) and protein (Okui *et al.*, 1968; Otaka *et al.*, 1969; Hikino *et al.*, 1969b) synthesis in mouse liver both *in vivo* and *in vitro*.

The mode of action of ecdysone has been intensively studied, mainly by Karlson's group. Reviews of the earlier work are available (Karlson and Sekeris, 1966a; Karlson, 1967c; Sekeris, 1967). The first indication that ecdysone might act in the nucleus was obtained by Clever and Karlson (1960) who showed that within 15 min after injection of amounts as small as 2×10^{-6} μg of ecdysone into the larva of the midge *Chironomus tentans*, a characteristic "puff" appears in the giant salivary gland chromosomes. The puffs representing active gene structures, are sites of messenger RNA synthesis. Based on these findings, Karlson (1963a) put forward the hypothesis that ecdysone may act directly in the nucleus by releasing genetic information in the form of mRNA, which is translated in the cytoplasm into specific proteins, such as enzymes. According to this hypothesis, ecdysone should eventually lead to enzyme induction. Every step of this reaction sequence has been experimentally demonstrated for ecdysone, by studying induction of the enzyme, DOPA decarboxylase (3,4-dihydroxy-L-phenylalanine carboxylyase), which is involved in the conversion of tyrosine to *N*-acetyldopamine, the sclerotizing agent (Karlson and Sekeris, 1966a; Sekeris, 1967). Karlson and Sekeris (1966b) suggest that it seems likely that ecdysone combines with some protein constituent of the chromosome, possibly a histone, which could well be a repressor in terms of the Jacob–Monod model. Other workers (Clever and Romball, 1966) have suggested that the control of specific RNA synthesis may be a primary effect of ecdysone or a consequence of preceding actions on other systems in the cell. Clever and Romball (1966) summarized the primary steps in the cellular response to ecdysone as follows:

$$\text{ecdysone} \rightarrow (?) \rightarrow \text{activation of specific genes} \rightarrow \text{mRNA} \rightarrow \text{protein} \rightarrow (?) \rightleftarrows \text{further gene activation(s)}$$

Kroeger (1966, 1968; Kroeger and Lezzi, 1966) also suggested that ecdysone indirectly controls gene activity by controlling the $Na^{+}:K^{+}$ concentration ratio within the nucleus so that the K^{+} concentration is increased. However, in recent studies (Congote *et al.*, 1969) with isolated fat body cell nuclei it has been demonstrated that potassium ions alone cannot effect stimulation of RNA polymerase, whereas ecdysone can. On the other hand, the fact that sodium ions could stimulate RNA synthesis in this *in vitro* system, led to the suggestion that it could be involved in the transport of nucleotides or other substances in the nuclei or in direct stimulation of gene loci in the chromosomes. In these experiments, the stimulation of RNA synthesis in isolated nuclei by ecdysone or juvenile hormone alone, could be partly inhibited by incubating the nuclei with both hormones together. Injection of a mixture of the two hormones has been reported (Patel and Madhavan, 1969) to produce similar effects on RNA synthesis in imaginal wing discs of the Ricini silkworm. The working hypothesis was, therefore, put forward (Congote *et al.*, 1970) that some genes can be regulated by *both* ecdysone and juvenile hormone at the same time (without excluding the possible existence of specific ecdysone- *or* juvenile hormone-dependent genes). One hormone alone would thus lead to activation of the ecdysone- *and* juvenile hormone-dependent genes, while both hormones together would inactivate them. In this way, the fluctuating concentrations of both ecdysone and juvenile hormone in the larvae could lead either to stimulation of RNA synthesis in some stages of development or to inhibition of this production in other stages. Clearly much work remains to be done on the mode of action of ecdysone and juvenile hormone as with other hormones. Kroeger (1968) has reviewed extensive evidence in support of the primary action of ecdysone in the control of intranuclear electrolyte balance, whereas recent work by Karlson's group has furnished strong support for the primary action of ecdysone in the control of specific RNA synthesis.

VII. Structure–Activity Correlations

A. Ecdysones

It is difficult to correlate exactly structural features with moulting activity, partly because different groups of workers have used different methods for bioassay (e.g. Kaplanis *et al.*, 1966a; Ohtaki *et al.*, 1967; Kobayashi *et al.*, 1967; Williams, 1968a; Sato *et al.*, 1968; Adelung and Karlson, 1969), but also because assays carried out at different times or in different laboratories cannot be compared. It is also important to note that different ecdysones exhibit different relative activities in different insects. For example, the ponasterone activities decrease in the order A,B,C in the *Musca* test, whereas the order is A,C,B in *Calliphora* and *Bombyx* (Kobayashi *et al.*, 1967; Hoffmeister *et al.*, 1968). Despite these shortcomings, some observations can be made. Although there have been reports (Karlson, 1956a; Hampshire and Horn, 1966) that ecdysone has approximately twice the activity of ecdysterone in the *Calliphora* bioassay, most workers report that ecdysterone has 1·25–2 times greater activity than ecdysone (e.g. Hocks *et al.*, 1966; Gibian, 1967; Hoffmeister *et al.*, 1968). Ecdysterone from *Manduca sexta* is reported to possess the same biological activity as ecdysone (Kaplanis *et al.*, 1966b). Hocks and co-workers (1966) have compared the activities in the *Calliphora* assay of the compounds shown in Fig. 4. Removal of the side-chain hydroxyl groups of ecdysone to give compound (31) also removes most of the activity, which is further reduced on removal of the 14α-hydroxyl group (32). Introduction of hydroxyl groups into the side-chain (48), however, restores substantial activity. Substantially higher activity has been reported in other work for compound (31) (Gibian, 1967).

The Australian group have shown that 22-deoxyecdysterone (36) has only $\frac{1}{50}$–$\frac{1}{100}$th the activity of ecdysterone in the *Calliphora* test, approximately the same as that of the synthetic ecdysone analogue (31) devoid of side-chain hydroxyl groups. It was, therefore, suggested that a 22-hydroxy group is essential for high biological activity in *Calliphora* and that the C-20 and C-25 hydroxyls probably contribute little to the biological activity of ecdysone.

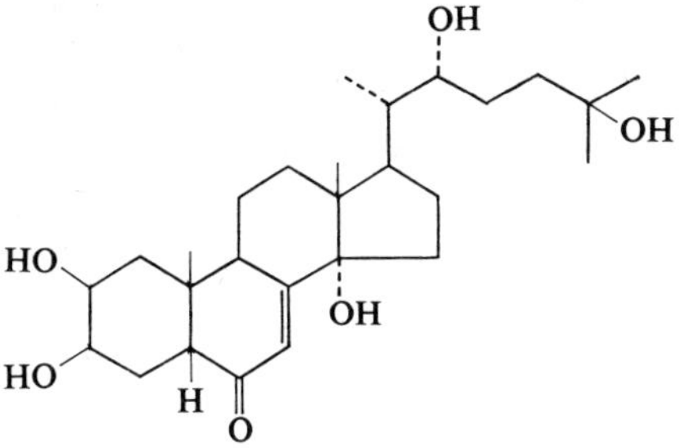

(1) Ecdysone
[1]

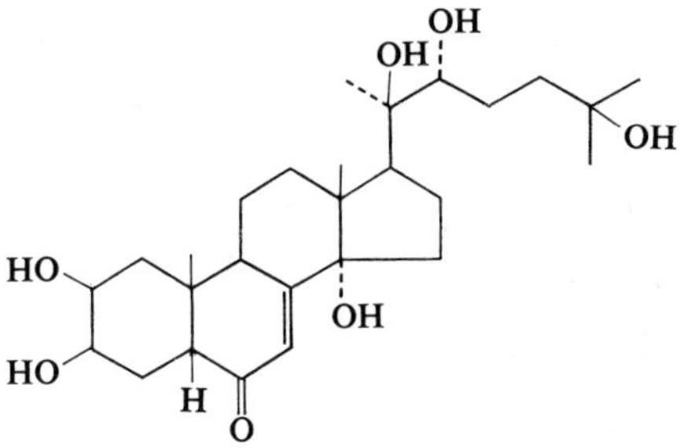

(2) Ecdysterone
[1·25]

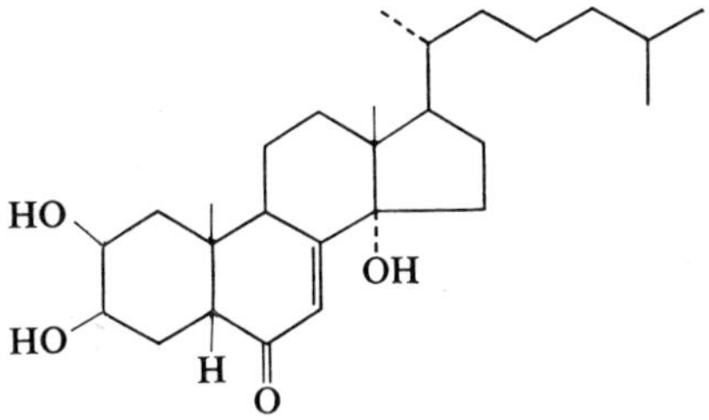

(31)
$\left[\frac{1}{50}\right]$

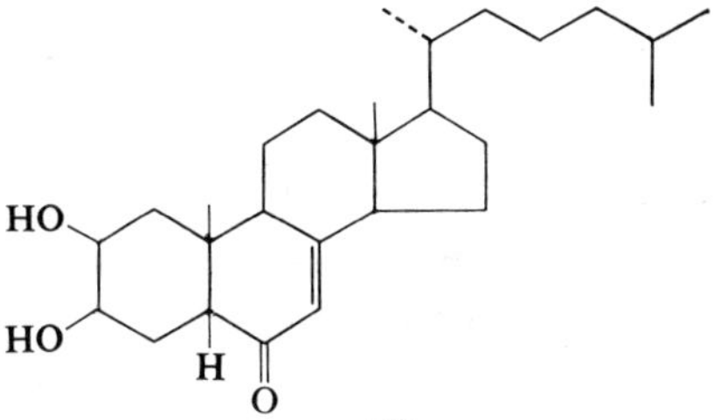

(32)
$\left[\frac{1}{80}\right]$

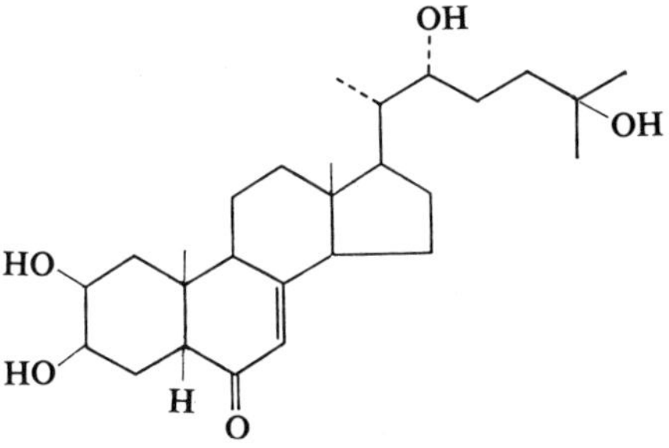

(48)
$\left[\frac{1}{15}\right]$

FIG. 4. Activities relative to ecdysone (in square brackets) of compounds in the *Calliphora* bioassay (Hocks *et al.*, 1966).

In actual fact, it has been reported (Gibian, 1967) that 25-deoxyecdysone (34) has one quarter the activity of ecdysone in *Calliphora*. It is also interesting that 22-epiecdysterone shows no activity in the *Calliphora* assay (Furlenmeier *et al.*, 1967). However, other workers (Harrison *et al.*, 1966) found that this compound showed one quarter to one eighth the activity of ecdysone when assayed in brainless pupae of the silkworm, *Samia cynthia*. Since the C-22 hydroxyl group, which seems to be important for biological activity, is masked in capitasterone (25), it is surprising that it shows high moulting hormone activity in *Sarcophaga peregrina* (Takemoto *et al.*, 1968j). However, it could possibly be metabolized to such a grouping in the test insect.

Turning now to a consideration of the ring structure, it is known that 5α-H compounds are inactive (Karlson, 1967b). Introduction of a 5β-hydroxyl group into ecdysterone results in four times the activity of ecdysone in *Calliphora* (Jizba *et al.*, 1967b). It seems that a 2-hydroxyl group is not essential, whereas compounds having 3β-hydroxyl groups, have higher activity than ones containing 3α-hydroxyl functions (Galbraith *et al.*, 1969b; Hoffmeister *et al.*, 1968). Podecdysone B (19) in which the Δ^7-6-keto grouping is absent, is noteworthy in that it has one fifth the activity of ecdysterone (Galbraith *et al.*, 1969c). The suggestion was made that this activity may be due to its *in vivo* isomerization to such a grouping. The inactivity of cheilanthone A (41) in the *Calliphora* assay illustrates the importance of the Δ^7 bond.

It is interesting that the methyl ketone (37), was inactive in the *Calliphora* test, but gave a positive response in induction of adult development of dauer pupae of the silkmoth (*Samia cynthia*) (Siddall *et al.*, 1967). However, rubrosterone (30) showed only weak activity in two different insect bioassays (Takemoto *et al.*, 1969a), but exhibited high stimulating effect upon protein synthesis in mouse liver, in common with other ecdysones. It was therefore suggested that for insect metamorphosis, a certain side-chain as well as nuclear structure is essential, whereas for stimulation of protein synthesis in mouse a certain nuclear structure only is required.

It is tempting to speculate that the unique features of rings A and B are responsible for moulting activity, which is enhanced by hydroxylations at C-14 and the side-chain. However, it is possible that the gross physico-chemical properties of the molecule may be an overriding factor.

B. SYNTHETIC STEROIDS SHOWING PHYSIOLOGICAL ACTIVITY IN INSECTS

The Prague group have synthesized a large number of cholestane, pregnane and androstane derivatives which seem to antagonize ecdysone in the insect, *Pyrrhocoris apterus* (Hora *et al.*, 1966; Velgová *et al.*, 1968; Lábler *et al.*, 1968; Velgová *et al.*, 1969). These compounds inhibit the hardening and sclerotization (tanning) of the cuticle in freshly moulted *Pyrrhocoris*, which later died. The relationship between the structure and activity of synthetic cholestane and pregnane derivatives has also been discussed (Velgová *et al.*, 1969). The presence of the cholestane side-chain is not essential for the antisclerotization activity, nor is the stereochemistry of the A/B ring junction important. At least one hydroxyl group in ring A is required for activity; it can be at either C-2 or C-3, although the C-2 hydroxy compounds are less active. The configuration of hydroxyl at C-3 is not critical, but a 6-keto group is important. The antisclerotization effects obtained in *Pyrrhocoris* do not have a general application for other insect species. For example, some of the most active compounds in *Pyrrhocoris* had no similar effects in pupae of *Galleria* or *Tenebrio*, although they inhibited the growth and development of their larvae. The mode of intervention of these synthetic steroids is as yet unknown. Many of these synthetic 6-keto cholestane and pregnane derivatives have a pronounced sterilizing activity in *Musca domestica*, but no obvious structure-activity correlation was observed (Řežábová *et al.*, 1968; Hora, 1969). The phenomena described in this section further illustrate the special effects which 6-keto steroids have in insects. They also emphasize the dangers of making generalizations from studies on a limited number of insect species.

ACKNOWLEDGEMENT

I wish to thank Professor T. W. Goodwin, F.R.S., for his continued interest and valuable suggestions and Drs N. J. de Souza, E. L. Ghisalberti and A. Willig for carrying out experimental work and for helpful discussions. I am grateful to Professor V. Herout for pre-publication information.

REFERENCES

Adelung, D. and Karlson, P. (1969). *J. Insect Physiol.* **15**, 1301.
Barritt, L. C. and Birt, L. M. (1970). *J. Insect Physiol.* **16**, 671.
Beck, S. D. and Kapadia, G. G. (1957). *Science, N.Y.* **126**, 258.
Berkoff, C. E. (1969). *Q. Rev., Lond.* **23**, 372.
Berreur, P. and Fraenkel, G. (1969). *Science, N.Y.* **164**, 1182.
Bode, C., Döpp, H. and Karlson, P. (1968). *Hoppe-Seyler's Z. Physiol. Chem.* **349**, 2.
Burdette, W. J. (1962). *Science, N.Y.* **135**, 430.
Burdette, W. J. and Bullock, M. N. (1963). *Science, N.Y.* **140**, 1311.
Butenandt, A. and Karlson, P. (1954). *Z. Naturf.* **9b**, 389.
Carlisle, D. B. (1965). *Gen. Comp. Endocr.* **5**, 366.
Carlisle, D. B. and Ellis, P. E. (1968a). *Science, N.Y.* **159**, 1472.
Carlisle, D. B. and Ellis, P. E. (1968b). *Nature, Lond.* **220**, 706.
Carlisle, D. B., Osborne, D. J., Ellis, P. E. and Moorhouse, J. E. (1963). *Nature, Lond.* **200**, 1230.
Caspi, E. and Hornby, G. M. (1968). *Phytochemistry* **7**, 423.
Clayton, R. B. (1964). *J. Lipid Res.* **5**, 3.
Clever, U. and Karlson, P. (1960). *Expl Cell. Res.* **20**, 623.
Clever, U. and Romball, C. G. (1966). *Proc. Natn. Acad. Sci. U.S.A.* **56**, 1470.
Congote, L. F., Sekeris, C. E. and Karlson, P. (1969). *Expl Cell. Res.* **56**, 338.
Congote, L. F., Sekeris, C. E. and Karlson, P. (1970). *Z. Naturf.* **25b**, 279.
Djerassi, C., Murray, D. H. and Villotti, R. (1961). *Proc. chem. Soc.* 450.
Djerassi, C., Knight, J. C. and Wilkinson, D. I. (1963). *J. Am. chem. Soc.* **85**, 835.
Djerassi, C., Knight, J. C. and Brockmann, H. (1964). *Chem. Ber.* **97**, 3118.
Dorfman, R. I. and Ungar, F. (1965). "Metabolism of Steroid Hormones", Academic Press, New York and London.
Emmerich, H. (1970). *J. Insect Physiol.* **16**, 725.
Etkin, W. and Gilbert, L. I., eds. (1968). "Metamorphosis, a Problem in Developmental Biology", North Holland Publishing Co., Amsterdam.
Faux, A., Horn, D. H. S., Middleton, E. J., Fales, H. M. and Lowe, M. E. (1969). *Chem. Commun.* 175.

Faux, A., Galbraith, M. N., Horn, D. H. S. and Middleton, E. J. (1970). *Chem. Commun.* 243.
Furlenmeier, A., Fürst, A., Langemann, A., Waldvogel, G., Hocks, P., Kerb, U. and Wiechert, R. (1967). *Helv. Chim. Acta* **50**, 2387.
Gabe, M., Karlson, P. and Roche, J. (1964). *In* "Comparative Biochemistry" (M. Florkin and H. S. Mason, eds.), Vol. 6, p. 245, Academic Press, London and New York.
Galbraith, M. N. and Horn, D. H. S. (1966). *Chem. Commun.* 905.
Galbraith, M. N. and Horn, D. H. S. (1969). *Aust. J. Chem.* **22**, 1045.
Galbraith, M. N., Horn, D. H. S., Hocks, P., Schulz, G. and Hoffmeister, H. (1967). *Naturwissenschaften* **54**, 1.
Galbraith, M. N., Horn, D. H. S., Middleton, E. J. and Hackney, R. J. (1968a). *Chem. Commun.* 83.
Galbraith, M. N., Horn, D. H. S., Porter, Q. N. and Hackney, R. J. (1968b). *Chem. Commun.* 971.
Galbraith, M. N., Horn, D. H. S., Middleton, E. J. and Hackney, R. J. (1969a). *Aust. J. Chem.* **22**, 1517.
Galbraith, M. N., Horn, D. H. S., Middleton, E. J. and Hackney, R. J. (1969b). *Aust. J. Chem.* **22**, 1059.
Galbraith, M. N., Horn, D. H. S., Middleton, E. J. and Hackney, R. J. (1969c). *Chem. Commun.* 402.
Galbraith, M. N., Horn, D. H. S., Middleton, E. J., Thomson, J. A., Siddall, J. B. and Hafferl, W. (1969d). *Chem. Commun.* 1134.
Galbraith, M. N., Horn, D. H. S., Thomson, J. A., Neufeld, G. J. and Hackney, R. J. (1969e). *J. Insect Physiol.* **15**, 1225.
Galbraith, M. N., Horn, D. H. S., Middleton, E. J. and Thomson, J. A. (1970). *Chem. Commun.* 179.
Gibian, H. (1967). *In* "Proceedings of the International Symposium on Drug Research", p. 176, Montreal.
Gilbert, L. I. (1964). *In* "The Hormones" (G. Pincus, K. V. Thimann and E. B. Astwood, eds.), Vol. 4, p. 67, Academic Press, New York and London.
Gilbert, L. I. and Schneiderman, H. A. (1961). *Am. Zool.* **1**, 11.
Goad, L. J. (1967). *In* "Terpenoids in Plants" (J. B. Pridham, ed.), p. 159, Academic Press, London and New York.
Goad, L. J. (1970). *In* "Natural Substances Formed Biologically from Mevalonic Acid" (T. W. Goodwin, ed.), Biochem. Soc. Symp. No. 29, p. 45, Academic Press, London and New York.
Hampshire, F. and Horn, D. H. S. (1966). *Chem. Commun.* 37.
Harrison, I. T., Siddall, J. B. and Fried, J. H. (1966). *Tetrahedron Letters* 3457.
Heed, W. B. and Kircher, H. W. (1965). *Science, N.Y.* **149**, 758.
Heinrich, G. and Hoffmeister, H. (1967). *Experientia* **23**, 995.
Heinrich, G. and Hoffmeister, H. (1968). *Tetrahedron Letters* 6063.
Herout, V. (1970). *In* "Progress in Phytochemistry" (L. Reinhold and Y. Liwschitz, eds), Vol. 2, John Wiley & Sons, New York.
Herout, V., Jizba, J. and Šorm, F. (1969). *Deut. Apotheker-Z.* **109** (41).
Hikino, H., Hikino, Y. and Takemoto, T. (1969a). *Tetrahedron* **25**, 3389.
Hikino, H., Nabetani, S., Nomoto, K., Arai, T., Takemoto, T., Otaka, T. and Uchiyama, M. (1969b). *Yakugaku Zasshi* **89**, 235.
Hikino, H., Nomoto, K. and Takemoto, T. (1969c). *Tetrahedron Letters* 1417.
Hikino, H., Kohama, T. and Takemoto, T. (1970). *Phytochemistry* **9**, 367.
Hocks, P. (1968). "Conference on Insect-Plant Interactions", Santa Barbara, 18–22 March (quoted in Takemoto *et al.*, 1968h).
Hocks, P. and Wiechert, R. (1966). *Tetrahedron Letters* 2989.

Hocks, P., Jäger, A., Kerb, U., Wiechert, R., Furlenmeier, A., Fürst, A. and Langemann, A. (1966). *Angew. Chem. Intern. Ed.* **5**, 673.
Hocks, P., Schulz, G., Watzke, E. and Karlson, P. (1967). *Naturwissenschaften* **54**, 44.
Hocks, P., Kerb, U., Wiechert, R., Furlenmeier, A. and Fürst, A. (1968). *Tetrahedron Letters* 4281.
Hoffmeister, H. (1966). *Angew. Chem.* **78**, 269.
Hoffmeister, H. (1970). *J. Am. Oil Chemists Soc.* **47**, 71A.
Hoffmeister, H. and Grützmacher, H. F. (1966). *Tetrahedron Letters* 4017.
Hoffmeister, H., Heinrich, G., Staal, G. B. and van der Burg, W. J. (1967a). *Naturwissenschaften* **54**, 471.
Hoffmeister, H., Grützmacher, H. F. and Duennebeil, K. (1967b). *Z. Naturf.* **22b**, 66.
Hoffmeister, H., Nakanishi, K., Koreeda, M. and Hsu, H. Y. (1968). *J. Insect Physiol.* **14**, 53.
Hora, J. (1969). *Collection Czech. Chem. Commun.* **34**, 344.
Hora, J., Lábler, L., Kasal, A., Černy, V. and Šorm, F. (1966). *Steroids* **8**, 887.
Horn, D. H. S., Middleton, E. J., Wunderlich, J. A. and Hampshire, F. (1966). *Chem. Commun.* 339.
Horn, D. H. S., Fabbri, S., Hampshire, F. and Lowe, M. E. (1968). *Biochem. J.* **109**, 399.
Huber, R. and Hoppe, W. (1965). *Chem. Ber.* **98**, 2403.
Huppi, G. and Siddall, J. B. (1967). *J. Am. chem. Soc.* **89**, 6790.
Huppi, G. and Siddall, J. B. (1968). *Tetrahedron Letters* 1113.
Imai, S., Fujioka, S., Nakanishi, K., Koreeda, M. and Kurokawa, T. (1967). *Steroids* **10**, 557.
Imai, S., Fujioka, S., Murata, E., Sasakawa, Y. and Nakanishi, K. (1968a). *Tetrahedron Letters* 3887.
Imai, S., Hori, M., Fujioka, S., Murata, E., Goto, M. and Nakanishi, K. (1968b). *Tetrahedron Letters* 3883.
Imai, S., Fujioka, S., Murata, E., Otsuka, K. and Nakanishi, K. (1969a). *Chem. Commun.* 82.
Imai, S., Murata, E., Fujioka, S., Koreeda, M. and Nakanishi, K. (1969b). *Chem. Commun.* 546.
Imai, S., Toyosato, T., Sakai, M., Sato, Y., Fujioka, S., Murata, E. and Goto, M. (1969c). *Chem. Pharm. Bull. (Tokyo)* **17**, 340.
Imai, S., Toyosato, T., Sakai, M., Sato, Y., Fujioka, S., Murata, E. and Goto, M. (1969d). *Chem. Pharm. Bull. (Tokyo)* **17**, 335.
Imai, S., Murata, E., Fujioka, S., Matsuoka, T., Koreeda, M. and Nakanishi, K. (1970). *Chem. Commun.* 352.
Ishii, S., Kaplanis, J. N. and Robbins, W. E. (1963). *Ann. Entomol. Soc. Am.* **56**, 115.
Jizba, J. and Herout, V. (1967). *Coll. Czech. Chem. Commun.* **32**, 2867.
Jizba, J., Herout, V. and Šorm, F. (1967a). *Tetrahedron Letters* 1689.
Jizba, J., Herout, V. and Šorm, F. (1967b). *Tetrahedron Letters* 5139.
Joly, R. A., Svahn, C. M., Bennett, R. D. and Heftmann, E. (1969). *Phytochemistry* **8**, 1917.
Kaplanis, J. N., Robbins, W. E. and Tabor, L. A. (1960). *Ann. Entomol. Soc. Am.* **53**, 260.
Kaplanis, J. N., Tabor, L. A., Thompson, M. J., Robbins, W. E., and Shortino, T. J. (1966a). *Steroids* **8**, 625.
Kaplanis, J. N., Thompson, M. J., Yamamoto, R. T., Robbins, W. E. and Louloudes, S. J. (1966b). *Steroids* **8**, 605.
Kaplanis, J. N., Thompson, M. J., Robbins, W. E. and Bryce, B. M. (1967). *Science, N.Y.* **157**, 1436.

Kaplanis, J. N., Robbins, W. E., Thompson, M. J. and Baumhover, A. H. (1969). *Science, N.Y.* **166**, 1540.
Karlson, P. (1956a). *In* "Vitamins and Hormones" (R. S. Harris, G. F. Marrian and K. V. Thimann, eds.), Vol. 14, p. 228, Academic Press, New York and London.
Karlson, P. (1956b). *Ann. Sci. Nat. Zool., Series II* **18**, 125.
Karlson, P. (1963a). *Perspect. Biol. Med.* **6**, 203.
Karlson, P. (1963b). *Angew. Chem. Intern. Ed.* **2**, 175. See also *Angew. Chem.* (1963) **75**, 257.
Karlson, P. (1966). *Naturwissenschaften* **53**, 445.
Karlson, P. (1967a). *In* "Proceedings of the Second International Congress on Hormonal Steroids", Milan, 1966 (L. Martini, F. Fraschini and M. Motta, eds.), p. 146, Excerpta Medica Foundation, Amsterdam.
Karlson, P. (1967b). *Rev. Pure Appl. Chem.* **14**, 75.
Karlson, P. (1967c). *Mem. Soc. Endocr.* No. 15, p. 67.
Karlson, P. (1970). *In* "Natural Substances Formed Biologically from Mevalonic Acid" (T. W. Goodwin, ed.), Biochem. Soc. Symp. No. 29, p. 45, Academic Press, London and New York.
Karlson, P. and Bode, C. (1969). *J. Insect Physiol.* **15**, 111.
Karlson, P. and Hoffmeister, H. (1963). *Z. Physiol. Chem.* **331**, 298.
Karlson, P. and Sekeris, C. E. (1964). *In* "Comparative Biochemistry" (M. Florkin and H. S. Mason, eds.), Vol. 6, p. 221, Academic Press, London and New York.
Karlson, P. and Sekeris, C. E. (1966a). *In* "Recent Progress Hormone Research (G. Pincus, ed.), Vol. 22, p. 473, Academic Press, London and New York.
Karlson, P. and Sekeris, C. E. (1966b). *Acta Endocr.* **53**, 505.
Karlson, P. and Skinner, D. M. (1960). *Nature, Lond.* **185**, 543.
Karlson, P., Hoffmeister, H., Hoppe, W. and Huber, R. (1963). *Ann. Chem.* **662**, 1.
Karlson, P., Hoffmeister, H., Hummel, H., Hocks, P. and Spiteller, G. (1965). *Chem. Ber.* **98**, 2394.
Kerb, U., Schulz, G., Hocks, P., Wiechert, R., Furlenmeier, A., Fürst, A., Langemann, A. and Waldvogel, G. (1966). *Helv. Chim. Acta* **49**, 1601.
Kerb, U., Wiechert, R., Furlenmeier, A. and Fürst, A. (1968). *Tetrahedron Letters* 4277.
King, D. S. and Siddall, J. B. (1969). *Nature, Lond.* **221**, 955.
Kircher, H. W. and Goodnight, K. (1970). *J. Am. Oil Chemists Soc.* **47**, 71A.
Knight, J. C. and Pettit, G. R. (1969). *Phytochemistry* **8**, 477.
Kobayashi, M. (1963). *Proc. Int. Congr. Zool.* **16**, 226.
Kobayashi, M. and Kimura, S. (1967). *J. Insect Physiol.* **13**, 545.
Kobayashi, M., Nakanishi, K. and Koreeda, M. (1967). *Steroids* **9**, 529.
Kok, L. T., Norris, D. M. and Chu, H. M. (1970). *Nature, Lond.* **225**, 661.
Koreeda, M. and Nakanishi, K. (1970). *Chem. Commun.* 351.
Krishnakumaran, A. and Schneiderman, H. A. (1968). *Nature, Lond.* **220**, 601.
Krishnakumaran, A. and Schneiderman, H. A. (1969). *Gen. Comp. Endocr.* **12**, 515.
Kroeger, H. (1966). *Expl Cell Res.* **41**, 64.
Kroeger, H. (1968). *In* "Metamorphosis, a Problem in Developmental Biology' (W. Etkin and L. J. Gilbert, eds.), p. 185, North-Holland Publishing Co., Amsterdam.
Kroeger, H. and Lezzi, M. (1966). *Ann. Rev. Entomol.* **11**, 1.
Kurata, H. (1968). *Bull. Japan Soc. Sci. Fisheries* **34**, 909.
Labler, L., Sláma, K. and Šorm, F. (1968). *Coll. Czech. Chem. Commun.* **33**, 2226.
Lowe, M. E., Horn, D. H. S. and Galbraith, M. N. (1968). *Experientia* **24**, 518.
Madhavan, K. and Schneiderman, H. A. (1969). *Biol. Bull.* **137**, 321.
Marmaras, V. J., Sekeris, C. E. and Karlson, P. (1966). *Acta Biochim. Polon.* **13**, 305.

Martin, M. M. and Carls, G. A. (1968). *Lipids* **3**, 256.
Monroe, R. E., Hopkins, T. L. and Valder, S. A. (1967). *J. Insect Physiol.* **13**, 219.
Mori, H. and Shibata, K. (1969). *Chem. Pharm. Bull.* **17**, 1970.
Mori, H., Shibata, K., Tsuneda, K. and Sawai, M. (1968). *Chem. Pharm. Bull.* (*Tokyo*) **16**, 563.
Moriyama, H. and Nakanishi, K. (1968). *Tetrahedron Letters* 1111.
Nakanishi, K. (1969). *Bull. Soc. Chim. Fr.* **10**, 3476.
Nakanishi, K., Koreeda, M., Sasaki, S., Chang, M. L. and Hsu, H. Y. (1966). *Chem. Commun.* 915.
Nakanishi, K., Koreeda, M., Chang, M. L. and Hsu, H. Y. (1968). *Tetrahedron Letters* 1105.
Oberlander, H. (1969a). *J. Insect Physiol.* **15**, 297.
Oberlander, H. (1969b). *J. Insect Physiol.* **15**, 1803.
Ohtaki, T., Milkman, R. D. and Williams, C. M. (1967). *Proc. Natn. Acad. Sci. U.S.A.* **58**, 981.
Ohtaki, T., Milkman, R. D. and Williams, C. M. (1968). *Biol. Bull.* **135**, 322.
Okui, S., Otaka, T., Uchiyama, M., Takemoto, T., Hikino, H., Ogawa, S. and Nishimoto, N. (1968). *Chem. Pharm. Bull.* (*Tokyo*) **16**, 384.
Otaka, T. and Uchiyama, M. (1969). *Chem. Pharm. Bull.* (*Tokyo*) **17**, 1883.
Otaka, T., Uchiyama, M., Takemoto, T. and Hikino, H. (1969). *Chem. Pharm. Bull.* (*Tokyo*) **17**, 1352.
Patel, N. and Madhavan, K. (1969). *J. Insect Physiol.* **15**, 2141.
Postethwait, J. H. and Schneiderman, H. A. (1970). *Biol. Bull.* **138**, 47.
Řežábová, B., Hora, J., Landa, V., Černý, V. and Šorm, F. (1968). *Steroids* **11**, 475.
Rimpler, H. (1969). *Tetrahedron Letters* 329.
Ritter, F. J. and Wientjens, W. H. J. M. (1967). *TNO-Niews* **22**, 381.
Robbins, W. E., Kaplanis, J. N., Monroe, R. E. and Tabor, L. A. (1961). *Ann. Entomol. Soc. Am.* **54**, 165.
Robbins, W. E., Thompson, M. J., Kaplanis, J. N. and Shortino, T. J. (1964). *Steroids* **4**, 635.
Robbins, W. E., Kaplanis, J. N., Thompson, M. J., Shortino, T. J., Cohen, C. F. and Joyner, S. C. (1968). *Science, N.Y.* **161**, 1158.
Sato, Y., Sakai, M., Imai, S. and Fujioka, S. (1968). *Appl. Ent. Zool.* **3**, 49.
Sauer, H. H., Bennett, R. D. and Heftmann, E. (1968). *Phytochemistry* **7**, 2027.
Schaefer, C. H., Kaplanis, J. N. and Robbins, W. E. (1965). *J. Insect Physiol.* **11**, 1013.
Sekeris, C. E. (1967). *In* "Regulation of Nucleic Acid and Protein Biosynthesis" (V. V. Koningsberger and L. Bosch, eds.), p. 388, Elsevier, Amsterdam.
Shaaya, E. (1969). *Z. Naturf.* **24b**, 718.
Shaaya, E. and Karlson, P. (1965a). *J. Insect Physiol.* **11**, 65.
Shaaya, E. and Karlson, P. (1965b). *Develop. Biol.* **11**, 424.
Shibata, K. and Mori, H. (1968). *Chem. Pharm. Bull.* (*Tokyo*) **16**, 1404.
Shimizu, K., Gut, M. and Dorfman, R. I. (1962). *J. biol. Chem.* **237**, 699.
Siddall, J. B., Cross, A. D. and Fried, J. H. (1966). *J. Am. chem. Soc.* **88**, 862.
Siddall, J. B., Horn, D. H. S. and Middleton, E. J. (1967). *Chem. Commun.* 899.
Slama, K. (1968). *In* Williams, C. M. and Robbins, W. E., *Bioscience* **18**, 791.
Slaytor, M. and Bloch, K. (1965). *J. biol. Chem.* **240**, 4598.
Souza, N. J. de, Ghisalberti, E. L., Rees, H. H. and Goodwin, T. W. (1969). *Biochem. J.* **114**, 895.
Souza, N. J. de, Ghisalberti, E. L., Rees, H. H. and Goodwin, T. W. (1970). *Phytochemistry* **9**, 1247.
Staal, G. B. (1967a). *Proc. Koninkl. Ned. Akad. Westenschap.* (Amsterdam) Ser. C. **70**, 410.

Staal, G. B. and van den Burg, W. J. (1967b). Paper read at the 19th Intern. Symp. over Fytofarmacie en Fytiatrie, Gent 1967, Mededelingen Rijksfaculteit Landbouwwetenschappen Gent, 1967b, XXXII nr 3/4, p. 393.

Stamm, M. D. (1959). *An. Real Soc. Esp. Fis. Quim.* (*Madrid*) **55B**, 171.

Svoboda, J. A., Hutchins, R. F. N., Thompson, M. J. and Robbins, W. E. (1969). *Steroids* **14**, 469.

Takemoto, T., Hikino, Y., Arai, T., Kawahara, M., Konno, C., Arihara, S. and Hikino, H. (1967a). *Chem. Pharm. Bull.* (*Tokyo*) **15**, 1816.

Takemoto, T., Hikino, Y., Nomoto, K. and Hikino, H. (1967b). *Tetrahedron Letters* 3191.

Takemoto, T., Ogawa, S. and Nishimoto, N. (1967c). *Yakugaku Zasshi* **87**, 1469.

Takemoto, T., Ogawa, S. and Nishimoto, N. (1967d). *Yakugaku Zasshi* **87**, 1474.

Takemoto, T., Ogawa, S., Nishimoto, N., Arihara, S. and Bue, K. (1967e). *Yakugaku Zasshi* **87**, 1414.

Takemoto, T., Ogawa, S., Nishimoto, N. and Hoffmeister, H. (1967f). *Z. Naturf.* **22B**, 481.

Takemoto, T., Ogawa, S., Nishimoto, N. and Tanigushi, S. (1967g). *Yakugaku Zasshi* **87**, 1478.

Takemoto, T., Arihara, S. and Hikino, H. (1968a). *Tetrahedron Letters* 4199.

Takemoto, T., Arihara, S., Hikino, Y. and Hikino, H. (1968b). *Chem. Pharm. Bull.* (*Tokyo*) **16**, 762.

Takemoto, T., Hikino, Y., Arai, T. and Hikino, H. (1968c). *Tetrahedron Letters* 4061.

Takemoto, T., Hikino, Y., Arai, T., Konno, C., Nabetani, S. and Hikino, H. (1968d). *Chem. Pharm. Bull.* (*Tokyo*) **16**, 759.

Takemoto, T., Hikino, Y., Arihara, S., Hikino, H., Ogawa, S. and Nishimoto, N. (1968e). *Tetrahedron Letters* 2475.

Takemoto, T., Hikino, Y., Hikino, H., Ogawa, S. and Nishimoto, N. (1968f). *Tetrahedron Letters* 3053.

Takemoto, T., Hikino, H., Jin, H., Arai, T. and Hikino, H. (1968g). *Chem. Pharm. Bull.* (*Tokyo*) **16**, 1636.

Takemoto, T., Hikino, Y., Okuyama, T., Arihara, S. and Hikino, H. (1968h). *Tetrahedron Letters* 6095.

Takemoto, T., Nomoto, K. and Hikino, H. (1968i). *Tetrahedron Letters* 4953.

Takemoto, T., Nomoto, K., Hikino, Y. and Hikino, H. (1968j). *Tetrahedron Letters* 4929.

Takemoto, T., Ogawa, S., Nishimoto, N., Hirayama, H. and Taniguchi, S. (1968k). *Yakugaku Zasshi* **88**, 1293.

Takemoto, T., Hikino, Y., Hikino, H., Ogawa, S. and Nishimoto, N. (1969a). *Tetrahedron* **25**, 1241.

Takemoto, T., Okuyama, T., Arihara, S., Hikino, Y. and Hikino, H. (1969b). *Chem. Pharm. Bull.* (*Tokyo*) **17**, 1973.

Thompson, M. J., Kaplanis, J. N., Robbins, W. E. and Yamamoto, R. T. (1967). *Chem. Commun.* 650.

Thomson, J. A., Siddall, J. B., Galbraith, M. N., Horn, D. H. S. and Middleton, E. J. (1969). *Chem. Commun.* 669.

Velgová, H., Labler, L., Černý, V., Šorm, F. and Sláma, K. (1968). *Coll. Czech. Chem. Commun.* **33**, 242.

Velgová, H., Černý, V., Šorm, F. and Sláma, K. (1969). *Coll. Czech. Chem. Commun.* **34**, 3354.

Williams, C. M. (1968a). *Biol. Bull.* **134**, 344.

Williams, C. M. (1968b). *In* Williams, C. M. and Robbins, W. E. *Bioscience* **18**, 791.

Wright, J. E. (1969). *Science, N.Y.* **163**, 390.

Zdarek, J. and Fraenkel, G. (1969). *Proc. Natn. Acad. Sci. U.S.A.* **64**, 565.

CHAPTER 8

Recent Progress in Carotenoid Chemistry

S. LIAAEN-JENSEN

Organic Chemistry Laboratories, Norwegian Institute of Technology, University of Trondheim, Trondheim, Norway

I. Introduction

Since only naturally occurring carotenoids are biochemically significant, we shall be primarily concerned with their chemistry in this review. Structural studies in the carotenoid field were discussed earlier in the 1966 symposium of the Phytochemical Society (Weedon, 1967a). Figure 1 gives the justification for reviewing a closely related topic so soon after its previous discussion and review. In an approximate manner the diagram plots the number of known, naturally occurring carotenoids against time (Zechmeister, 1934; Karrer and Jucker, 1948; Liaaen-Jensen and Jensen, 1965; Straub, 1971). Around 300 natural carotenoids are now recognized and more or less correct structures have been ascribed to about 200 of these (Straub, 1971). During the last few years a large number of new carotenoids have been isolated and several novel structural features have been revealed, as indicated to the right of the curve on Fig. 1. Thus variations both in the carbon skeleton, as well as the substitution pattern, have been disclosed. The main reason for this progress is the application of new chemical and physical methods, as indicated to the left of the curve in Fig. 1.

In this paper selected topics where considerable progress has recently been made will be discussed.

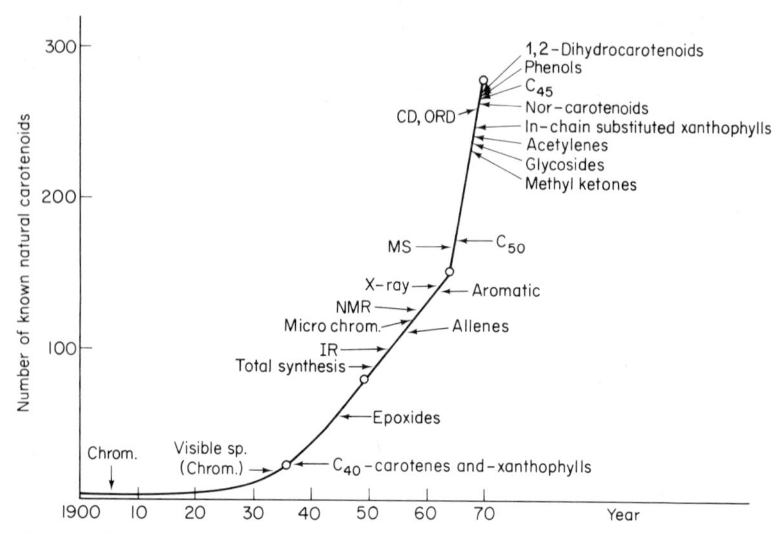

FIG. 1. Historical data concerning the study of naturally occurring carotenoids.

II. NATURAL CAROTENOIDS—NOVEL STRUCTURES

Let us first consider representatives for the new structural types that have come to light in recent years. According to the traditional definition of a carotenoid its skeleton is formally built up of eight isoprenoid units (Karrer and Jucker, 1948). By the original definition carotenoids were yellow-red pigments, their colour of course being primarily caused by their characteristic polyene chain. Later the colourless biosynthetic C_{40}-precursors of the coloured carotenoids were included, but now we must also include blue representatives. Apo-carotenoids formally derived by oxidative cleavage of the chain were formerly included. Now one must also allow extension of the carbon skeleton beyond C_{40}.

Decaprenoxanthin (1) was the first C_{50}-carotenoid discovered (Liaaen Jensen and Weeks, 1966; Liaaen Jensen, 1967; Liaaen-Jensen *et al.*, 1968; Schwieter and Liaaen-Jensen, 1969). Some ten C_{50}-carotenoids have been characterized during the last few years from non-photosynthetic bacteria (Weeks *et al.*, 1969; Liaaen-Jensen, 1970; Kelly *et al.*, 1970; Norgård *et al.*, 1970). The C_{50}-carotenoids so far encountered represent variations of the same general theme (2), either aliphatic tertiary carotenols or cyclic representatives with α- or β-type rings. The two extra C_5-units are attached to 2,2'-positions.

Three C_{45}-carotenoids have also been isolated, one aliphatic (3) and two cyclic (4, 5) representatives (Norgård and Liaaen-Jensen, 1969; Weeks *et al.*, 1969). Again these structures agree with the hypothetical addition of isoprenoid C_5-units to the isopropylidene double bond of a traditional C_{40}-carotenoid

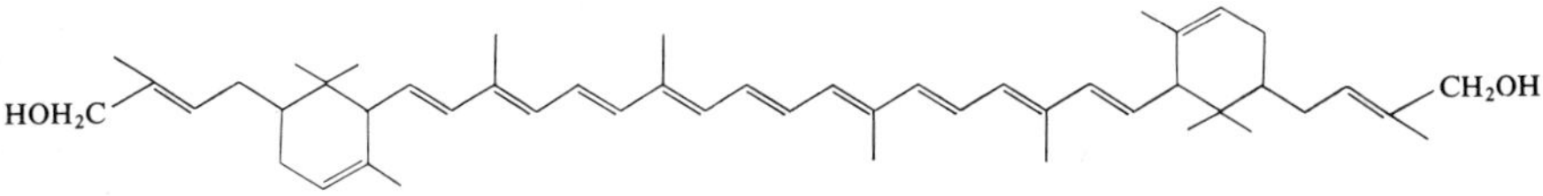

(1) Decaprenoxanthin

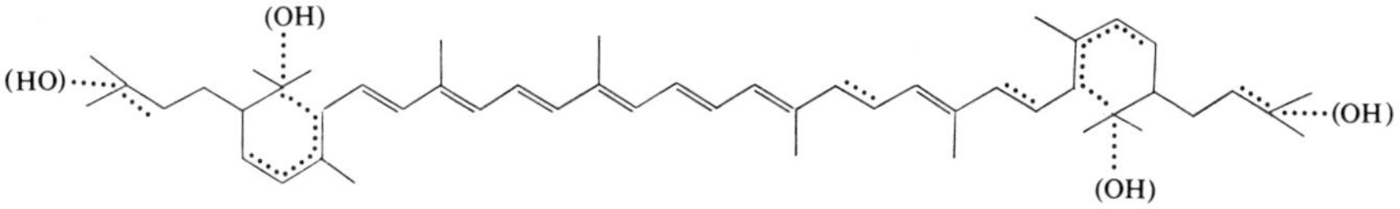

(2) General carbon skeleton

(Liaaen-Jensen *et al.*, 1968). The biosynthesis of these higher isoprenologs is discussed elsewhere (Weeks, 1971).

Turning from the representatives with supernumerary carbon skeletons to apo-type carotenoids with deficient carbon skeletons, methyl ketones have in recent years proved common. Some examples are given below. The structure of citranaxanthin (6), the first member of this series, was established by Yokoyama and White (1965a), as was that of sintaxanthin (7) (Yokoyama and White, 1965b). Paracentrone (8) from sea-urchins is a recent representative (Galasko *et al.*, 1969). A chemical conversion of fucoxanthin (9) into paracentrone (8) has recently been achieved (Hora *et al.*, 1970). By Oppenauer oxidation of (9) the 3-hydroxy group was converted to a ketone, causing

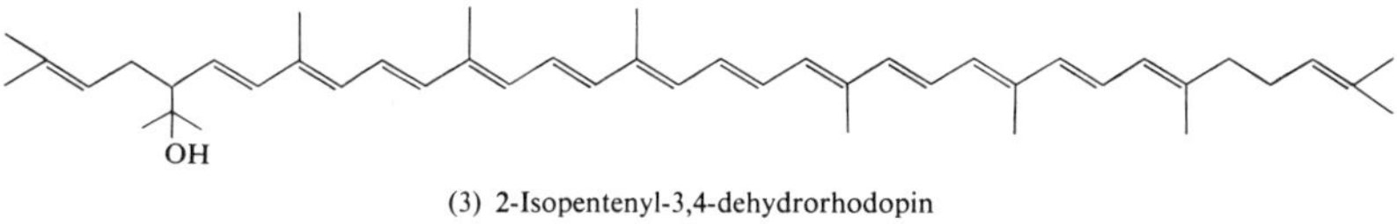

(3) 2-Isopentenyl-3,4-dehydrorhodopin

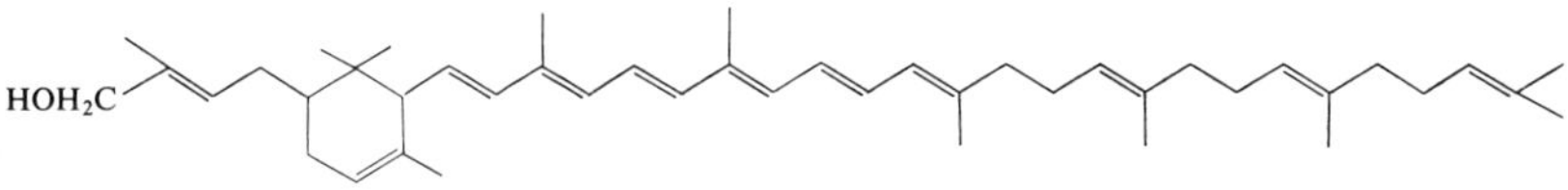

(4) Nonaprenoxanthin

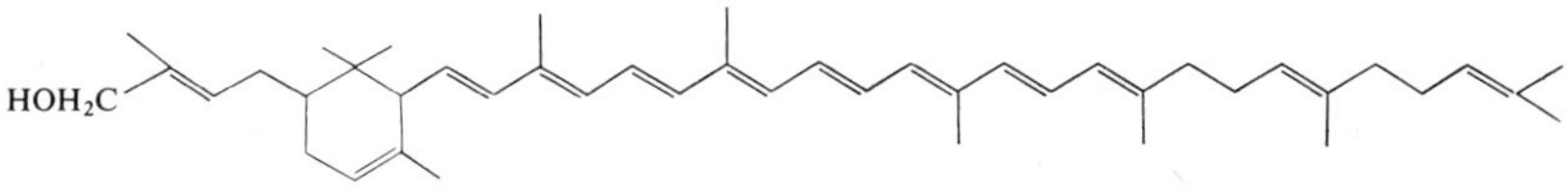

(5) 11′,12′-Dehydrononaprenoxanthin

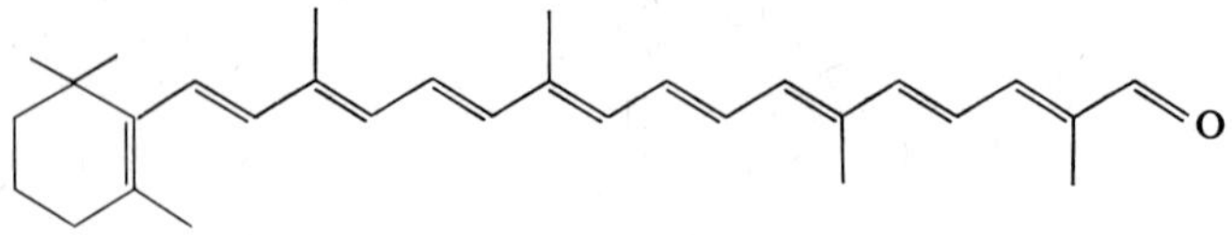

(10) β-Apo-8′-carotenal

Retro-aldol cleavage ↑ $-H_3C-\underset{\underset{O}{\|}}{C}-CH_3$

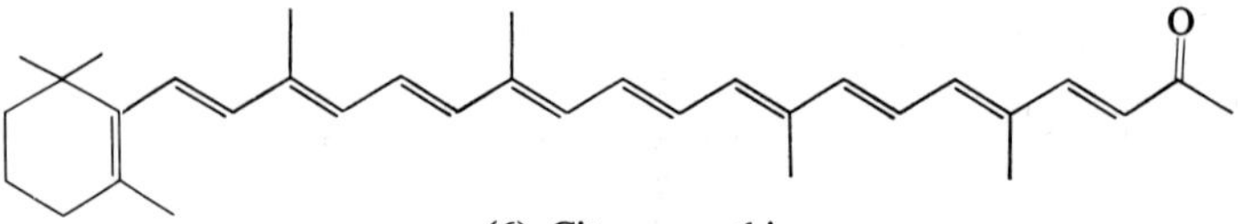

(6) Citranaxanthin

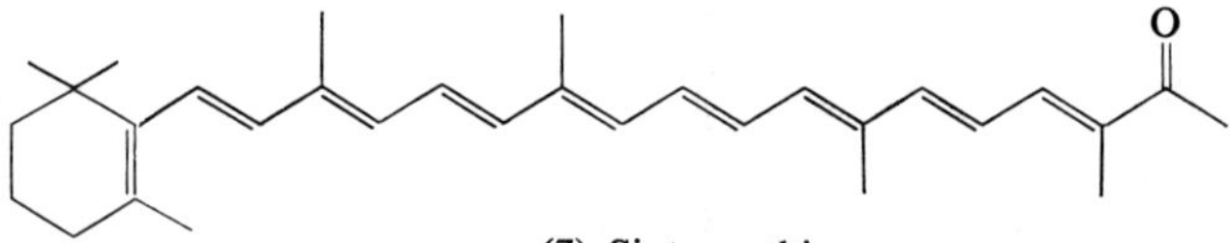

(7) Sintaxanthin

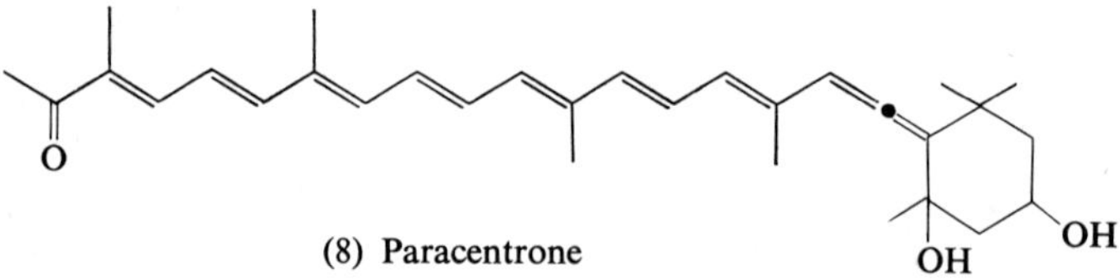

(8) Paracentrone

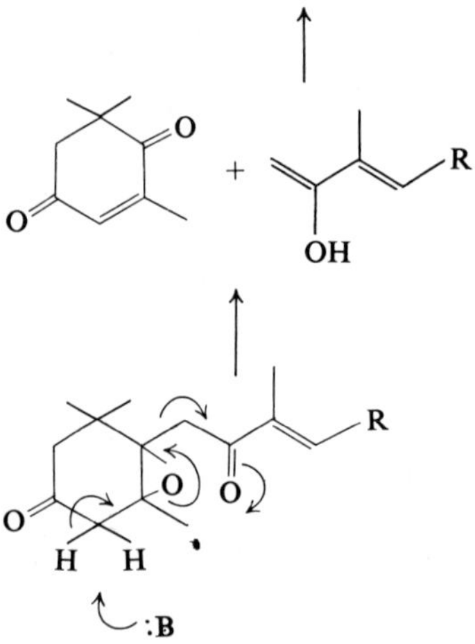

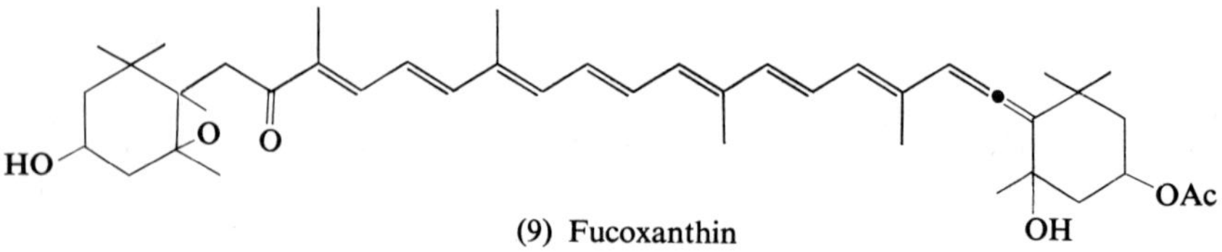

(9) Fucoxanthin

activation of the 4-position. In the presence of an electron donor electron displacement resulted in paracentrone (8). The acetoxy group was hydrolysed under mild alkaline conditions in the final step. This interesting conversion may have a biochemical counterpart. The methyl ketones (6) and (7) of citrus fruits may be formed biosynthetically by some type of *retro*-aldol cleavage. Chemically the methyl ketones undergo such cleavage in the presence of alkali with loss of acetone, thus citranaxanthin (6) gives β-apo-18′-carotenal (C_{30}) (10).

Another new class of carotenoids with carbon skeletons deficient in carbon atoms is the nor-carotenoids. Actinioerythrin (11) with two cyclopentenone rings, obtained from sea anemones, is the first representative of this class (Lederer, 1933; Hertzberg and Liaaen-Jensen, 1968; Hertzberg *et al.*, 1969). Actinioerythrin (C_{38}) provides on careful alkali treatment the blue violerythrin (Heilbron *et al.*, 1935), a cyclopentenedione. Violerythrin (12) is the darkest blue coloured carotenoid reported, but is not a natural carotenoid. We have postulated that actinioerythrin (C_{38}) is formed *in vivo* from astaxanthin (13) via a hypothetic intermediary triketone with ring contraction by benzilic acid rearrangement (Hertzberg and Liaaen-Jensen, 1968). A chemical conversion of astacin (14) into violerythrin (12), presumably by the same mechanism, has been achieved (Holzel *et al.*, 1969).

Also carotenoids with C_{40}-skeletons have lately revealed several structural variations. The oxygen substitution on the polyene chain is interesting, first demonstrated in the rhodopinal series (Aasen and Liaaen-Jensen, 1967b; Francis and Liaaen-Jensen, 1970), typical of some photosynthetic sulphur bacteria (Schmidt *et al.*, 1965). To the rhodopinal series belong allylic alcohols such as rhodopinol (15) and aldehydes of the rhodopinal (16) type. The cross-

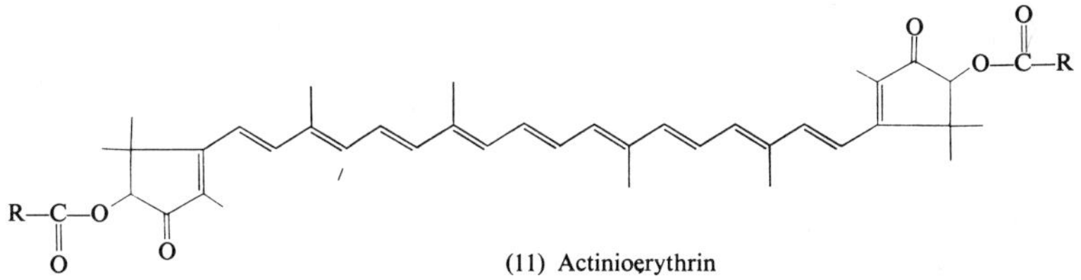

O O—C—R O R—C—O O O

(11) Actinioerythrin

KOH, O_2

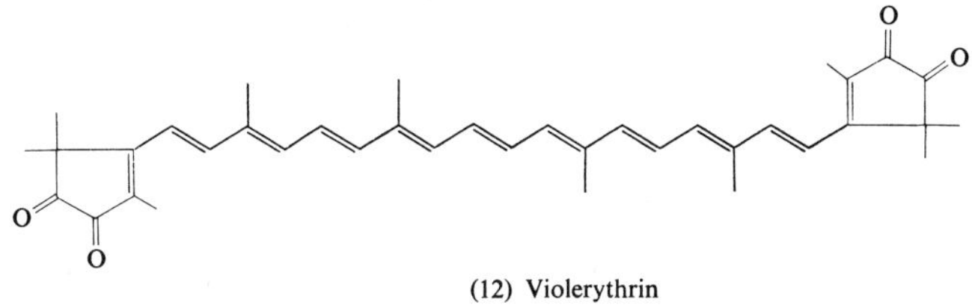

O O O O

(12) Violerythrin

Hypothetical *in vivo* conversion

(13) Astaxanthin ⇒ ⇒ $\xrightarrow{-CO_2}$ ⇒ (11) Actinioerythrin

In vitro conversion

(14) Astacene $\xrightarrow{MnO_2}$ [ ] → $\xrightarrow{MnO_2}$ (12) Violerythrin

conjugated aldehyde groups cause a remarkable bathochromic shift in the visible light spectrum. Of more recent date are loroxanthin (17) (Aitzetmüller *et al.*, 1969) and pyrenoxanthin (18) (Yamamoto *et al.*, 1969) from green algae. Pyrenoxanthin may be identical with loroxanthin.

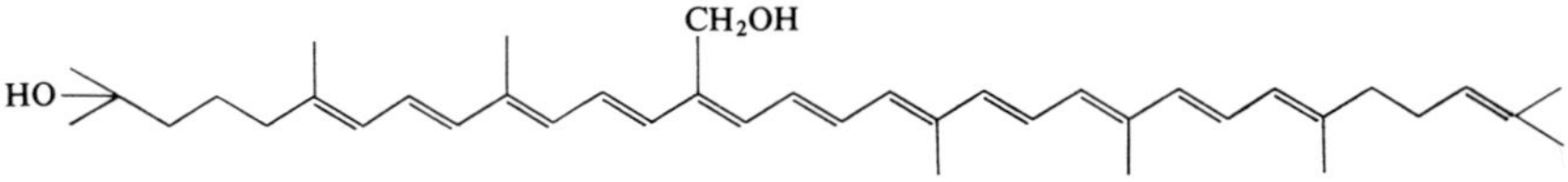

CH_2OH

HO

(15) Rhodopinol

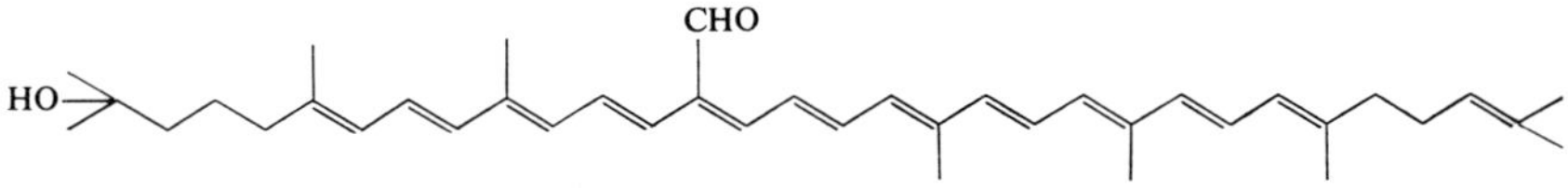

CHO

HO

(16) Rhodopinal

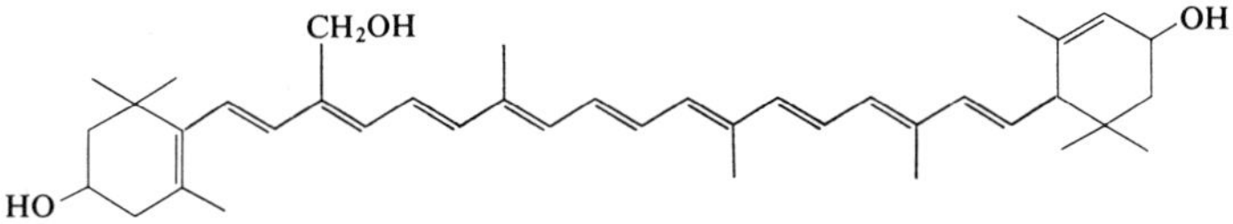

CH_2OH OH

HO

(17) Loroxanthin

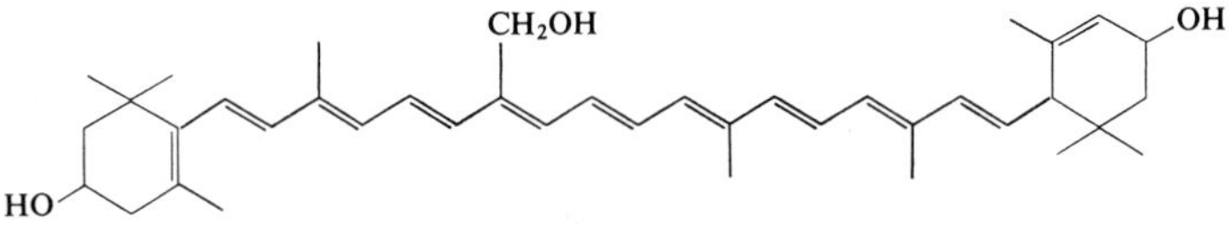

CH_2OH OH

HO

(18) Pyrenoxanthin

Carotenoid epoxides have been well established by Karrer's school, but were until recently restricted to β-rings with epoxy groups in 5,6-position (19) and their furanoid rearrangement products, the 5,8-oxides (20) (Karrer and Jucker, 1948). 1,2-Epoxyphytoene (21) has recently been isolated from tomatoes (Britton and Goodwin, 1969) and represents a new structural element. We have a related oxidic C_{50}-carotenoid under investigation (S. Norgård and S. Liaaen-Jensen, unpublished).

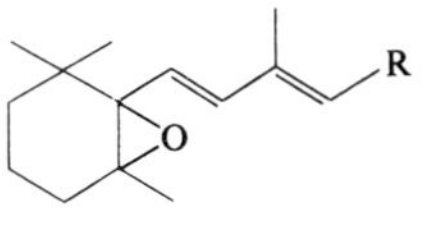

R O

(19)

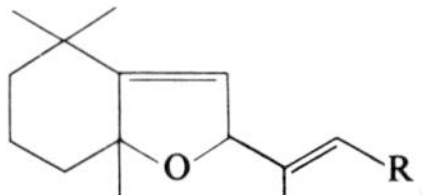

O R

(20)

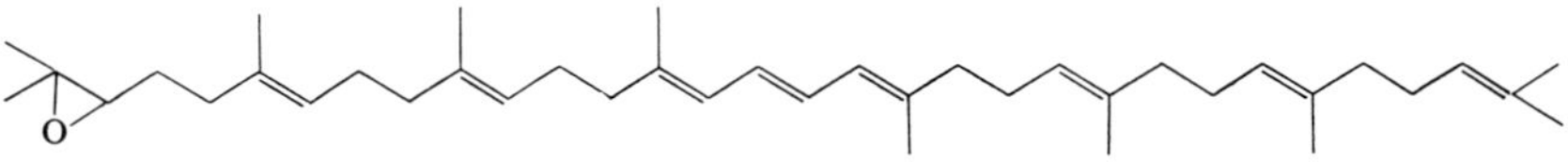

O

(21) 1,2-Epoxyphytoene

As a general rule the methyl groups of natural carotenoids have been tertiary, exhibiting sharp singlets in the proton magnetic resonance spectrum. An exception to this rule is the recently reported 1,2-dihydro carotenes, 1,2-dihydro-neurosporene (22), 1,2-dihydrolycopene (23) and 1,2-dihydro-3,4-dehydrolycopene (24) obtained from a photosynthetic bacterium (Malhotra *et al.*, 1970). 1,2-Dihydrolycopene has been prepared by total synthesis (Kjøsen and Liaaen-Jensen, 1970).

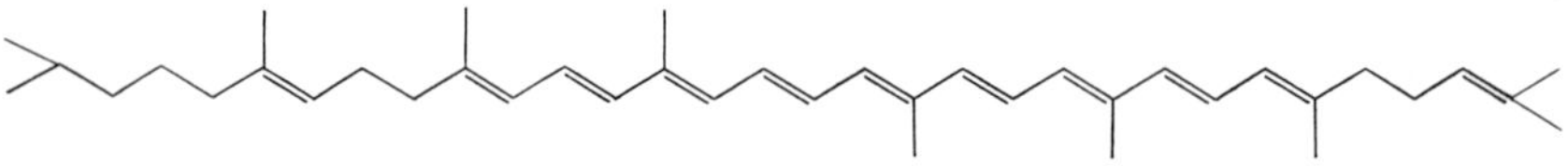

(22) 1,2-Dihydroneurosporene

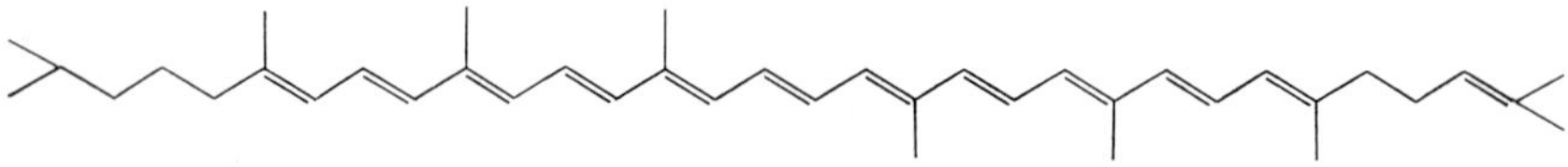

(23) 1,2-Dihydrolycopene

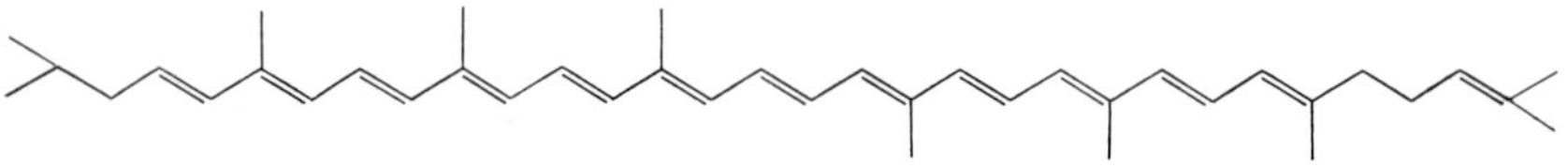

(24) 1,2-Dihydro-3,4-dehydrolycopene

Terminal methylene groups are common in other terpenoids, but have not been encountered previously in carotenoids. We have evidence for such an arrangement in aleuriaxanthin (36) of fungal origin—the position of the hydroxy group is not fully established (Liaaen Jensen, 1965; Arpin *et al.*, 1971a).

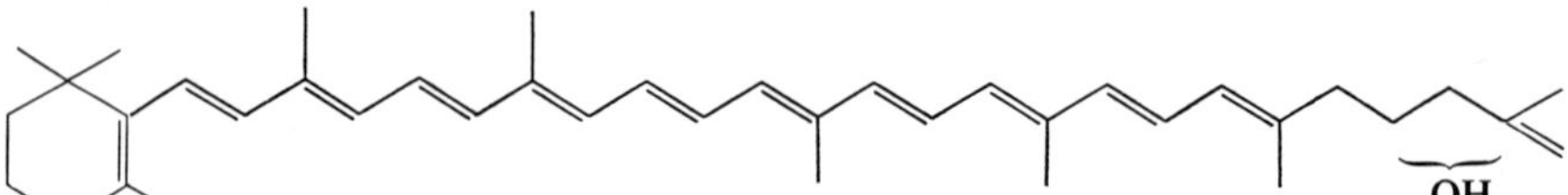

OH

(36) Aleuriaxanthin

Aryl carotenoids have been known for some time (for a review see Liaaen Jensen, 1966). Phenolic ones with characteristic properties were recently isolated from a *Streptomyces* sp. The structures 3-hydroxyisorenieratene (25) and 3,3′-di-hydroxyisorenieratene (26) were unequivocally established by degradation and total synthesis (Arcamone *et al.*, 1966). On oxidation the diol (26) gives an unstable, dark-blue carotenoid, presumably the corresponding quinone (27) (Nybraathen and Liaaen-Jensen, 1970).

Further variations in the substitution pattern have recently been demonstrated. A great number of carotenol fatty acid esters have been studied. These can be efficiently separated in thin-layer partition systems on paraffin-impregnated kieselguhr (Egger, 1964, 1968). Considerable variation is found

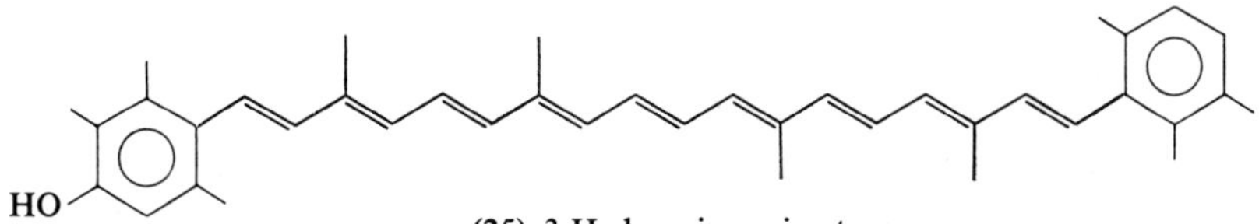

HO

(25) 3-Hydroxyisorenieratene

OH

HO

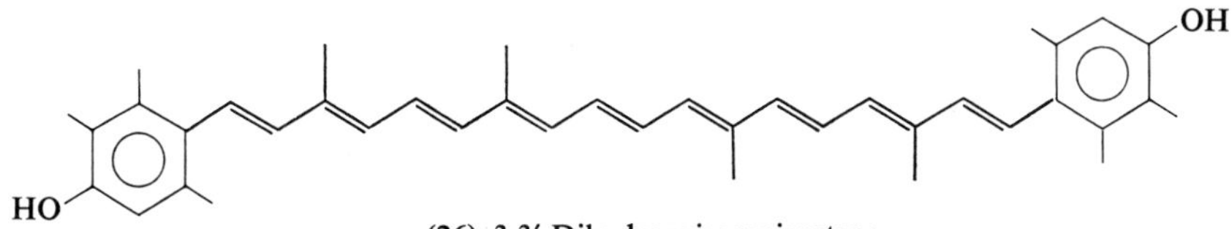

(26) 3,3′-Dihydroxyisorenieratene

Ox.

O

O

?

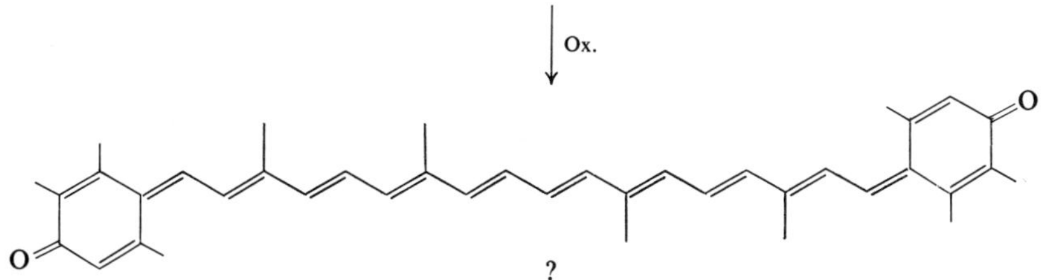

(27)

in the fatty acid substituents ranging from acetic to caprylic, capric, lauric, myristic, palmitic, stearic, palmitoleic, oleic and linoleic acids (Kuhn *et al.*, 1930, 1931; Booth, 1964; Jensen, 1964; Egger and Kleinig, 1966; Kleinig, 1967; Kleinig and Egger, 1967a,b; Kleinig and Nietsche, 1968; Egger, 1968; Vacheron *et al.*, 1969; Hertzberg *et al.*, 1969). However, the fatty acids esterified with carotenols are so far restricted to straight chain saturated or unsaturated fatty acids.

Carotenoid glycosides were not detected until a few years ago (Hertzberg and Liaaen-Jensen, 1967b); crocin (28) (Karrer and Miki, 1929) in reality is an ester. We now know several representatives of tertiary (29, 30, 31) and secondary (32, 33) carotenoid glycosides with considerable variation in both carbohydrate and aglycone. Glucosides (Hertzberg and Liaaen-Jensen, 1967b; Norgård *et al.*, 1970), a mannoside (Aasen *et al.*, 1969), rhamnosides (Hertzberg and Liaaen-Jensen, 1969a,b) and O-methyl desoxy hexosides related to (32) and (33) (Francis *et al.*, 1970) are demonstrated. Well-known carotenoids such as myxoxanthophyll (32), oscillaxanthin (33) and aphanizophyll (Tischer, 1938; Hertzberg and Liaaen-Jensen, 1971) of blue-green algae belong to this class. The special properties of the strongly polar carotenoid glycosides pose interesting problems concerning their function and location in the living cell.

Protein complexes have long been known in the carotenoid series (Kuhn and Lederer, 1933; Wald *et al.*, 1948). Considerable progress has been made

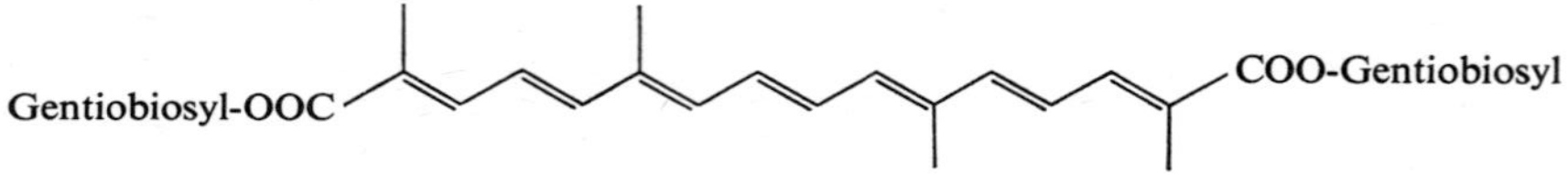

Gentiobiosyl-OOC COO-Gentiobiosyl

(28) Crocin

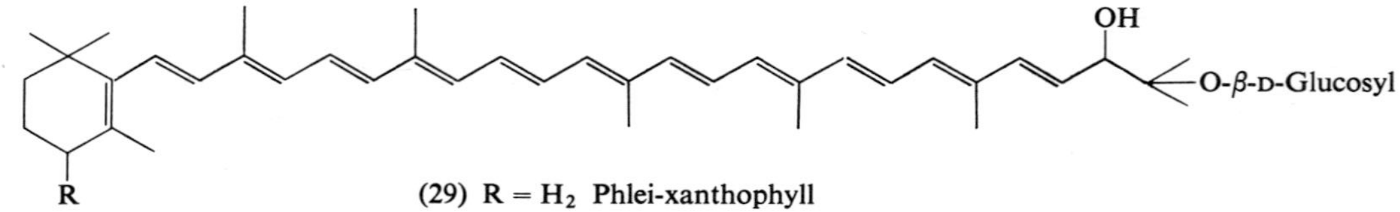

R OH O-β-D-Glucosyl

(29) R = H_2 Phlei-xanthophyll

(30) R = O 4-Keto-phlei-xanthophyll

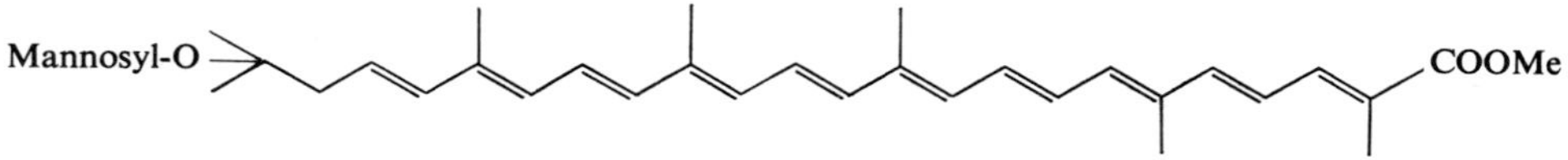

Mannosyl-O COOMe

(31) Methyl 1-Mannosyl-1,2-dihydro-3,4-didehydro-apo-8′-lycopenoate

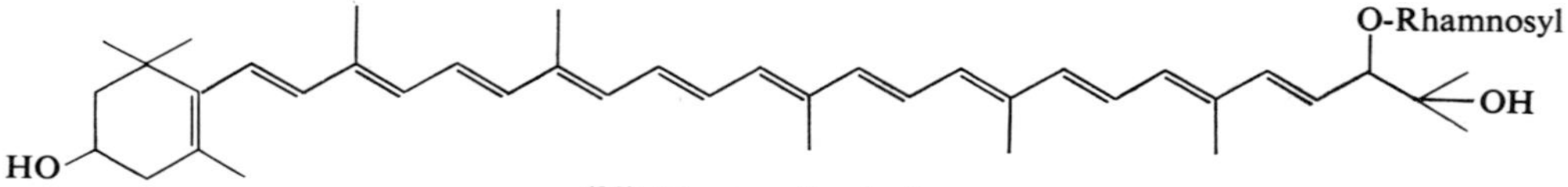

HO O-Rhamnosyl OH

(32) Myxoxanthophyll

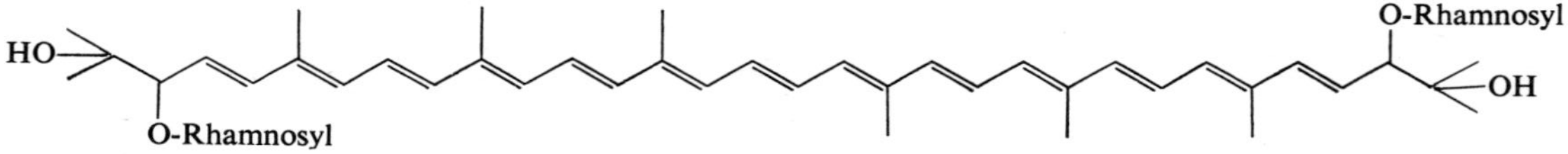

HO O-Rhamnosyl O-Rhamnosyl OH

(33) Oscillaxanthin

in the purification and characterization of these macromolecules (Cheesman *et al.*, 1967). The most closely examined carotenoprotein crustacyanin of lobster shells, where astaxanthin is the prosthetic group, has according to molecular weight determination by analytical ultracentrifugation, light scattering and osmometry a molecular weight of *ca* 320,000. α-Crustacyanin contains according to investigations by Kuhn and Kuehn (1967) 16 and according to Cheesman *et al.* (1966) *ca* 10 astaxanthin molecules. In neutral solutions of low ionic strength irreversible dissociation occurs into 8 sub-units of β-crustacyanin with molecular weight *ca* 38,000, containing 2 astaxanthin molecules (see Fig. 2). Reversible dissociation of α-crustacyanin in acid, alkaline or urea-containing solutions into colourless and pigmented aggregates has been demonstrated. On neutralization or dialysis the blue chromoprotein is regenerated. The various coloured complexes all contain extractable astaxanthin. They exhibit different colours and the colour and quaternary protein structure as a function of the degree of solvation of the astaxanthin molecule have been discussed (Kuhn and Kuehn, 1967).

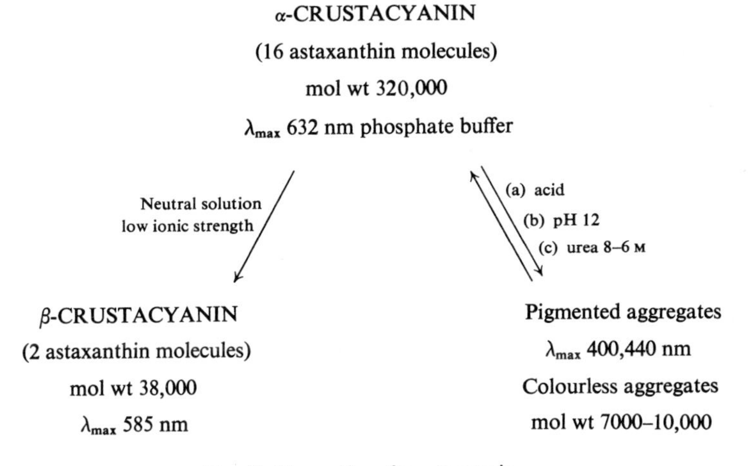

FIG. 2. Properties of crustacyanin.

III. MASS SPECTROMETRY

As the slope of the curve in Fig. 1 indicates, mass spectrometry is the most important recent tool for investigating carotenoid structures. The first study in this field was published by Isler's group five years ago (Schwieter *et al.*, 1965), and the application of mass spectrometry to the carotenoid field has since received much attention (e.g. Baldas *et al.*, 1966, 1969; Enzell *et al.*, 1969; Francis, 1969; Schwieter *et al.*, 1970; Enzell, 1970; Liaaen-Jensen, 1970).

Let me summarize some characteristic features of carotenoid mass spectra. For the vast majority of natural carotenoids, or a suitable derivative thereof, molecular ions may be observed. Hence high resolution measurements permit establishment of unambiguous molecular formulae. In carotenoids containing the normal central polyene chain the molecular ion is confirmed by losses of 92, 106, 158 and often 79 mass units (Schwieter *et al.*, 1965; Enzell *et al.*, 1968, 1969). The oxygen functions are often revealed, e.g. by losses of 18 (water from an alcohol) and 60 (acetic acid from an acetate) mass units, and the fragmentation pattern provides further structural information. Studies of deuterated (Schwieter *et al.*, 1970; Kjøsen *et al.*, 1971) and other derivatives have been used to devise mechanisms for the fragmentations observed. The currently entertained hypothesis for the elimination of toluene, xylene and dimethylcyclodecapentaene from the polyene chain, advanced by Schwieter *et al.* (1970) and based on a mechanism proposed by Edmunds and Johnstone (1965) for a corresponding thermal reaction, is given below. The mechanism involves a four-membered ring intermediate and rationalizes the essential results of the electron impact induced fragmentation of the common carotenoids. By this mechanism cleavage of double bonds is implied. If oxygen

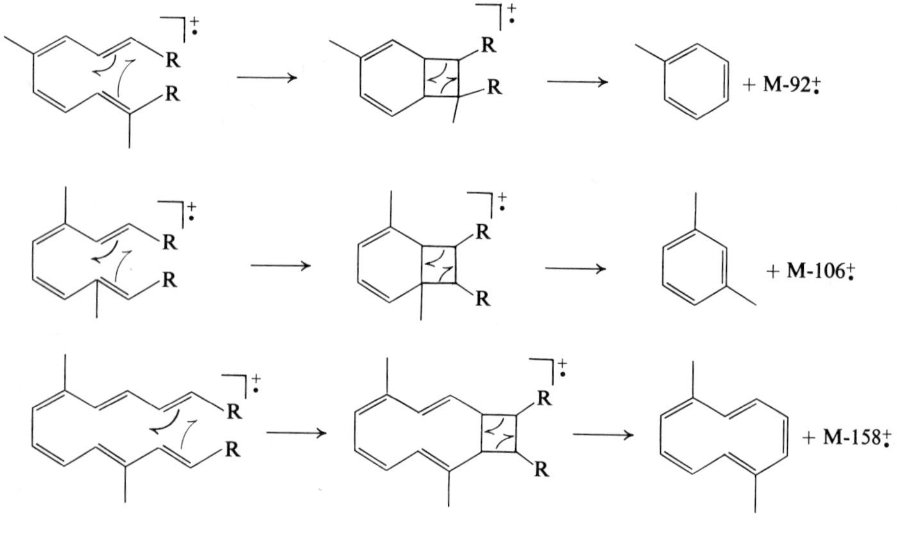

+ M-92$^{+\cdot}$

+ M-106$^{+\cdot}$

+ M-158$^{+\cdot}$

substituents are present in a methyl group attached to the central polyene chain, the same type of elimination still occurs (Aitzetmüller *et al.*, 1969; Francis and Liaaen-Jensen, 1970; Enzell and Liaaen-Jensen, 1971), and the extra substituent is revealed by the mass shift of the losses observed. It has been demonstrated with deuterated carotenes that in bicyclic carotenoids toluene originates from the central part of the molecule (only the 20 or 20′ methyl groups are present in the toluene eliminated), whereas xylene is derived from the C-8 to C-13 and C-8′ to C-13′ part of the chain (Schwieter *et al.*, 1970; Kjøsen *et al.*, 1971). Consequently it is possible to differentiate between oxygen substitution in C-19 (C-19′) and C-20 (C-20′) in a bicyclic carotenoid on the basis of any mass shift observed for the common losses of 92 and 106 mass units. For aliphatic carotenoids this relationship is more complex and of less diagnostic value.

In order to illustrate the application of mass spectrometry to specific problems I shall discuss a few examples from our own research.

The crucial assignment of attachment of the extra C_5-units of decaprenoxanthin (1) followed from a *retro* Diels-Alder fragmentation resulting in loss of 140 mass units from the molecular ion. The corresponding diacetate and dialdehyde showed analogous fragmentation (Liaaen-Jensen *et al.*, 1968). Recently loss of 124 mass units has been taken as evidence for the presence of the corresponding hydrocarbon structural element (Weeks *et al.*, 1969). However, $106 + 18 = 124$ and the M-124 peak may alternatively represent a combination of loss of xylene and water from a hydroxy group present elsewhere in the molecule. The latter fragmentation therefore calls for confirmation by high resolution measurements.

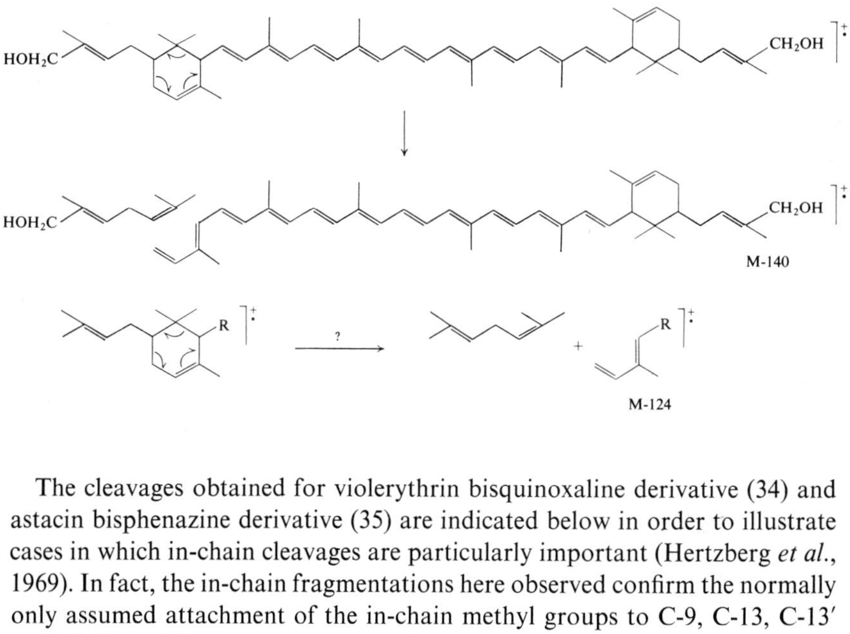

HOH₂C CH₂OH

HOH₂C CH₂OH M-140

R ? R M-124

The cleavages obtained for violerythrin bisquinoxaline derivative (34) and astacin bisphenazine derivative (35) are indicated below in order to illustrate cases in which in-chain cleavages are particularly important (Hertzberg *et al.*, 1969). In fact, the in-chain fragmentations here observed confirm the normally only assumed attachment of the in-chain methyl groups to C-9, C-13, C-13′ and C-9′ positions. Proton magnetic resonance spectroscopy is not discriminating at this point.

The mass spectra of glycoside acetates have proved very informative (Hertzberg and Liaaen-Jensen, 1969a,b; Aasen *et al.*, 1969; Francis *et al.*, 1970). As an example the fragmentations of a peracetate of a C_{50}-D-glucoside from *Sarcina lutea* are given below. The formation of the tetraacetoxyoxonium ion

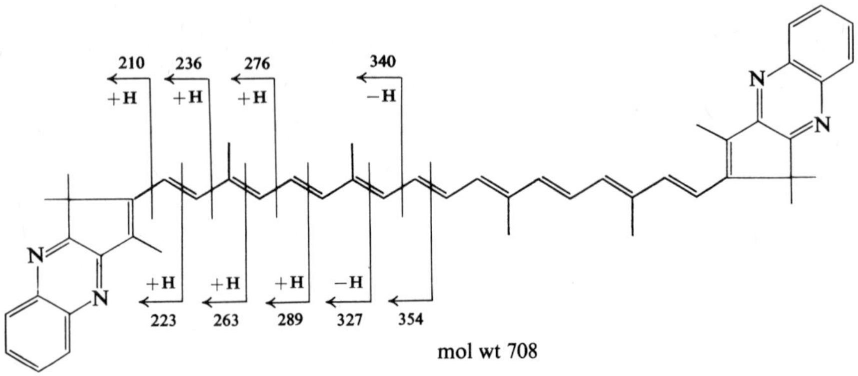

mol wt 708

(34) Violerythrin *bis*quinoxaline derivative

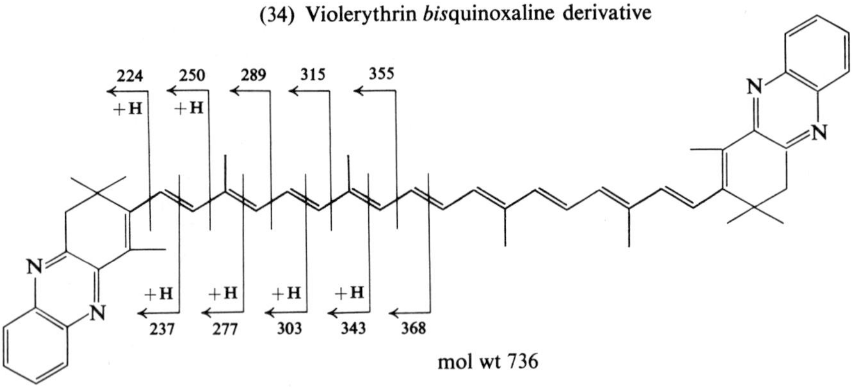

mol wt 736

(35) Astacin *bis*phenacine derivative

and its further fragmentation in known manner (Biemann *et al.*, 1963) revealed the hexoside formulation, subsequently confirmed by chemical evidence (Norgård *et al.*, 1970a). This glucoside occurs together with the unsymmetrical C_{50}-diol sarcinaxanthin (Liaaen-Jensen, 1970; Arpin *et al.*, 1971b). Assuming that the glucoside is sarcinaxanthin glucoside, alternative locations of the extra hydroxy group and the glucoside were further examined. If allocation of the glucoside to C-18 (18′) is not favoured for steric reasons, the problem is restricted to deciding between the location of the extra hydroxy group at either 18 or 18′. The in-chain cleavage observed for the acetylated glucoside favoured the 18′ assignment. However, the accurately determined ions supporting these in-chain cleavages could alternatively be interpreted as caused by combined losses of a hydrocarbon-type end group and a 58 mass units fragment (CH_2COO) from the other acetylated end group. The latter possibility was ruled out from the mass spectrum of the per (trideutero) acetate. Had the formulation of one hydrocarbon end group been correct, a shift of 2 mass units only should be observed for the in-chain cleavages, whereas these in fact all shifted 3 units, according to the formulation of the

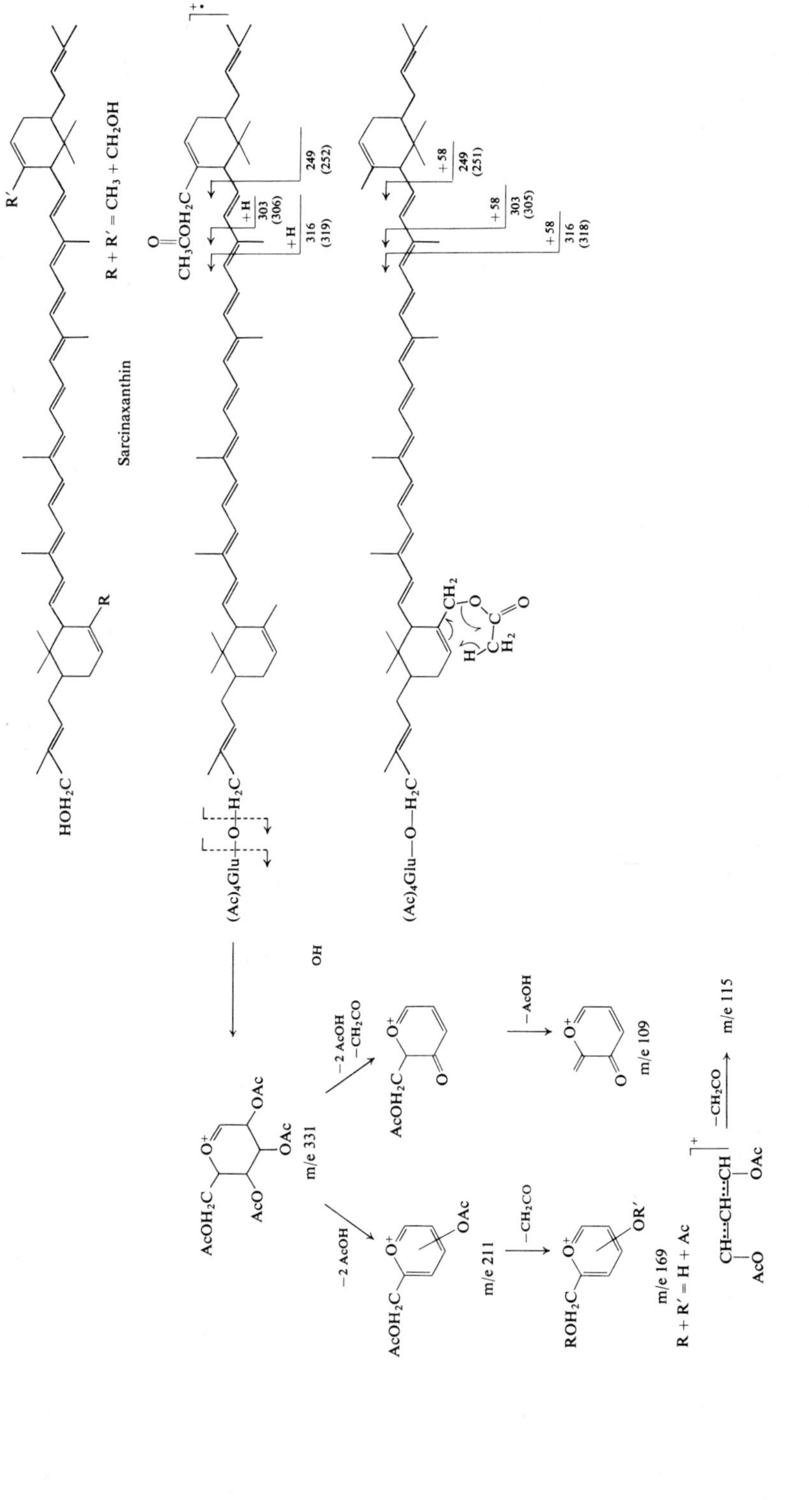

HOH$_2$C R R′

Sarcinaxanthin

$R + R' = CH_3 + CH_2OH$

$(Ac)_4Glu-O-H_2C$

CH_3COH_2C (C=O)

+H 249 (252)

+H 303 (306)

+H 316 (319)

$(Ac)_4Glu-O-H_2C$

CH_2 O C O H C H_2

+58 249 (251)

+58 303 (305)

+58 316 (318)

$AcOH_2C$ O^+ AcO OAc OAc

m/e 331

−2 AcOH, $-CH_2CO$ → $AcOH_2C$ O^+ O — −AcOH → O^+ O m/e 109

−2 AcOH → $AcOH_2C$ O^+ OAc m/e 211 — $-CH_2CO$ → ROH_2C O^+ OR′ m/e 169

$R + R' = H + Ac$

$CH\cdots CH\cdots CH$ (AcO, OAc)$^+$ — $-CH_2CO$ → m/e 115

glucoside and the extra acetoxy group hence present on different sides. If the steric argument is irrelevant, the hydroxy and glucoside groups may change positions. The implication of these findings—one oxygenated substituent on each side of the molecule—for the structure of sarcinaxanthin is being further studied.

IV. Synthesis

At a time when one is very enthusiastic about the potentiality of spectrometric methods, it should be stressed that no matter how obvious a structure may appear from spectral evidence, unequivocal proof must be sought by partial or total synthesis.

A. Partial Synthesis

Chemical conversions of carotenoids are considered as partial synthesis.

One reaction that has proved very useful lately is reductive opening of 5,6-epoxides or 5,8-epoxides with lithium aluminium hydride, providing the tertiary alcohol and by unknown mechanism the corresponding olefin (Grob and Siekmann, 1965; Cholnoky *et al.*, 1966, 1967, 1969; Schimmer and Krinsky, 1966; Hertzberg and Liaaen-Jensen, 1967b; Bonnet *et al.*, 1969). The latter reaction is at present the most convenient method for de-epoxidation.

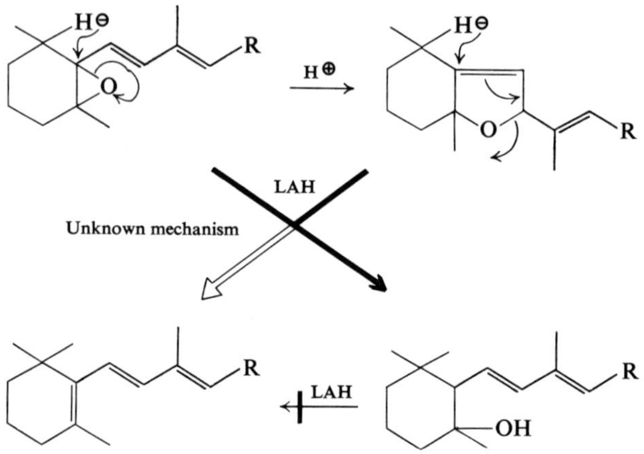

Of related type is the hydrogenolytic elimination of allylic glycosides. As an example is given the reaction of myxoxanthophyll (32) with lithium aluminium hydride, providing saproxanthin (37) and anhydrosaproxanthin (38) (Hertzberg and Liaaen-Jensen, 1969a).

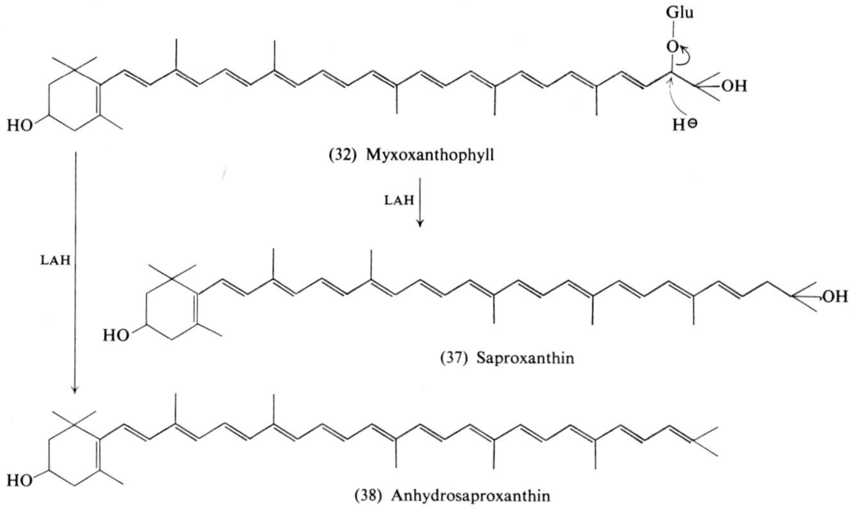

Glu
OH
HO
H⊖
(32) Myxoxanthophyll
LAH
LAH
HO
OH
(37) Saproxanthin
HO
(38) Anhydrosaproxanthin

The acid-catalysed conversion claimed by Egger *et al.* (1969) of the allenic epoxide neoxanthin (77) into the acetylenic diadinochrome, involves the conversion given below. This most interesting reaction needs further evidence.

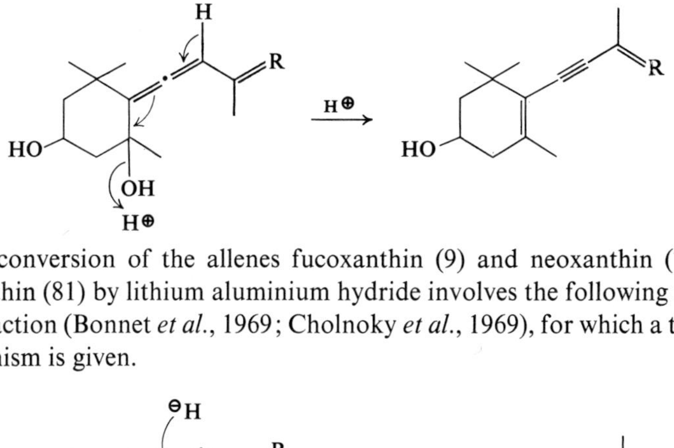

H
R
HO
OH
H⊕
H⊕
R
HO

The conversion of the allenes fucoxanthin (9) and neoxanthin (77) into zeaxanthin (81) by lithium aluminium hydride involves the following remarkable reaction (Bonnet *et al.*, 1969; Cholnoky *et al.*, 1969), for which a tentative mechanism is given.

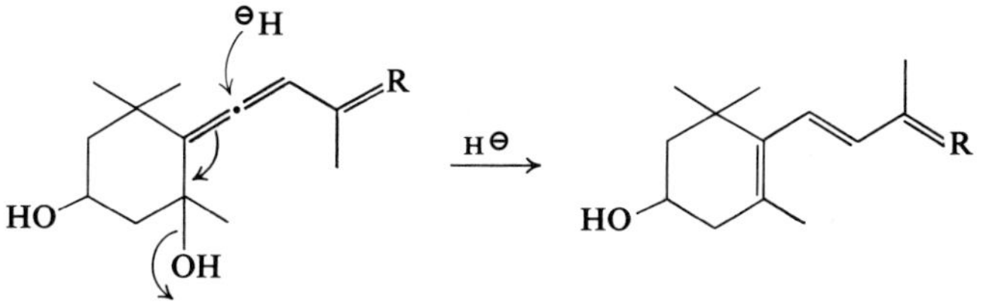

⊖H
R
HO
OH
H⊖
R
HO

Other new reaction types in partial synthesis have not really been developed or adapted recently. However, several examples of extensive use of chemical reactions in structure determination are reported. The reactions are readily carried out in micro scale with the efficient separation methods and spectro-

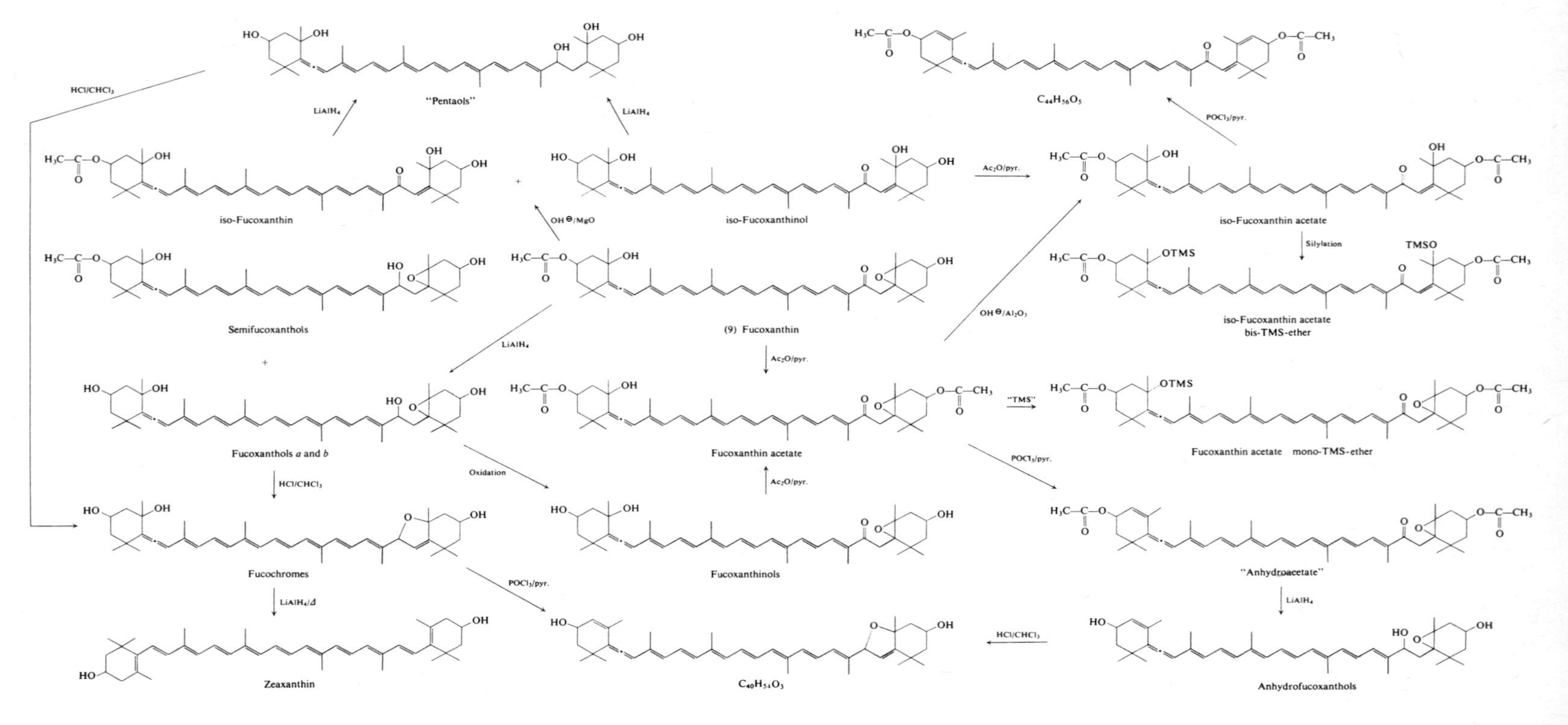

FIG. 3. Reactions of fucoxanthin (9).

scopic techniques available. Chemical reactions of fucoxanthin (9), mainly studied by Weedon's group (Bonnet *et al.*, 1969) and my husband (Jensen, 1966), cited in Fig. 3, serve to demonstrate this point.

Supporting evidence for the structure advanced for actinioerythrin (11) by our group (Hertzberg *et al.*, 1969), was also obtained by preparation of some thirty derivatives, Fig. 4.

B. TOTAL SYNTHESIS

As an example of the recent achievements in this field the first total synthesis of rhodoxanthin (49), performed by Mayer *et al.* (1967) will be discussed, Fig. 5. Selective ketal formation of the cyclohexenedione (40), followed by addition of lithium ethoxy acetylide gave the acetylenic ketal (41) in high yield. Selective hydrogenation with Lindlar catalyst gave the *cis* enol ether (42), which on treatment with phosphorus tribromide, followed by acid hydrolysis gave the ketoaldehyde (43). The carbon chain was elongated in a Wittig reaction with carboethoxy ethylidene triphenyl phosphorane, providing the keto ester (44). The keto group was again protected by ketal formation and the ester group reduced by lithium aluminium hydride to the allylic alcohol (45). The ketal was submitted to acid hydrolysis, and the resulting keto alcohol converted in the usual manner to the corresponding Wittig salt (46), the phosphorane (47) of which was condensed with the C_{12}-dial (48) to rhodoxanthin (49). The required C_{12}-dial (48) was obtained from 3-methyl-2-penten-4-yn-1-ol (49a) by Glaser dimerization to the diol (50), which was oxidized to the corresponding dial (51) and partly hydrogenated by means of Lindlar catalyst and isomerized to the *trans* C_{12}-dial (48). Rhodoxanthin (49), eschscholtzxanthin (the corresponding diol) and eschscholtzxanthone (the ketol; Bodea *et al.*, 1964) are the only natural carotenoids with *retro* structure.

As another example let us consider the recent stereochemically controlled synthesis of the methyl ester (52) of natural bixin by Weedon's school (Pattenden *et al.*, 1970), Fig. 6. In natural bixin (53) the carbon–carbon double bond number two from the carbomethoxy group has *cis* configuration (Barber *et al.*, 1960b). In the synthesis the *cis* aldehyde (55) was the key intermediate. Its synthesis proceeded from the lactol (56) by a Horner condensation to the *cis* carboxylic acid (57), which via the acid chloride (58), followed by reduction with tri-*tertiary*-butoxy aluminium hydride gave the aldehyde (55) with complete retention of stereochemistry. In passing it should be commented that the Horner reaction used in this step has proved very useful in total synthesis of carotenoids for reacting ketones where the Wittig reaction fails (Manchand *et al.*, 1965; Aasen and Liaaen-Jensen, 1967a; Kjøsen and Liaaen-Jensen, 1970). The desired *cis* ester was obtained by condensing the phosphorane (59) with the *cis* aldehyde (55). The phosphorane (59) was obtained by the usual method from the hydroxy ester (60), starting with the well-known C_{10}-dial (61).

The stereochemically controlled synthesis of methyl natural bixin leads directly to the last topic to be discussed.

(11) Actinioerythrin

(a) Ac.
(b) Silylation
(c) CH_2N_2
(d) Acetal formation

(a) Ac.
(b) Silylation
(c) CH_2N_2

KOH

O_2

$NaBH_4$

Ac.

LAH

Ox.

NH_2OH

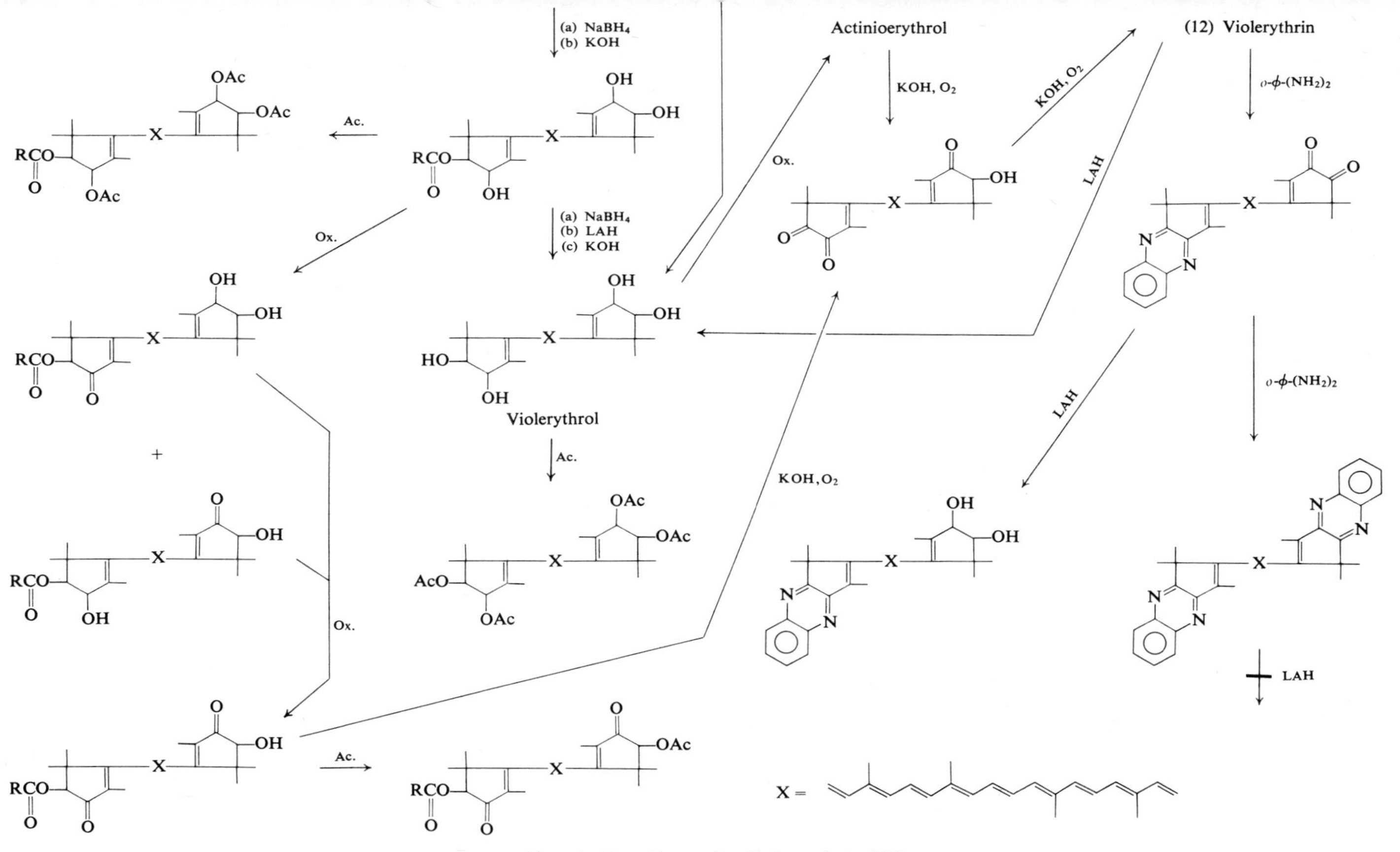

FIG. 4. Reactions of actinioerythrin (11).

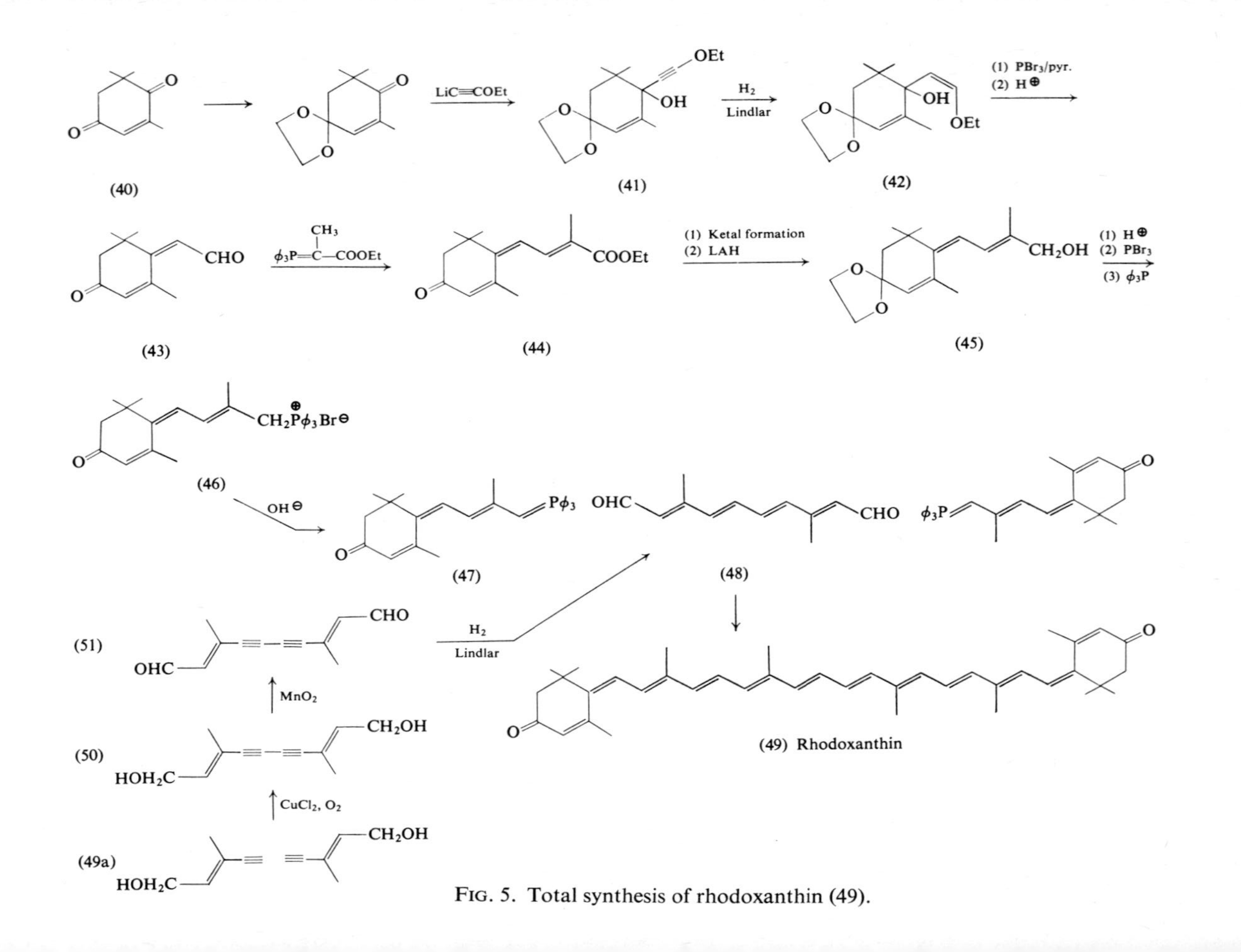

FIG. 5. Total synthesis of rhodoxanthin (49).

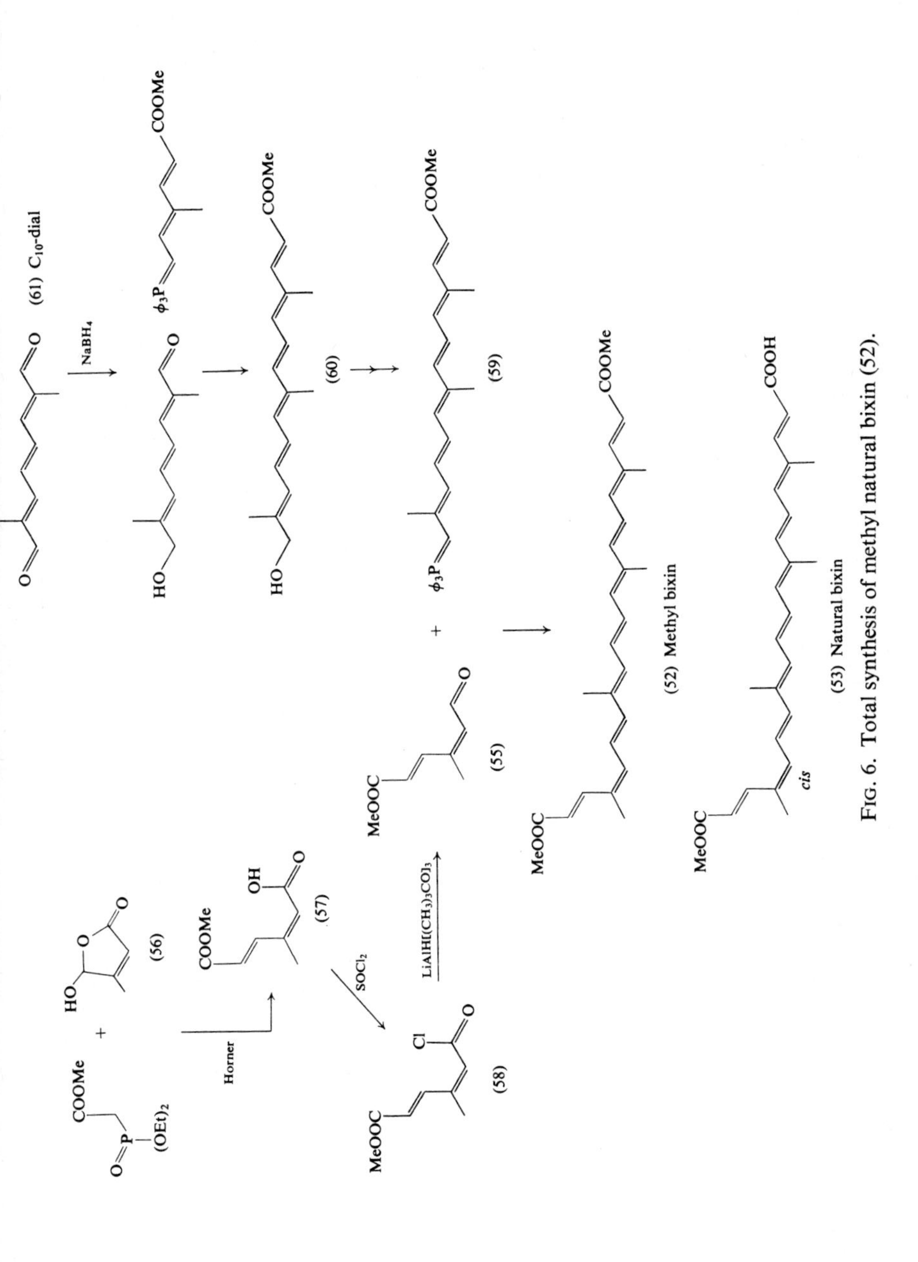

FIG. 6. Total synthesis of methyl natural bixin (52).

V. Stereochemistry

A. *cis-trans* isomerism

The phenomenon of *cis-trans* isomerism of carotenoids was extensively studied by Zechmeister's school (Zechmeister, 1962). Although much information about the isomerization reaction, including its reversibility, and the chromatographic and spectral properties in visible light of *cis* carotenoids was obtained at that time, the exact allocation of *cis* bonds hitherto has depended on synthetic methods or proton magnetic resonance spectroscopy. *Cis* carotenoids are usually recognized as such by their hypsochromic colour shift relative to the parent *trans* compound and in most cases by the presence of a *cis* peak in the near-visible region. This relation is not expected to be valid for terminal *cis* bonds, the presence of which was recently demonstrated by Brown and Weedon (1968) in gazaniaxanthin (62), a natural *cis* isomer of rubixanthin (63). In recent years chromatographic separation of gazaniaxanthin (62) and rubixanthin (63) has not been achieved (cf. Schön, 1938). These carotenoids exhibit the same visible light absorption spectra and analogous optical rotatory dispersion (ORD) and circular dichroism (CD) curves, but different melting point (Brown and Weedon, 1968; Arpin and Liaaen-Jensen, 1969). Gazaniaxanthin (62) could be converted into the higher melting *trans* isomer, rubixanthin (63), on iodine-catalysed stereomutation.

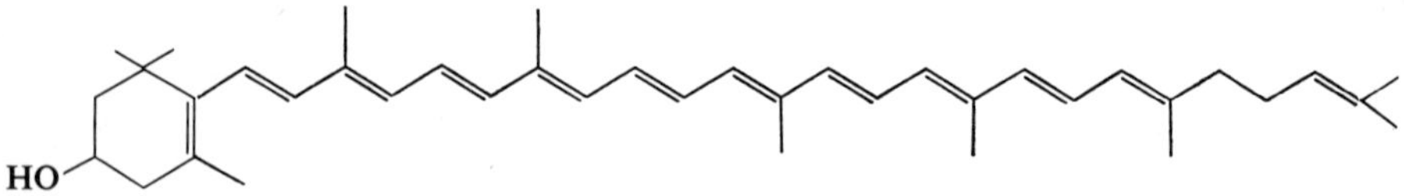

HO

(63) Rubixanthin

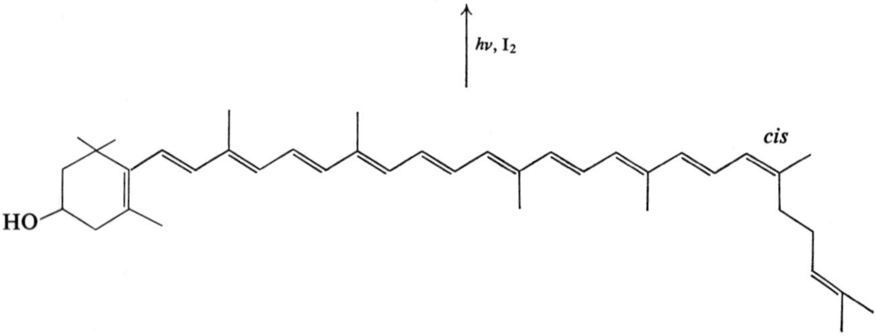

$h\nu, I_2$

cis

HO

(62) Gazaniaxanthin

Cis isomers are usually less stable than the *trans* compounds. Exceptions to this rule are found for the cross-conjugated carotenals of the rhodopinal (15) type (Aasen and Liaaen Jensen, 1967b; Ke *et al.*, 1970). Acetylenic carotenoids also have rather stable *cis* isomers (Weedon, 1970; Francis *et al.*, 1970).

B. RELATIVE AND ABSOLUTE CONFIGURATION

Turning to the absolute stereochemistry of carotenoids with asymmetric carbon atoms, Karrer's school obtained the first information (Entschel and Karrer, 1960; Faigle and Karrer, 1961a,b) on capsanthin (64) and capsorubin (65), and the subject was further pursued by Weedon's group (Barber *et al.*, 1960a, 1961b). Ozonolysis gave a trimethyl-hydroxy-cyclopentane carboxylic acid (66) which on chromic acid oxidation afforded L(−)-camphoronic acid (67) identical with the one obtained from (+)-camphor (68), thus giving the absolute stereochemistry of capsanthin (64) at C-5. *Trans* configuration between the hydroxy group and the carbonyl group in capsanthin (64) was subsequently claimed from synthetic work (Cooper *et al.*, 1962), and confirmed by Faigle *et al.* (1964) for the cyclopentane carboxylic acid (66) by thorough lactonization experiments and proton magnetic resonance spectroscopy. An unambiguous synthesis of the cyclopentane carboxylic acid (66) has since been achieved from (+)-camphor (68) (Weedon, 1967b).

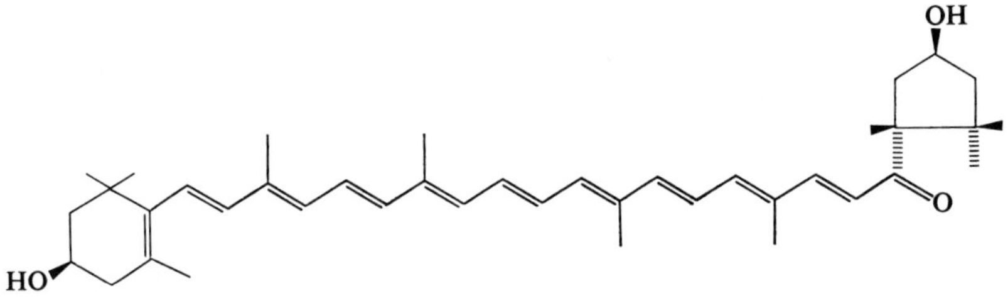

(64) Capsanthin

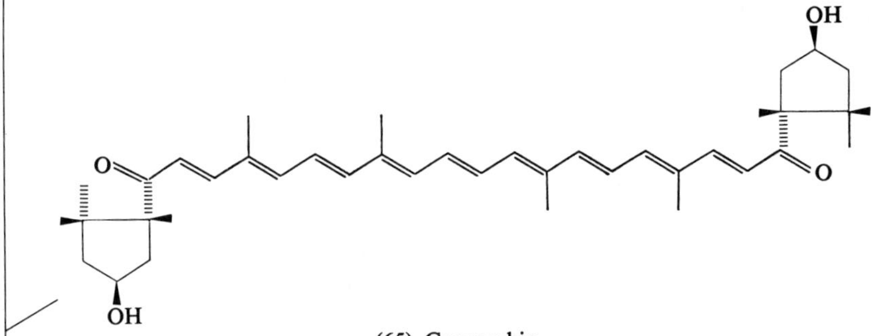

(65) Capsorubin

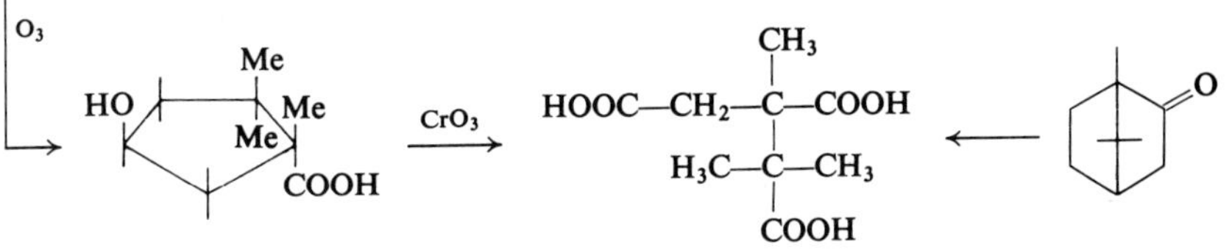

O_3

(66)

CrO_3

HOOC—CH$_2$—C(CH$_3$)(COOH)—C(CH$_3$)$_2$—COOH

(67) L-(−)-Camphoronic acid

(68) (+)-Camphor

More recently the absolute configuration of α-carotene (69) has been reported by Eugster *et al.* (1969), Fig. 7. (+)- and (−)-α-Cyclogeranic acid (70) were separately converted into (+)- and (−)-α-ionone (71) by reactions that retain the absolute stereochemistry at C-6. The total synthesis of the enantiomeric α-carotenes (69 and 72) and ε-carotenes (73 and enantiomer) were performed by means of the pure α-ionone enantiomers. The chirality of the enantiomeric α-ionones used was established by correlation with (+)-manool (74). (−)-α-Ionone (71) was partly hydrogenated to (−)-dihydro-α-ionone (75), which was converted into (−)-α-cyclogeranyl acetic acid which was cyclized to the (−)-*trans*-δ-lactone (76), also obtained from (+)-manool (74). It was concluded that the naturally occurring (+)-α-carotene (69) has the configuration 6*R*.

Fucoxanthin (9), neoxanthin (77) and violaxanthin (78) have been converted into zeaxanthin (79) by reactions that retain the stereochemistry at C-3 and C-3′ (Bonnet *et al.*, 1969; Cholnoky *et al.*, 1967, 1969), Fig. 8. The products all have the same relative configuration at these positions according to their ORD curves (Bartlett *et al.*, 1969). The allenic ketone (80), which is a degradation product of fucoxanthin (9), was converted into the corresponding *p*-bromobenzoate. X-ray analysis of the latter by de Ville *et al.* (1969) showed the given absolute configuration and it may thus be inferred that zeaxanthin (79), neoxanthin (77), fucoxanthin (9) and violaxanthin (78) also have the *R*-configuration at C-3 and C-3′. ORD measurements by Bartlett *et al.* (1969) suggest that cryptoxanthin, rubixanthin (63), gazaniaxanthin (62), β-citraurin and reticulataxanthin have the same *R*-configuration at C-3. Moreover, the perhydro derivative (81) of the acetylenic alloxanthin (82) gives the same perhydro derivative (81) as zeaxanthin (79) with the same ORD curves. Hence the same configuration at C-3 and C-3′ is also inferred for alloxanthin. The ORD correlation of the diacetylenic alloxanthin (82) and the monoacetylenic analogue, diatoxanthin, is in further agreement at this point (Bartlett *et al.*, 1969).

Natural violaxanthin (78) has ORD properties different from those of the main products obtained by *in vitro* epoxidation of zeaxanthin (79) diacetate and subsequent hydrolysis. Models indicate that epoxidation by monoperphthalic acid is favoured.* On this basis the epoxy group and hydroxy group in violaxanthin (78) and neoxanthin (77), and probably also in fucoxanthin (9), are assumed to possess *trans* configurations (Cholnoky *et al.*, 1969; Bartlett *et al.*, 1969) (Fig. 8).

The absolute configuration of the asymmetric centres at C-3 (C-3′) and or C-6 (C-6′) in the more common C_{40}-carotenoids has thus been established during the last year by English and Swiss efforts, whereas the configuration at C-2 (C-2′) in C_{50}-carotenoids is still unknown. Natural bacterioruberin (83)

* *Cis* to the C-3 oxygen function.

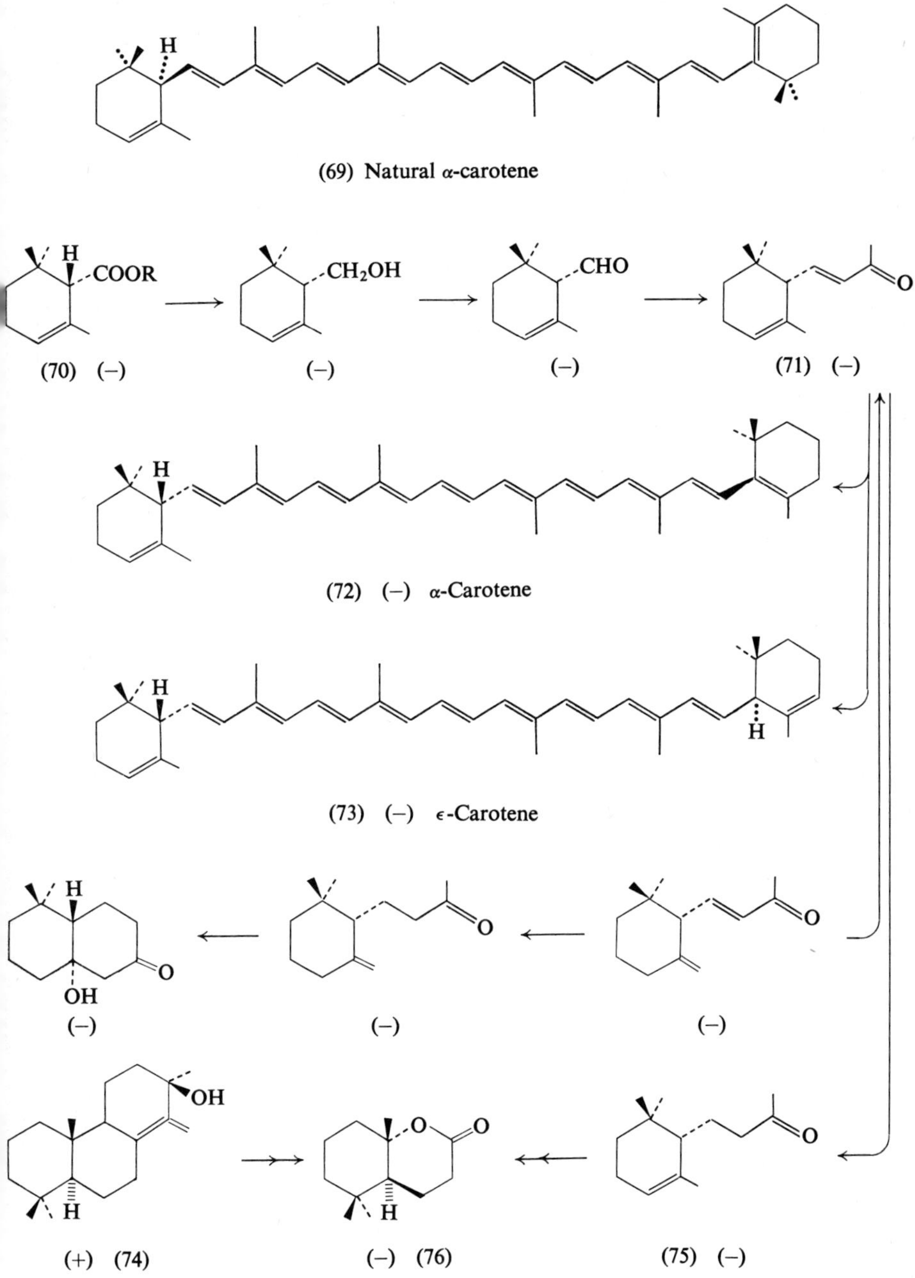

(69) Natural α-carotene

COOR CH$_2$OH CHO O

(70) (−) (−) (−) (71) (−)

(72) (−) α-Carotene

(73) (−) ε-Carotene

OH O O O

(−) (−) (−)

OH O O

(+) (74) (−) (76) (75) (−)

FIG. 7. Stereochemistry of α-carotene (69).

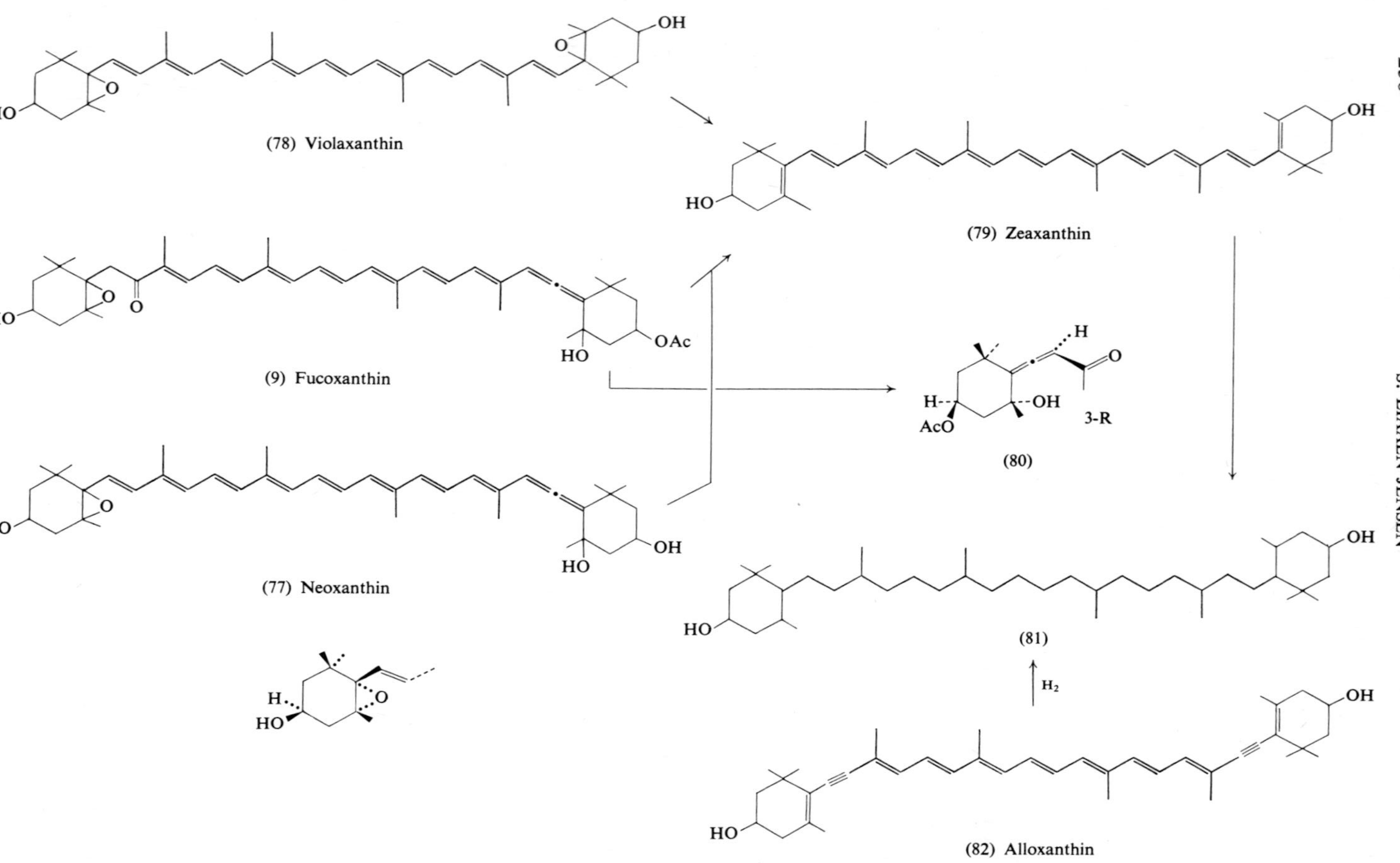

FIG. 8. Relative and absolute configuration of various carotenoids.

HO OH OH OH

(83) Bacterioruberin

OH OH

(84) Bisanhydrobacterioruberin

(Kelly *et al.*, 1970) from *Halobacterium salinarium* and natural bisanhydrobacterioruberin (84) (Norgård *et al.*, 1970b) from *Corynebacterium poinsettiae* exhibit CD properties which suggest the same relative configuration at C-2 and C-2′ (Borch and Liaaen-Jensen, unpublished). This reveals, as expected, that the extra C_5-units in C_{50}-carotenoids are added in a stereospecific manner. However, establishment of the absolute configuration of natural C_{50}-carotenoids, as well as the chemical synthesis of such carotenoids, are among the problems that are left for the future.

REFERENCES

Aasen, A. J. and Liaaen Jensen, S. (1967a). *Acta Chem. Scand.* **21**, 371–377.

Aasen, A. J. and Liaaen Jensen, S. (1967b). *Acta Chem. Scand.* **21**, 2185–2204.

Aasen, A. J., Francis, G. W. and Liaaen-Jensen, S. (1969). *Acta Chem. Scand.* **23**, 2605–2615.

Aitzetmüller, K., Strain, H. H., Svec, W. A., Grandolfo, M. and Katz, J. (1969). *Phytochemistry* **8**, 1761–1770.

Arcamone, F., Camerino, B., Cotta, E., Franseschi, G., Grein, A., Penco, S. and Spala, C. (1966). *Experientia* **25**, 241–242.

Arpin, N. and Liaaen-Jensen, S. (1969). *Phytochemistry* **8**, 185–193.

Arpin, N., Francis, G. W., Hertzberg, S. and Liaaen-Jensen, S. To be published.

Arpin, N., Norgård, S., Francis, G. W. and Liaaen-Jensen, S. To be published.

Baldas, J., Porter, Q. N., Cholnoky, L., Szabolcs, J. and Weedon, B. C. L. (1966). *Chem. Commun.* 852–854.

Baldas, J., Porter, Q. N., Leftwick, A. P., Holzel, R., Weedon, B. C. L. and Szabolcs, J. (1969). *Chem. Commun.* 415–416.

Barber, M. S., Jackman, L. M., Warren, C. K. and Weedon, B. C. L. (1960a). *Proc. chem. Soc.* 19.

Barber, M. S., Jackman, L. M. and Weedon, B. C. L. (1960b). *Proc. Chem. Soc.* 23.

Barber, M. S., Hardisson, A., Jackman, L. M. and Weedon, B. C. L. (1961a). *J. chem. Soc.* 1625–1630.

Barber, M. S., Jackman, L. M., Warren, C. K. and Weedon, B. C. L. (1961b). *J. chem. Soc.* 4019–4024.

Bartlett, L., Klyne, W., Mose, W. P., Scopes, P. M., Galasko, G., Mallams, A. K. and Weedon, B. C. L. (1969). *J. Chem. Soc.* 2527.

Biemann, K., De Jongh, D. C. and Schnoes, H. K. (1963). *J. Am. chem. Soc.* **85**, 1763–1771.
Bodea, C., Nicoara, E. and Salontai, T. (1964). *Rev. Roum. Chem.* **9**, 517–521.
Bonnett, R., Mallams, A. K., Spark, A. A., Tee, J. L., Weedon, B. C. L. and McCormick, A. (1969). *J. chem. Soc.*, 429–454.
Booth, V. H. (1964). *Phytochemistry* **3**, 229–234.
Britton, G. and Goodwin, T. W. (1969). *Phytochemistry* **8**, 2257–2258.
Brown, B. O. and Weedon, B. C. L. (1968). *Chem. Commun.* 382–384.
Cheesman, D. F., Zagalsky, P. F. and Ceccaldi, H. J. (1966). *Proc. r. Soc.* Ser. B. **164**, 130–151.
Cheesman, D. F., Lee, W. L. and Zagalsky, P. T. (1967). *Biol. Rev.* **42**, 131–160.
Cholnoky, L., Györgyfy, K., Szabolcs, J. and Weedon, B. C. L. (1966). *Chem. Commun.* **13**, 404–405.
Cholnoky, L., Szabolcs, J. and Toth, G. (1967). *Annalen* **708**, 218–223.
Cholnoky, L., Györgyfy, K., Ronai, A., Szabolcs, J., Toth, G., Galasko, G., Mallams, A. K., Waight, E. C. and Weedon, B. C. L. (1969). *J. chem. Soc.* C 1256–1263.
Cooper, R. D. G., Jackman, L. M. and Weedon, B. C. L. (1962). *J. chem. Soc.* 215.
De Ville, T. E., Hursthouse, M. B., Russell, S. W. and Weedon, B. C. L. (1969). *Chem. Comm.* 754.
Edmunds, F. S. and Johnstone, R. A. W. (1965). *J. Chem. Soc.* 2892–2897.
Egger, K. (1964). *Ber.* **77**, 10–15.
Egger, K. (1968). *Z. Naturf.* **23b**, 733–735.
Egger, K. and Kleinig, H. (1966). *Z. Pflanzenphysiol.* **55**, 224–228.
Egger, K. and Kleinig, H. (1967). *Phytochemistry* **6**, 437–440.
Egger, K., Dabbagh, A. G. and Nietsche, N. (1969). *Tetrahedron Letters* 2995–2998.
Entschel, R. and Karrer, P. (1960). *Helv. Chim. Acta* **43**, 89–94.
Enzell, C. R. (1969). *Pure Appl. Chem.* **20**, 497–516.
Enzell, C. R., Francis, G. W. and Liaaen-Jensen, S. (1968). *Acta Chem. Scand.* **22**, 1054–1055.
Enzell, C. R., Francis, G. W. and Liaaen-Jensen, S. (1969). *Acta Chem. Scand.* **23**, 727–750.
Enzell, C. R. and Liaaen-Jensen, S. (1971). *Acta Chem. Scand.* **25**. (In press.)
Eugster, C. H., Buchecker, R., Tscharner, C., Uhde, G. and Ohloff, G. (1969). *Helv. Chim. Acta* **52**, 1729–1731.
Faigle, H. and Karrer, P. (1961a). *Helv. Chim. Acta* **44**, 1257–1261.
Faigle, H. and Karrer, P. (1961b). *Helv. Chim. Acta* **44**, 1904–1907.
Faigle, W., Müeller, H., v. Philipsborn, W. and Karrer, P. (1964). *Helv. Chim. Acta* **47**, 741–745.
Francis, G. W. (1969). *Acta Chem. Scand.* **23**, 2916–2918.
Francis, G. W. and Liaaen-Jensen, S. (1970). *Acta Chem. Scand.* **24**, 2705–2713.
Francis, G. W., Hertzberg, S., Andersen, K. and Liaaen-Jensen, S. (1970). *Phytochemistry* **9**, 629.
Francis, G. W., Upadhyaj, R. R. and Liaaen-Jensen, S. (1971). *Acta Chem. Scand.* **25**, 3050–3053.
Galasko, G., Hora, J., Toube, T. P., Weedon, B. C. L., André, D., Barbier, M., Lederer, M. and Villanueva, V. R. (1969). *J. chem. Soc.* C 1264–1265.
Grob, E. C. and Siekmann, W. (1965). *Helv. Chim. Acta* **48**, 1199–1203.
Heilbron, J. M., Jackson, H. and Jones, R. N. (1935). *Biochem. J.* **29**, 1384–1388.
Hertzberg, S. and Liaaen-Jensen, S. (1967a). *Acta Chem. Scand.* **21**, 15–41.
Hertzberg, S. and Liaaen-Jensen, S. (1967b). *Phytochemistry* **6**, 1119–1126.
Hertzberg, S. and Liaaen-Jensen, S. (1968). *Acta Chem. Scand.* **22**, 1714–1716.
Hertzberg, S. and Liaaen-Jensen, S. (1969a). *Phytochemistry* **8**, 1259–1280.

Hertzberg, S. and Liaaen-Jensen, S. (1969b). *Phytochemistry* **8**, 1281–1292.

Hertzberg, S. and Liaaen-Jensen, S. (1971). *Phytochemistry*. To be published.

Hertzberg, S., Liaaen-Jensen, S., Enzell, C. R. and Francis, G. W. (1969). *Acta Chem. Scand.* **23**, 3290–3312.

Holzel, R., Leftwick, A. P. and Weedon, B. C. L. (1969). *Chem. Commun.* 128–129.

Hora, J., Toube, T. P. and Weedon, B. C. L. (1970). *J. chem. Soc.* C 241–284.

Jensen, A. (1964). *Acta Chem. Scand.* **18**, 840–841.

Jensen, A. (1966). *Norw. Inst. Seaweed Res. Report* No. 31, Tapir, Trondheim.

Karrer, P. and Jucker, E. (1948). "Carotinoide". Birkhäuser, Basel.

Karrer, P. and Miki, K. (1929). *Helv. Chim. Acta* **12**, 985–986.

Ke, B., Imsgard, F., Kjøsen, H. and Liaaen-Jensen, S. (1970). *Biochim. biophys. Acta* **210**, 139–152.

Kelly, M., Norgård, S. and Liaaen-Jensen, S. (1970). *Acta Chem. Scand.* **24**, 2259–2260.

Kjøsen, H. and Liaaen-Jensen, S. (1970). *Acta Chem. Scand.* **24**. (In press.)

Kjøsen, H., Liaaen-Jensen, S. and Enzell, C. R. (1971). *Acta Chem. Scand.* **25**. (In press.)

Kleinig, H. (1967). *Z. Naturf.* **22b**, 977–979.

Kleinig, H. and Egger, K. (1967a). *Phytochemistry* **6**, 611–619.

Kleinig, H. and Egger, K. (1967b). *Phytochemistry* **6**, 1681–1686.

Kleinig, H. and Nietsche, H. (1968). *Phytochemistry* **7**, 1171–1175.

Kuhn, R. and Kuehn, H. (1967). *Europ. J. Biochem.* **2**, 349–360.

Kuhn, R. and Lederer, E. (1933). *Ber.* **66**, 488–495.

Kuhn, R., Winterstein, A. and Kaufmann, W. (1930). *Ber.* **63**, 1489–1497.

Kuhn, R., Winterstein, A. and Lederer, E. (1931). *Hoppe Seyler's Z. Physiol. Chem.* **197**, 141–160.

Lederer, E. (1933). *Compt. rend. Soc. Biol., Paris* **113**, 1391.

Liaaen Jensen, S. (1965). *Phytochemistry* **4**, 925–936.

Liaaen Jensen, S. (1966). *In* "Biochemistry of Chloroplasts" (T. W. Goodwin, ed.), pp. 437–441, Academic Press, London and New York.

Liaaen Jensen, S. (1967). *Acta Chem. Scand.* **21**, 1972.

Liaaen-Jensen, S. (1969). *Pure Appl. Chem.* **20**, 421–448.

Liaaen-Jensen, S. and Jensen, A. (1965). *In* "Progress Chemistry Fats and other Lipids" (R. T. Holman, ed.), Vol. VIII, Part 2, pp. 133–212.

Liaaen Jensen, S. and Weeks, O. B. (1966). *Norw. J. Chem. Mining Met.* **26**, 130.

Liaaen-Jensen, S., Hertzberg, S., Weeks, O. B. and Schwieter, U. (1968). *Acta Chem. Scand.* **22**, 1171–1186.

Manchand, P. S., Rüegg, R., Schwieter, U., Siddons, P. T. and Weedon, B. C. L. (1965). *J. chem. Soc.* 2019–2026.

Malhotra, H. C., Britton, G. and Goodwin, T. W. (1970). *Chem. Commun.* 127–128.

Mayer, H., Montavon, M., Rüegg, R. and Isler, O. (1967). *Helv. Chim. Acta* **50**, 1606–1618.

Norgård, S. and Liaaen-Jensen, S. (1969). *Acta Chem. Scand.* **23**, 1463.

Norgård, S., Francis, G. W., Jensen, A. and Liaaen-Jensen, S. (1970a). *Acta Chem. Scand.* **24**, 1460–1462.

Norgård, S., Aasen, A. J. and Liaaen-Jensen, S. (1970b). *Acta Chem. Scand.* **24**, 2183–2197.

Nybraathen, G. and Liaaen-Jensen, S. (1970). *Acta Chem. Scand.* **24**. (In press.)

Pattenden, G., Way, J. E. and Weedon, B. C. L. (1970). *J. Chem. Soc.* C 235–242.

Schimmer, B. P. and Krinsky, N. I. (1966). *Biochem.* **5**, 3649–3657.

Schmidt, K., Pfennig, N. and Liaaen-Jensen, S. (1965). *Arch. Mikrobiol.* **52**, 132–146.

Schwieter, U. and Liaaen-Jensen, S. (1969). *Acta Chem. Scand.* **23**, 1057–1058.

Schwieter, U., Bolliger, H. R., Chopard-dit-Jean, L. H., Englert, G., Kofler, M., König, A., v. Planta, C., Rüegg, R., Vetter, W. and Isler, O. (1965). *Chimia* **19**, 294–302.
Schwieter, U., Englert, G., Rigassi, N. and Vetter, W. (1969). *Pure Appl. Chem.* **20**, 365–420.
Schön, K. (1938). *Biochem. J.* **32**, 1566–1570.
Straub, O. (1971). *In* "Carotenoids—a Progress Report" (O. Isler, ed.), Birkhäuser, Basel.
Tischer, J. (1938). *Hoppe Seyler's Z. physiol. Chem.* **251**, 109–128.
Vacheron, M. J., Guilly, R., Arpin, N. and Michel, G. (1969). *Phytochemistry* **8**, 897–903.
Wald, G., Nathanson, N., Jencks, W. P. and Tarr, E. (1948). *Biol. Bull. mar. biol. Woods Hole* **95**, 249–250.
Weedon, B. C. L. (1967a). *In* "Terpenoids in Plants" (J. B. Pridham, ed.), pp. 119–128, Academic Press, London and New York.
Weedon, B. C. L. (1967b). *Chem. Br.* 424–432.
Weedon, B. C. L. (1970). *Rev. Pure Appl. Chem.* **20**, 51–66.
Weeks, O. B. (1971). *In* "Biochemistry of the Terpenoids" (T. W. Goodwin, ed.), Academic Press, London and New York.
Weeks, O. B., Andrewes, A. G., Brown, B. O. and Weedon, B. C. L. (1969). *Nature, Lond.* **224**, 879–882.
Yamamoto, H., Yokoyama, H. and Boettger, H. (1969). *J. Org. Chem.* **34**, 4207–4208.
Yokoyama, H. and White, M. (1965a). *J. org. Chem.* **30**, 2481–2482.
Yokoyama, H. and White, M. (1965b). *J. org. Chem.* **30**, 3994–3996.
Zechmeister, L. (1934). "Carotinoide, ein biochemischer Bericht über pflantzliche und tierische Polyenfarbstoffe", Springer, Berlin.
Zechmeister, L. (1962). "*Cis-trans* isomeric Carotenoids, Vitamins A and Arylpolyenes", Springer, Wien.

CHAPTER 9

General Aspects of Carotenoid Biosynthesis

G. BRITTON

Department of Biochemistry, The University of Liverpool, Liverpool, England

I. INTRODUCTION

The recent great advance in knowledge of the chemistry of the carotenoid pigments which has been made possible by the development of new purification methods, and by the harnessing of several physical techniques, e.g. proton magnetic resonance and mass spectrometry, optical rotatory dispersion and X-ray crystallography (see Jensen, 1971; Weedon, 1970), has not yet been followed by similar advances in knowledge of carotenoid biosynthesis. The elucidation of the structures of many of the 300–350 presently known naturally occurring carotenoids has rapidly been followed by the postulation of hypothetical, though chemically reasonable, schemes for their biosynthesis. In most cases, no biochemical investigation of these postulated pathways has been attempted. In other cases, despite much work, no definite conclusions may justifiably be drawn about the mechanisms of the reactions involved. Thus, although the main outlines of the pathways by which carotenoids are biosynthesized are generally accepted, very little detail is known of the individual steps involved.

Various aspects of the biosynthesis of carotenoids have been the subject of several recent reviews (Goodwin, 1965; Liaaen-Jensen, 1965; Chichester, 1967; Chichester and Nakayama, 1967; Czygan, 1967; Porter and Anderson, 1967). The present report will therefore concentrate mainly on work published in the last 3–4 years, i.e. 1966 to early 1970.

II. Carotenoid Biosynthesis in Higher Plants, Fungi and Non-Photosynthetic Bacteria

A. General Outline of Carotene Biosynthesis

Since about 1950, studies of the inheritance of pigmentation in tomatoes, studies of the effects of mutations on pigment synthesis in microorganisms, and studies of the inhibition of carotenoid synthesis in microorganisms by diphenylamine, together with the determination of the structures of the compounds involved, have led to the postulation of a number of schemes involving sequential desaturation and cyclization, which could lead to the formation of all common naturally occurring carotene hydrocarbons. The evidence on which such proposals were based has been discussed in the reviews previously cited. The idea of sequential formation has been widely accepted.

Figure 1 illustrates such a scheme for carotene biosynthesis in higher plants, fungi and non-photosynthetic bacteria. According to this scheme, two molecules of geranylgeranyl pyrophosphate (GGPP, 1) (synthesized from mevalonic acid (MVA) by the normal terpenoid pathway) condense to give phytoene (2) as the first C_{40} compound formed. This then undergoes successive losses of two hydrogen atoms to give phytofluene (3), ζ-carotene (4), neurosporene (5) and lycopene (6), with cyclization of neurosporene or lycopene occurring to give the bicyclic α-carotene (11), β-carotene (12) and ε-carotene (13) via the monocyclic α-zeacarotene (7), β-zeacarotene (8), δ-carotene (9) and γ-carotene (10).

Some of the proposed steps in this scheme will be considered in more detail. Discussion will be limited to those reactions specific to carotenoid biosynthesis; the early biogenetic stages common to all terpenoids have been extensively reviewed.

B. Biosynthesis of Phytoene

1. *General Aspects*

Early suggestions (Grob *et al.*, 1961) that the first C_{40} intermediate in carotenoid biosynthesis was lycopersene (14) formed from GGPP in a manner analogous to the formation of the triterpene squalene from farnesyl pyrophosphate (FPP) are now largely discounted. It is now generally accepted that

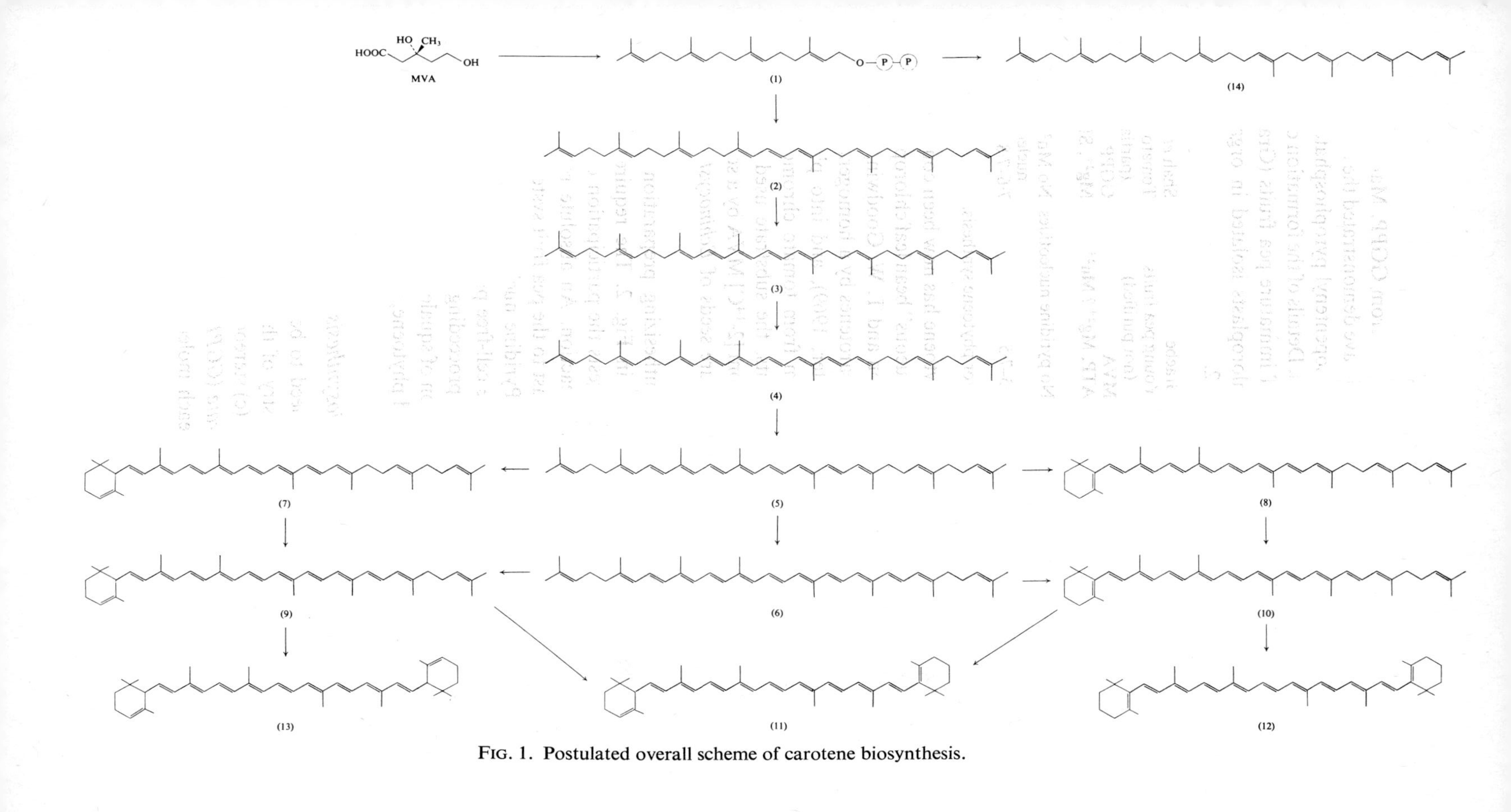

FIG. 1. Postulated overall scheme of carotene biosynthesis.

phytoene is the first C_{40} compound formed from GGPP. Many studies with intact systems and cell-free preparations have demonstrated the incorporation of terpenoid precursors such as MVA, isopentenyl pyrophosphate (IPP) and FPP into phytoene and other carotenoids. Details of the formation of phytoene from MVA by cell-free preparations of immature pea fruits (Graebe, 1968) and by a preparation of bean leaf chloroplasts isolated in organic media (Charlton *et al.*, 1967) are given in Fig. 2.

	Charlton *et al.*	Graebe	Shah *et al.*
Source of enzyme	Bean leaf chloroplasts (not purified)	Young pea fruits (not purified)	Tomato plastids (partially purified)
Substrate	MVA	MVA	GGPP
Co-factor requirements	ATP, either Mg^{2+} or Mn^{2+}	ATP, Mg^{2+} ? $\mathbf{Mn^{2+}}$	$\mathbf{Mg^{2+}}$, SH protector
	No pyridine nucleotides	No pyridine nucleotides	No Mn^{2+} or pyridine nucleotides
pH optimum	7·0–7·5	6·5–7·5	7·6–7·8

FIG. 2. Cell-free systems for phytoene synthesis.

The role of GGPP as precursor of phytoene has now been confirmed by its conversion into phytoene by the "non-aqueous" bean leaf chloroplast preparation (M. J. Buggy, J. R. Vose, G. Britton and T. W. Goodwin, unpublished results, 1969), into phytoene and other carotenes by a homogenate of *Phycomyces blakesleeanus* (Lee and Chichester, 1969), and into phytoene by a partially purified soluble enzyme system from tomato chromoplasts (Shah *et al.*, 1968). In all of these experiments, the substrate used was all-*trans* [4,8,12,16-^{14}C] GGPP biosynthesized from [2-^{14}C] MVA by a soluble enzyme fraction from the endosperm of immature seeds of *Echinocystis macrocarpa* (Oster and West, 1968).

Details of the soluble phytoene-synthesizing preparation from tomato plastids (Shah *et al.*, 1968) are given in Fig. 2. The requirement for the sulphydryl reagent, dithiothreitol, suggests the participation of a sulphydryl enzyme group in the condensation reaction. An absolute requirement for Mg^{2+} was also demonstrated, in contrast to the pea fruit system, in which the absolute requirement was for Mn^{2+}. Pyridine nucleotides were not required for phytoene synthesis by any of these cell-free preparations, thus providing further evidence against the reaction proceeding via lycopersene, formation of which, by analogy with the formation of squalene, should require NADPH. No intermediates between GGPP and phytoene, enzyme-bound or otherwise, have been isolated.

2. *Stereochemistry of Phytoene Biosynthesis*

Three details of stereochemistry need to be defined in a consideration of phytoene synthesis; (a) stereochemistry of the GGPP substrate; (b) stereochemistry of the phytoene product; (c) stereochemical course of the reaction.

(*a*) *Stereochemistry of the substrate* (*GGPP*). In the formation of GGPP, one hydrogen atom from C-4 of each molecule of MVA is lost during the

IPP-DMAPP (dimethylallyl pyrophosphate) isomerization, or the prenyl transferase condensations. It is now established that in the formation of *trans*-prenols, the *pro*-4*R** hydrogen atom of MVA is retained, and the *pro*-4*S* hydrogen atom lost at each stage, whereas in the formation of *cis*-prenols, the reverse is true (Archer *et al.*, 1966; Cornforth *et al.*, 1966; Hemming 1970).

Phytoene, biosynthesized from [2-^{14}C,(4*R*)-4-^{3}H$_1$] MVA by *Phycomyces blakesleeanus* and by tomato slices has been shown to retain all the tritium label whereas all the tritium label from [2-^{14}C,(4*S*)-4-^{3}H$_1$] MVA is lost (Goodwin and Williams, 1966), thus indicating that phytoene biosynthesis proceeds via the all-*trans* prenol route, and all-*trans* GGPP. This has now been confirmed by conversion of all-*trans* GGPP into phytoene by cell-free preparations, as described above.

(*b*) *Stereochemistry of the product, phytoene.* Although almost all naturally occurring carotenoids have the all-*trans* configuration, it has been established that phytoene isolated from carrot oil has the 15-*cis* configuration, the other double bonds being *trans* (Rabourn *et al.*, 1954; Rabourn and Quackenbush, 1956; Jungalwala and Porter, 1965; Davis *et al.*, 1966). Weeks (1971), however, has suggested that all-*trans* phytoene is the isomer isolated from *Flavobacterium dehydrogenans*. The stereochemistry of the C-15,15′ double bond in the phytoene produced by the previously mentioned cell-free systems has not been determined, and it is important not only for studies of this reaction, but also for a description of later stages of carotenoid biosynthesis that this stereochemical point is established.

(*c*) *Stereochemical course of the reaction.* In the formation of phytoene, two hydrogen atoms are lost, one from C-1 of each molecule of GGPP (initially C-5 of MVA). GGPP will have the *pro*-5*R* and *pro*-5*S* hydrogen atoms of MVA (designated H_{5R} and H_{5S} respectively) as shown in Fig. 3, by analogy with the established labelling pattern in FPP (Popják *et al.*, 1961; Cornforth *et al.*, 1966; Donninger and Popják, 1966).

It has been shown that all the tritium from [2-^{14}C,(5*R*)-5-^{3}H$_1$] MVA is retained in the formation of phytoene (configuration of the C-15 double bond not determined) in tomato slices and fruits of *Physalis alkekengii*, the hydrogen atoms lost thus being the *pro*-5*S* hydrogen atoms of MVA, i.e. the *pro*-1*S* hydrogen atoms of GGPP (Fig. 3), (Williams *et al.*, 1966, 1967; Walton *et al.*, 1969). This has recently been confirmed in the cell-free preparation of "non-aqueous" bean leaf chloroplasts (Buggy *et al.*, 1969).

3. *Mechanism of Phytoene Formation*

Several reaction mechanisms have been proposed for the formation of phytoene from GGPP (Goodwin, 1965; Chichester, 1967; Williams *et al.*,

* The term "*pro*-4*R* hydrogen atom of MVA" refers to that hydrogen atom which, when labelled with deuterium or tritium, confers the *R* configuration at C-4 of MVA.

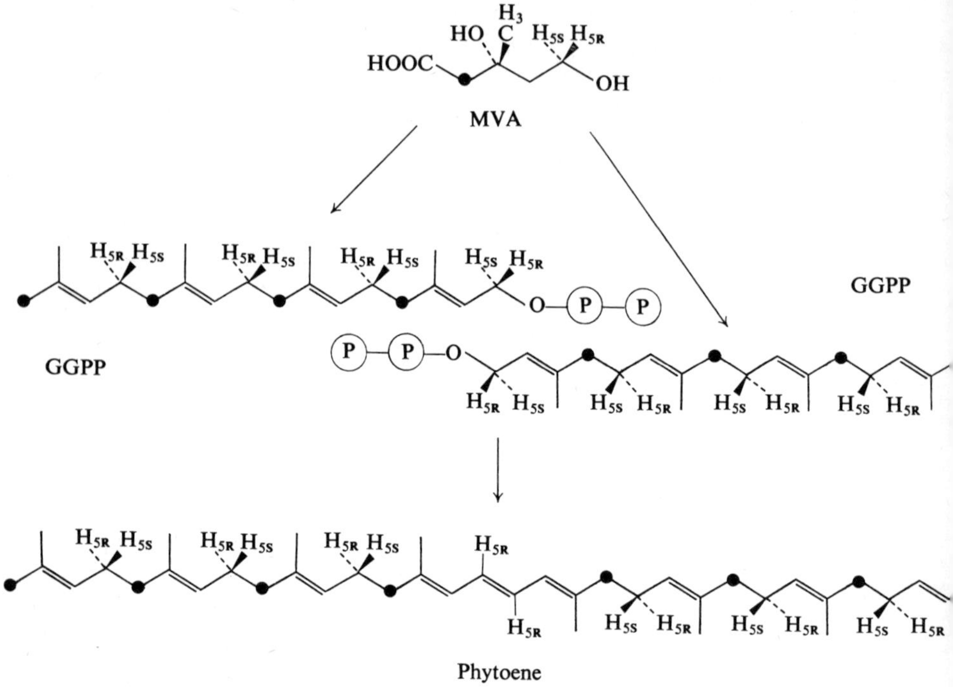

FIG. 3. Stereochemistry of phytoene formation.

1967). However, since the configuration of the phytoene produced is not established, and no information is available about possible intermediates in the reaction, the postulation of any mechanism from the evidence available hardly seems justified.

C. DESATURATION

1. General Observations

Whilst there can be no doubt that in several carotenogenic systems studied, an enzyme is present which will convert GGPP into phytoene, the position of phytoene as a possible precursor of the other, less-saturated, carotenoids is not so well-defined. It is obvious that the coloured carotenoids are formed from a more saturated precursor, i.e. GGPP, and this has been demonstrated experimentally, but some doubt exists as to whether, under normal conditions, phytoene also is an intermediate in the formation of the more unsaturated carotenoids. Several pieces of evidence are available which have been interpreted as indicating that free phytoene may not be such an intermediate.

2. *The Possible Role of Phytoene as an Intermediate in the Biosynthesis of Other Carotenes*

Genetic experiments, e.g. studies of pigment inheritance in tomatoes, and studies of the pigmentation in mutants of several microorganisms indicate the likelihood of the coloured carotenoids being formed from a more saturated precursor. This idea gains further support from experiments on the inhibitory effect of diphenylamine (DPA) and 2-hydroxybiphenyl on carotenogenesis in several microorganisms. In the presence of these inhibitors, synthesis of the normal carotenoids is prevented, and more saturated carotenoids, especially phytoene, accumulate. In some experiments, removal of the inhibitor was shown to allow synthesis of the normal pigments to proceed, though in several cases this synthesis did not seem to occur at the expense of the accumulated saturated compounds, particularly phytoene (Goodwin and Osman, 1954; Jensen *et al.*, 1959; Villoutreix, 1960). These experiments are discussed in the reviews previously cited.

Recent experiments have demonstrated the incorporation of [^{14}C] GGPP into carotenes, e.g. into phytoene, phytofluene, ζ-carotene, neurosporene and lycopene by a homogenate of a mutant strain of *Phycomyces blakesleeanus* (Lee and Chichester, 1969) and into phytoene, lycopene, β-carotene and δ-carotene by tomato slices (J. R. Vose, G. Britton and T. W. Goodwin, unpublished results, 1969). This therefore shows that the coloured carotenes are formed from a more saturated precursor, GGPP, but the nature of the intermediate which undergoes desaturation is not clear. Thus it has not conclusively been established whether desaturation normally occurs at the C_{20} stage or the C_{40} stage, or whether the reaction sequence involves enzyme-bound intermediates or proceeds via free phytoene.

Very few direct conversions of phytoene into other carotenes have been reported, e.g. into β-carotene by extracts of *Sporobolomyces shibatanus* (Kakutani *et al.*, 1964; Kakutani, 1966), into δ-carotene by extracts of *Staphylococcus aureus* (Suzue, 1960), and into phytofluene by a tomato chromoplast preparation (Beeler and Porter, 1962).

Subbarayan *et al.* (1970) have now reported the conversion of [^{14}C] phytoene into phytofluene and lycopene in the light by a spinach chloroplast enzyme system. Synthesis of phytofluene and lycopene was stimulated by NADP, and FAD appeared to be an absolute requirement for lycopene synthesis. These authors also report similar conversions of phytoene into other carotenes by a tomato enzyme system (Kushwaha *et al.*, 1970).

Many experiments have been performed in which the time courses of pigment synthesis following removal of an inhibitor or change in nutrient conditions, and of incorporation of radioactive substrates into carotenoids have been followed. From the results of this type of experiment, several workers have concluded that phytoene did not behave as a precursor of other carotenes (Purcell *et al.*, 1959; Krzeminski and Quackenbush, 1960; Villoutreix, 1960; Yokoyama *et al.*, 1961; Purcell, 1964; Karunakaran *et al.*, 1966). Other workers, however, from the results of similar experiments, have deduced that phytoene is a precursor of other carotenoids (Beeler and Porter, 1963; Davies *et al.*, 1963; Nusbaum-Cassuto *et al.*, 1967; Harding *et al.*, 1969).

The position of phytoene is thus in many ways anomalous, and it is quite possible that free phytoene as such is not a normal intermediate in the biosynthesis of other carotenes. Detailed work at the enzyme level is clearly necessary to resolve this problem.

3. *Stereochemistry of Desaturation*

(*a*) *Stereochemistry of the intermediates.* As previously explained, phytoene isolated from carrot oil has the 15-*cis* configuration. Phytofluene from the same source is also apparently the 15-*cis* isomer, whereas ζ-carotene, neurosporene and lycopene are all-*trans* (Davis *et al.*, 1966). Phytoene isolated from other sources may also have the all-*trans* structure (Weeks, 1971). It is therefore necessary to establish whether the intermediates in the formation of the other carotenes are the all-*trans* or 15-*cis* isomers of phytoene and phytofluene. If the 15-*cis* isomers are involved, then an isomerization step, in addition to desaturation, must occur in the formation of the coloured carotenes. The participation of the all-*trans* isomers would seem to be more reasonable. This problem also must await enzyme studies.

(*b*) *Stereochemistry of hydrogen loss.* Whatever the stage at which desaturation occurs, and whichever isomers of phytoene and phytofluene (if any) may be involved, the introduction of each double bond requires the loss of two hydrogen atoms, one of which originates from C-2 of MVA, the other from C-5. Figure 4 shows the distribution of these hydrogen atoms in MVA, GGPP and phytoene, and their probable location in the lycopene molecule. The incorporation of [2-^{14}C,(5*R*)-5-$^{3}H_1$] MVA and [2-^{14}C,5-$^{3}H_2$] MVA into phytoene and other carotenes in tomato slices (Williams *et al.*, 1966, 1967) has shown that what was originally the *pro*-5*R* hydrogen atom of MVA is lost in the introduction of each double bond, and the *pro*-5*S* hydrogen atom retained.

It is thus of interest to determine the stereochemistry of hydrogen loss from the other positions (C-8,12,8′,12′, originally C-2 of MVA), and thereby ascertain whether the hydrogen atoms lost bear a *cis* or *trans* relationship. A series of incubations of [2-^{14}C,(2*R*)-2-$^{3}H_1$] MVA and [2-^{14}C,(2*S*)-2-$^{3}H_1$] MVA with tomato slices gave results which suggested that the stereospecificity of the tritium label had been lost (R. J. H. Williams, G. Britton and T. W. Goodwin, unpublished results, 1967). A reasonable explanation for this and related observations lies in the earlier stages of terpenoid biosynthesis (Goad, 1970), in which an equilibrium exists between IPP and DMAPP, the equilibrium being strongly in favour of DMAPP (Shah *et al.*, 1965). The stereospecificity of the tritium label at C-2 of MVA is retained during conversion of MVA into IPP (Cornforth *et al.*, 1966), but IPP is in equilibrium with DMAPP, in which the tritium now labels one of the hydrogen atoms of a methyl group. Reversal of the equilibrium converts DMAPP back into IPP,

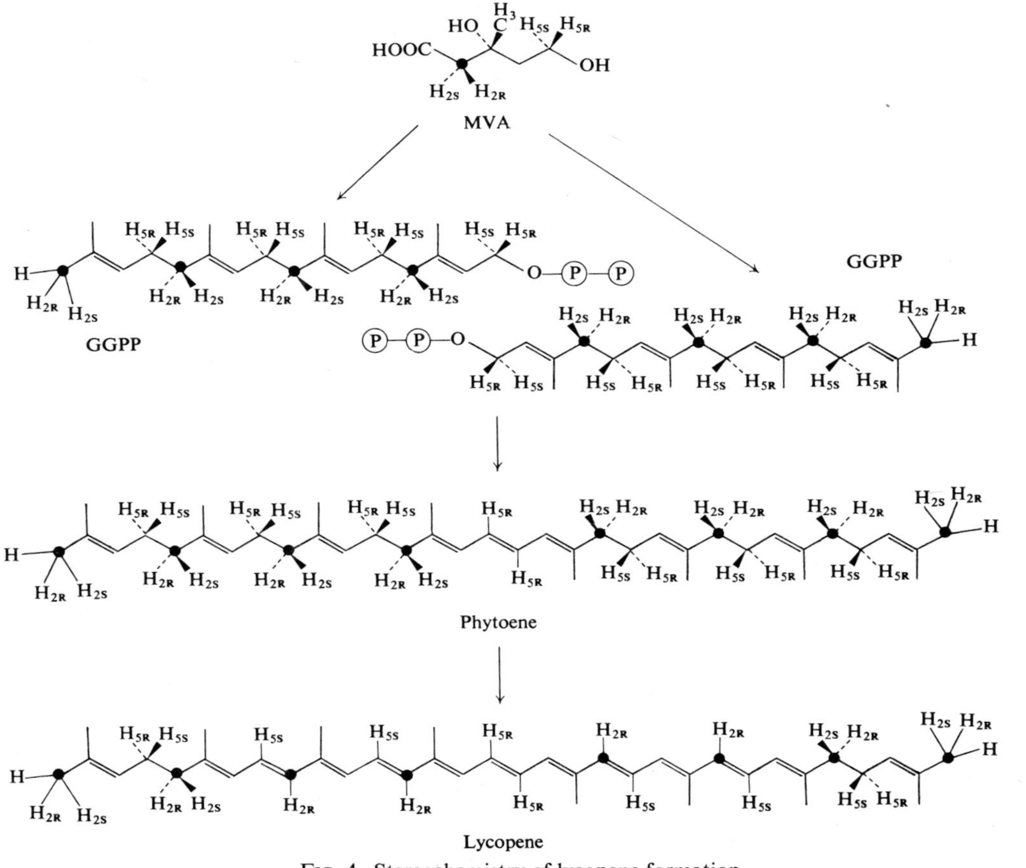

FIG. 4. Stereochemistry of lycopene formation.

in which the two tritium-labelled species (15) and (16) are equally probable, (Fig. 5). The stereospecificity of labelling is thus lost.

If synthesis of higher terpenyl pyrophosphates is not sufficiently rapid, this randomization of the tritium label by the isomerization will mean that the substrates of the later enzymes are no longer stereospecifically labelled. Thus no valid results were obtained for the incorporation of [2-^{14}C,(2*R*)-2-$^{3}H_1$] MVA and [2-^{14}C,(2*S*)-2-$^{3}H_1$] MVA into carotenes by tomato slices over 24–48 h.

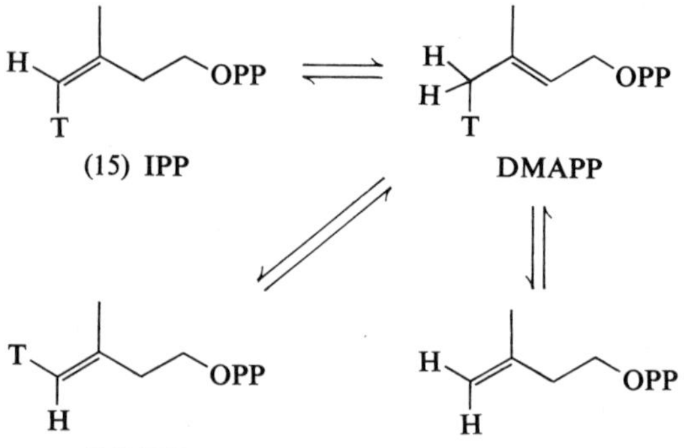

FIG. 5. Randomization of ^{3}H-label during IPP–DMAPP isomerization.

Attempts were therefore made to overcome this problem by using as substrate GGPP prepared from [2-^{14}C,(2*R*)-2-$^{3}H_1$] MVA or [2-^{14}C,(2*S*)-2-$^{3}H_1$] MVA by an enzyme extract of *Echinocystis macrocarpa* (Oster and West, 1968). GGPP is synthesized very rapidly by this system, so that appreciable randomization of label by the isomerization should not occur. Preliminary results of such experiments (J. R. Vose, G. Britton and T. W. Goodwin, unpublished results, 1969) indicate that in the desaturation reaction of carotenoid biosynthesis, the hydrogen atoms lost from C-8,12,8′,12′, are those which were originally the *pro*-2*S* hydrogen atoms of MVA, the *pro*-2*R* hydrogen atoms being retained. This indicates an overall *trans* loss of hydrogen atoms in the introduction of the double bonds (Fig. 4). Further work is in progress to confirm this finding.

4. *Mechanism of Desaturation*

No conclusions can be drawn about the mechanism of desaturation. The only relevant facts available are that an electron acceptor seems to be required (Rilling, 1962), that the reaction appears to involve a *trans* loss of hydrogen atoms and that FAD appears to be an absolute requirement. No enzyme studies have been reported, and nothing is known of cofactor requirements. The desaturation could be simply a dehydrogenation, could involve a hydroxylation–dehydration mechanism, or may follow some entirely different course. Much further work on this problem is obviously necessary.

D. CYCLIZATION

1. General Aspects

It is evident that the cyclic carotenes must be formed from some acyclic precursor. Much controversy and argument has been centred around the stage at which cyclization occurs, i.e. which acyclic compound is cyclized and is thus the precursor of the cyclic carotenes. Most of the argument has concentrated on the rival claims of neurosporene and lycopene as the branch point. Evidence which has been used to support these claims has been summarized by other reviewers. There has also been argument about whether α- and β-rings (17, 18 respectively) are formed sequentially or independently. This problem at least seems to be well in hand.

The natural occurrence of β-zeacarotene and probably of α-zeacarotene, which may be considered to be cyclic derivatives of neurosporene, shows that cyclization can occur at or before the neurosporene level of desaturation, but does not necessarily indicate that this is the normal route for the formation of cyclic carotenes. No direct conversion of neurosporene into any cyclic carotenoid has been reported.

Earlier reports of the conversion of lycopene into β-carotene by carrot leaf chloroplasts (Decker and Uehleke, 1961) and into γ-, δ-, and β-carotenes by isolated tomato plastids or spinach chloroplasts (Wells *et al.*, 1964) have now been confirmed by Hill and Rogers (1969) who obtained a small incorporation of [^{14}C] lycopene into β-carotene in bean leaf chloroplasts and tomato plastids, and this work has been extended by Kushwaha *et al.* (1969) in experiments with enzyme systems from spinach chloroplasts and tomato plastids. In this work, incorporation of [15,15′-$^{3}H_2$] lycopene into α-, β-, γ-, δ- and neo-β-carotenes by spinach chloroplasts and into β- and neo-β-carotenes by a soluble preparation of these plastids, and similar conversions into α-, β-, γ-, δ- and neo-β-carotenes by plastids of the high-beta and high-delta tomato fruit, and by soluble preparations from these plastids are reported. In the experiment with the soluble preparation of spinach chloroplasts, FAD was an absolute requirement, and formation of β-carotene was aided by the presence of NADP.

These findings indicate that cyclization of lycopene can occur. They do not necessarily show that formation of the cyclic carotenes normally occurs via cyclization of lycopene.

2. Independent Synthesis of α- and β-Rings

In several postulated schemes of carotene biosynthesis, it has been suggested that only one type of ring (e.g. β-) is formed by cyclization of an acyclic precursor, the other ring type (α-) being formed by isomerization of the β-ring, i.e. β-carotene has been postulated as an intermediate in the formation of α-carotene (Porter and Lincoln, 1950; Porter and Anderson, 1962). The alternative view of α-carotene as a precursor of β-carotene has also been proposed (Rabourn, 1957; Hocking, 1967).

Another possibility is that α- and β-rings are formed independently by cyclization of an acyclic precursor. Studies by Tomes (1967) on the competition effect of the β- and δ-carotene genes on α- and β-ring formation in the tomato provide evidence which suggests independent formation of the two ring types, with competition for a common precursor.

The independent parallel synthesis of α- and β-carotenes and of δ- and γ-carotenes has been proved in a series of labelling experiments by Williams

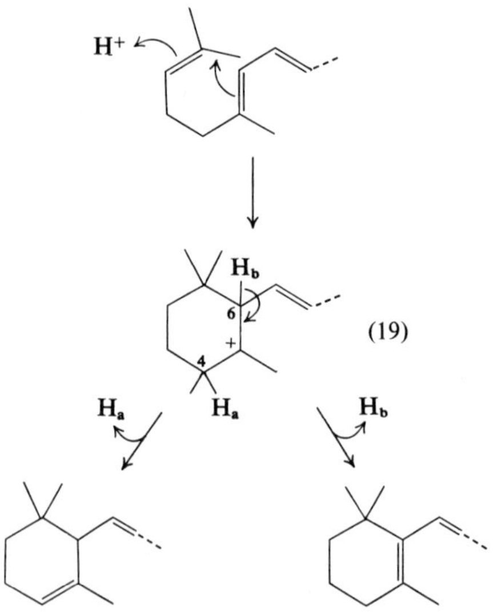

(19)

(17) α-ring (18) β-ring

FIG. 6. Independent synthesis of α- and β-rings.

et al. (1967). The basis of these experiments is outlined in Fig. 6. It is usually accepted that the cyclization reaction proceeds via an intermediate such as a carbonium ion (19). In this proposed intermediate, the hydrogen atom at C-6, which would be lost in formation of the β-ring, arises from C-4 of MVA. Labelling experiments with [2-^{14}C,(4*R*)-4-$^{3}H_1$] MVA have shown that the hydrogen atoms from the C-6,6′ positions are lost in the formation of β-carotene (Goodwin and Williams, 1965a; Williams *et al.*, 1967). The hydrogen atom at C-6′ of α-carotene would therefore be unlabelled if this compound arose by isomerization of β-carotene. In fact experiments have shown that tritium is retained at this position in α-carotene from carrot roots (Goodwin and Williams, 1965b), tomato slices (Williams *et al.*, 1967) and maize leaves (Goodwin *et al.*, 1968) showing that β-carotene is not an intermediate in α-carotene formation.

Similar experiments with [2-^{14}C,2-$^{3}H_2$] MVA, which would label the hydrogen atoms at C-4 of a proposed carbonium ion intermediate, have shown that although one labelled hydrogen atom is lost in the formation of α-carotene, all the tritium is retained at these positions in β-carotene, indicating that β-carotene is not formed from α-carotene (Williams *et al.*, 1967). The two ring types thus arise independently.

3. *Stereochemistry of Cyclization*

There are several possibilities for the stereochemistry of formation of the cyclic carotenes, depending on whether the acyclic precursor is folded in a chair or boat conformation, and on which face proton attack to initiate cyclization occurs. Four possibilities are illustrated in Fig. 7. In order to define

Chair folding

(A)

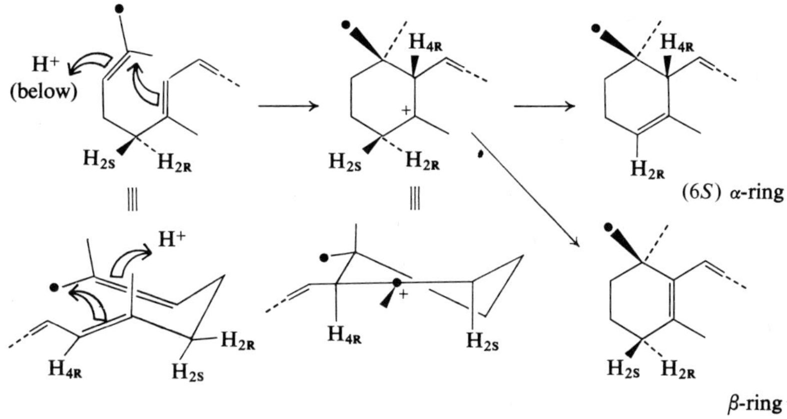
H+
(below)
H2S
H2R
H4R
H2R
(6S) α-ring
H+
H2R
H4R
H2S
H4R
H2S
H2S H2R
β-ring

(B)

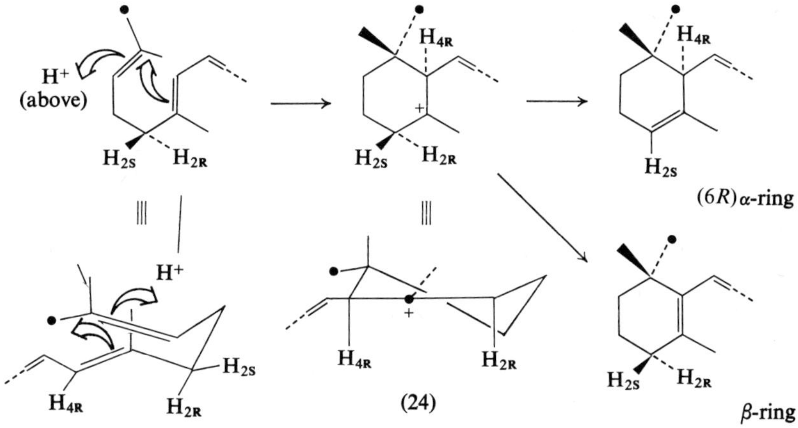
H+
(above)
H2S
H2R
H4R
H2S
(6R)α-ring
H+
H2S
H4R
H2R
H4R
H2R
H2S H2R
β-ring

(24)

Boat folding

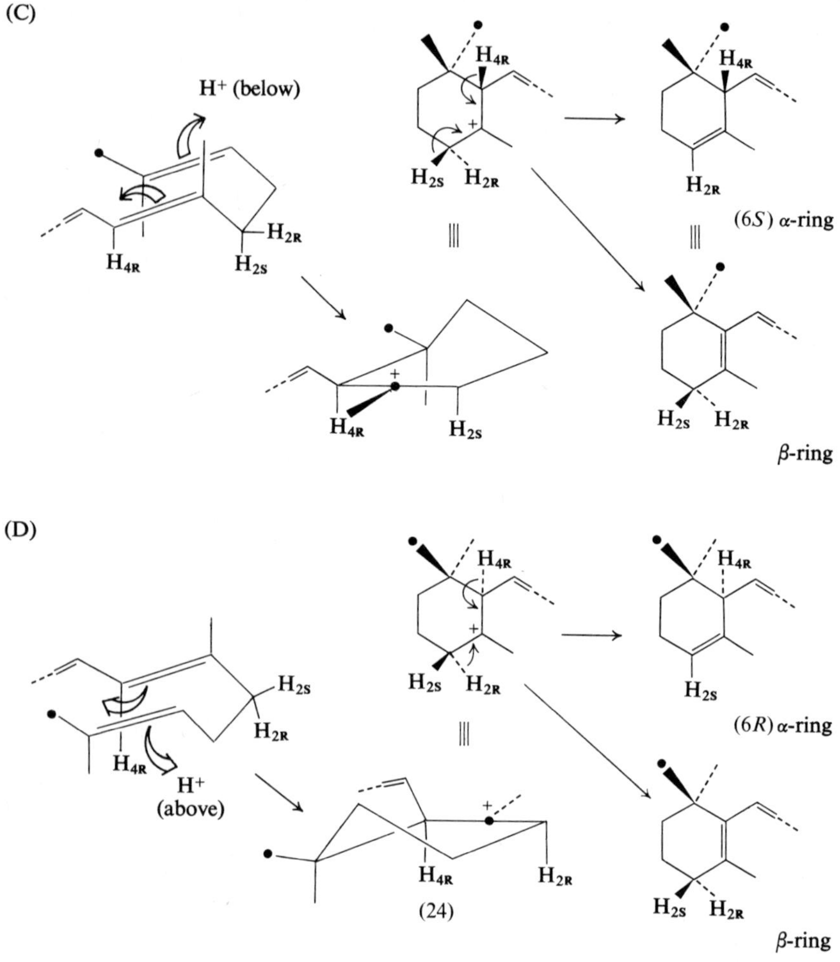

(C)

(D)

FIG. 7. Possibilities for the stereochemical course of cyclization.

the stereochemical course of the cyclization reaction, four facts must be ascertained. These are (i) the absolute configuration of C-6′ of the α-ring, (ii) which of the hydrogen atoms is stereospecifically lost from C-4′ in formation of the α-ring, (iii) the respective stereochemical positions at C-1 in the cyclic carotenes which the C-1 methyl groups of the acyclic precursor adopt, (iv) whether covalently bonded enzyme-substrate intermediates are involved. Some progress has been made towards a solution to these problems.

The absolute configuration of α-carotene has recently been determined by Eugster *et al.* (1969) to be as shown (20), i.e. 6′*R*. This eliminates examples

(A) and (C) (Fig. 7) as possible cyclization mechanisms for formation of the α-ring. Preliminary studies on the incorporation of GGPP biosynthesized from [2-^{14}C,(2*R*)-2-$^{3}H_1$] MVA and [2-^{14}C,(2*S*)-2-$^{3}H_1$] MVA into δ-carotene by the high-delta tomato mutant have indicated that the hydrogen atom lost in formation of the α-ring is that which was originally the *pro*-2*S* hydrogen atom of MVA (J. R. Vose, G. Britton and T. W. Goodwin, unpublished results, 1969).

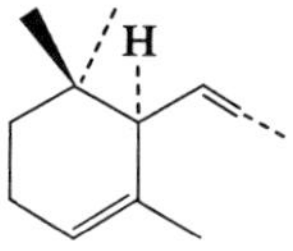

(20)

Some information has recently been reported which is relevant to the problem of the orientation of the C-1 methyl groups in the formation of β-carotene. It is generally assumed that incorporation of [2-^{14}C] MVA into acyclic carotenes results in a stereospecific labelling of one of the C-1 methyl groups, as shown in Fig. 8. This has now been tested by Tefft *et al.* (1970)

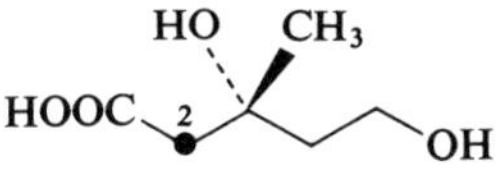

MVA

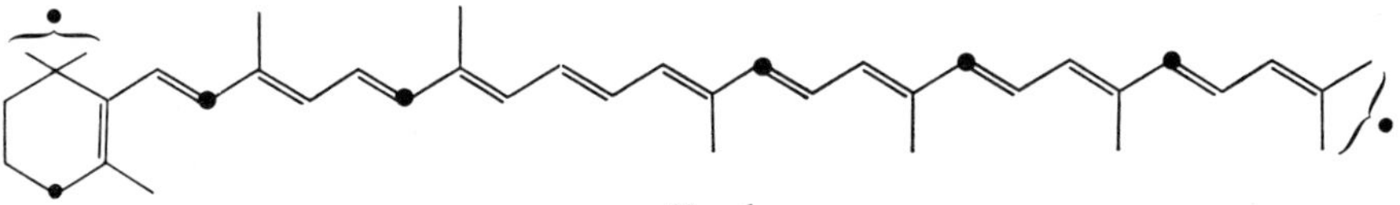

Torulene

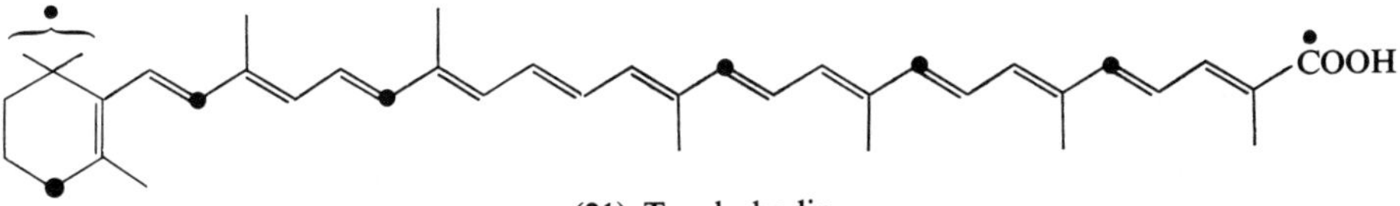

(21) Torularhodin

FIG. 8. Biosynthesis of torularhodin.

for the acyclic end group of torularhodin (21) from the yeast *Rhodotorula rubra*. In this carotenoid, one of the C-1 methyl groups is modified into a carboxyl group, and it was shown that the carboxyl group contained the label from [2-^{14}C] MVA, the methyl group being inactive. This was the first demonstration of the stereospecific labelling of the C-1 substituents of a carotenoid.

Now, recent work by Bu'Lock *et al.* (1970) on the absolute configuration, and the biosynthesis, of trisporic acid (22) has given results which have been used to deduce by extrapolation which of the C-1 methyl groups of β-carotene originates from C-2 of MVA. In this work the absolute configuration of trisporic acid was shown to be as in (22), and it was also shown that label from [2-^{14}C] MVA was introduced into the methyl group at C-1 (carotene numbering), the carboxyl group being unlabelled. Since other experiments had shown the formation of trisporic acid from β-carotene, it was deduced that the *pro-R* methyl group at C-1 of β-carotene is labelled from [2-^{14}C] MVA (Fig. 9).

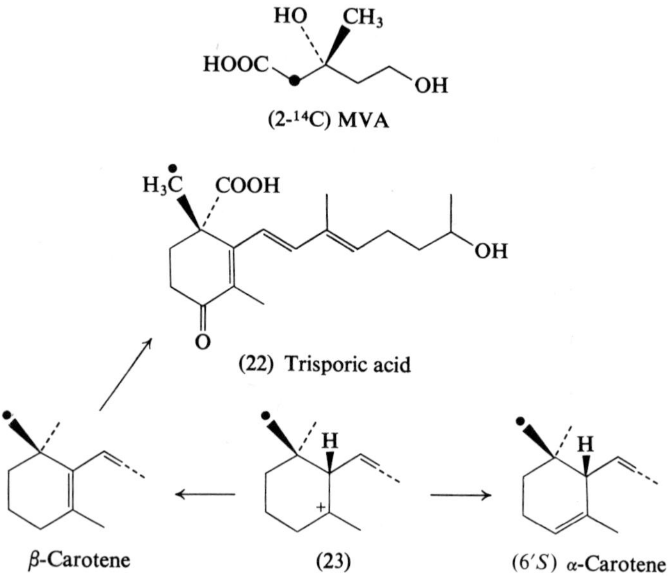

FIG. 9. Stereochemistry of trisporic acid and β-carotene biosynthesis.

This suggested that the β-ionone ring must have been formed via an intermediate such as (23) which, by loss of a proton from C-4 would have given rise to an α-ring with the 6*S* configuration, which is opposite to the naturally occurring enantiomer. Boat-folding of the acyclic precursor (type D, Fig. 7) could give rise to the correctly labelled β-carotene, and to the correct optical isomer of α-carotene, if the branch-point in the biosynthesis of the two ring types is at the carbonium ion intermediate stage. Alternative possibilities are that the divergence of the pathway to α- and β-ionone rings occurs at an earlier stage than at first thought, i.e. the ring type may be determined by the initial folding of the acyclic precursor, or that the stereochemistry of folding and cyclization may be different in different organisms.

It appears that the hydrogen atom lost from C-4 in formation of the α-ring is probably not the axial H_{2R} which would be expected if the conformation of the intermediate approaches that shown (24, Fig. 7). However, there is no reason to suppose that the conformation of such an intermediate would be a strict boat or chair, and loss of the unexpected hydrogen atom can readily be accommodated.

Further work is necessary to determine whether labelling of the C-1 methyl groups in the α-ring is the same as that suggested for the β-ring, to establish whether enzyme-bound intermediates are involved, and to determine if the stereochemistry of cyclization is the same in all organisms.

E. OVERALL PICTURE OF CAROTENE BIOSYNTHESIS

GGPP is now established as the C_{20} precursor of phytoene and other carotenes. There are, however, several different opinions about the rest of the pathway. Some workers believe that the sequence

GGPP → phytoene → neurosporene or lycopene → cyclic carotenes

is established. Others feel that the position of phytoene as a precursor is in doubt. A further proposal has been that desaturation and cyclization occur at the C_{20} level, to give the series of intermediates shown in Fig. 10, combinations

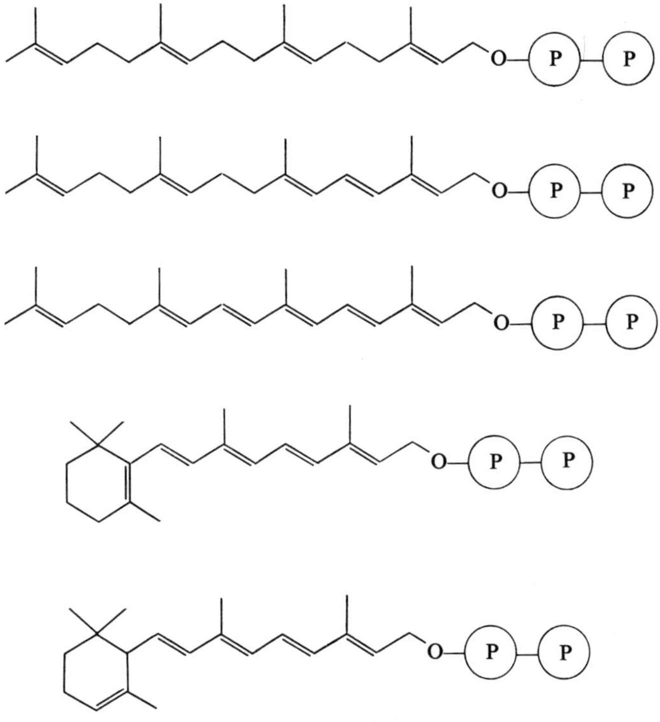

FIG. 10. Postulated C_{20} intermediates in carotene biosynthesis.

of which can give rise to the whole range of carotenes (Quackenbush, 1965, quoted by Porter and Anderson, 1967). Perhaps too much emphasis has been placed on arguments about whether phytoene is an intermediate, whether neurosporene or lycopene is the precursor of the cyclic carotenes, etc. The cyclizing enzyme, for example, may simply be capable of cyclizing any compound with the appropriate end group (25).

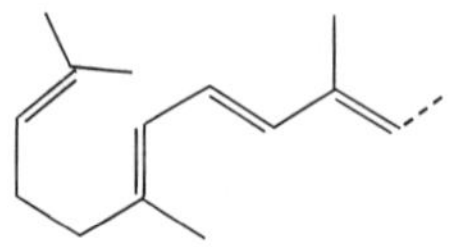

(25)

The possibility of a multi-enzyme complex capable of converting GGPP into carotenes with no free intermediates must also be considered. The existence of such an enzyme complex containing desaturating and cyclizing enzymes which could convert GGPP into carotenes could account for some experimental findings difficult to explain on the basis of other ideas. The apparently anomalous position of phytoene in some systems, for example, could be explained if synthesis of other carotenes occurred via enzyme-bound intermediates, by-passing the pool of free phytoene.

The overall picture obviously requires clarification. Much more work is necessary, especially at the enzyme level before any conclusions can justifiably be drawn.

F. BIOSYNTHESIS OF XANTHOPHYLLS

1. *General Aspects*

Many naturally occurring carotenoids are known which contain oxygen functions, e.g. hydroxy-, oxo- and epoxy-groups. Little is known of the biosynthesis of these xanthophylls. It is generally accepted that the introduction of oxygen functions into carotenoids occurs at a late stage in the biosynthesis, e.g. zeaxanthin is thought to be formed by hydroxylation of β-carotene. There has been no experimental verification of this, and no reliable evidence of direct conversions is available.

It has been established by ^{18}O experiments, that the oxygen of hydroxy- and epoxy-substituents in the xanthophylls of *Chlorella* and *Phaseolus lunatus* comes from molecular oxygen rather than from water (Yamamoto *et al.*, 1962a,b,c; Yamamoto and Chichester, 1965).

Experiments with [2-^{14}C,(4*R*)-4-$^{3}H_1$] MVA and [2-^{14}C,2-$^{3}H_2$] MVA similar to those described for α- and β-carotenes have demonstrated that, in maize leaves, no interconversion of α- and β-rings in xanthophylls occurs (Goodwin *et al.*, 1968; T. J. Walton, G. Britton and T. W. Goodwin, unpublished results, 1968, 1970).

2. *Stereochemistry of Hydroxylation*

The absolute configuration (*R*) at C-3 in β-cryptoxanthin and zeaxanthin has now been established (27) (De Ville *et al.*, 1969). This has enabled the stereochemical course of the hydroxylation to be examined. C-3 of β-carotene arises from C-5 of MVA, and the hydrogen atoms from C-5 of MVA would be orientated as shown (26). Studies of the incorporation of [2-^{14}C,(5*R*)-5-$^{3}H_1$] MVA and [2-^{14}C,5-$^{3}H_2$] MVA into lutein, β-cryptoxanthin and zeaxanthin in maize leaves, *Physalis alkekengii*, and a *Flavobacterium* species, have shown that the *pro*-5*R* hydrogen atom is lost and the *pro*-5*S* hydrogen atom retained during hydroxylation (Walton *et al.*, 1969; J. C. B. McDermott, G. Britton and T. W. Goodwin, unpublished results, 1969, 1970). This shows that ketonic intermediates are not involved in the hydroxylation and hydroxylation occurs

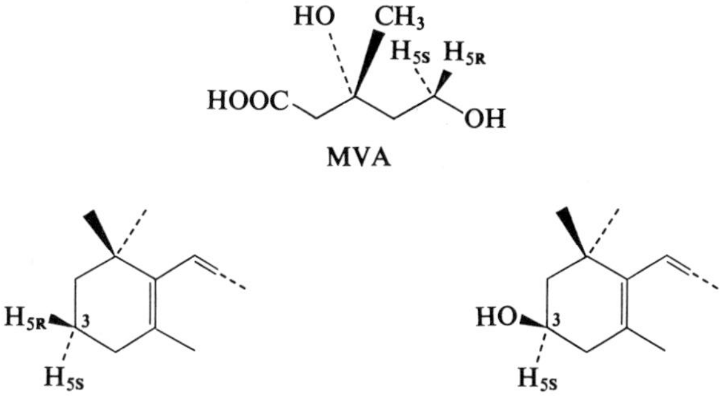

HO CH$_3$ H_{5S} H_{5R} HOOC OH

MVA

H_{5R} 3 H_{5S}

(26) β-Carotene

HO 3 H_{5S}

(27) (3R)-Zeaxanthin

FIG. 11. Xanthophyll biosynthesis; stereochemistry of hydroxylation at C-3.

with retention of configuration (Fig. 11). This is typical of the mixed function oxidase type of hydroxylation well known in the biosynthesis of sterols and other lipid compounds (Hayano, 1962). This work has also shown that it is unlikely that any conversion of xanthophylls into carotenes occurs.

3. *Conclusions*

There has been much speculation about the biosynthesis of xanthophylls, mechanisms having been proposed for the biosynthesis of almost every known naturally occurring compound. There is, however, a great shortage of knowledge about the pathways of biosynthesis. There are very many problems which may be solved by the application of specific labelling techniques, and especially by enzyme studies.

III. CAROTENOID BIOSYNTHESIS IN PHOTOSYNTHETIC BACTERIA

A. GENERAL OUTLINE

The photosynthetic bacteria are characterized by their ability to synthesize carotenoids of structural types not usually found in other organisms. In particular, synthesis of a range of acyclic carotenoids, generally with tertiary hydroxy- and methoxy-groups is usual, e.g. spirilloxanthin (28) and related compounds. In the presence of diphenylamine (DPA), several photosynthetic bacteria are unable to synthesize their normal pigments, the spirilloxanthin series, and accumulate in their place more saturated carotenoids, especially phytoene (Goodwin and Osman, 1954). Because of this, it has usually been assumed that the early stages of biosynthesis, via GGPP, phytoene and the usual desaturation series, are similar to those occurring in other organisms.

Present knowledge of structures and distribution of carotenoids in photosynthetic bacteria clearly reveals that these organisms are capable of performing biosynthetic reactions different from those of higher plants. The main biochemical interest has therefore been in the (presumably late) biogenetic steps by which these novel structural features are elaborated.

B. ATHIORHODACEAE

1. *Biosynthesis of Spirilloxanthin*

A common feature of the carotenoids of the photosynthetic bacteria is the presence of tertiary hydroxy- and methoxy-groups in compounds of the spirilloxanthin (28) and spheroidene (29) types. The biosynthesis of these compounds in the Athiorhodaceae has been studied by Stanier, Liaaen-Jensen and co-workers (Stanier, 1960; Liaaen-Jensen, 1962, 1963). These workers cultured *Rhodospirillum rubrum* anaerobically in the presence of diphenylamine, and then studied the kinetics of the disappearance of the accumulated saturated carotenoids, and appearance of the normal compounds of the spirilloxanthin series when the DPA inhibition was removed. This allowed them to propose a scheme (Fig. 12) for the biosynthesis of spirilloxanthin from lycopene by a series of hydration, methylation and desaturation reactions (Liaaen-Jensen *et al.*, 1958, 1961). Similar work suggested the formation by a series of similar reactions of spheroidene from neurosporene in *Rhodopseudomonas spheroides* (Liaaen-Jensen *et al.*, 1961; Liaaen-Jensen, 1963) and an extension of this pathway to spirilloxanthin in *R. gelatinosa* was also proposed (Eimhjellen and Liaaen-Jensen, 1964) (Fig. 13).

A recent reinvestigation involving determination of the structures of other compounds present in DPA-inhibited cultures of *Rhodospirillum rubrum* has allowed Davies (1970a) to suggest the operation of other pathways of spirilloxanthin biosynthesis. The presence of spheroidene and 1′-hydroxy-1′,2′-dihydrospheroidene (Davies *et al.*, 1969; Davies and Holmes, 1969a; Malhotra *et al.*, 1969; Davies, 1970a) suggested the operation, in *R. rubrum* of the pathway from neurosporene via spheroidene, as described by Eimhjellen and Liaaen-Jensen (1964) for *Rhodopseudomonas gelatinosa*.

Davies (1970a) has also proposed the functioning of another novel pathway on the basis of the discovery that the ζ-carotene present in DPA-inhibited cultures of *Rhodospirillum rubrum* was not the normal (in higher plants and fungi) symmetrical 7,8,7′,8′-tetrahydrolycopene (4) but was the unsymmetrical isomer, 7,8,11,12-tetrahydrolycopene (30) (Davies *et al.*, 1969; Davies and Holmes, 1969b; Davies, 1970b). Compounds related to this were also identi-

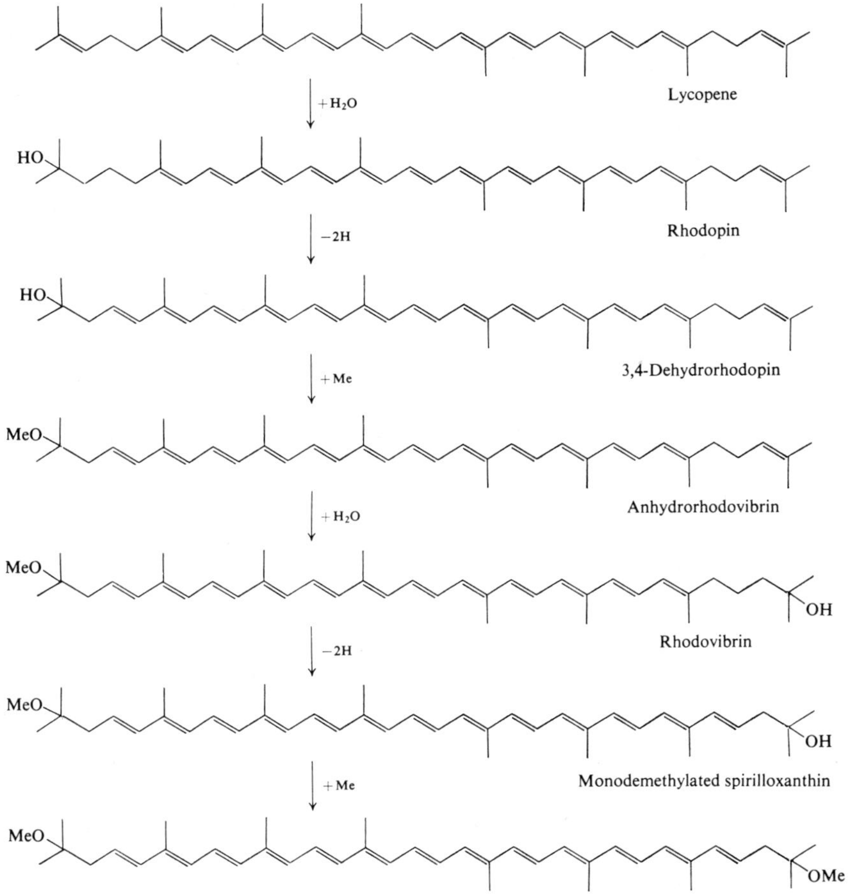

Lycopene

$+H_2O$

HO

Rhodopin

$-2H$

HO

3,4-Dehydrorhodopin

$+Me$

MeO

Anhydrorhodovibrin

$+H_2O$

MeO OH

Rhodovibrin

$-2H$

MeO OH

Monodemethylated spirilloxanthin

$+Me$

MeO OMe

(28) Spirilloxanthin

FIG. 12. Postulated scheme for the biosynthesis of spirilloxanthin from lycopene.

fied, and a scheme for the conversion of 7,8,11,12-tetrahydrolycopene into spheroidene and hence spirilloxanthin was postulated (Davies 1970a) (Fig. 14).

Further investigations of DPA-inhibited cultures of *R. rubrum*, however, have revealed the presence of many other minor carotenoids, including hydroxy- and methoxy-derivatives of phytoene and phytofluene (probably 1-hydroxy-1,2-dihydrophytoene (31) and 1-hydroxy-1,2-dihydrophytofluene (32) and the corresponding mono-*O*-methyl ethers, (33, 34) (Malhotra *et al.*, 1970a), and a "dimethoxy-ζ-carotene" (3,4,7,8,11,12,3′,4′-octahydro-spirilloxanthin (35) (Malhotra *et al.*, 1970b)). The occurrence of these compounds, and a wide range of other 1-hydroxy-1,2-dihydro- carotenoids

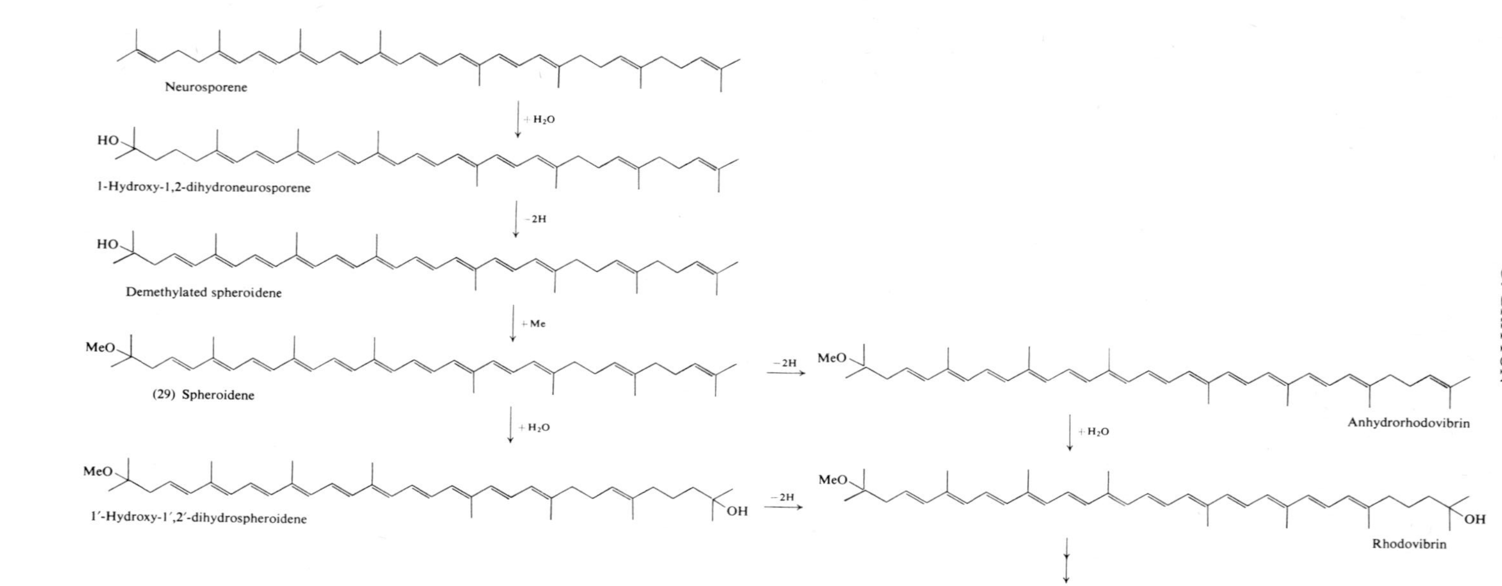

FIG. 13. Postulated scheme for the biosynthesis of spheroidene and spirilloxanthin from neurosporene.

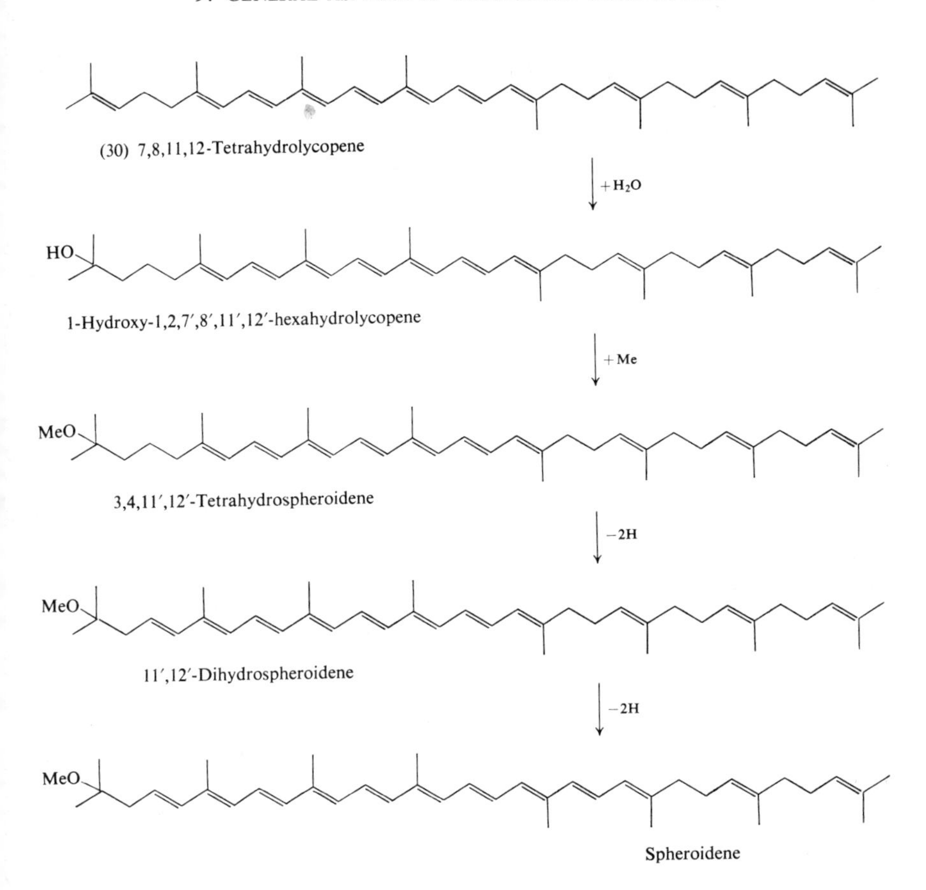

FIG. 14. Postulated scheme for the biosynthesis of spheroidene from 7,8,11,12-tetrahydrolycopene.

(B. H. Davies, personal communication, 1970) suggests that, under conditions of DPA inhibition, the enzymes responsible for hydration of the C-1,2 double bond have a relatively low specificity, and can act at all levels of desaturation. Any attempt to deduce pathways by which spirilloxanthin is biosynthesized under normal conditions from the structures of compounds present in DPA inhibited cultures must be made, if at all, with great caution.

2. *Biosynthesis of 1,2-Dihydrocarotenes by* Rhodopseudomonas viridis

Rhodopseudomonas viridis is an unusual *Rhodopseudomonas* species in that, in place of the spheroidene–spirilloxanthin series, it appeared to accumulate neurosporene and lycopene as major pigments (K. E. Eimhjellen, quoted by Thornber *et al.*, 1969). Recent work by Malhotra *et al.* (1970c) however, has shown that, although neurosporene and lycopene are present in anaerobic

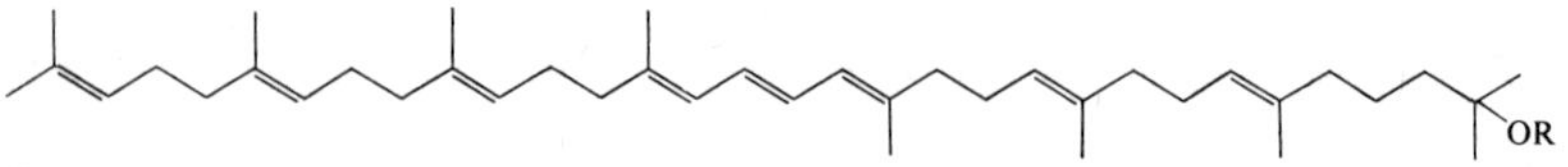

(31) (R = H) (32) (R = CH_3)

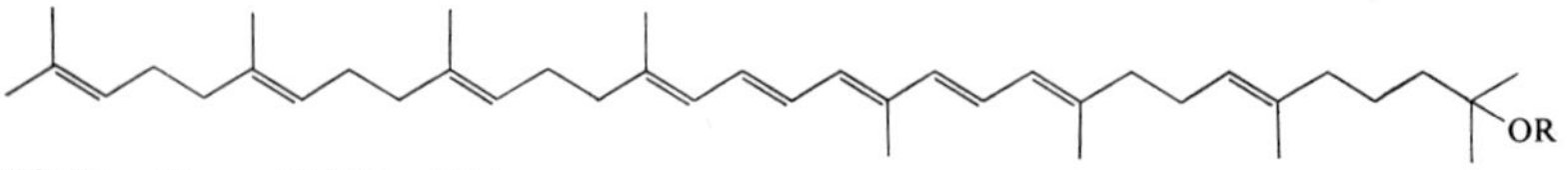

(33) (R = H) (34) (R = CH_3)

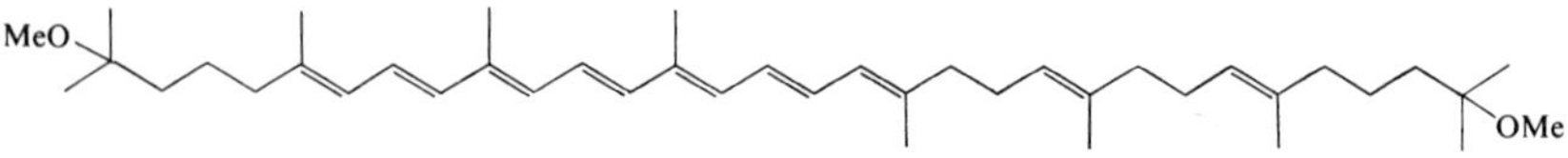

(35)

cultures of *R. viridis*, the main carotenoids have the novel 1,2-dihydro- end group, viz. 1,2-dihydroneurosporene (36), 1,2-dihydrolycopene (37) and 1,2-dihydro-3,4-didehydrolycopene (38). Further work (Malhotra *et al.*, 1970d) on normal and DPA-inhibited cultures has shown the presence of phytoene and phytofluene and their 1,2-dihydro-derivatives (39, 40). (The effect of DPA on carotenogenesis by this organism was small.) A 1,2-dihydro-derivative of unsymmetrical ζ-carotene (1,2,7,8,11,12-hexahydrolycopene, 41) was also isolated, together with ζ-carotene, which, unusually in the Athiorhodaceae, was the symmetrical 7,8,7′,8′-tetrahydrolycopene (4). The co-occurrence of symmetrical ζ-carotene and the unsymmetrical dihydro-ζ-carotene is intriguing.

Two 1,2,1′,2′-tetrahydro-compounds were also isolated, 1,2,1′,2′-tetrahydroneurosporene (42) and 1,2,1′,2′-tetrahydrolycopene (43). It seems likely that the enzyme responsible for the hydrogenation is rather unspecific, although the possibility of the hydrogenation occurring at an early stage (GGPP or phytoene) to give a dihydro-precursor of the other carotenoids cannot be ruled out. No biochemical studies of the biosynthetic pathway have yet been attempted.

Preliminary work (A. Ben-Aziz, H. C. Malhotra, G. Britton and T. W. Goodwin, unpublished results, 1970) indicates the presence in *R. viridis* of "cross-conjugated" aldehydes related to the rhodopinal (50) series in *Chromatium warmingii* (Aasen and Liaaen-Jensen, 1967), but with the 1,2-dihydro end group. Such compounds have not previously been found in the Athiorhodaceae.

3. *Introduction of 2-Oxo-Groups in* Rhodopseudomonas spheroides *and* R. gelatinosa

When grown photosynthetically under anaerobic conditions, *R. spheroides* and *R. gelatinosa* accumulate carotenoids of the spheroidene and spirilloxanthin series. On transfer to aerobic conditions, however, an almost quantitative

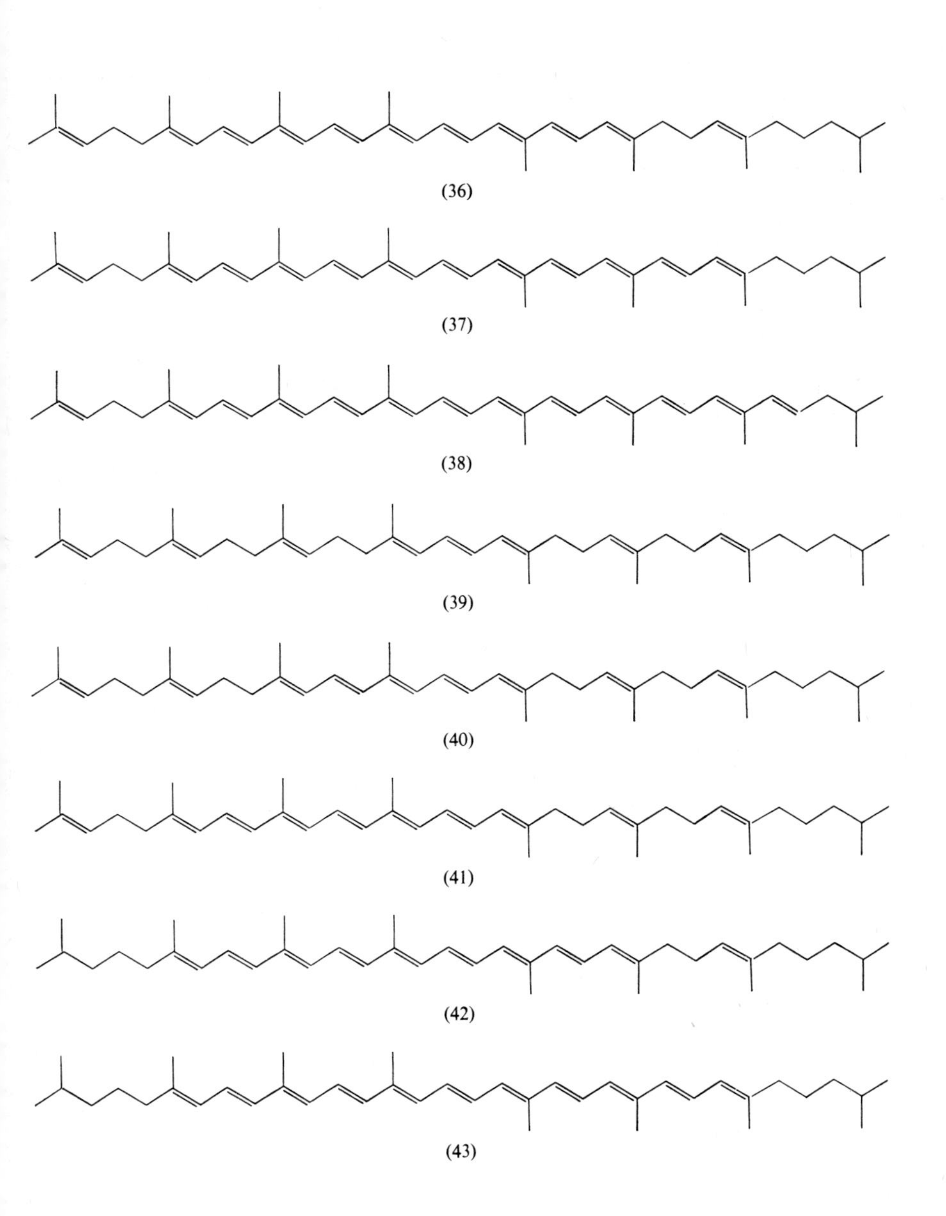
(36)
(37)
(38)
(39)
(40)
(41)
(42)
(43)

conversion of these compounds into their 2-oxo derivatives occurs (Fig. 15) (Van Niel, 1947; Goodwin *et al.*, 1955, 1956; Shneour, 1962a).

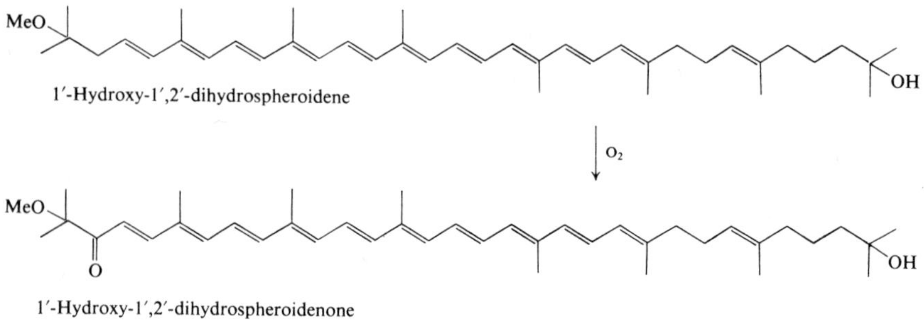

MeO

OH

1′-Hydroxy-1′,2′-dihydrospheroidene

O_2

MeO

O

OH

1′-Hydroxy-1′,2′-dihydrospheroidenone

FIG. 15.

4. *Formation of Cyclic Carotenoids by* Rhodomicrobium vannielii

Studies of the pigmentation of *R. vannielii*, an organism whose taxonomic position has been in some doubt, but which is now placed in the Athiorhodaceae (Van Niel, 1963) revealed the presence, in addition to members of the spirilloxanthin series, of a compound which appeared to be β-carotene (Volk and Pennington, 1950; Conti and Benedict, 1962). This identification has now been confirmed (Ryvarden and Liaaen-Jensen, 1964; A. Ben-Aziz, G. Britton and T. W. Goodwin, unpublished results, 1970). Other compounds also identified in anaerobic cultures were 1′-methoxy-1′,2′-dihydro-3′,4′-didehydro-γ-carotene (46), and, surprisingly, a compound which could not be distinguished from β-cryptoxanthin (47) (A. Ben-Aziz, G. Britton and T. W. Goodwin, unpublished results, 1970).

This is the only known case of the formation of cyclic carotenoids in the Athiorhodaceae, and the only example of the presence of β-carotene and related compounds in the photosynthetic bacteria. The presence of β-cryptoxanthin in *anaerobic* cultures of *R. vannielii* is interesting, because synthesis

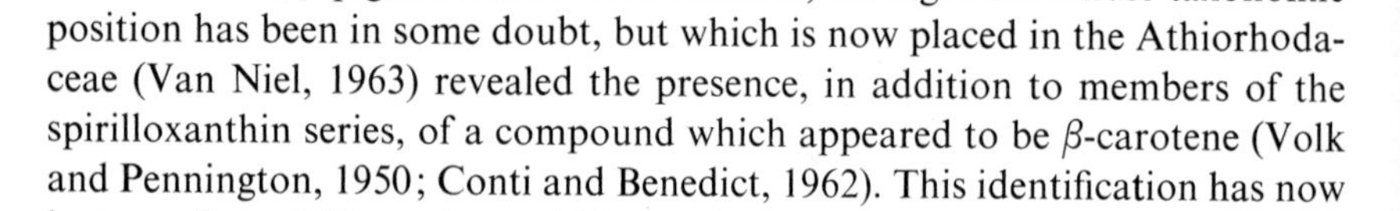

OMe

(46)

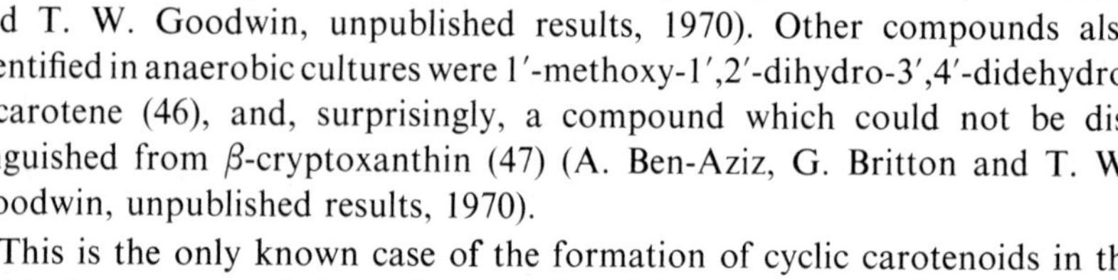

OH

(47)

of this compound by higher plants and non-photosynthetic bacteria appears to involve an NADPH-molecular oxygen mixed function-oxidase type of enzyme.

C. THIORHODACEAE AND CHLOROBACTERIACEAE

Several members of the Thiorhodaceae and Chlorobacteriaceae, like the Athiorhodaceae, synthesize carotenoids of the spirilloxanthin series, but in addition some species have the ability to elaborate carotenoids with novel structural features, especially aromatic end groups, e.g. chlorobactene (48) from *Chlorobium* species (Chlorobacteriaceae) (Liaaen-Jensen *et al.*, 1964), okenone (49) from *Chromatium okenii* (Thiorhodaceae) (Schmidt *et al.*, 1963; Liaaen-Jensen, 1966, 1967). Nothing is known about the mechanism of formation of the aromatic ring.

Another bacterium, *C. warmingii* (Thiorhodaceae), is able to oxidize an in-chain methyl group to an aldehyde group, under anaerobic conditions, e.g. in rhodopinal (50) (Aasen and Liaaen-Jensen, 1967), the naturally occurring isomer of which is 13-*cis*. It remains to be determined whether this isomerization is a consequence of the oxidation reaction, or whether these two novel features are introduced independently.

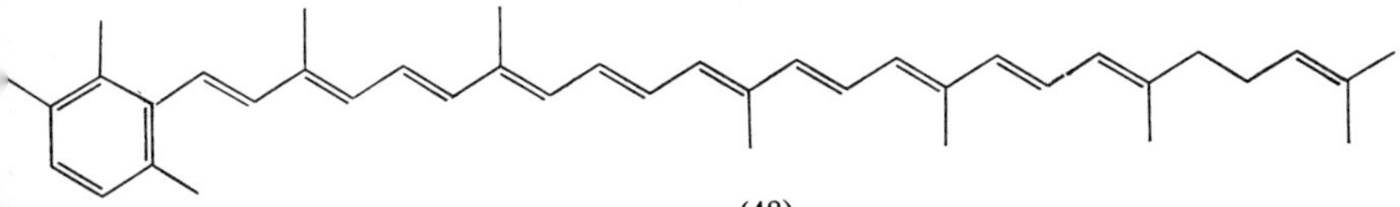

(48)

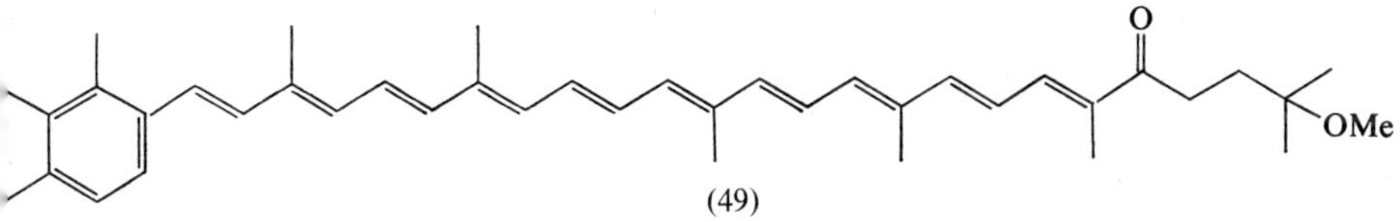

O

OMe

(49)

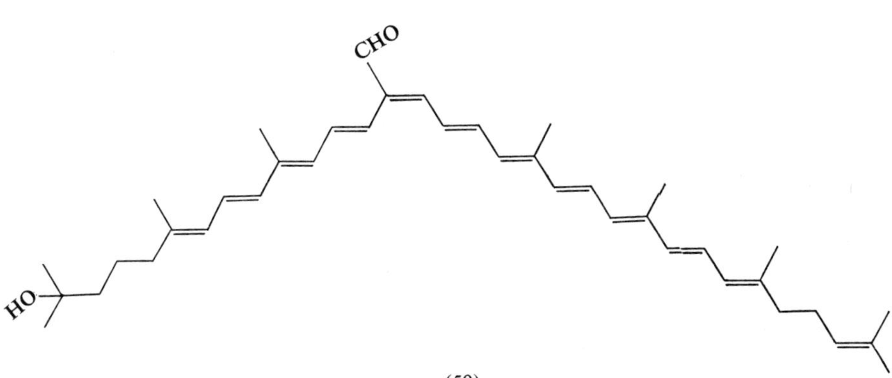

CHO

HO

(50)

D. CAROTENOID BIOSYNTHESIS IN PHOTOSYNTHETIC BACTERIA. GENERAL OBSERVATIONS

The above brief and incomplete survey of structural types of carotenoids and their distribution in photosynthetic bacteria does not solve any biogenetic problems, but rather emphasizes the wide range of problems which must be tackled in this field. No enzyme or co-factor studies have been undertaken, no direct conversions of one carotenoid into another have been achieved, and no information is yet available about the stereochemical course of any of the reactions. All that can be given is an outline of the types of reactions which photosynthetic bacteria are able to perform, which differ from those found in higher plants and other organisms. These reactions show the presence of a further range of enzymes which must be studied.

E. TYPES OF REACTION

[It will be assumed that formation of the first C_{40} intermediate (phytoene ?) from GGPP occurs in the same way as in higher plants. This obviously needs to be confirmed.]

1. *Desaturation (C-7,8;11,12)*

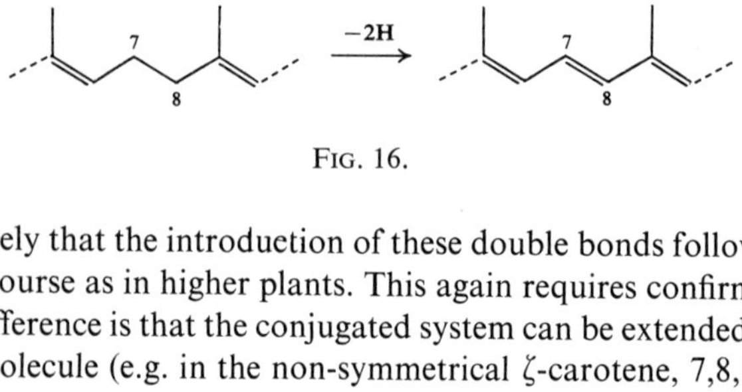

−2H

FIG. 16.

It seems likely that the introduction of these double bonds follows the same mechanistic course as in higher plants. This again requires confirmation. One interesting difference is that the conjugated system can be extended to one side only of the molecule (e.g. in the non-symmetrical ζ-carotene, 7,8,11,12-tetrahydrolycopene (30) rather than the introduction of double bonds occurring alternately on either side of the molecule. No other information is available.

2. *Hydration (C-1,2)*

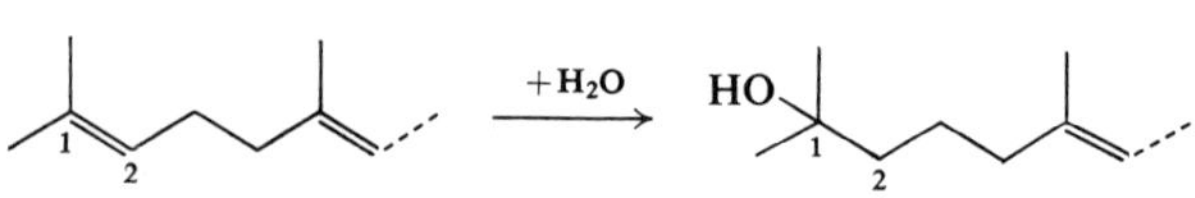

$+H_2O$

FIG. 17.

Formation of the 1-hydroxy-1,2-dihydro- end group takes place under anaerobic conditions, and it seems reasonable to expect a hydration type of mechanism.

3. O-*Methylation*

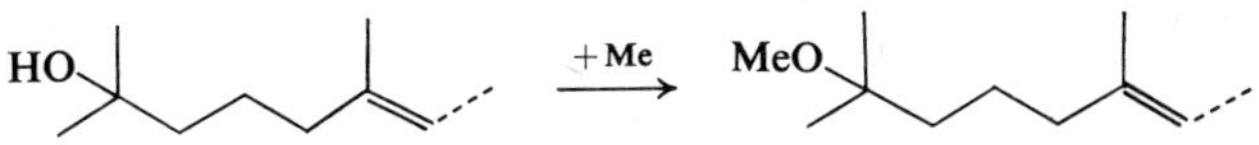

HO ... $\xrightarrow{+Me}$ MeO ...

FIG. 18.

It has been demonstrated (Benedict and Beckman, 1964) that S-adenosylmethionine is the direct source of the methyl group in compounds of the spheroidenone series. Incorporation of the labelled methyl group of S-adenosylmethionine or methionine into spirilloxanthin and related compounds was also obtained with a cell-free preparation of chromatophores of *Rhodospirillum rubrum*.

4. *Desaturation* (*C-3,4*)

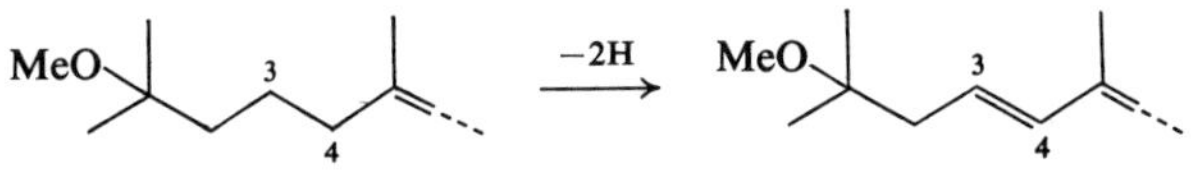

MeO 3 4 $\xrightarrow{-2H}$ MeO 3 4

FIG. 19.

The introduction of the C-3,4 double bond, a common feature in carotenoids of photosynthetic bacteria, is not inhibited by diphenylamine to the same extent as the other desaturation reactions. This suggests that a different enzyme and different mechanism may be involved. It is not yet known if the stereochemistry of hydrogen loss is the same in both types of desaturation.

The presence of an adjacent 1,2-dihydro- or substituted 1,2-dihydro- end group seems to be a requirement for desaturation at C-3,4 to occur.

5. *Hydrogenation*

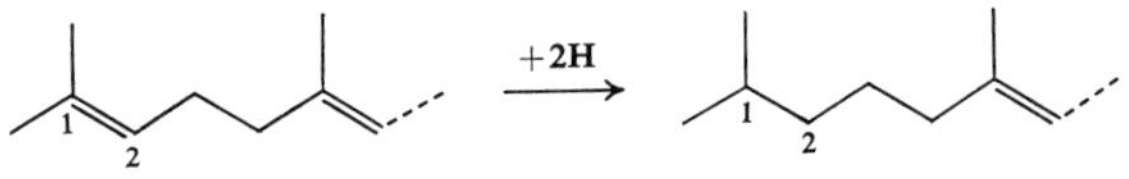

1 2 $\xrightarrow{+2H}$ 1 2

FIG. 20.

Nothing is known about the mechanism of this reaction.

6. *Aerobic Introduction of Oxo-groups* (*C-2*)

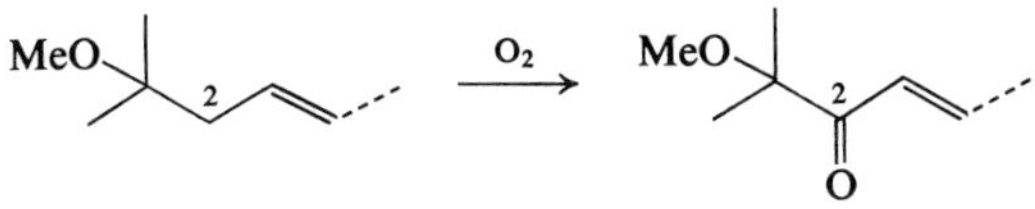

MeO 2 $\xrightarrow{O_2}$ MeO 2 O

FIG. 21.

Introduction of the 2-oxo substituent appears to require the presence of a double bond at C-3,4, the introduced oxo- group then being in conjugation with the main chromophore. It has been demonstrated by ^{18}O experiments that the oxygen atom of the oxo- group comes from molecular oxygen (Shneour, 1962b). The mechanism of the reaction is not known.

7. *Cyclization*

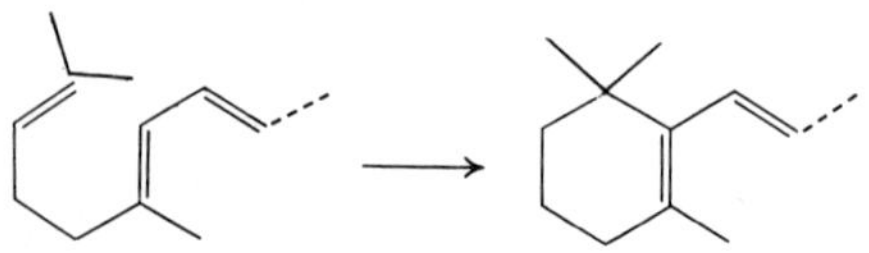

FIG. 22.

Formation of the β-ionone ring is limited to one species, *Rhodomicrobium vannielii*. It is not known if the cyclization reaction follows the same course as in higher plants.

8. *Hydroxylation* (*C-3*)

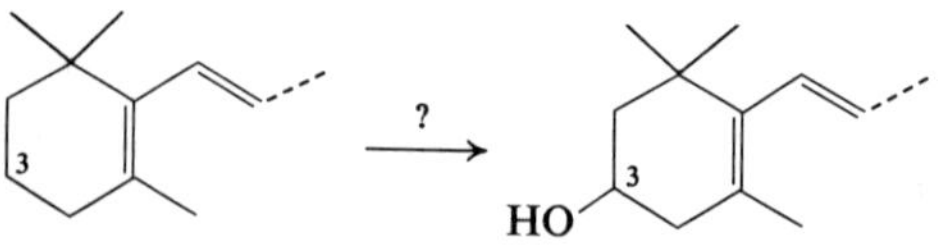

FIG. 23.

In *Rhodomicrobium vannielii*, introduction of the hydroxyl group at C-3 of β-cryptoxanthin apparently occurs under anaerobic conditions, whereas in higher plants and non-photosynthetic bacteria, molecular oxygen is required. No information is available about the mechanism or stereochemistry of this reaction.

9. *Aromatization*

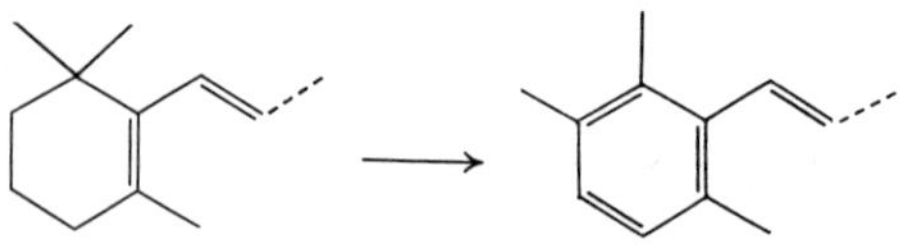

FIG. 24.

Formation of the aromatic ring is thought to arise via a β-ring (Liaaen-Jensen, 1966). Nothing is known of the mechanism.

10. Introduction of Oxo-Functions (Anaerobic)

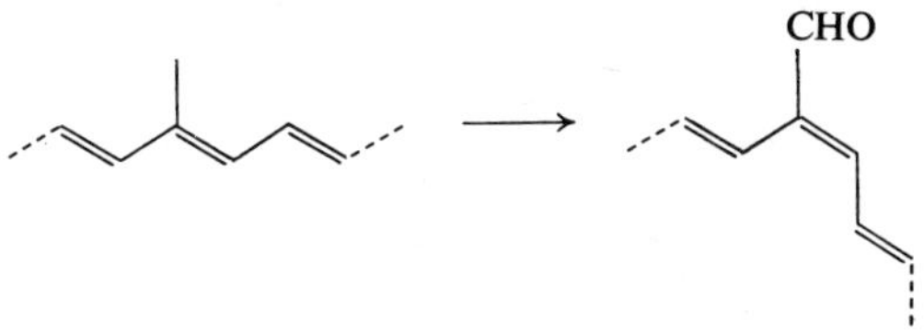

CHO

FIG. 25.

Formation of the aldehyde group in rhodopinal (50) and the 2-oxo group of okenone (49) occur under anaerobic conditions. A possible mechanism involving hydroxylation and oxidation of the allylic hydroxyl group has been proposed by Liaaen-Jensen (1963).

F. COMPARISON OF REACTION TYPES

Comparison of the structures of the 1-hydroxy-1,2-dihydro- and 1,2-dihydro- end groups and the β-ring suggests the possibility that these may represent alternative end products from a common precursor. Such a possibility has been considered by Liaaen-Jensen (1963). As illustrated in Fig. 26, proton attack on an acyclic precursor could give rise to a carbonium ion (51) which could be stabilized in three ways; (i) addition of H^- (from NADPH?) to give the 1,2-dihydro- end group; (ii) addition of OH^- to give the normal 1-hydroxy-1,2-dihydro end group; (iii) cyclization.

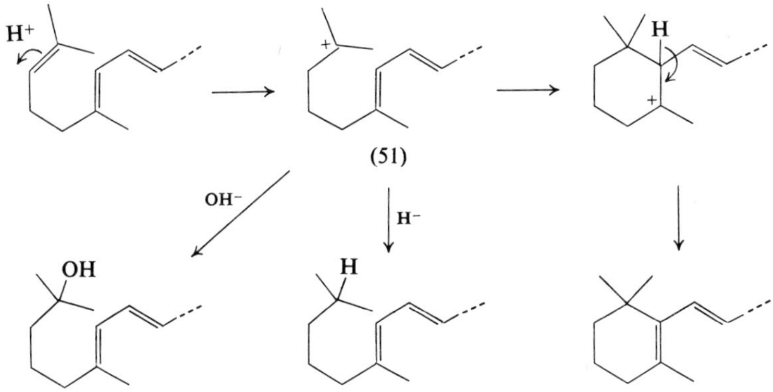

H+
H
(51)
OH−
H−
OH
H

FIG. 26. Scheme for the formation of alternative end-groups from a common precursor.

G. CONCLUSIONS

Much chemical work with physical techniques has led to the determination of the structures of many compounds present in the photosynthetic bacteria. This has made it clear that these organisms are capable of carrying out several reactions different from those performed by higher plants. Very little biochemical work on these problems has yet been undertaken, and no details are available about the enzymes responsible, co-factor requirements, the stereochemical course of the reactions or the reaction mechanisms. Again the possibility of the reactions taking place on a multi-enzyme complex must be considered. Very many problems are waiting to be tackled.

ACKNOWLEDGEMENTS

The author wishes to express his sincere gratitude to all those co-workers in the Departments of Biochemistry at U.C.W. Aberystwyth, and Liverpool University who have contributed much of the experimental work reported, and especially to Professor T. W. Goodwin, F.R.S., who introduced him to the field of carotenoid biosynthesis, and who has always given advice and encouragement during this work. Much of the work described was supported by research grants from the Science Research Council and Roche Products Ltd.

REFERENCES

Aasen, A. J. and Liaaen-Jensen, S. (1967). *Acta Chem. Scand.* **21**, 2185.
Archer, B. L., Barnard, D., Cockbain, E. G., Cornforth, J. W., Cornforth, R. H. and Popják, G. (1966). *Proc. r. Soc.* B **163**, 519.
Beeler, D. A. and Porter, J. W. (1962). *Biochem. biophys. Res. Commun.* **8**, 367.
Beeler, D. A. and Porter, J. W. (1963). *Archs Biochem. Biophys.* **100**, 167.
Benedict, C. R. and Beckman, L. D. (1964). *Pl. Physiol.* **39**, 726.
Buggy, M. J., Britton, G. and Goodwin, T. W. (1969). *Biochem. J.* **114**, 641.
Bu'Lock, J. D., Austin, D. J., Snatzke, G. and Hruban, L. (1970). *J. chem. Soc.* D 255.
Charlton, J. M., Treharne, K. J. and Goodwin, T. W. (1967). *Biochem. J.* **105**, 205.
Chichester, C. O. (1967). *Pure appl. Chem.* **14**, 215.
Chichester, C. O. and Nakayama, T. O. M. (1967). *In* "Biogenesis of Natural Compounds", 2nd Ed. (P. Bernfeld, ed.), p. 641, Pergamon Press, London.
Conti, S. F. and Benedict, C. R. (1962). *J. Bacteriol.* **83**, 929.
Cornforth, J. W., Cornforth, R. H., Donninger, C. and Popják, G. (1966a). *Proc. r. Soc.* B **163**, 492.
Cornforth, J. W., Cornforth, R. H., Popják, G. and Yengoyan, L. (1966b). *J. biol. Chem.* **241**, 3970.
Czygan, F. C. (1967). *Ber. dtsch. Bot. Ges.* **80**, 627.
Davies, B. H. (1970a). *Biochem. J.* **116**, 93.
Davies, B. H. (1970b). *Biochem. J.* **116**, 101.
Davies, B. H. and Holmes, E. A. (1969a). *Biochem. J.* **113**, 33P.
Davies, B. H. and Holmes, E. A. (1969b). *Biochem. J.* **113**, 35P.
Davies, B. H., Villoutreix, J., Williams, R. J. H. and Goodwin, T. W. (1963). *Biochem. J.* **89**, 96P.
Davies, B. H., Holmes, E. A., Loeber, D. E., Toube, T. P. and Weedon, B. C. L. (1969). *J. chem. Soc.* C 1266.
Davis, J. B., Jackman, L. M., Siddons, P. T. and Weedon, B. C. L. (1966). *J. chem. Soc.* C 2154.
Decker, K. and Uehleke, H. (1961). *Hoppe-Seyler's Z. physiol. Chem.* **323**, 61.
De Ville, T. E., Hursthouse, M. B., Russell, S. W. and Weedon, B. C. L. (1969). *J. chem. Soc.* D 1311.
Donninger, C. and Popják, G. (1966). *Proc. r. Soc.* B **163**, 465.
Eimhjellen, K. E. and Liaaen-Jensen, S. (1964). *Biochim. biophys. Acta* **82**, 21.

Eugster, C. H., Buchecker, R., Tscharner, C., Uhde, G. and Ohloff, G. (1969). *Helv. Chim. Acta* **52**, 1729.
Goad, L. J. (1970). *In* "Natural Substances Formed Biologically from Mevalonic Acid" (T. W. Goodwin, ed.), p. 45, Academic Press, London and New York.
Goodwin, T. W. (1965). *In* "Chemistry and Biochemistry of Plant Pigments" (T. W. Goodwin, ed.), p. 143, Academic Press, London and New York.
Goodwin, T. W. and Osman, H. G. (1954). *Biochem. J.* **56**, 222.
Goodwin, T. W. and Williams, R. J. H. (1965a). *Biochem. J.* **94**, 5C.
Goodwin, T. W. and Williams, R. J. H. (1965b). *Biochem. J.* **97**, 28C.
Goodwin, T. W. and Williams, R. J. H. (1966). *Proc. r. Soc.* B **163**, 515.
Goodwin, T. W., Land, D. G. and Osman, H. G. (1955). *Biochem. J.* **59**, 491.
Goodwin, T. W., Land, D. G. and Sissins, M. E. (1956). *Biochem. J.* **64**, 486.
Goodwin, T. W., Britton, G. and Walton, T. J. (1968). *Pl. Physiol.* **43**, (Suppl.) S-46.
Graebe, J. E. (1968). *Phytochemistry* **7**, 2003.
Grob, E. C., Kirschner, K. and Lynen, F. (1961). *Chimia* **15**, 308.
Harding, R. W., Huang, P. C. and Mitchell, H. K. (1969). *Archs Biochem. Biophys.* **129**, 696.
Hayano, M. (1962). *In* "Oxygenases" (O. Hayaishi, ed.), p. 181, Academic Press, London and New York.
Hemming, F. W. (1970). *In* "Natural Substances Formed Biologically from Mevalonic Acid" (T. W. Goodwin, ed.), p. 105, Academic Press, London and New York.
Hill, H. M. and Rogers, L. J. (1969). *Biochem. J.* **113**, 31P.
Hocking, D. (1967). *Can. J. Microbiol.* **13**, 859.
Jungalwala, F. B. and Porter, J. W. (1965). *Archs Biochem. Biophys.* **110**, 291.
Kakutani, Y. (1966). *J. Biochem. (Tokyo)* **59**, 135.
Kakutani, Y., Suzue, G. and Tanaka, S. (1964). *J. Biochem. (Tokyo)* **56**, 195.
Karunakaran, A., Karunakaran, M. E. and Quackenbush, F. W. (1966). *Archs Biochem. Biophys.* **114**, 326.
Krzeminski, L. F. and Quackenbush, F. W. (1960). *Archs Biochem. Biophys.* **88**, 287.
Kushwaha, S. C., Subbarayan, C., Beeler, D. A. and Porter, J. W. (1969). *J. biol. Chem.* **244**, 3635.
Kushwaha, S. C., Suzue, G., Subbarayan, C. and Porter, J. W. (1970). *J. biol. Chem.* **245**, 4708.
Lee, T.-C. and Chichester, C. O. (1969). *Phytochemistry* **8**, 603.
Liaaen-Jensen, S. (1962). *Kgl. Norske Videnshab. Selskabs. Skrifter* **8**, 5.
Liaaen-Jensen, S. (1963). *In* "Bacterial Photosynthesis" (H. Gest, A. San Pietro and L. P. Vernon, eds.), p. 19, Antioch Press, Yellow Springs, Ohio.
Liaaen-Jensen, S. (1965). *Ann. Rev. Microbiol.* **19**, 163.
Liaaen-Jensen, S. (1966). *In* "Biochemistry of Chloroplasts", Vol. I (T. W. Goodwin, ed.), p. 437, Academic Press, London and New York.
Liaaen-Jensen, S. (1967). *Acta Chem. Scand.* **21**, 961.
Liaaen-Jensen, S. (1971). *In* "Aspects of Terpenoid Chemistry and Biochemistry" (T. W. Goodwin, ed.), p. 223, Academic Press, London and New York.
Liaaen-Jensen, S., Cohen-Bazire, G., Nakayama, T. O. M. and Stanier, R. Y. (1958). *Biochim. biophys. Acta* **29**, 477.
Liaaen-Jensen, S., Cohen-Bazire, G. and Stanier, R. Y. (1961). *Nature, Lond.* **192**, 1168.
Liaaen-Jensen, S., Hegge, E. and Jackman, L. M. (1964). *Acta Chem. Scand.* **18**, 1703.
Malhotra, H. C., Britton, G. and Goodwin, T. W. (1969). *Phytochemistry* **8**, 1047.

Malhotra, H. C., Britton, G. and Goodwin, T. W. (1970a). *FEBS Letters* **6**, 334.
Malhotra, H. C., Britton, G. and Goodwin, T. W. (1970b). *Phytochemistry* **9**, 2369.
Malhotra, H. C., Britton, G. and Goodwin, T. W. (1970c). *J. chem. Soc.* D 127.
Malhotra, H. C., Britton, G. and Goodwin, T. W. (1970d). *Int. J. Vit. Res.* **40**, 315.
Nusbaum-Cassuto, E., Villoutreix, J. and Malengé, J.-P. (1967). *Biochim. biophys. Acta* **136**, 459.
Oster, M. O. and West, C. A. (1968). *Archs Biochem. Biophys.* **127**, 112.
Popják, G., Goodman, DeW. S., Cornforth, J. W., Cornforth, R. H. and Ryhage, R. (1961). *J. biol. Chem.* **236**, 1934.
Porter, J. W. and Anderson, D. G. (1962). *Archs Biochem. Biophys.* **97**, 520.
Porter, J. W. and Anderson, D. G. (1967). *A. Rev. Pl. Physiol.* **18**, 197.
Porter, J. W. and Lincoln, R. E. (1950). *Archs Biochem. Biophys.* **27**, 390.
Purcell, A. E. (1964). *Archs Biochem. Biophys.* **105**, 606.
Purcell, A. E., Thompson, G. A., Jr. and Bonner, J. (1959). *J. biol. Chem.* **234**, 1081.
Quackenbush, F. W. (1965). *Japanese-American Conference on Biosynthesis of Carotenes, Kyoto, Japan.* (Unpublished results.)
Rabourn, W. J. (1957). *Am. Chem. Soc. 132nd Meeting, New York.* Abstr. 88C.
Rabourn, W. J. and Quackenbush, F. W. (1956). *Archs Biochem. Biophys.* **61**, 111.
Rabourn, W. J., Quackenbush, F. W. and Porter, J. W. (1954). *Archs Biochem. Biophys.* **48**, 267.
Rilling, H. C. (1962). *Biochim. biophys. Acta* **65**, 156.
Ryvarden, L. and Liaaen-Jensen, S. (1964). *Acta Chem. Scand.* **18**, 643.
Schmidt, K., Liaaen-Jensen, S. and Schlegel, H. G. (1963). *Arch. Mikrobiol.* **46**, 117.
Shah, D. H., Cleland, W. W. and Porter, J. W. (1965). *J. biol. Chem.* **240**, 1946.
Shah, D. V., Feldbruegge, D. H., Houser, A. R. and Porter, J. W. (1968). *Archs Biochem. Biophys.* **127**, 124.
Shneour, E. A. (1962a). *Biochim. biophys. Acta* **62**, 534.
Shneour, E. A. (1962b). *Biochim. biophys. Acta* **65**, 510.
Stanier, R. Y. (1960). *In* "The Harvey Lectures", 1958–1959, p. 215, Academic Press, New York and London.
Subbarayan, C., Kushwaha, S. C., Suzue, G. and Porter, J. W. (1970). *Archs Biochem. Biophys.* **137**, 547.
Suzue, G. (1960). *Biochim. biophys. Acta* **45**, 616.
Tefft, R. E., Goodwin, T. W. and Simpson, K. L. (1970). *Biochem. J.* **117**, 921.
Thornber, J. P., Olson, J. M., Williams, D. M. and Clayton, M. L. (1969). *Biochim. biophys. Acta* **172**, 351.
Tomes, M. L. (1967). *Genetics* **56**, 227.
Van Niel, C. B. (1947). *Antonie van Leeuwenhoek, J. Microbiol. Serol.* **12**, 156.
Van Niel, C. B. (1963). *In* "Bacterial Photosynthesis" (H. Gest, A. San Pietro and L. P. Vernon, eds), p. 459, Antioch Press, Yellow Springs, Ohio.
Villoutreix, J. (1960). *Biochim. biophys. Acta* **40**, 434.
Volk, W. A. and Pennington, D. (1950). *J. Bacteriol.* **59**, 169.
Walton, T. J., Britton, G. and Goodwin, T. W. (1969). *Biochem. J.* **112**, 383.
Weedon, B. C. L. (1970). *F. Chem. Org. Naturst.* **27**, 81.
Weeks, O. B. (1971). *In* "Aspects of Terpenoid Chemistry and Biochemistry" (T. W. Goodwin, ed.), p. 291, Academic Press, London and New York.
Wells, L. W., Schelbe, W. J. and Porter, J. W. (1964). *Fedn. Proc.* **23**, 426.
Williams, R. J. H., Britton, G. and Goodwin, T. W. (1966). *Biochem. J.* **101**, 7P.
Williams, R. J. H., Britton, G., Charlton, J. M. and Goodwin, T. W. (1967a). *Biochem. J.* **104**, 767.
Williams, R. J. H., Britton, G. and Goodwin, T. W. (1967b). *Biochem. J.* **105**, 99.

Yamamoto, H. Y. and Chichester, C. O. (1965). *Biochim. biophys. Acta* **109**, 303.

Yamamoto, H. Y., Chichester, C. O. and Nakayama, T. O. M. (1962a). *Archs Biochem. Biophys.* **96**, 645.

Yamamoto, H. Y., Chichester, C. O. and Nakayama, T. O. M. (1962b). *Photochem. Photobiol.* **1**, 53.

Yamamoto, H. Y., Nakayama, T. O. M. and Chichester, C. O. (1962c). *Archs Biochem. Biophys.* **97**, 168.

Yokoyama, A., Nakayama, T. O. M. and Chichester, C. O. (1961). *J. biol. Chem.* **237**, 681.

CHAPTER 10

Biosynthesis of C_{50} Carotenoids

OWEN B. WEEKS

New Mexico State University, Arts and Sciences Research Center, Las Cruces, New Mexico, U.S.A.

I. Introduction

The first example of a naturally occurring fifty-carbon (C_{50}) atom carotenoid was dehydrogenans-P439 (Liaaen-Jensen and Weeks, 1966; Weeks and Garner, 1967) which now has the trivial name decaprenoxanthin (Weeks *et al.*, 1969). The carotenoid is the principal pigment of a Gram-positive bacterium *Flavobacterium dehydrogenans*. Decaprenoxanthin (Fig. 1) has been studied extensively and details of its structure and properties published (Weeks and Garner, 1967; Liaaen-Jensen *et al.*, 1968; Andrewes, 1969; Schwieter and Liaaen-Jensen, 1969). Since 1966 additional C_{50} carotenoids have been characterized and some of these already were known and had tacitly been given the usual C_{40} basic structure. The list includes (Liaaen-Jensen, 1970): sarcinaxanthin, an isomer of decaprenoxanthin; bacterioruberin and related minor carotenoids from *Halobacterium* species; bisanhydro-bacterioruberin, the principal carotenoid of *Corynebacterium poinsettiae* which Starr and Saperstein (1953) thought might be spirilloxanthin, and other carotenoids from the same bacterium. Recently corynexanthin (Hodgkiss *et al.*, 1954) has been characterized as a monoglucoside of decaprenoxanthin (Weeks and Andrewes, 1970). Thus some C_{50} carotenoids are familiar names whose true nature has only recently become known. The ease with which C_{50} carotenoids conceal their distinguishing feature, and the importance of mass spectrometry in revealing it, is well documented (Liaaen-Jensen *et al.*, 1968; Andrewes, 1969; Enzell, 1969; Liaaen-Jensen, 1969). There are two generalizations which may be made about the C_{50} structure and its distribution in nature.

The basic C_{50} structure as it is known now is a branched, polyene chain in which branching results from substitution of an isopentenyl function at positions 2 and 2′, of the carotenoid molecule. Three examples illustrate this: bacterioruberin which is an acyclic carotenoid; the *C. poinsettiae* carotenoid,

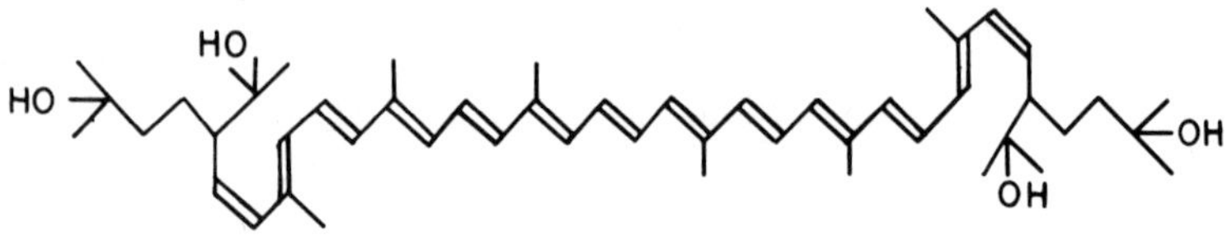

Bacterioruberin

$C_{50}H_{76}O_4$

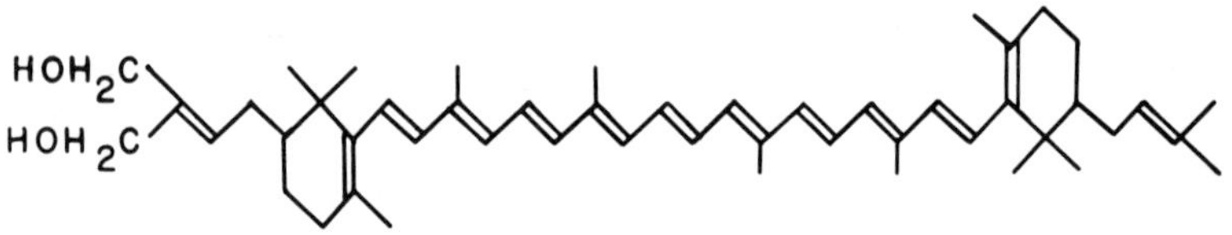

C. poinsettiae P 450

$C_{50}H_{72}O_2$

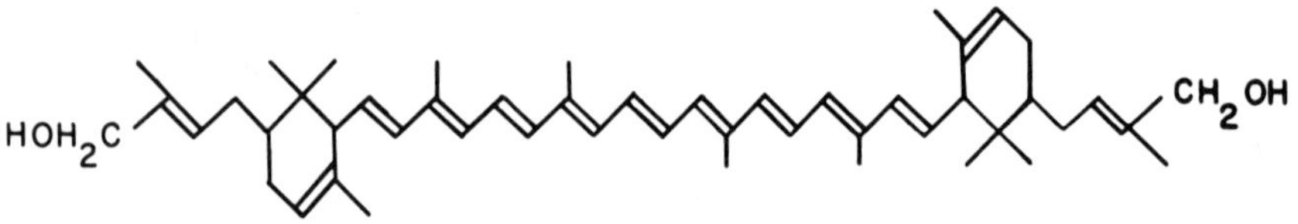

Decaprenoxanthin

$C_{50}H_{72}O_2$

FIG. 1. Selected examples of decapreno carotenoids.

C.p. 450, which has a β-cyclic structure; and decaprenoxanthin with its α cyclic rings (Fig. 1). The additional structural features are analogous to those already known for C_{40} carotenoids with the exception of substitution on the isopentenyl side chains which now seems limited to formation of hydroxyl groups.

The second generalization has to do with distribution of C_{50}-carotenoids in nature. These carotenoids have been found only in bacteria and especially in Gram-positive, aerobic forms. All are non-photosynthetic. It can be argued that *Halobacterium*, the source of bacterioruberin, is a Gram-negative bacterium and similar to the pseudomonads. The bacterium has an unusual cell wall and possibly it has more in common with cell walls of Gram-positive bacteria than Gram-negative. Pigmented, aerobic, Gram-positive bacteria may be arranged into three seemingly natural taxonomic groups: the coryne-bacteria, the brevibacteria and a group designated *Cellulomonas*-dehydro-genans-xanthe (Hester and Weeks, 1969). The groups are readily distinguished taxometrically and each has a distinctive range of nucleotide base ratio values.

The *Cellulomonas*-dehydrogenans-xanthe group includes at least thirty bacterial strains among which are the so-called *Pseudomonas xanthe* (Weeks, 1955) which is Gram-positive, various species of *Cellulomonas*, the fourteen strains of *F. dehydrogenans* described by Ferrari and Zannini (1958), strains of *Corynebacterium poinsettiae* and *C. mediolanum* which are indistinguishable, and unnamed, Gram-positive flavobacteria (D. J. Hester and O. B. Weeks, unpublished studies). All strains of *F. dehydrogenans*, the species of *Cellulo-monas* and *P. xanthe*, produce a decaprenoxanthin-like carotenoid which has been confirmed as decaprenoxanthin by mass spectrometry in most cases. This taxometric group has a nucleotide base ratio range between 70–76·5% guanine plus cytosine (% GC) which is near the upper limit reported for bacteria (Hill, 1966). It is of seeming interest that *Sarcina* from which the C_{50} sarcinaxanthin is obtained (Liaaen-Jensen *et al.*, 1967; Liaaen-Jensen, 1969) has a GC ratio above 70%. It is premature to decide that C_{50}-carotenoids are restricted to Gram-positive bacteria having nucleotide base ratios above 70%, *Halo-bacterium* species are reported to be between 60 and 69% GC and to be Gram-negative, but it is among these bacteria such carotenoids occur commonly.

Biosynthesis of C_{50} carotenoids may be assumed to be that which is accepted for carotenoids generally (Goodwin, 1965a; Porter and Anderson, 1967) unless the branched chain were to arise as a pre-phytoene event. Structural studies of the carotenoid pigments of *F. dehydrogenans* suggest the C_{50} structure arises during post-phytoene carotenogenesis (Weeks and Garner, 1967; Andrewes, 1969; Weeks *et al.*, 1969). It is assumed that the major details of terminal carotenogenesis in *F. dehydrogenans* may be reconstructed from the known carotenoid structures and that the pattern which emerges will serve as a model for biosynthesis of C_{50} carotenoids generally.

II. Carotenoid System of *Flavobacterium dehydrogenans*

The carotenoid system of *F. dehydrogenans* has been studied (Weeks and Garner, 1967) and structures for most of the constituent carotenoids reported (Liaaen-Jensen *et al.*, 1968; Andrewes, 1969; Weeks *et al.*, 1969). Quantitative and qualitative variations in carotenoids of the bacterium occur and depend upon light and cultural conditions (Fig. 2).

Carotenogenesis in *F. dehydrogenans* is controlled by light and populations grown in absence of light (dark-grown) contain little or no carotenoid.

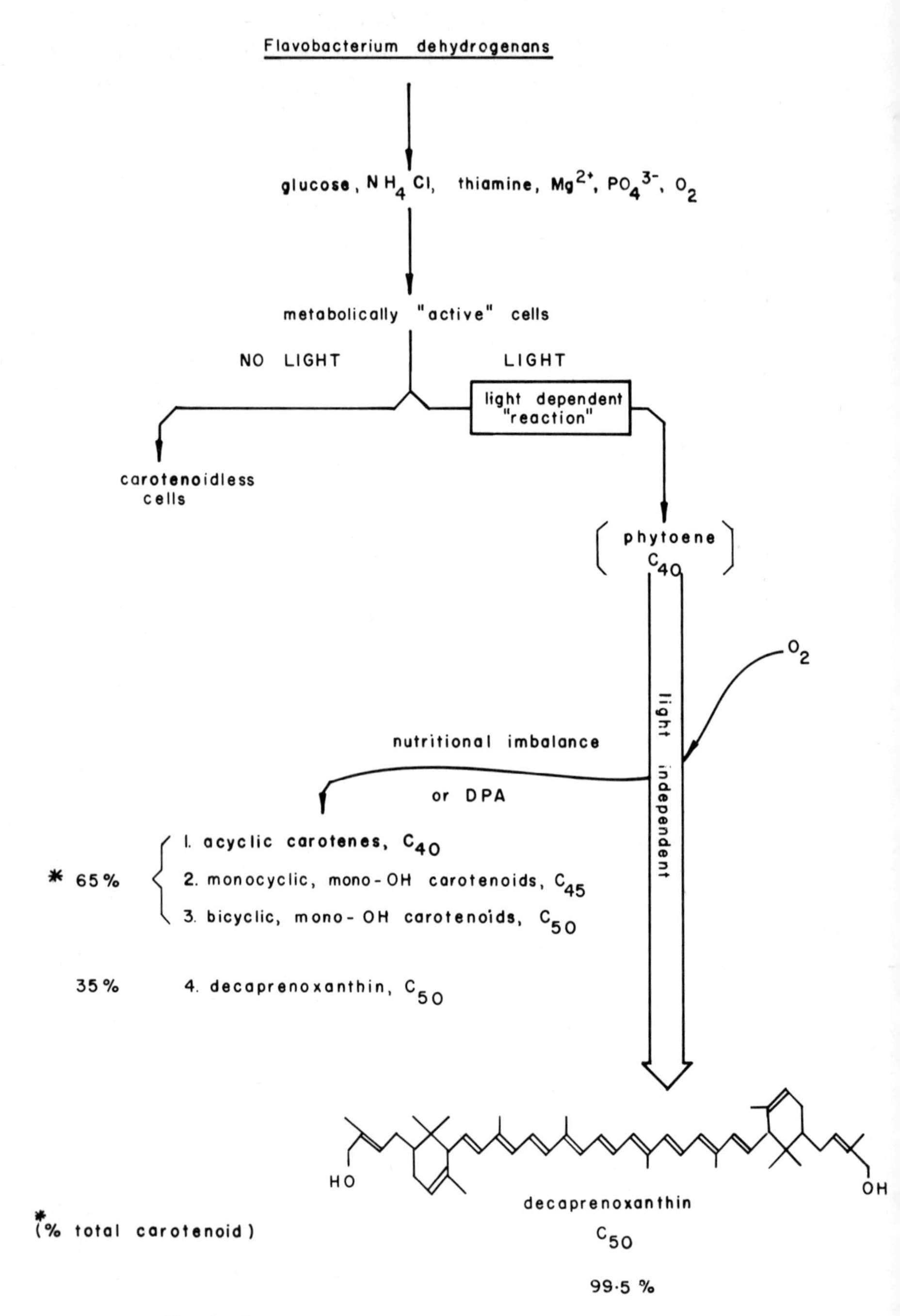

FIG. 2. Carotenogenic system of *Flavobacterium dehydrogenans*.

Cultures grown with continuous light contain 3 to 12 μg total carotenoid/mg bacterial cell nitrogen (cell-N), while dark-grown cultures have between 0·003 and 0·03 μg, a difference of three orders of magnitude in carotenoid content of light and dark cultures. The amount of carotenoid is related to time populations are incubated in light (Fig. 3). Illumination of dark-grown cells

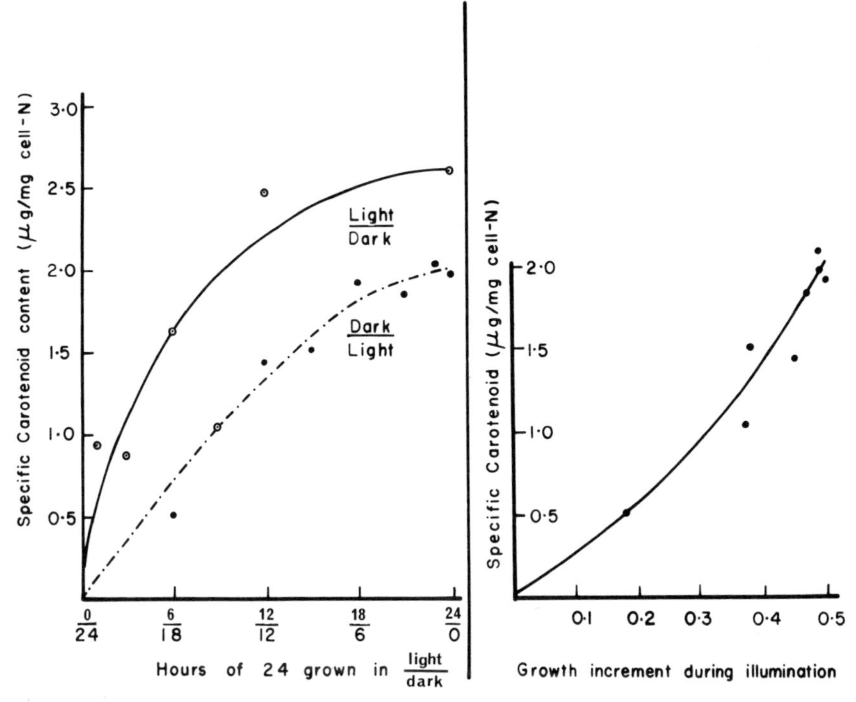

FIG. 3. Influence of light on biosynthesis of carotenoids in *Flavobacterium dehydrogenans*. Light/dark populations were grown in light initially and then in the dark for the remainder of the 24 h period. Dark/light was a reversal of the experimental conditions. Growth increment in light is the increase in optical density (580 nm) during incubation of cultures in light. Growth medium AGY, continuous shaking during incubation at 30° (Weeks and Garner, 1967).

for 1 min, 0°, induces carotenogenesis during one reproductive cycle of the bacterium when the cells are incubated in the dark subsequent to photoactivation (Fig. 4). Further study has shown light-induction of carotenogenesis is temperature independent, O_2-dependent and requires protein biosynthesis following the photochemical event (O. B. Weeks *et al.*, unpublished studies). Light is not necessary for post-phytoene aspects of the biosynthesis but oxygen is.

Nutritional requirements for growth and carotenogenesis in *F. dehydrogenans* are not exceptional and may be satisfied by NH_4Cl (or asparagine), glucose and a small amount of yeast extract. Large volume culturing requires added magnesium and the systems are buffered to pH 7·0 with phosphates

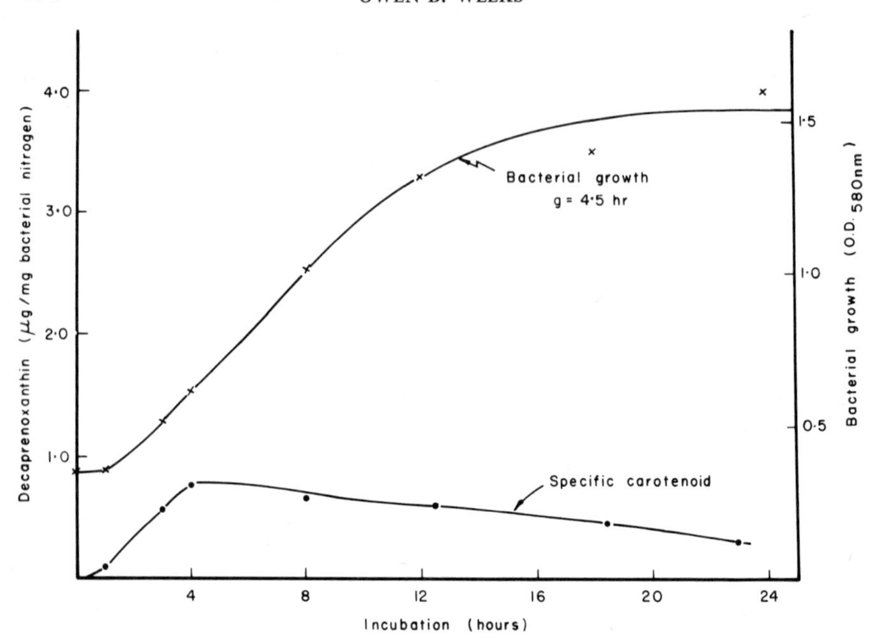

FIG. 4. Specific decaprenoxanthin (μg/mg cell-N) biosynthesized by *Flavobacterium dehydrogenans* photo-activated (1500-ft candles) for 2 min, 0°, and then allowed to grow in the dark with continuous shaking during incubation at 30°. Specific carotenoid amounts after various incubation times are compared with bacterial growth. Carotenoid increased during the initial generation (g) of growth and declined subsequently (O. B. Weeks *et al.*, unpublished study).

(Weeks and Garner, 1967). Yeast extract may be replaced with a mixture of thiamine, riboflavin, biotin, pyridoxine, folic acid and *p*-amino benzoic acid. The only absolute requirement is for thiamine but the additional B-vitamins increase both bacterial cell and specific carotenoid yields. The amount of yeast extract necessary varies with the commercial product and the conditions under which, and the time, a given yeast extract has been stored. Welch (1965) studied this circumstance in *F. dehydrogenans* and related the effectiveness of a yeast extract to its thiamine content. Increasing the amount of yeast extract above that required for growth influences the quantity and quality of carotenoids in *F. dehydrogenans* (Table I). Similar variations occur when diphenylamine is added to the bacterial culture during its early development, i.e. during the first 8 h incubation, 30° (Table II).

The third cultural condition which produces similar variations in the carotenoid product is growth in a peptone, beef extract medium (Table II). Decaprenoxanthin is the principal carotenoid in each medium and essentially the only carotenoid in the synthetic medium or media in which the B-vitamins are replaced by small quantities of yeast extract. In each instance inhibited carotenogenesis is partially overcome by increasing the aeration during

TABLE I

Carotenoids separated from methanol extracts of *Flavobacterium dehydrogenans* grown in several media (from Weeks and Garner, 1967)

		% Total each carotenoid							
		Medium AGY limited aeration[a]				Medium AGY maximum aeration[b]			
	Synthetic	% yeast extract				% yeast extract			
Carotenoid designation	medium[c]	0·001	0·01	0·1	1·0	0·01	0·025	0·05	0·1
Dehydrogenans-									
phytoene	0	0	6·3	26	29	0	0	56·7	60·2
phytofluene	0	0	4·2	9	7	0	tr	4·8	5·0
ζ-carotene	0	0	0·8	1	0	0	tr	0·3	0·3
P420 complex	0	0	6·0	4·5	4	0·5	2·4	1·9	1·9
neurosporene	0·1	1·4	0·1	0	0	0·5	0	0	0
P439	99·9	98·6	82·3	59	60	99·0	97·6	36·4	32·6
lycopene	0	0	0	0	0	0	tr	tr	tr
Total carotenoid, μg	729	159	532	190	135	9697	8810	17298	13800
Total cell-nitrogen, mg	99	40	76	52	79	1710	2576	2184	1794
Total carotenoid, μg/mg cell nitrogen	7·4	4	6·9	3·7	1·7	5·7	3·4	7·9	7·7

[a] Cultural conditions: 24 h incubation; 30°; continuous illumination; continuous agitation, 1·5-litre system.
[b] Cultural conditions: 24 h incubation; 30°; continuous illumination; continuous agitation plus aeration; 12-litre systems.
[c] Synthetic medium and Medium AGY (Weeks and Garner, 1967).

TABLE II

Carotenoids separated from methanol extracts of *Flavobacterium dehydrogenans* grown in nutrient broth medium and medium AGY plus diphenylamine[a] (from Weeks and Garner, 1967)

Carotenoid designation	% Total each carotenoid[b]					
	Culture system: Nutrient broth, no DPA				Culture system: Medium AGY plus DPA	
	Limited aeration[c]		Maximum aeration[d]		Limited aeration[e]	Maximum aeration[d]
	24 h	48 h	12 h	24 h	24 h	24 h
Dehydrogenans-						
phytoene	39·6	28·4	1·4	0	66	46
phytofluene	9·6	0	1·9	0	10	6
ζ-carotene	1·1	0	0	0	0	0
ζ-carotene diOH(?)	1·7	2·9	0	0	0	0
P420 complex	9·4	8·9	10·5	0	20	12
neurosporene	0	tr	0	1·6	0	0
P439	38·9	59·9	85·1	98·3	4	35
lycopene	0	tr	0·4	0	0	tr
Total carotenoid, μg	1469	1532	6833	4717	58	3212
Total cell-nitrogen, mg	354	336	1717	828	—	731
Total carotenoid, μg/mg cell nitrogen	4·2	4·6	3·9	5·7	—	4·4

[a] Cultural conditions: 30°, constant agitation (limited aeration) or constant agitation plus aeration (maximum aeration); continuous illumination. See original reference for composition of nutrient broth and Medium AGY (0·01 % yeast extract). Diphenylamine (DPA) added to a concentration of 6×10^{-5} M, 2 h after inoculation.

[b] tr = trace; 0 = no compound detected. [c] 3-litre systems.

[d] 12-litre systems. [e] 1·5-litre system.

incubation (Tables I and II). Metabolic events which explain inhibition of carotenogenesis are not yet known but the result has been useful both to obtain the minor carotenoids for structural studies and for studies of post-phytoene carotenogenesis.

III. Structures of Carotenoids from *Flavobacterium dehydrogenans*

Investigation of methanol extracts of *F. dehydrogenans* which had been grown in light in cultural conditions which interfered with post-phytoene carotenogenesis, showed the carotenoid mixture (Tables I and II) included compounds which were similar, if not identical, to the carotenes of the well-known Porter-Lincoln series (Porter and Lincoln, 1950). The remaining carotenoids were classed as mono- and dihydroxy compounds of unknown structure (Weeks and Garner, 1967). Structural details of the principal carotenoid, decaprenoxanthin, were established first (Liaaen-Jensen *et al.*, 1968) and later for most of the minor carotenoids (Andrewes, 1969; Weeks *et al.*, 1969). The carotenoid system of *F. dehydrogenans* is shown in Table III and some individual structures in Figs 1 (decaprenoxanthin) and 8.

Structural studies of the hydrolycopenes and lycopene from *F. dehydrogenans* have shown these are authentic counterparts of well known carotenoids (Andrewes, 1969; Weeks *et al.*, 1969). Tentative identification by electronic

Table III
Carotenoids from *Flavobacterium dehydrogenans*

Preliminary designation[a]	Identity[b,c]
Dehydrogenans—	
phytoene	phytoene, $C_{40}H_{64}$
phytofluene	phytofluene, $C_{40}H_{62}$
ζ-carotene	ζ-carotene, $C_{40}H_{60}$
[ζ-carotene, asymmetrical][b]	[7′,8′,11′,12′-tetrahydrolycopene][d]
[neurosporene][d]	[neurosporene][d]
lycopene	lycopene, $C_{40}H_{56}$
P373	nonaprenoxanthin = 2-(3-hydroxymethyl-but-2-enyl)-7′,8′,11′,12′-tetrahydro-δ-carotene
P422	11′,12′-dehydrononaprenoxanthin = 2-(3-hydroxy-methyl-but-2-enyl)-7′,8′-dihydro-δ-carotene
[P452][d]	[7′,8′,11′,12′-dehydrononaprenoxanthin][d]
P439-mono-OH	deshydroxy decaprenoxanthin = 2-(3-hydroxy-methyl-but-2-enyl)-2′-(3-methyl-but-2-enyl)-ϵ-carotene
P439[e]	decaprenoxanthin = 2′,2′-di(3-hydroxymethyl-but-2-enyl)-ϵ-carotene

[a] Weeks and Garner, 1967; [b] Andrewes, 1969; [c] Weeks *et al.*, 1969; [d] [structure presumed, not established]; [e] Liaaen-Jensen *et al.*, 1968.

absorption spectrometry and chromatographic adsorptivity was confirmed by mass spectrometry after preliminary mass spectral study of authentic hydrolycopenes and lycopene. Authentic phytoene, phytofluene, ζ-carotene, neurosporene and lycopene each showed a strong molecular ion and characteristic fragmentation pattern (Fig. 5). Mass spectral fragmentation occurred

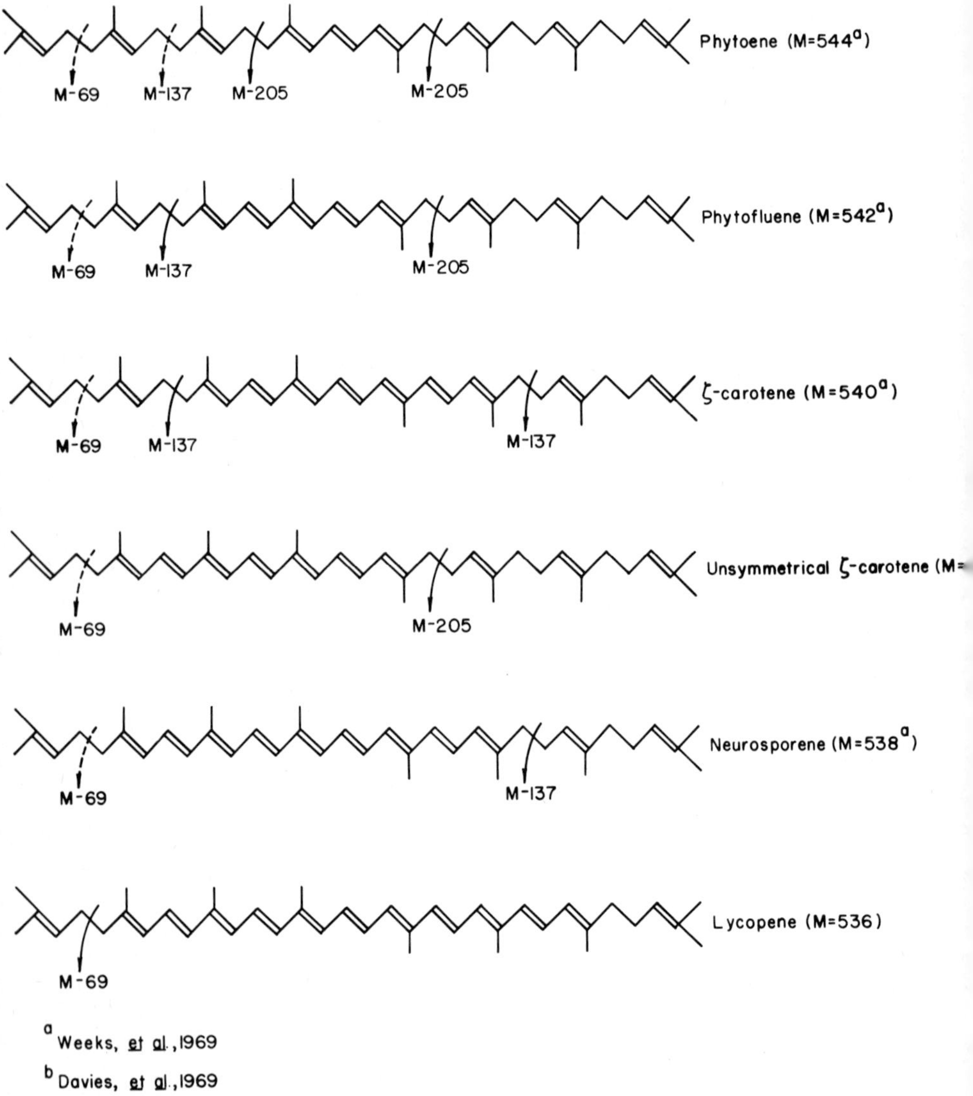

Phytoene (M=544[a])
M-69 M-137 M-205 M-205

Phytofluene (M=542[a])
M-69 M-137 M-205

ζ-carotene (M=540[a])
M-69 M-137 M-137

Unsymmetrical ζ-carotene (M=
M-69 M-205

Neurosporene (M=538[a])
M-69 M-137

Lycopene (M=536)
M-69

[a] Weeks, *et al.*, 1969
[b] Davies, *et al.*, 1969

FIG. 5. Characteristic mass spectral fragmentation patterns of authentic hydrolycopenes and lycopene (Data from Weeks *et al.*, 1969).

preferentially at *bis*-allylic bonds adjacent to the chromophore which not only established the length of each chromophore but also its position in the central polyene chain (Weeks *et al.*, 1969).

Dehydrogenans phytoene, phytofluene, ζ-carotene and lycopene are C_{40} in nature. Dehydrogenans-neurosporene has not been obtained in sufficient quantity for more than tentative identification but there seems no reason to question that it too has an authentic C_{40}-structure.

Dehydrogenans phytoene and phytofluene have always been obtained from the bacterium as predominately all-*trans* isomers (Andrewes, 1969) whereas most naturally occurring phytoene and phytofluene are said to be central (15,15′) mono *cis* isomers (Jungalwala and Porter, 1965; Davis *et al.*, 1966).

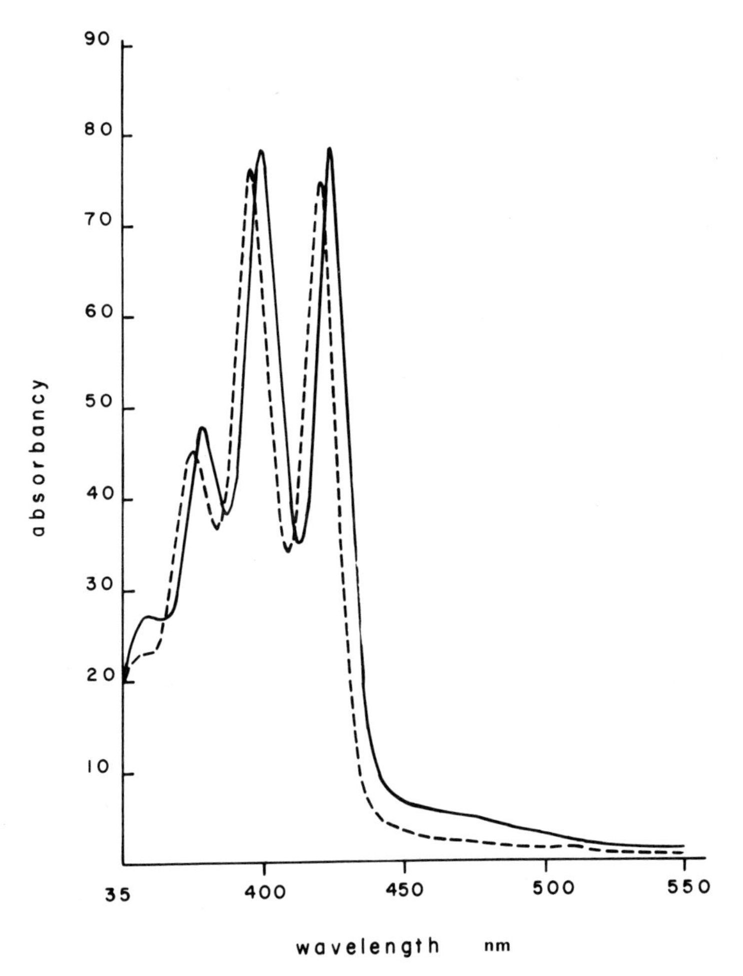

FIG. 6. Electronic absorption spectra of ζ-carotene (—) and unsymmetrical ζ-carotene (---) from *Flavobacterium dehydrogenans* (Andrewes, 1969).

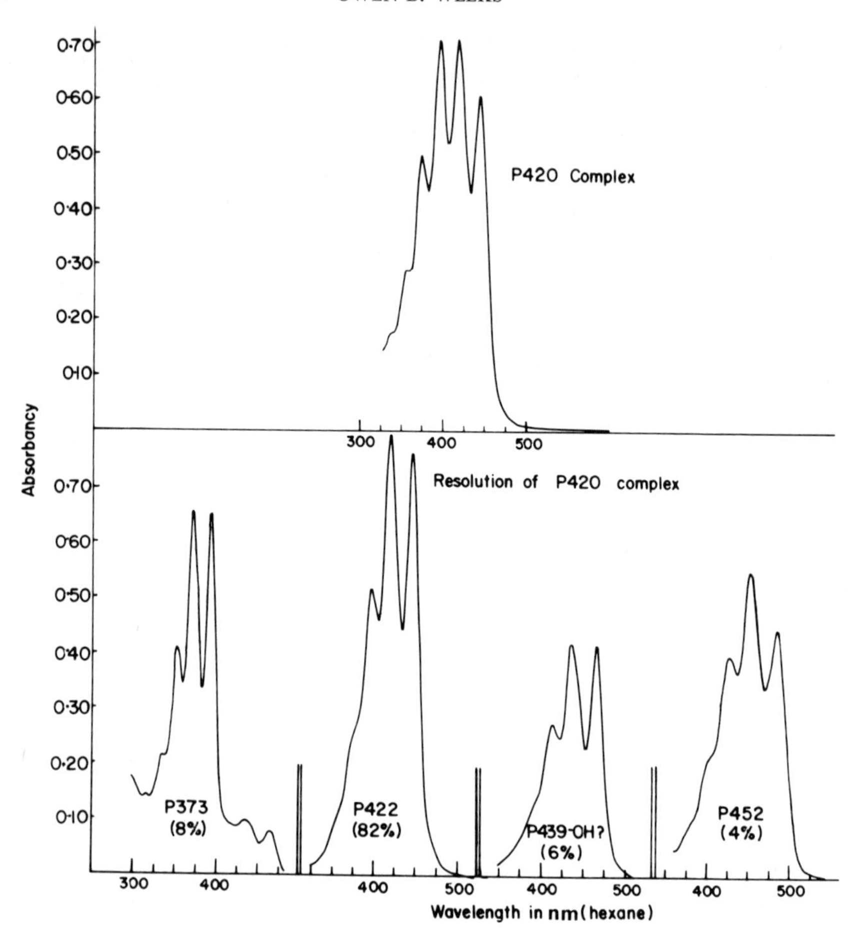

FIG. 7. Electronic absorption spectra of unresolved P420 complex and the constituent carotenoids from *Flavobacterium dehydrogenans* after resolution on magnesia:Celite (1:1) columns using petroleum ether containing 7·5% acetone as eluant (Weeks and Garner, 1967).

It has been assumed that condensation of 2 moles of geranylgeranyl pyrophosphate produces mono-*cis* phytoene and that the *cis* configuration is maintained until all-*trans* ζ-carotene is formed two reaction steps later in the Porter-Lincoln biosynthetic system (Porter and Anderson, 1967). Occurrence of all-*trans* phytoene and phytofluene in *F. dehydrogenans* suggests stereospecific differences will be found in different carotenogenic systems.

The second unusual feature in the hydrolycopene series of carotenes from *F. dehydrogenans* is the occurrence of two isomers of ζ-carotene (Andrewes, 1969). The one is authentic ζ-carotene which has a symmetrical molecule, and the second was presumed to be asymmetrical ζ-carotene, i.e. 7′,8′,11′,12′-tetrahydrolycopene (Davies *et al.*, 1969). Electronic absorption spectra of the two carotenes are shown in Fig. 6. Absorption maxima of the unsymmetrical

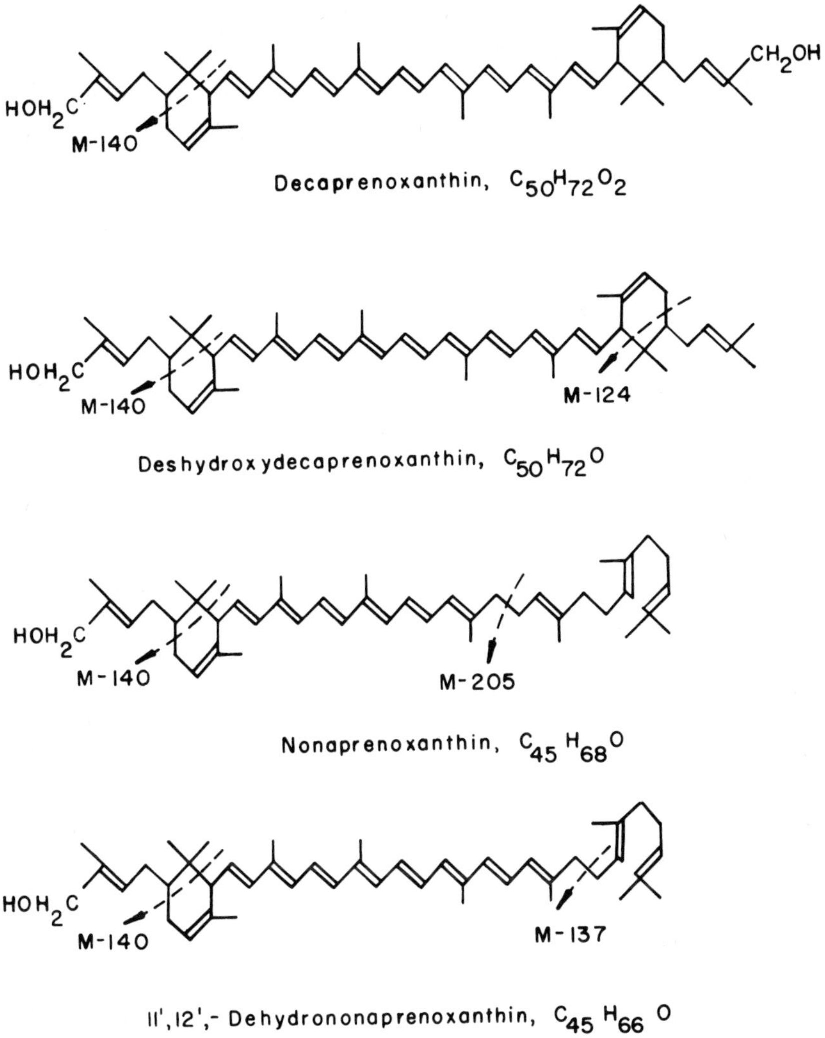

HOH$_2$C CH$_2$OH
M-140

Decaprenoxanthin, $C_{50}H_{72}O_2$

HOH$_2$C
M-140 M-124

Deshydroxydecaprenoxanthin, $C_{50}H_{72}O$

HOH$_2$C
M-140 M-205

Nonaprenoxanthin, $C_{45}H_{68}O$

HOH$_2$C
M-140 M-137

11',12',- Dehydrononaprenoxanthin, $C_{45}H_{66}O$

FIG. 8. Nonapreno and decapreno carotenoids from *Flavobacterium dehydrogenans* showing characteristic mass spectral fragmentation (Andrewes, 1969).

ζ-carotene (375, 396, 420 nm, hexane) are shifted hypsochromically from those of ζ-carotene (378, 400, 425 nm, hexane). The hypsochromic shift does not appear to be the result of *cis*-isomerization since comparison of extinction values at 396 and 420 nm indicates the presumed asymmetrical ζ-carotene is in all-*trans* form (Andrewes, 1969).

Culture conditions which allow the acyclic dehydrogenans carotenes to accumulate (Tables I and II) result also in accumulation of the P420 complex (Weeks and Garner, 1967). This mixture of carotenoids is chromatographically

TABLE IV
Quantitative resolution of P420-complex from *Flavobacterium dehydrogenans* (Weeks and Garner, 1967)

Carotenoid designation	% Total each carotenoid Growth medium[b] Nutrient broth	AGY (0·025)	AGY (0·05)	AGY (0·10)
Dehydrogenans-				
P373	8	8	16	13
P439 (mono-OH)	6	20	7	10
P422	82	66	76	76
P452	4	6	2	2
Total P420 complex, μg	276	223	139	319

[a] Chromatographic resolution used magnesia: celite (1:1) columns and petroleum ether plus 7·5% acetone as developer and eluant.
[b] P420-complex was obtained from the bacteria grown in nutrient broth or Medium AGY containing different amounts (%) of yeast extract; see original reference.

homogeneous during initial separation on alumina, activity grade 3, columns. Once it has been separated the mixture may be resolved into three, sometimes four, components using magnesia:celite (1:1) columns (Fig. 7). Ultimate

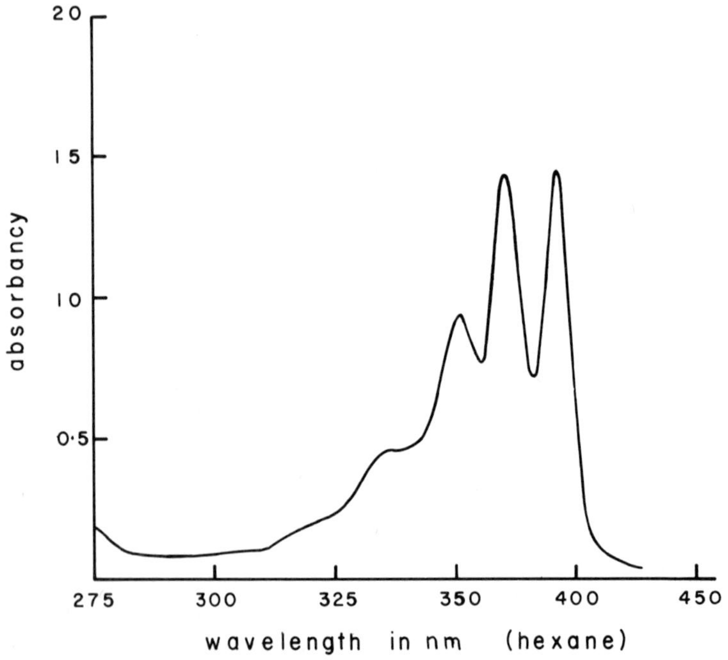

FIG. 9. Electronic absorption spectrum of nonaprenoxanthin from *Flavobacterium dehydrogenans*. Solvent, n-hexane. (Weeks *et al.*, 1969).

purification is done with thick-layer chromatography. The four carotenoids were designated P373, P422, P452 and monohydroxy-P439. Structures are known for P373, P422 and monohydroxy-P439 but that of P452 is not. The latter carotenoid is not only the least abundant constituent of the complex (Table IV) but also the molecule seems unusually labile. Electronic absorption spectra of P373, P422 and P452 suggested chromophores of 6, 8, and 10 double bonds and once the structure of P439 (decaprenoxanthin) was known, it was assumed these carotenoids were similar in their structures but different in having only one end of the molecules cyclized. Chromophoric differences were assumed to represent the degree of development of double bond systems toward the acyclic end of the molecules. The presumed structures were confirmed (Fig. 8) first for P422 (Weeks and Andrewes, 1968) and later for P373 and monohydroxy-P439 (Andrewes, 1969; Weeks *et al.*, 1969). The trivial name nonaprenoxanthin was given to P373 and thus P422 would be 11′,12′-dehydrononaprenoxanthin. Monohydroxy-P439 is deshydroxy-decaprenoxanthin. The structure presumed for P452 is 7′,8′,11′,12′-dehydrononaprenoxanthin. Electronic absorption spectra for all-*trans* P373 and P422 are shown in Figs 9 and 10.

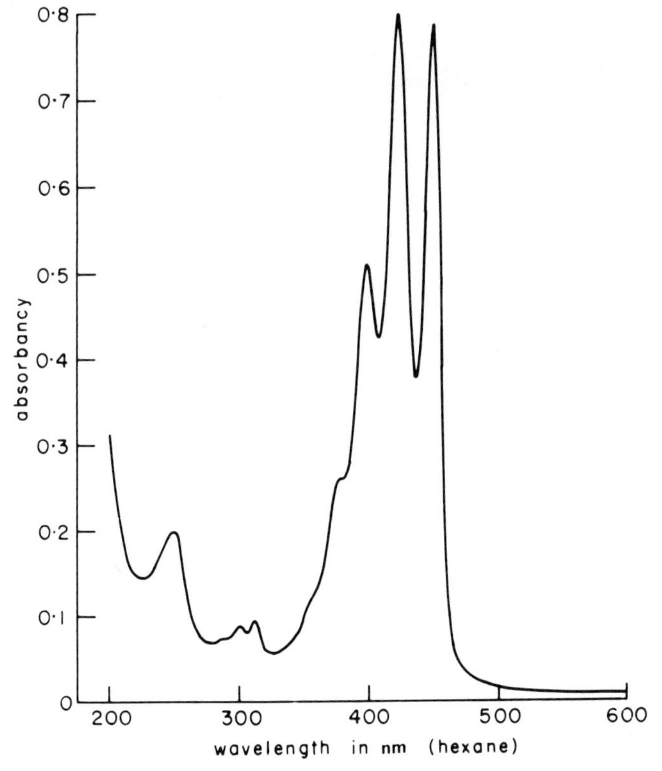

FIG. 10. Electronic absorption spectrum of 11′,12′-dehydrononaprenoxanthin from *Flavobacterium dehydrogenans*. Solvent, n-hexane (Andrewes, 1969).

IV. Biosynthesis of Decaprenoxanthin

Dark-grown *Flavobacterium dehydrogenans* is carotenoidless and data indicate that the entire carotenoid system is not functioning (Weeks and Garner, 1967). Dark-grown cells, however, are competent to make carotenoid when activated by light and when experimental conditions allow protein biosynthesis. Such bacterial cells will produce carotenoid in the absence of appreciable increase in cell mass. Experiments in which chloramphenicol was used to prevent protein biosynthesis at various time intervals following photo-activation showed the carotenogenic system was fully competent 2 min after exposure of dark-grown *F. dehydrogenans* to light. It is possible, therefore, to study biosynthesis of C_{50} carotenoids in this bacterium without the intrusion of biochemical reactions relating to growth.

It is possible also to produce populations of *F. dehydrogenans* in which pre-phytoene carotenogenesis has occurred but post-phytoene events have been altered from the usual pattern (Tables I and II). Culture conditions which seem optimum for carotenogenesis result in decaprenoxanthin being essentially the only carotenoid formed. If a nutritional imbalance is created or diphenylamine added to the medium, other carotenoids appear. Such bacterial populations may be washed (phosphate buffer, pH 7) to remove whatever it is in the medium that causes unbalanced carotenogenesis, i.e. diphenylamine or ingredients of the culture medium. Subsequent incubation of washed cells in phosphate buffer results in decreases in amounts of hydrolycopenes and increases in decaprenoxanthin (Table V) (Weeks and Garner, 1967). Use of

Table V
Carotenogenesis in populations of *Flavobacterium dehydrogenans*

Carotenoid	Percent of total carotenoid: Medium AGY	Medium AGY plus DPA[a]	DPA-cells reincubated in buffer[b]
Phytoene	0	29	18
Phytofluene	0	5	0
ζ-Carotene	0	1	0
Neurosporene	1·4	1	0
Lycopene	0	2	0·3
P420 complex	0	23	10·0
Decaprenoxanthin	98·6	39	70·0
Total carotenoid (μg/l)	159	77	110
Total specific carotenoid (μg/mg cell-N)	4	4	7

[a] Growth medium (%) = asparagine (0·2), glucose (0·5), yeast extract (0·001), Mg^{2+}, phosphate buffer (pH 7·0). Incubation = 30°, 24 h, continuous shaking and illumination. Diphenylamine (DPA) concentration in medium, 6×10^{-5} M.

[b] DPA-inhibited cells washed free of DPA with phosphate buffer (pH 7·0), resuspended in phosphate buffer and then incubated an additional 24 h.

these experimental systems and information on structures of the carotenoids makes it possible to arrange the carotenoids of *F. dehydrogenans* to represent a biosynthetic scheme explaining at least the broad features of post-phytoene carotenogenesis which culminate in decaprenoxanthin.

Carotenogenesis in *F. dehydrogenans* is illustrated in Fig. 11 (Andrewes, 1969). The acyclic carotenoids are arranged as a Porter-Lincoln series (Porter and Lincoln, 1950). Andrewes (1969) has proposed that for *F. dehydrogenans* this series differs in two of its features: the all-*trans* isomers of phytoene and phytofluene are shown rather than the central mono-*cis* isomers which Porter and Anderson (1967) have suggested and the unsymmetrical isomer of ζ-carotene (Davies *et al.*, 1969) is included. The nonaprenocarotenoids have also been arranged into a series in which each higher homologue has two fewer hydrogen atoms. The latter compounds are a conspicuous feature in biosynthesis of decaprenoxanthin and, perhaps, in biosynthesis of C_{50} carotenoids generally.

The nonapreno-carotenoids could originate from asymmetrical ζ-carotene and successive dehydrogenations produce the two higher homologues. Alternately each nonapreno-carotenoid could originate from a different carotene: nonaprenoxanthin from asymmetrical ζ-carotene; 11′,12′-dehydrononaprenoxanthin from neurosporene; P452, which presumably is 7′,8′,11′,12′-dehydrononaprenoxanthin, from lycopene. There is at present no basis for deciding between the two origins nor does one preclude the other. Irrespective of the origin, the question may be asked whether the nonapreno-carotenoids are obligatory intermediates in biosynthesis of decapreno-carotenoids. Jensen (1970) has also found a nonapreno-carotenoid in *Corynebacterium poinsettiae*. Biochemical evidence from studies of *F. dehydrogenans* suggests the C_{45} carotenoids need not be obligatory intermediates in biosynthesis of C_{50} carotenoids but could represent an alternate, perhaps incidental and competing, pathway.

Data (Table V) show that when carotenogenesis in *F. dehydrogenans* was inhibited by diphenylamine (DPA) the carotenes and P420 complex accounted for 60 % of total carotenoid and decaprenoxanthin only 40 %. Non-inhibited systems, by contrast, contained almost no carotenoid except decaprenoxanthin (Table V, Medium AGY). When DPA inhibited populations were washed to remove the DPA and then reincubated in phosphate buffer, decaprenoxanthin nearly doubled in amount and there was concomitant decrease in the hydrolycopenes and P420 complex. There seems little doubt that the latter carotenoids are converted to decaprenoxanthin but any attempt to show stoichiometric relationship is made difficult by the increase in total carotenoid in the cell suspension reincubated after washing to remove DPA (Table V). *F. dehydrogenans* grown in nutrient broth showed the carotenes and P420 complex that were present after 12 h incubation had disappeared after 24 h and the disappearance of these minor carotenoids coincided with an increase in decaprenoxanthin (Table II, nutrient broth, maximum aeration). *F.*

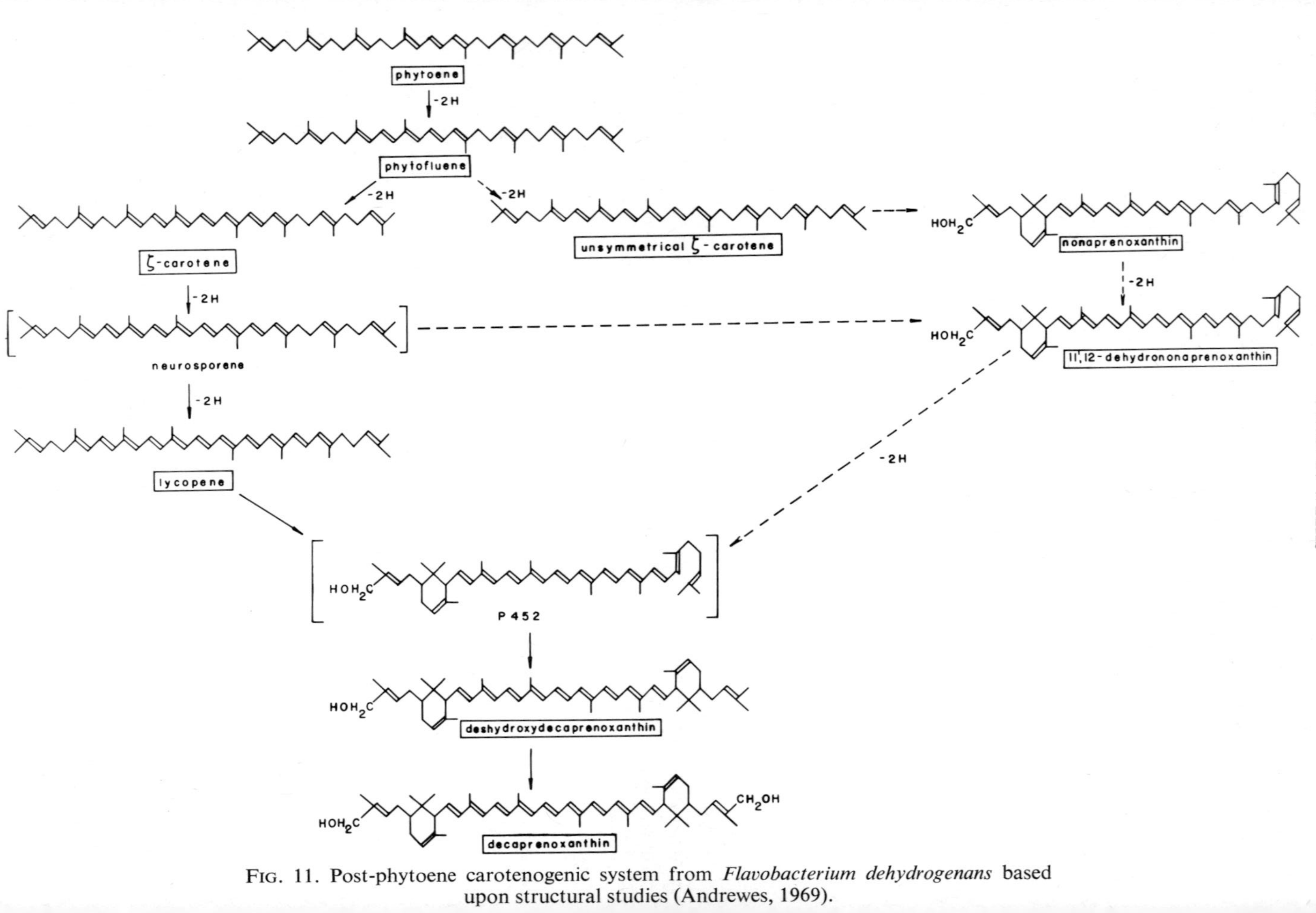

FIG. 11. Post-phytoene carotenogenic system from *Flavobacterium dehydrogenans* based upon structural studies (Andrewes, 1969).

dehydrogenans, however, may be grown so that the carotenes and P420 complex do not appear (Table I). The presence of the minor carotenoids when culture conditions are modified could be the result of metabolic feedback regulation slowing post-phytoene carotenogenesis and making apparent intermediate reactions which otherwise are inconspicuous. Alternately, inhibition of a biochemical mainstream could allow fortuitous reactions and cause compounds such as the nonaprenoxanthins to be formed. These then would be either end-products or, as seems more likely in *F. dehydrogenans*, represent an alternate biosynthetic route to decaprenoxanthin. Such fortuitous reactions would not be improbable if the different precursor molecules had a common structural feature prerequisite to formation of the cyclic end group which the nonaprenoxanthins and decaprenoxanthins of *F. dehydrogenans* share. Cyclization, however, seems secondary to the principal reaction which characterizes biosynthesis of decapreno carotenoids, i.e. the addition of an isopentenyl side chain at positions 2,2′ of the C_{40} carotenoid precursor.

The two general mechanisms which have been proposed by Porter and Anderson (1967) to explain formation of cyclic end groups of carotenoids are shown in Fig. 12. These mechanisms are based upon investigations by Goodwin and his colleagues (Goodwin, 1965b; Goodwin and Williams, 1965). In each illustration cyclization is preceded by nucleophilic attack at position 2 and if the nucleophile is a proton the result is a β-ring or an α-ring depending whether a proton is removed from position 6 or 4. If an isopentenyl group replaces the proton the characteristic nonapreno or decapreno carotenoid would result. It has been suggested by Andrewes (1969) that cyclization of an acyclic carotenoid requires it to have a 1,5,7-triene terminus (Fig. 12) and it is not impossible that a 1,3,5,7-tetraene terminus (Fig. 13, IIIa) would not cyclize. The 1,5,7-triene terminal structure is common to one end of asymmetrical ζ-carotene and neurosporene and to both ends of lycopene (Fig. 11). This would require lycopene to be the C_{40} acyclic precursor of decaprenoxanthin possibly with P452 being a transient, monocyclic intermediate. On the same basis the nonapreno carotenoids of *F. dehydrogenans* would originate separately from asymmetrical ζ-carotene and neurosporene and could not become bicyclic, i.e. additional dehydrogenations to form the second 1,5,7-triene structure would be necessary. This scheme would place the nonapreno carotenoids of *F. dehydrogenans*, except for P452, in a minor biosynthetic position and justify the concept that their origin is from fortuitous reactions and because the enzyme system catalysing cyclization can promote the reaction in several acyclic precursors with a common structural feature.

Extension of the scheme proposed to explain carotenogenesis in *F. dehydrogenans* to decapreno carotenoids generally, is possible if addition of the isopentenyl side chains and cyclization are accepted as independent biochemical reactions. Stereospecific removal of a proton from position 4 after nucleophilic addition of the isopentenyl side chain (Fig. 12) would produce a decapreno carotenoid such as decaprenoxanthin (Fig. 13; Ia,b); from position

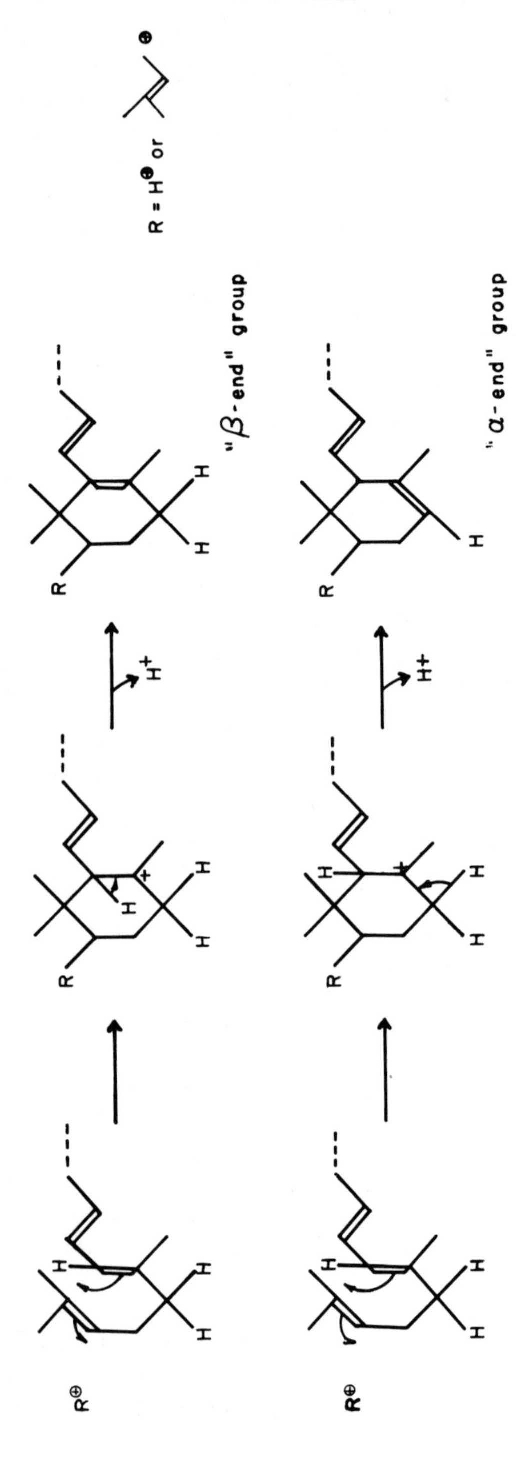

FIG. 12. Cyclization mechanisms for carotenoids (Porter and Anderson, 1967).

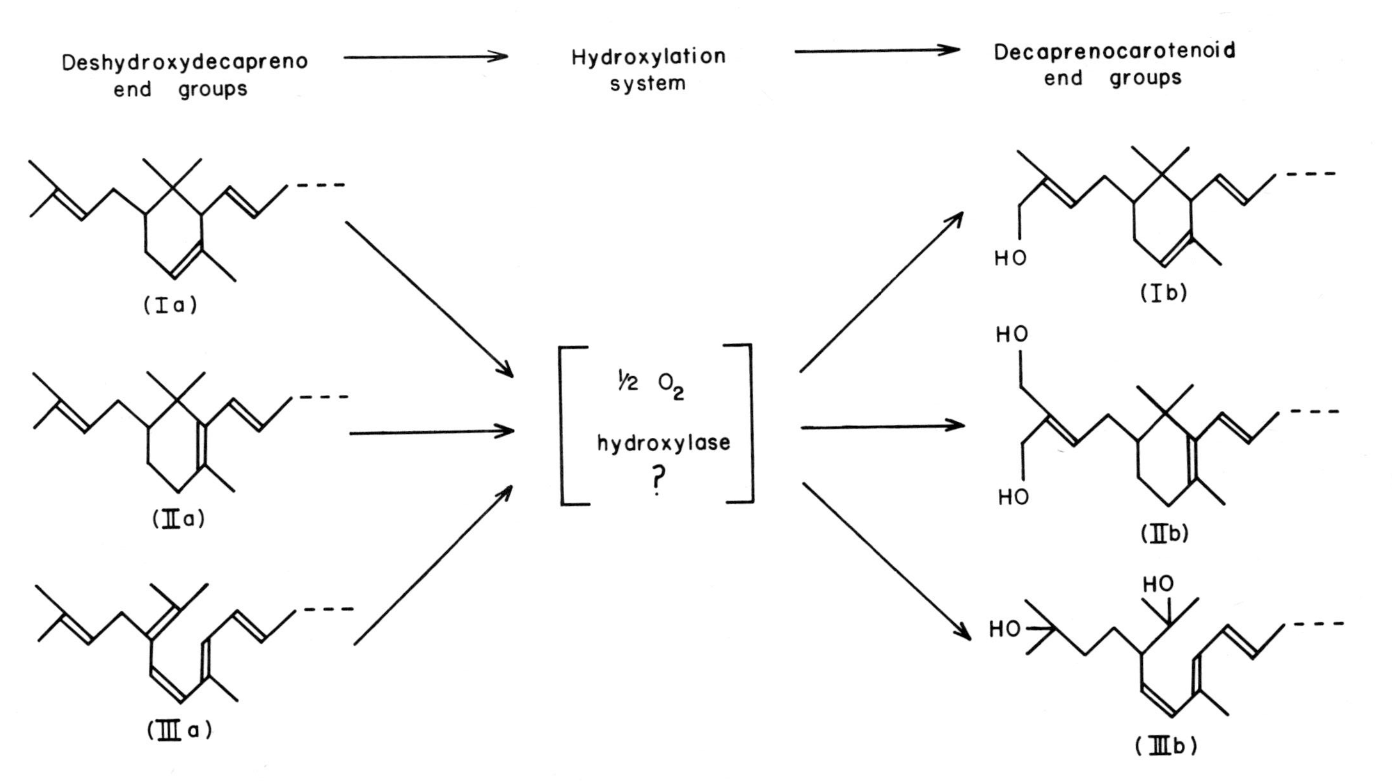

FIG. 13. Possible origin of characteristic end groups of decapreno carotenoids. Ib = decaprenoxanthin from *Flavobacterium dehydrogenans*; IIb = C.p. 450 from *Corynebacterium poinsettiae*; IIIb = bacterioruberin from *Halobacterium* species.

6, a carotenoid such as C.p. 450 (Fig. 13; IIa,b) and the 1,3,5,7-tetraene terminus, exemplified by 3,4,3′,4′-dehydrolycopene, would restrict biosynthesis to side chain addition (Fig. 13; IIIa,b) and result in a carotenoid such as bacterioruberin (Fig. 1).

Hydroxyl group formation is regarded as the final biosynthetic event in formation of both nonapreno and decapreno carotenoids (Fig. 13). Primary hydroxyl groups on one or both of the isopentenyl side chains are known for the cyclic decapreno carotenoids (Fig. 1). The acyclic examples, by contrast, are characterized by tertiary hydroxyl groups which may be on the isopentenyl side chains and at the 1-position of the molecule (Fig. 13, IIIb) or, for bis-anhydro-bacterioruberin, the 1-position only (Liaaen-Jensen, 1969). It seems reasonable to suppose that the catalyst is a mixed function oxidase (hydroxyl-ase, Massart and Vercautern, 1959) but this supposition is based upon analogy and requires investigation for proof. The hydroxyl groups occur in positions on the molecule which suggest the reaction mechanism requires the isopropyli-dene group for insertion of oxygen.

ACKNOWLEDGEMENTS

The contribution of A. G. Andrewes, R. A. Berry, Fayek Saleh, R. J. Garner and James Bryan, New Mexico State University, and B. O. Brown, Queen Mary College, University of London, to an understanding of carotenoid biosynthesis in *Flavobacterium dehydrogenans* is gratefully acknowledged.

REFERENCES

Andrewes, A. G. (1969). "Carotenoids of *Flavobacterium dehydrogenans* Arnaudi", Ph.D. Diss. New Mexico State University, Las Cruces, New Mexico, U.S.A.

Davis, J. B., Jackman, J. M., Siddons, P. T. and Weedon, B. C. L. (1966). *J. chem. Soc.* (C), 2154.

Davies, B. H., Holmes, E. A., Loeber, D. E., Taube, T. P. and Weedon, B. C. L. (1969). *J. chem. Soc.* (C), 1266.

Enzell, C. (1969). *Pure appl. Chem.* **20**, 479.

Ferrari, A. and Zannini, E. (1942). *Ann. Microbiol. Enzymol.* **8**, 138.

Goodwin, T. W. (1965a). *In* "Chemistry and Biochemistry of Plant Pigments" (T. W. Goodwin, ed.), Academic Press, London and New York.

Goodwin, T. W. (1965b). *Biochem. J.* **96**, 2P.

Goodwin, T. W. and Williams, R. J. H. (1965). *Biochem. J.* **94**, 5C.

Hester, D. J. and Weeks, O. B. (1969). *Bacteriol. Proc.* **19** (G5).

Hill, L. R. (1966). *J. gen. Microbiol.* **44**, 419.

Hodgkiss, W., Liston, J., Goodwin, T. W. and Jamikorn, M. (1954). *J. gen. Microbiol.* **11**, 438.

Jungalwala, F. B. and Porter, J. W. (1965). *Archs Biochem. Biophys.* **110**, 291.

Liaaen-Jensen, S. (1969). *Pure appl. Chem.* **20**, 421.

Liaaen-Jensen, S. and Weeks, O. B. (1966). *Norwegian J. Chem. Mining, Metall.* **N26**, 130.

Liaaen-Jensen, S., Weeks, O. B., Strang, R. H. C. and Thirkell, D. (1967). *Nature, Lond.* 214.

Liaaen-Jensen, S., Hertzberg, S., Weeks, O. B. and Schwieter, U. (1968). *Acta Chem. Scand.* **22**, 1171.
Massart, L. and Vercautern, R. (1959). *A. Rev. Biochem.* **28**, 527.
Porter, J. W. and Lincoln, R. W. (1950). *Archs Biochem. Biophys.* **27**, 380.
Porter, J. W. and Anderson, D. G. (1967). *A. Rev. Pl. Physiol.* **18**, 197.
Schwieter, U. and Liaaen-Jensen, S. (1969). *Acta Chem. Scand.* **23**, 1057.
Starr, M. P. and Saperstein, S. (1953). *Archs Biochem. Biophys.* **43**, 157.
Weeks, O. B. (1955). *J. Bacteriol.* **69**, 649.
Weeks, O. B. and Andrewes, A. G. (1968). *Bacteriol. Proc.* **23** (G29).
Weeks, O. B. and Andrewes, A. G. (1970). *Archs Biochem. Biophys.* **937**, 284.
Weeks, O. B. and Garner, R. J. (1967). *Archs Biochem. Biophys.* **121**, 35.
Weeks, O. B., Andrewes, A. G., Brown, B. O. and Weedon, B. C. L. (1969). *Nature, Lond.* **224**, 879.
Welch, L. J. (1965). "Physiological and Nutritional Factors Influencing Carotenogenesis in *Flavobacterium dehydrogenans*". M.S. Thesis, University of Idaho, Moscow, Idaho.

CHAPTER 11

Algal Carotenoids

T. W. GOODWIN

Biochemistry Department, The University of Liverpool, Liverpool, England

I. Nature and Distribution

A. Chlorophyta

1. General

Under normal circumstances carotenoids are found only in the chloroplasts of the Chlorophyta, but as will be discussed later they can accumulate outside the chloroplast under certain adverse nutritional conditions. The general distribution of carotenoids in the unicellular green algae and the colonial forms, except the Siphonales and Prasinophyceae, is similar to that observed in the green leaves of higher plants; the main components are β-carotene (1), lutein (2), violaxanthin (3) and neoxanthin (4). Small amounts of α-carotene (5) are also occasionally encountered and the usual $\alpha:\beta$ ratio is about 1:20; however, one green alga, *Chlamydomonas agloeformis*, contains "much α-carotene" (Strain, 1958). The xanthophyll:carotene ratio can vary from about 1:3 as in *C. reinhardi* (Sagar and Zalokar, 1958) to about 5·5:1 as in *Nannochloris atomus* (Jeffrey, 1961). Various *cis*-isomers of neoxanthin are reported in *Chlorella ellipsoidea* (Tsukida *et al.*, 1968). The various green algae which have been examined and found to fit in with this generalization are listed in Table I.

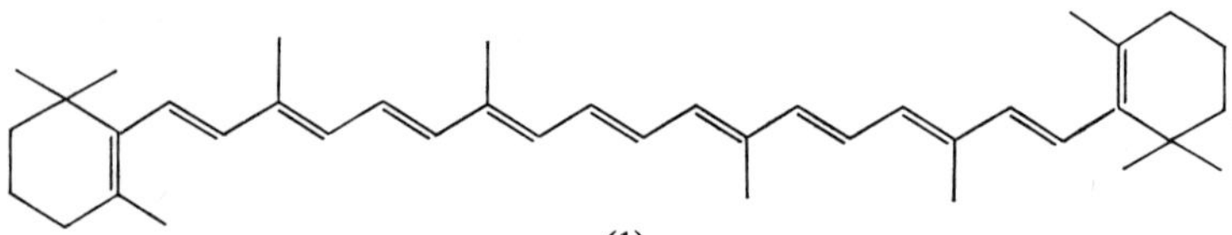

(1)

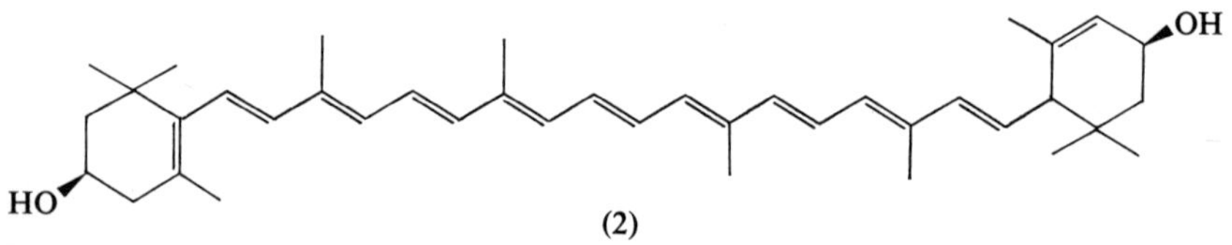

HO OH

(2)

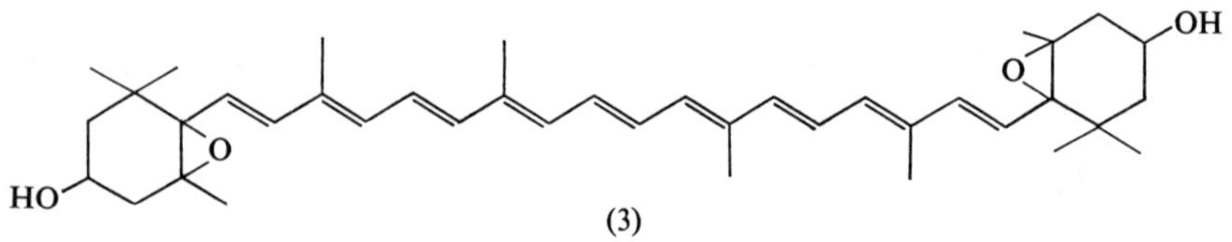

HO O O OH

(3)

HO OH O OH

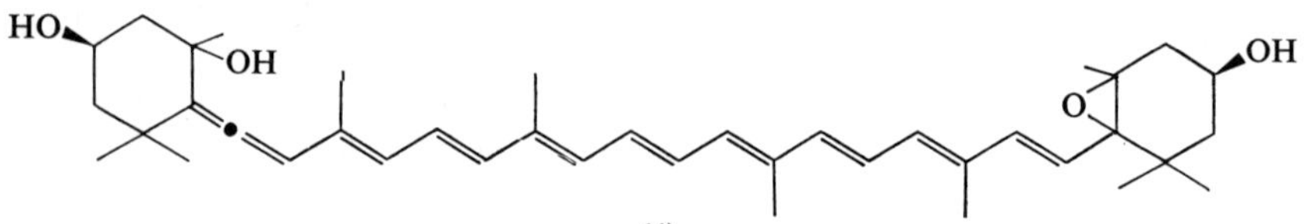

(4)

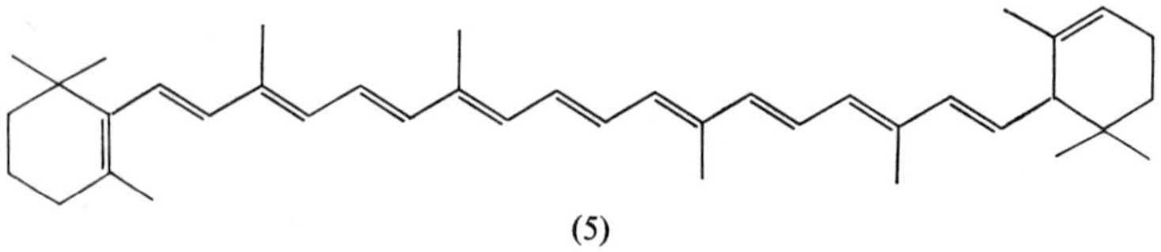

(5)

TABLE I

Green algae in which the chloroplast pigments are the same as those in higher plants. (Strain, 1958, 1966, unless otherwise stated; species marked with an asterisk contain α- as well as β-carotene)

CHLOROCOCCALES

CHLOROCOCCACEAE
Chlorococcum sp.*

COELASTRACEAE
*Coelastrum proboscideum**

OOCYSTACEAE
*Ankistrodesmus braunii**[1]
*A. falcatus**
*Chlorella pyrenoidosa**[a]
*C. vulgaris**[2]

PROTOSIPHONACEAE
*Protosiphon botryoides**[3]

SCENEDESMACEAE
*Scenedesmus bifulgatus**
*S. brasiliensis**
*S. obliquus**[a]

CLADOPHORALES

CLADOPHORACEAE
*Chaetomorpha aerea**[b]
*C. antennina**
*Cladophora crispata**
*C. fascicularis**
*C. glomerata**
*C. graminea**
*C. membranacea**
C. sp.*
*C. trichotoma**[a]
Rhizoclonium implexuum
*Spongomorpha coalita**

DASYCLADALES

DASYCLADACEAE
Acetabularia clavata
A. crenulatum
A. mobii
Bornetella sphaerica
Neomeris annulata

SIPHONOCLADALES

ANADYOMENACEAE
Microdictyon setchellianum

BOODLEACEAE
*Boodlea kaeneana**
Struvea sp.*

VALONIACEAE
Cladophoropsis herpestica[5]
*Dictyosphaeria cavernosa**
D. favulosa
D. versluysii
*Valonia fastigiata**
*Valoniopsis pachynema**[5]

TETRASPORALES

COCCOMYXACEAE
*Coccomyxa elongata**[10]
*C. simplex**
*C. viridis**

ULOTRICHALES

CHAETOPHORACEAE
*Chaetophora incrassata**
*Draparnaldia glomerata**
Draparnaldia sp.*
Stigeoclonium sp.*

PROTOCOCCACEAE
Protococcus sp.*

ULOTRICHACEAE
*Hormidium flaccidum**
Hormidium sp.*
*Stichococcus subtilis**
Ulothrix sp.*

UVALES

Enteromorpha (? *clathrata*)
E. compressa
E. (? *intestinalis*)
Enteromorpha sp.*
E. tubulosa
Monostroma sp.*
*Ulva fasciata**
*U. lactuca**[4,5]
U. latissima[4b]
*U. linza**
*U. reticulata**
Ulva sp.

VOLVOCALES

CHLAMYDOMONADACEAE
*Chlamydomonas agloeformis**
C. reinhardii[2,6,c]

POLYBLEPHARIDACEAE
*Dunaliella salina**
D. primolecta[8]
*D. tertiolecta**[5,7]

ZYGNEMATALES

MESOTAENIACEAE
*Mestanium caldariorum**

TABLE I—*continued*

ZYGNEMATACEAE	*Cosmarium debaryi*
Spirogyra sp.*	*Euastrum verrucosum*
Zygnema pectinatum[d]	*Micrasterias americana*
DESMIDIACEAE[9]	*M. denticulata*
Closterium acerosum	*M. papillifera*
C. ehrenbergii	*M. pinnatifida*
C. littorale	*M. truncata*

References: [1] Dersch (1960); [2] Krinsky and Levine (1964); [3] Kleinig and Czygan (1969); [4] de Nicola and Furnari (1954); [5] de Nicola (1961); [6] Sager and Zalokar (1958); [7] Bunt (1964); [8] Riley and Wilson (1965); [9] Herrmann (1968); [10] Whittle and Castleton (1969).

[a] Also contains loroxanthin (6).

[b] In one investigation neoxanthin was not detected, but flavochrome (7) and auroxanthin (8) were found (de Nicola, 1961); this requires confirmation.

[c] Luteoxanthin, trollein and P-460 [possibly a *cis* isomer of γ-carotene (9)], also present (Krinsky and Levine (1964).

[d] Fucoxanthin (p. 328) was reported in a very early investigation (Heilbron, 1942); this was probably due to contamination of the specimen with diatoms.

Certain *Chlorella* species contain in addition, small amounts of a unique xanthophyll (loroxanthin, 6) in which one of the in-chain methyl groups has been oxidized to a hydroxymethyl group (Aitzetmüller *et al.*, 1969). The known distribution of loroxanthin is given in Table II.

It is interesting that the free-living *Trebouxia humicula* and the phycobiont *T. decolorans*, which is the symbiont of the lichen *Xanthoria parietina*, both contain the same "green plant" carotenoids (de Nicola and Tomaselli, 1961; de Nicola and di Benedetto, 1962).

TABLE II
Green algae in which loroxanthin has been detected
(Aitzetmüller *et al.*, 1969)

Chlorella vulgaris	*Scenedesmus obliquus*
Cladophora trichotomata	*Ulva rigida*
C. ovoidea	

2. *Siphonales*

The Siphonales are unique in synthesizing siphonaxanthin, a carotenoid not found in any other algae (Strain, 1949, 1951, 1958, 1966). In addition siphonein, an ester of siphonaxanthin also accumulates. Usually xanthophylls found in photosynthetic tissues are unesterified. The structure of siphonaxanthin is (10) (Walton *et al.*, 1970). The esterifying fatty acid in siphonein from *Caulerpa prolifera* is lauric acid (Kleinig and Egger, 1967a); in *Codium*

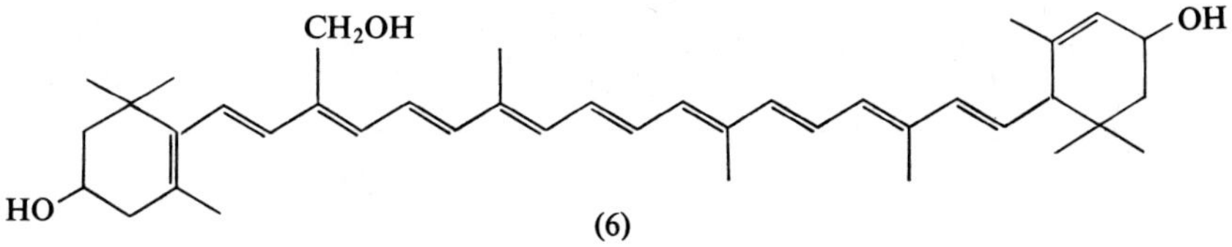

CH_2OH OH HO

(6)

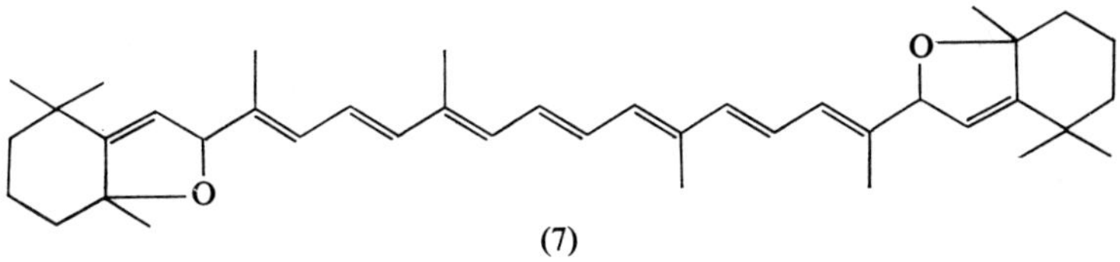

O O

(7)

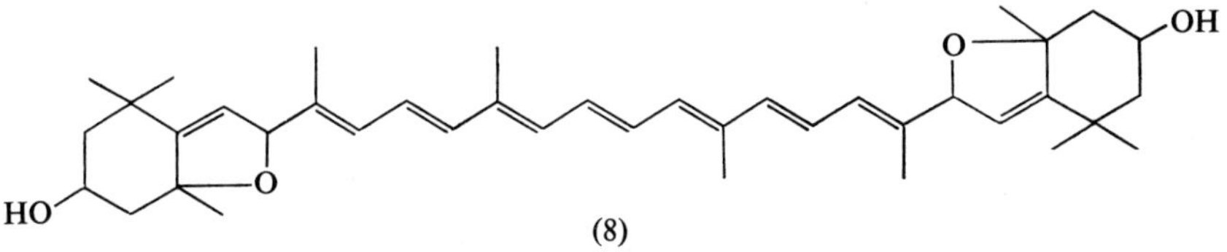

O OH HO O

(8)

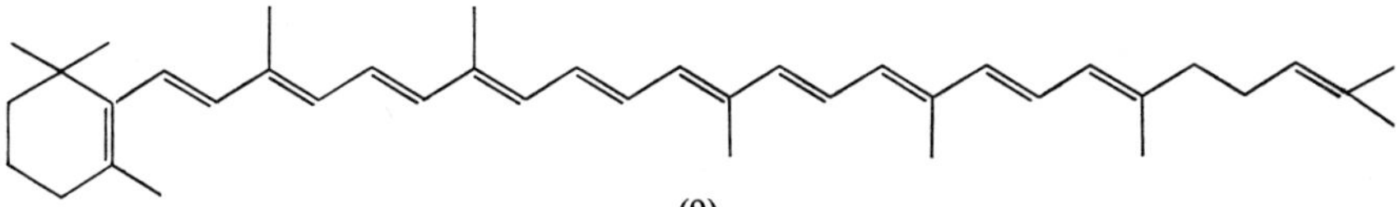

(9)

CH_2OH OH HO O

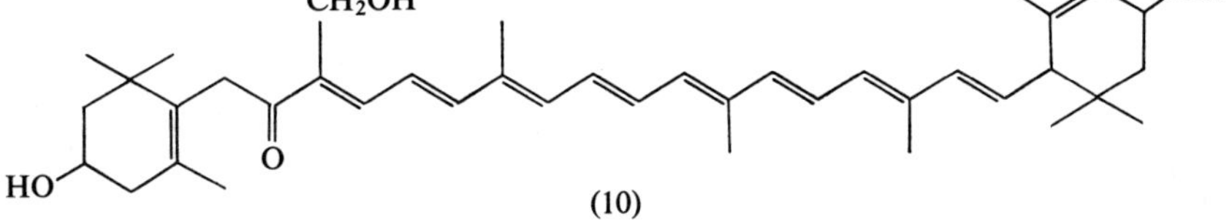

(10)

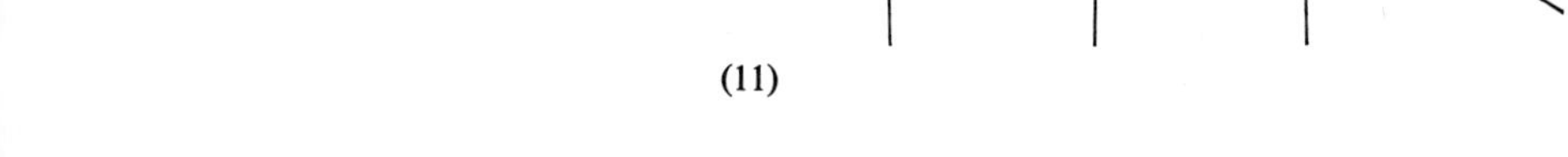

(11)

TABLE III
The Siphonalean green algae which contain Siphonaxanthin (from Strain, 1958, 1965, 1966)

BRYOPSIDACEAE	*C. duthieae*
Bryopsis spp.	*C. fragile*
B. corticulans[a]	*C. lucasii*
B. hypnoides	*C. muelleri*
CAULERPACEAE	*C. spongiosum*
Caulerpa cupressoides	*Halimeda discoidea*
C. distichophylla	*Halimeda opuntia*
C. filiformis	*Halimeda tuna*
C. lentillifera	*Penicillus capitatus*
C. prolifera	*Udotea flabellum*
C. racemosa	DICHOTOMOSPHONACEAE
C. serrulata	*Dichotomosiphon tuberosus*
C. sertularioides	HALICYSTIDACEAE
CODIACEAE	*Derbesia lamourouxii*
Chlorodesmis comosa	*D. vaucheriaeformis*
Codium coronatum	*Halicystis ovalis*

[a] Also contains the rare ϵ-carotene (30).

fragile it is certainly not lauric acid (Walton *et al.*, 1970) and it remains to be seen what further variations in the fatty acid component are encountered as additional species are examined. The algae known to contain siphonaxanthin and siphonein are listed in Table III. They all contain the other xanthophylls characteristic of the green algae (Strain, 1965) but most are characteristically different in containing more α-carotene than β-carotene; an exception is *Dichotomosiphon tuberosus*, collected from Lake Michigan (Strain, 1958), which is the only fresh water species so far examined. On the other hand one

TABLE IV
The Prasinophyceae which contain normal chlorophycean carotenoids (Ricketts, 1970)

Haematococcus sp.	*P. subcordiformis*
Heteromastix rotunda	*P. striata*
Heteromastix spp.	*Prasinocladus lubricus*
Mesostigma viride	*P. marinus*
Monomastix minuta	*Prasinocladus* sp.
Pedinomonas minor[a]	*Pyramimonas obovata*
Platymonas chuii	*P. urceolata*
P. tetrathele	*P. grossii*
	Spermatozopsis exsultans

[a] Also synthesizes γ-carotene and lycopene in significant amounts (Ricketts, 1967).

marine species *Caulerpa filiformis* contained no siphonein or siphonaxanthin although α-carotene preponderated over β-carotene (Strain, 1965).

3. *Charophyceae*

Chara fragilis contains β-carotene and the usual chlorophycean carotenoids; in addition it produces small amounts of γ-carotene (16) and lycopene (11) but no α-carotene (Strain, 1958).

4. *Prasinophyceae*

A number of these flagellates contain the usual green algal carotenoids (Table IV) (Ricketts, 1970), but some are unique. *Micromonas pusilla*, *M. squamata* and *Nephroselmis gilva* produce a keto-carotenoid micronone, which has not yet been fully characterized (Ricketts, 1966, 1967). Others within this class which are significantly different from normal are *Asteromonas propulsa* (usual pigments plus siphonein), and *Heteromastix longifilis*, *Pyramimomonas amylifera*, *Pachysphaera* sp., and *Pterosperma* sp. (no lutein but two unidentified pigments K_1 and K_2 (Ricketts, 1970).

5. *Mutants*

A number of X-ray mutants of *Chlorella vulgaris* have been described in which chlorophyll synthesis is almost completely blocked and carotenoid synthesis interrupted (Claes, 1954, 1956); similar mutants have been obtained from *C. pyrenoidosa* (Kessler and Czygan, 1966). The carotenoids synthesized by these mutants are listed in Table V (see also p. 348). A carotenoid-deficient

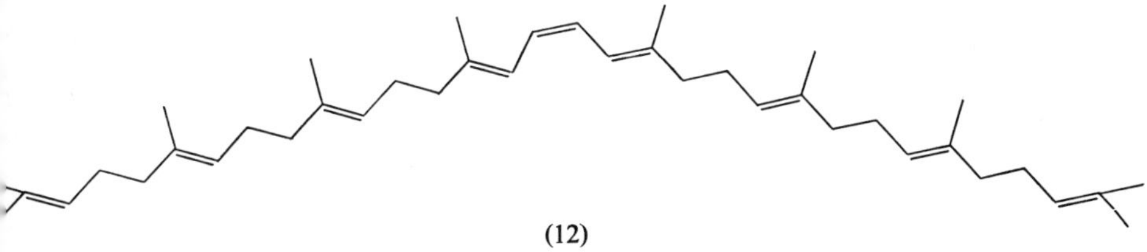

(12)

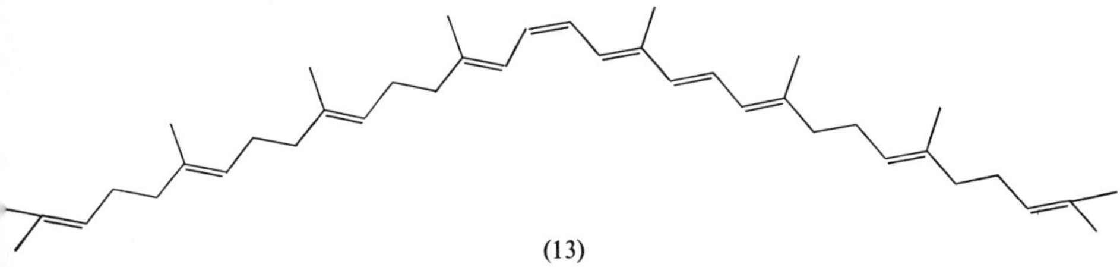

(13)

TABLE V
Polyenes synthesized by some mutant strains of *Chlorella vulgaris* (Claes, 1954, 1956) and *C. pyrenoidosa* (Kessler and Czygan, 1966)

Mutant	Polyenes synthesized
Chlorella vulgaris	
5/871[a]	Phytoene (12) only
5/518[a]	Phytoene, phytofluene (13), ζ-carotene (14) (no xanthophylls)
9a[a]	Phytoene, phytofluene, ζ-carotene, β-zeacarotene (15); no β-carotene but normal xanthophylls
5/520	*In light*: normal synthesis
	In darkness: phytoene, phytofluene, ζ-carotene, proneurosporene, prolycopene; no β-carotene or xanthophylls
C. pyrenoidosa	
G34	Phytoene only
G41	Usual pigments
G44	Phytoene, ζ-carotene; mixture of 9 unidentified xanthophylls

[a] A combination of light and oxygen is lethal to this mutant.

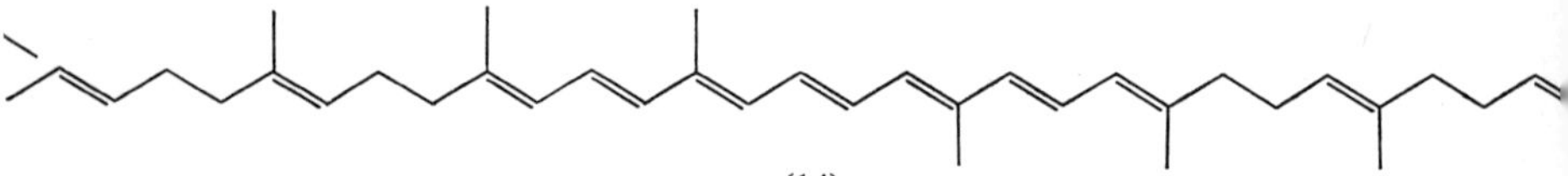

(14)

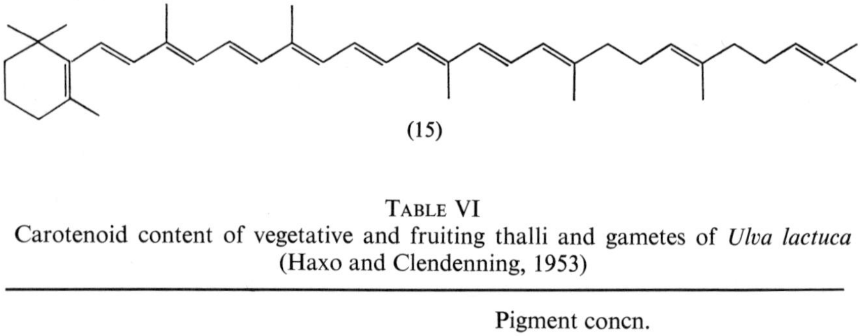

(15)

TABLE VI
Carotenoid content of vegetative and fruiting thalli and gametes of *Ulva lactuca* (Haxo and Clendenning, 1953)

Part	Pigment concn. (mg/g dry wt) Carotenes[a]	Xanthophylls
Vegetative thallus	0·34	0·54
Yellow fruiting thallus ♂	1·35	0·68
Green fruiting thallus ♀	1·49	0·64
Gametes ♂	8·20	4·68
Gametes ♀	4·70	4·20

[a] In vegetative thalli the carotene fraction, being chloroplastidic, is almost entirely β-carotene; in other cases γ-carotene predominates.

mutant of *Chlamydomonas reinhardii* synthesizes only carotenes, mainly β-carotene, at a concentration 200–500 times less than that of the wild-type (Sager and Zalokar, 1958).

6. *Extra-plastidic Accumulation*

The carotenoid content of the fruiting areas of certain colonial green algae such as *Ulva lobata* (Strain, 1951) and *U. lactuca* (Haxo and Clendenning, 1953) is increased owing to the increased carotene synthesis in these areas. The concentration is even higher (4–6 times) in the gametes themselves (Table VI). The major increase is in γ-carotene (43% of the total), and an unidentified carotene (22% of the total). A similar accumulation of γ-carotene and lycopene occurs in the antherida of *Chara ceratophylla* and *Nitella syncarpa* (Karrer *et al.*, 1943).

Under unfavourable cultural conditions, such as nitrogen deficiency many green algae become yellow or red, owing to the formation of large amounts of carotenoids outside the chloroplasts. The pigments concerned are generally β-carotene and its ketonic derivatives echinenone (4-oxo-β-carotene, 16), astaxanthin (3,3′-dihydroxy-4,4′-dioxo-β-carotene, 17) and canthaxanthin (3,3′-dioxo-β-carotene, 18). In the old literature before the complexity of the

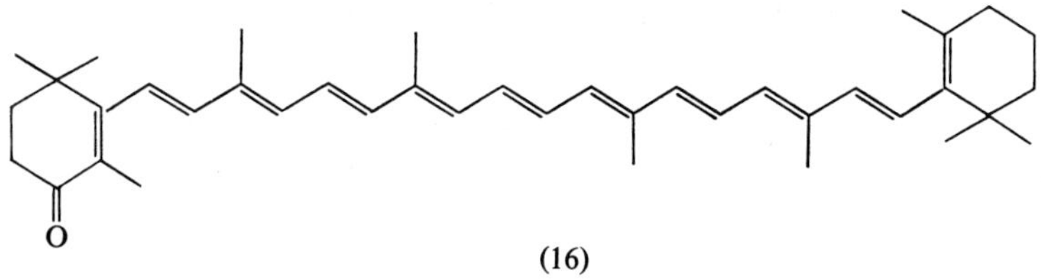

O

(16)

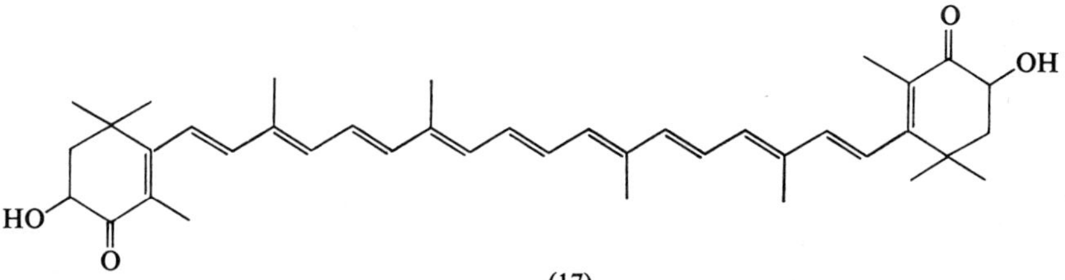

O

OH

HO

O

(17)

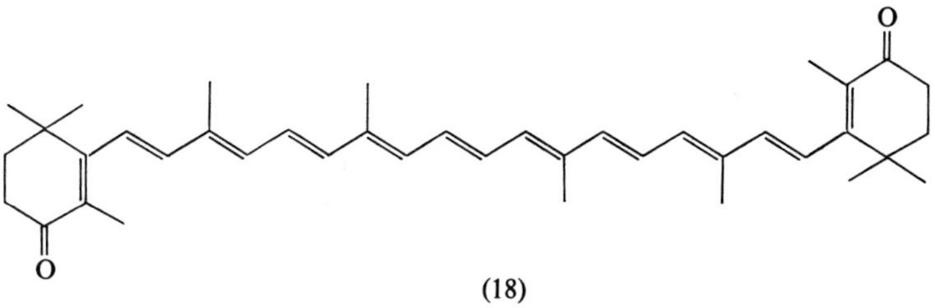

O

O

(18)

problem was realized, the "pigment" was named haematochrome. Today these pigments are sometimes called "secondary carotenoids", which is also not a happy choice. The additional pigments are found outside the chloroplast, for example in the aplanospores in *Haematococcus* sp. (Czygan and Kessler, 1967) and in intracytoplasmic deposits with no limiting membrane in *Protosiphon botryoides* (Fig. 1) (Berkaloff, 1967). The term extra-plastidic carotenoids might be an appropriate term to distinguish these pigments from those regularly found associated only with chloroplasts. A list of the algae in which synthesis of these pigments can be induced is given in Table VII. In *Ankistrodesmus braunii* the freeze etching technique indicates that the pigments accumulate in lipid vacuoles (Mayer and Czygan, 1969), whilst in

TABLE VII
Green algae which are known to produce extra-plastidic carotenoids

Algae	Pigments	Reference
Protosiphon botryoides[b] (5 strains)	1, 2, 3,[a] 4,[a] 5,[a] 6, 7	Kleinig and Czygan (1969)
Haematococcus pluvialis	1,[a] 7, 8	Goodwin and Jamikorn (1954); Droop (1955); Sestak and Baslerova (1963); Tischer (1941); Czygan (1966); Czygan (1968)
Nannochloris atomus	7	Jeffrey (1961)
Chlorella pyrenoidosa (= *fusca*)	1, 2	Czygan (1964)
Dictyococcus cinnabarinus	1, 2, 9[f]	Dentice di Accadia *et al.* (1966, 1968); Kessler and Czygan (1965); Czygan (1968)
Crucigenia apiculata	1[a]	Czygan (1964)
Scenedesmus brasiliensis	1	Czygan (1964)
Scenedesmus spp.[d]	1, 2, 7	Czygan (1964); Sestak and Baslerova (1963); Kessler and Czygan (1967)
Chlorella fusca v. *rubescens*[c]	1, 2, 7	Kessler *et al.* (1968)
Ankistrodesmus spp.[e]	1, 2, 7	Kessler and Czygan (1967)
Chlorella zofingiensis	1, 2, 7	Dersch (1960); Sestak and Baslerova (1963); Kessler and Czygan (1965)
Chlorococcum wimmeri	1, 2, 7, 8	Czygan and Kessler (1967); Czygan (1968)
Brachiomonas simplex	7	(see Goodwin, 1952)
Sphaeroplea		Kleinig (1967)
Acetabularia mediterranea		Richter (1958); Kleinig and Egger (1967)

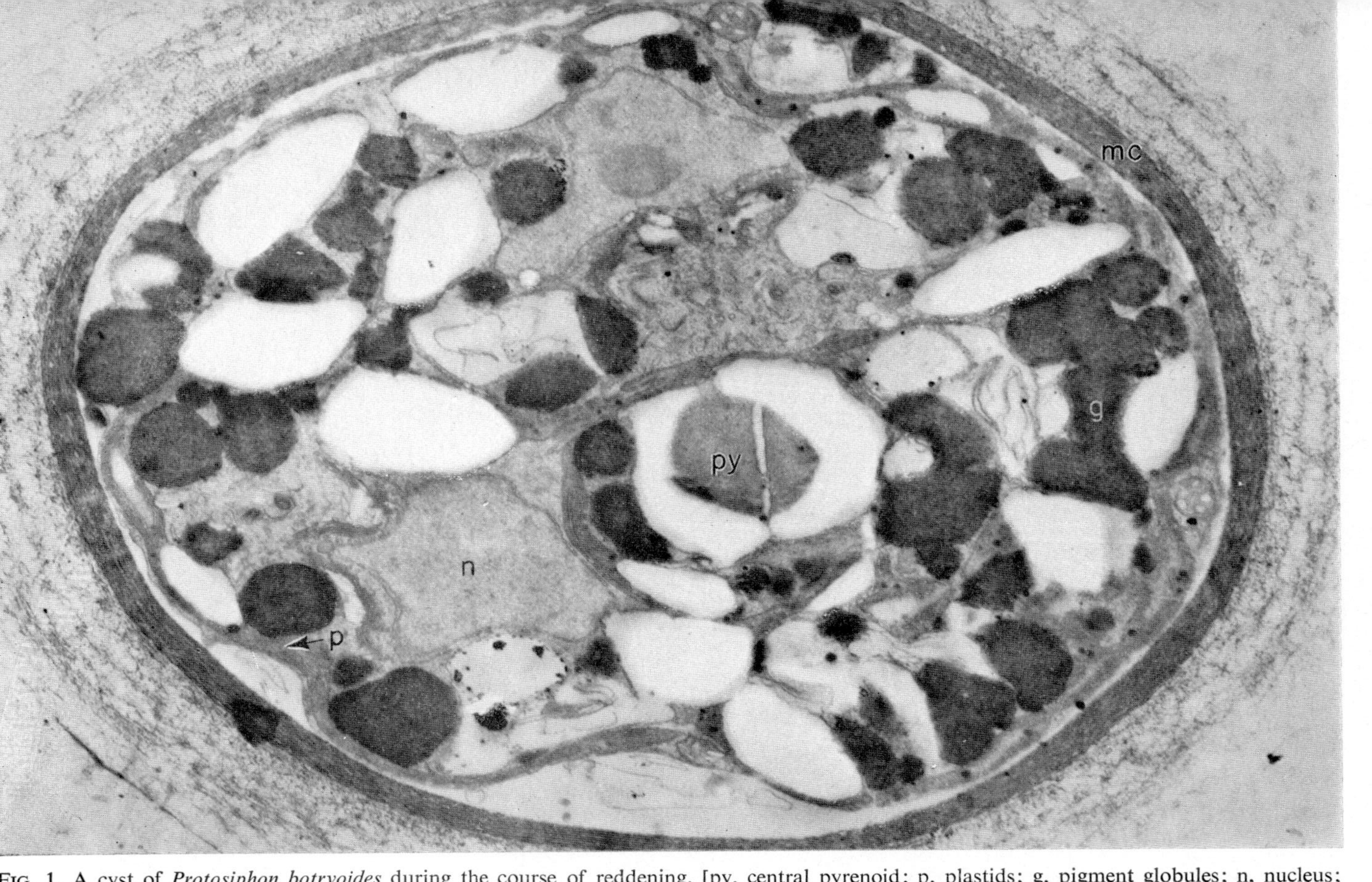

FIG. 1. A cyst of *Protosiphon botryoides* during the course of reddening. [py, central pyrenoid; p, plastids; g, pigment globules; n, nucleus; mc, pectocellulose membrane; magnification × 20,000.] (Electron micrograph kindly supplied by Dr C. Berkaloff.) Note the accumulation of carotenoids as intracytoplasmic deposits with no limiting membrane.

TABLE VII—*continued*

Algae	Pigments	Reference
Chlamydomonas nivalis	1, (2), 7	Czygan (1968); Villela (1966); Bold and Parker (1962); Sestak and Baslerova (1963)
Chromochloris cinnabarina		Sestak and Baslerova (1963)
Hydrodictyon reticulatum	1, 7	Dersch (1960); Czygan (1968)
Spongiochloris typica		McLean (1967)
Sphaeroplea		Kleinig (1967)
Haematococcus droebakensis	1, 2, 7	Czygan (1968)
Chlorococcum infusionum	1, 2, 7	Czygan (1968)
C. multinucleatum	1, 2, 7	Czygan (1968)
C. oleofaciens	1, (2), 7	Czygan (1968)
Coelastrum proposcideum v. *dilatatum*	1, 2, (7)	Czygan (1968)
Crucigenia apiculata	1, 2, 7	Czygan (1968)
Trentepohlia aurea	10	Czygan and Kalb (1966)
Scotiella sp.	1, 2, 7	Czygan (1968)

1. Echinenone (16); 2. Canthaxanthin (18); 3. 4′-Hydroxyechinenone (19); 4. 4-Hydroxy, 3′,4′-diketo-β-carotene (20); 5. 3-Hydroxy-4,4′-diketo-β-carotene (21); 6. 3-Hydroxy-3′,4,4′-triketo-β-carotene (22); 7. Astaxanthin (17); 8. 3,4,4′-trihydroxy-β-carotene (23); 9. 3,4-diketo-β-carotene (24); 10. β-Carotene (1).

[a] Not present in all strains.

[b] The hydroxy pigments are esterified.

[c] Previously *Halochlorella rubescens*.

[d] Present in 27 strains.

[e] Present in 25 strains.

[f] Tentative identification.

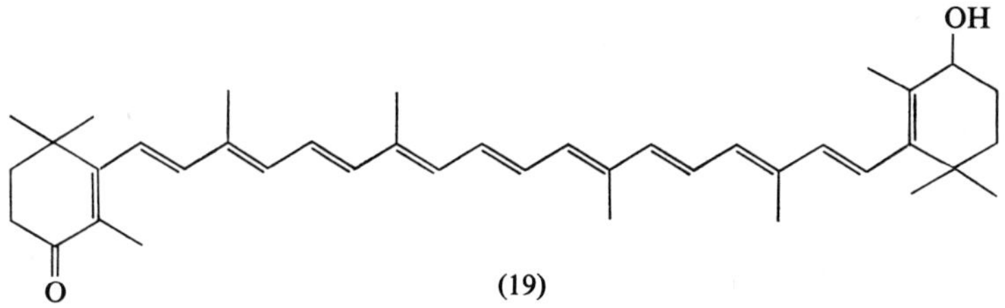

OH

O

(19)

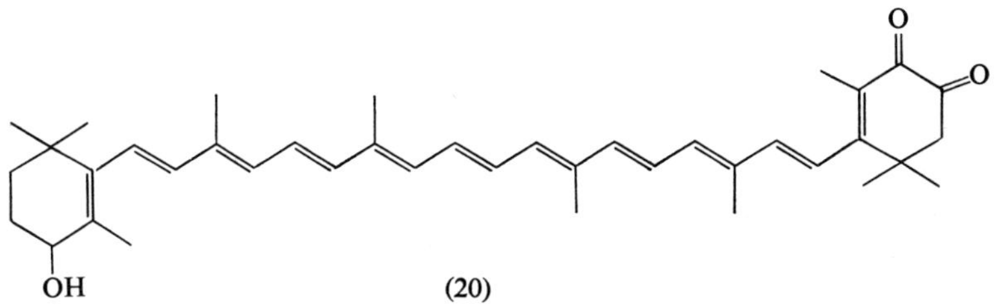

O

O

OH

(20)

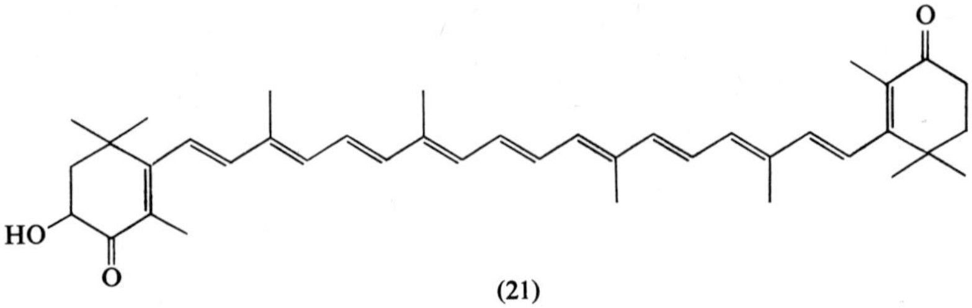

HO O O

(21)

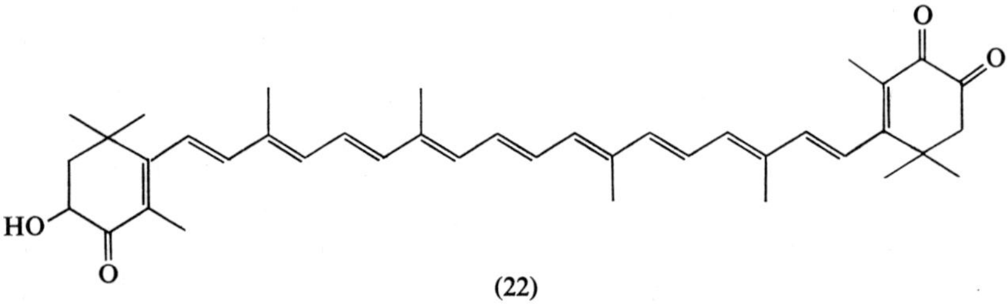

HO O O O

(22)

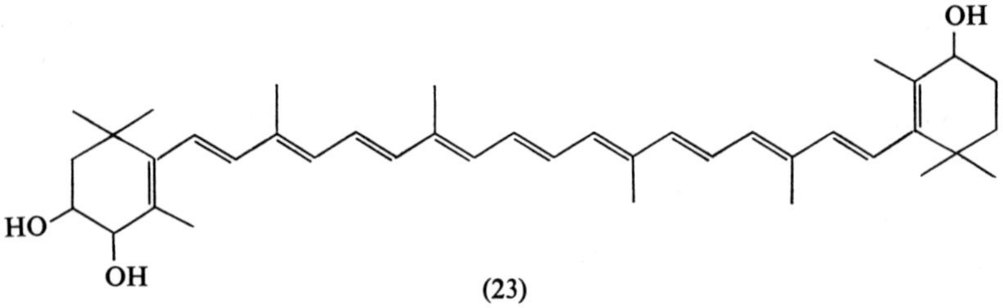

HO OH OH

(23)

O O

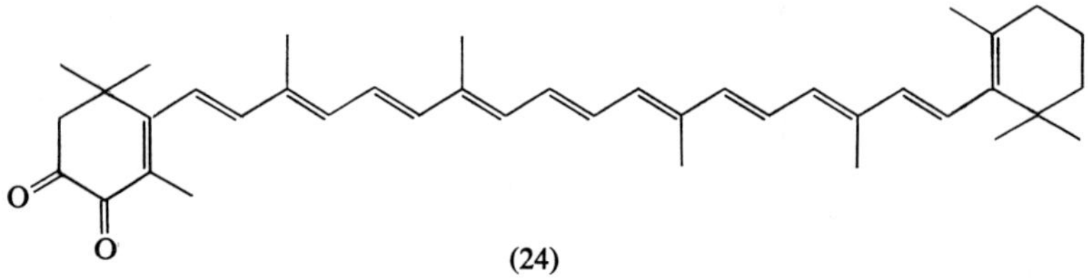

(24)

Haematococcus astaxanthin accumulates in "perinuclear cytoplasm characterized by a network of ribosome-coated endoplasmic reticulum segments, free ribosomes, dictysomes in active stages of vesicle formation, and mitochondria". They do not exist within any organelle or vesicle (Lang, 1968).

A large number of mutants of *Chlorella pyrenoidosa* lose their ability to synthesize canthaxanthin on nitrogen-poor media; however, they frequently tend to revert to the wild-type as indicated by their ability to form extraplastidic carotenoids (Kessler and Czygan, 1966).

B. PHAEOPHYTA

1. Heterokontae (Xanthophyceae)

Xanthophylls were first reported in *Chlorobotrys stellata* (Poulton, 1930) and the pigment was thought to be flavoxanthin (lutein 5,8-epoxide) (Kuhn and Brockmann, 1932); the same pigment was reported in *Botrydium granulatum* (Carter *et al.*, 1939) but later investigations have revealed β-carotene and 3–4 unidentified xanthophylls (Strain, 1958; Allen *et al.*, 1964; Thomas and Goodwin, 1965; Mattox and Williams, 1965; Kleinig and Egger, 1967; Strain *et al.*, 1968; Egger *et al.*, 1969). The exact identification of the pigments is still not settled. However, it now seems that the major xanthophylls in all heterokonts examined except *Pleurochloris commutata* are heteroxanthin ($C_{40}H_{56}O_4$), esters of vaucheriaxanthin ($C_{40}H_{56}O_5$), diatoxanthin (25), diadinoxanthin (26) (Strain *et al.*, 1968; Egger *et al.*, 1969; Stransky and Hager, 1970) and neoxanthin (Stransky and Hager, 1970). The structures of heteroxanthin and vaucheriaxanthin are not yet known. The former is the pigment found in a number of heterokonts and once thought to be a derivative of trollixanthin (Thomas and Goodwin, 1965). It is also clear that the pigment in the nine heterokonts examined by Thomas and Goodwin (1965) is diadinoxanthin and not antheraxanthin. Their pigment was identified by comparison with "antheraxanthin" from *Euglena* which has recently been shown to be diadinoxanthin (see p. 336). Similarly *Botrydium* sp. contains diadinoxanthin (Egger *et al.*, 1969) and not antheraxanthin (27) as first thought (Kleinig and Egger, 1967b). *P. commuta* on the other hand contains violaxanthin, antheraxanthin and zeaxanthin and both free and esterified vaucheriaxanthin although

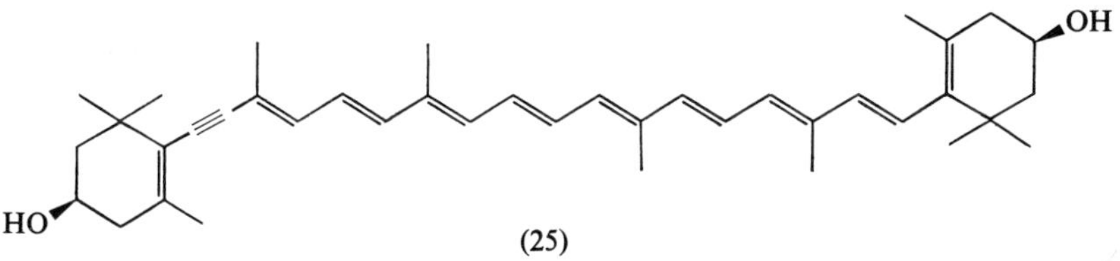

HO OH

(25)

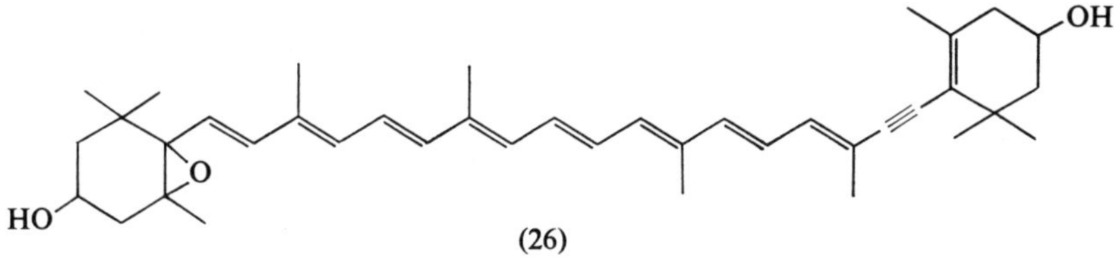

HO O OH

(26)

HO O OH

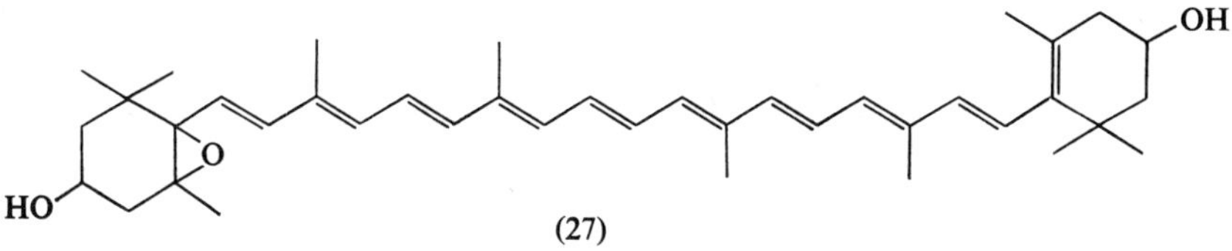

(27)

Botrydiopsis alpina which also belongs to the Pleurochloridaceae contains acetylenic pigments (Stransky and Hager, 1970). Heterokonts which have been examined for carotenoids are listed in Table VIII.

TABLE VIII
Heterokonts which have been examined for carotenoids

Alga	Reference
Botrydium becherianum	Thomas and Goodwin (1965)
B. granulatum	Stransky and Hager (1970)
Botrydiopsis alpina	Thomas and Goodwin (1965); Stransky and Hager (1970)
Bumilleria exilis	Thomas and Goodwin (1965)
B. sicula	Stransky and Hager (1970)
Bumilleriopsis brevis	Thomas and Goodwin (1965)
B. filiformis	Stransky and Hager (1970)
Chloridella neglecta	Thomas and Goodwin (1965)
Heterococcus caespitosus	Stransky and Hager (1970)
H. fuorensis	Thomas and Goodwin (1965)
Heterothrix debilis	Strain (1958); Thomas and Goodwin (1965); Stransky and Hager (1970)
Mischococcus sphaerocephalus	Whittle and Casselton (1969)
Monocilia (*Heterococcus*) *chodatii*	Strain (1958)
Ophiocytum majus	Thomas and Goodwin (1965)
Pleurochloris commutata	Whittle and Casselton (1969); Stransky and Hager (1970)
P. magna	Whittle and Casselton (1969)
Polyedriella helvetica	Strain (1958); Whittle and Casselton (1969)
Tribonema aequale	Thomas and Goodwin (1965); Falk and Kleinig (1968); Whittle and Casselton (1969); Stransky and Hager (1970); Strain *et al.* (1970)
T. bombycinum	Strain (1958); Falk and Kleinig (1968)
T. minus	Strain (1958); Falk and Kleinig (1968)
Vaucheria tenestris	Stransky and Hager (1970)
Vaucheria spp.	Strain (1958); Strain *et al.* (1968); Egger *et al.* (1969)
Vischeria sp.	Allen *et al.* (1964); Thomas and Goodwin (1965)
V. stellata	Strain (1958)

2. *Chrysophyceae*

In 1942 Heilbron reported β-carotene and fucoxanthin (28) in a mixed culture of *Apistonema carteri*, *Thallochrysis litoralis* and *Gloeschrysis maritima*, but it was not until almost twenty years later that this was confirmed on a pure culture of the chrysophytes *Ochromonas danica* and *Prymnesium parvum* (Allen *et al.*, 1960). Fucoxanthin represented some 75% of the total carotenoids and the remainder is β-carotene with traces of diatoxanthin and a pigment

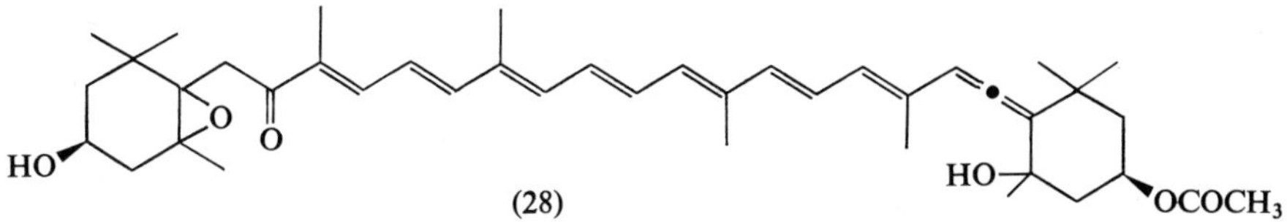

(28)

more strongly adsorbed than fucoxanthin. Other chrysophytes which contain fucoxanthin are listed in Table IX. Diatoxanthin is also present in *Isochrysis galbana* and *Sphaleromantis* sp., and diadinoxanthin in the former and dinoxanthin (structure still in doubt) in the latter (Jeffrey, 1961). There is still argument about the presence of diadinoxanthin in *I. galbana* (Dales, 1960; Jeffrey, 1961).

TABLE IX
Chrysophyceae known to synthesize fucoxanthin

Alga	Reference
Chrysochromulina ericina	Dales (1960)
Coccolithus huxleyi	Jeffrey and Allen (1964); Riley and Wilson (1965)
Dictateria inomata	Dales (1960)
Isochrysis galbana	Dales (1960); Jeffrey (1961); Bunt (1964); Riley and Wilson (1965)
Monochrysis lutheri	Parsons (1961)
Ochromonas danica	Allen *et al.* (1960)
Pavlova gyrans	Dales (1960)
Phaeaster type	Dales (1960)
Phaeocystis pouchetti	Dales (1960)
Prymnesium parvum	Allen *et al.* (1960)
Pseudopedinella sp.	Dales (1960)
Sphaleromantis sp.	Jeffrey (1961)
Syracosphera carteri	Parsons (1961)

β-Carotene was the only carotene found in *O. danica* and it is the major carotene in *P. parvum* (Allen *et al.*, 1960). Most of the other algae quoted in Table IX also contained α- and γ-carotene (Dales, 1960). The xanthophyll: carotene ratio is high for photosynthetic tissues, ranging from 4·8:1 in *O. danica* (Allen *et al.*, 1960) to 25·5:1 in *Sphaleromantis* sp. (Dales, 1960). The carotenoid concentration in *P. parvum* (10·57 mg/g dry wt) makes it probably the richest known algal source of carotenoids (Allen *et al.*, 1960).

It is interesting that a high concentration of carotenoids (and chlorophylls) have been found in a finely coiled lamellar system which is distinct from the chloroplast in *Hymenomonas* sp. *huxleyii* (Olson *et al.*, 1967).

3. *Phaeophyceae*

Recent careful investigations have revealed that the main pigments of the Phaeophyceae are β-carotene (1), violaxanthin (3) and fucoxanthin (28) (Jensen, 1966). No α-carotene was detected in genera from Norway (Jensen, 1966) or Australia (Strain, 1966). Some Norwegian genera contained zeaxanthin (29) (Jensen, 1966), but it was not possible to decide whether it was not the result of *post mortem* changes as was earlier suggested (Heilbron and Phipers, 1933). The naturally occurring fucoxanthin is the all *trans* form (Jensen, 1966) and the previously described *cis*-fucoxanthin (Strain, 1958, 1966; Strain *et al.*, 1944; Strain and Manning, 1943) is probably an artefact.

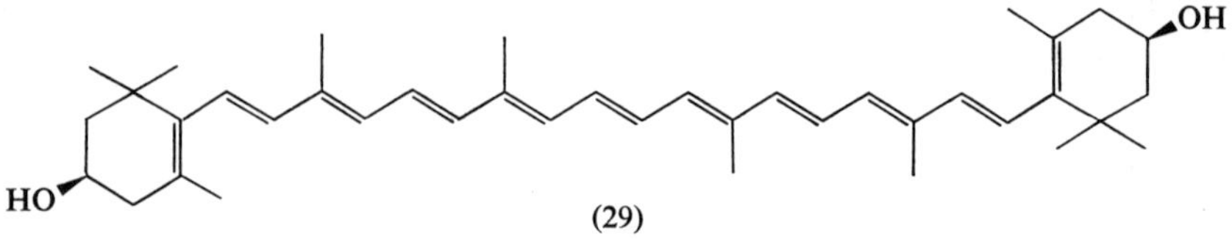

OH

HO

(29)

Traces of diatoxanthin and diadinoxanthin are occasionally encountered (Strain *et al.*, 1944; Strain and Manning, 1943; Jensen, 1966) and are considered not due to contamination of the specimens with diatoms. Lutein has never been encountered in recent studies (Jensen, 1966) although it was earlier reported in *Ectocarpus* (Strain and Manning, 1943). The Phaeophyceae which have been examined for carotenoids are listed in Table X, and some quantitative data are given in Table XI. These values can vary considerably with environmental and cultural conditions.

Differential distribution of extra-plastidic carotenoids occurs in *Fucus serratus*, *F. vesiculosus* and *Ascophyllum nodosum*. The bright orange-yellow of the male gametes is due almost entirely to β-carotene and the olive-green of the ova to a mixture of fucoxanthin and chlorophyll (Carter *et al.*, 1948).

4. *Bacillariophyceae* (*Diatoms*)

The pure cultures of diatoms so far examined (Table XII) are characterized by the presence of the acetylenic carotenoids diatoxanthin and diadinoxanthin in addition to β-carotene and fucoxanthin (Strain, 1951, 1958, 1966; Strain *et al.*, 1944). The recent demonstration that *cis*-fucoxanthin in the Phaeophyceae is an artefact, probably also applies to the reports of this pigment in diatoms, and the earlier reports of lutein and zeaxanthin may have been due to misidentification of dinoxanthin and diatoxanthin. ε-Carotene (30) is reported in traces in *Nitzschia closterium* (Strain *et al.*, 1944).

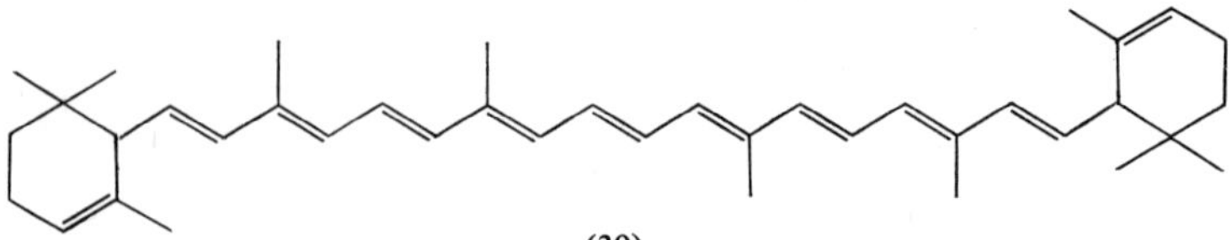

(30)

TABLE X
Phaeophyceae which have been examined for carotenoids (excluding those listed in Table XI)

Alga	Reference
Alaria esculenta	Jensen (1966)
Ascophyllum nodosum	Lunde (1937); Seybold and Egle (1938); Owen (1954); Jensen (1966)
Bachelotia fulvescens	Strain (1958)
Colpomenia sinuosa	Strain (1958, 1966)
Cystophyllum muricatum	Strain (1966)
Cystoseria osmundacea	Strain (1958)
Desmarestia aculeata	Willstatter and Page (1914); Jensen (1966)
Dictyopteris acrostichoides	Strain (1966)
Dictosiphon foeniculaceus	Jensen (1966)
Dictyota spp.	Willstätter and Page (1914); Seybold and Egle (1938); Strain (1958, 1966)
Ecklonia radiata	Strain (1966)
Egregia menziesii	Strain (1958)
Fucus spp.	Willstätter and Page (1914); Lunde (1937); Seybold and Egle (1938); Owen (1954); Liaaen and Sörensen (1956); Strain (1958, 1966); Jensen (1966)
Halidrys siliquosa	Seybold and Egle (1938)
Haplogloia andersonii	Strain (1958)
Hesperophycus harveyanus	Strain (1958)
Heterochordaria abietina	Strain (1958)
Hormosira banksii	Strain (1966)
Laminaria spp.	Willstätter and Page (1914); Lunde (1937); Seybold and Egle (1938); Owen (1954); Strain (1958, 1966); Jensen (1966)
Leathesia difformis	Jensen (1966)
Macrocystis integrifolia	Strain (1958)
Myriogloia (*Haplogloia*) *sciurus*	Strain (1966)
Nemacystus decipens	Strain (1966)
Padina spp.	Strain (1958, 1966)
Pelvetia canaliculata	Lunde (1937); Jensen (1966)
Pelvetiopsis limitata	Strain (1958)
Petalonia fascia	Jensen (1966); Strain (1966)
Petrospongium rugosum	Strain (1958, 1966)
Phyllospora comosa	Strain (1966)
Pocockiella nigrescens	Strain (1966)
Postelsia palmaeformis	Strain (1958)
Pterygophora californica	Strain (1958)
Pylaiella litoralis	Jensen (1966)
Sargassum spp.	Strain (1958, 1966)
Scaberia agardhii	Strain (1966)
Scytosiphon lomentaria	Strain (1966)
Spatoglossum sp.	Strain (1966)
Turbinaria ornata	Strain (1958, 1966)
Zonaria crenata	Strain (1966)

TABLE XI
Quantitative distribution of major carotenoids in some Norwegian Phaeophyceae (I) (mg/kg dry matter) (Jensen, 1966)

Alga	Date of collection	Carotene	Violaxanthin	Fucoxanthin
Alaria esculenta	29/3	210	267	1700
Ascophyllum nodosum	23/2	74	140	280
Desmarestia aculeata	10/6	190	197	1750
Dictyosiphon foeniculaceus	20/6	150	77	411
Fucus serratus	17/2	135	180	960
F. vesiculosus	16/2	80	162	340
Laminaria digitata	16/3	63	110	468
Leathesia difformis	10/6	230	200	390
Pelvetia canaliculata	19/2	100	294	487
Petalonia fascia	12/5	183	164	1060
Pylaiella litoralis	22/4	590	2500	5100
Scytosiphon lomentarius	19/4	170	182	643

TABLE XII
Diatoms examined for carotenoids

Species	Reference
Fragilaria sublinearis	Bunt (1964)
Isthmia nervosa	Strain *et al.* (1944)
Navicula torquatum	Strain *et al.* (1944)
Nitzschia closterium	Strain *et al.* (1944)
N. closterium f. *minutissima*	Strain *et al.* (1944); Strain (1966)
N. dissipata	Wassink and Kersten (1946)
N. palea	Strain *et al.* (1944)
Phaeodactylum tricornutum	Jeffrey (1961)
Stephanopyxis turris	Strain *et al.* (1944)
Thalassiosera gravida	Strain *et al.* (1944)

C. RHODOPHYTA

A large number of red algae has now been examined from both classes Bangiodeae and Florideae and it is clear that α- and β-carotene and their derivatives lutein and zeaxanthin are widely distributed (Strain, 1958, 1966; Allen *et al.*, 1964). The general pattern is thus relatively simple but considerable quantitative variations occur. Although the amounts of carotenoids in red algae and the relative distribution between carotenes and xanthophylls are similar to those encountered in green algae, there the similarity stops. For example zeaxanthin frequently predominates over lutein and in a number of cases, e.g. *Polysiphonia collinsii* (Strain, 1958), *Porphyridium aerugineum* and

Asterocytis ramosa (Chapman, 1966) lutein is apparently completely absent. Similarly α-carotene frequently occurs in greater amounts than β-carotene and this is particularly characteristic of the Delessericeae (Strain, 1958, 1966). Although α-carotene is frequently absent there is only one report of a red algae (*Phycodrysa sinuosa*) from which β-carotene is to be completely absent (Larsen and Haug, 1956). The red algae examined for carotenoids are listed in Table XIII.

The comparatively simple pattern of carotenoid distribution is occasionally disrupted. A pigment thought possibly to be neoxanthin was detected in one specimen of *Nemalion multifidum* (Allen *et al.*, 1964) and antheraxanthin was reported as the main pigment two red algae from Hawaii (*Acantophora spicifera* and *Gracilaria lichenoides*) (Aihara and Yamamoto, 1968). However, no antheraxanthin was reported in *G. sjoestedtii* isolated in California (Strain, 1958) or in *G. edulis* from Australia (Strain, 1966). Both the Hawaian algae contained β-cryptoxanthin (31) whilst *Lenormandia prolifera* (Saenger and Rowan, 1968) synthesizes α-cryptoxanthin (32). Violaxanthin was reported in *Halosaccion glandiforma* (Strain, 1968).

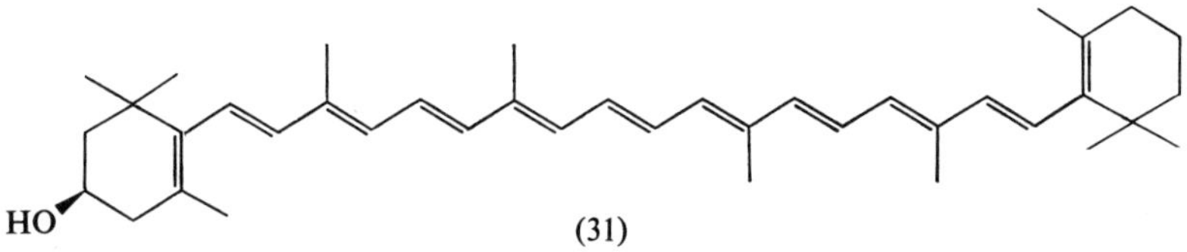

HO

(31)

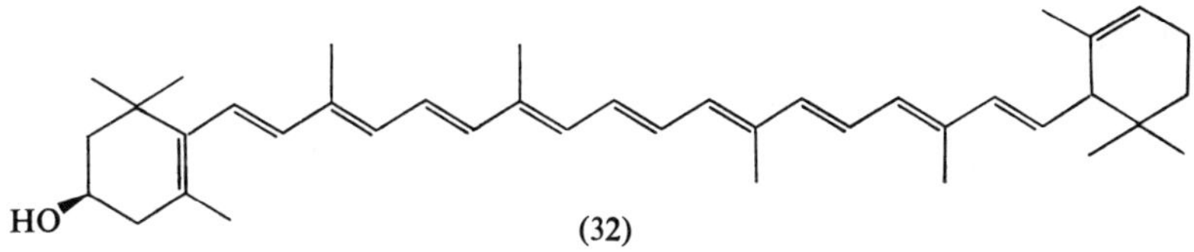

HO

(32)

D. PYRROPHYTA (DINOPHYCEAE)

The situation with regard to this class is somewhat confusing. In the fresh water *Peridinium cinctum* the characteristic pigment is peridinin (Kylin, 1927) which is said to be identical with sulcatoxanthin first isolated from *Anemonia sulcata* (Heilbron *et al.*, 1935; Seybold *et al.*, 1941); its structure is still unknown. Peridinin has also been reported in *Gymnodinium* sp. (Jeffrey, 1961), *Amphidinium klebsii* (Bunt, 1964), *Gonyaulax polyedra* (Sweeney *et al.*, 1959), *Amphidinium carteri* (Parsons and Strickland, 1963); *Prorocentrum micans* (Pinckard *et al.*, 1953) and *Anthopleura xanthogrammica* (Strain *et al.*, 1944). Associated with peridinin in these algae are diadinoxanthin and dinoxanthin and certain unidentified pigments in small quantities, some of which may be *cis*-isomers of the pigments just enumerated. The symbiotic dinoflagellates

TABLE XIII

Red algae which have been examined for carotenoids (From Strain, 1958, 1966, unless otherwise stated)

BANGIALES

Asterocytis ramosa βZ
Erythrotrichia carnea β*, Z
Porphyra naiadum β*
P. perforata β

PORPHYRIDIALES

Porphyridium aerugineum[8] β* Z*
P. cruentum β*, Z
Rhodosorus marinus β*, Z

NEMALIONALES

Asparagopsis armata β, L
A. taxiformis α, L
Batrachospermum sp. β, L
Cumagloia andersonii β, L
Galaxaura sp. β, L
Galaxaura umbellata α, L
Gloiophloea confusa α, L
Nemalion multifidum[1,2] β, L*
Rhodochorton rothii α, L

CERAMIALES

Amansia dietrichiana α, L
Antithamnion plumula[1,2] β, L*
Botryoglossum farlowianum α, L
Callithamnion californianum α, L
C. pikeanum β, L
Callophyllis marginifructa α, L
Centroceras clavatum β, L
C. davulata α, Z
Ceramium eatonianum β, L
Chondria sp. β*, Z
Cryptopleura lobulifera α, L
C. violacea α, L
Griffithsia pacifica α, L
Griffithsia sp. α, L
Hymenena flabelligera α, L
H. kylinii α, L
H. multiloba α, L
Laurencia heteroclada β*, Z
L. nidifica β*, Z
L. obtusa β, Z
L. pacifica β*, Z
L. rigida β*, L
Laurencia sp. β* (L = Z)
L. spectabilis β, Z
Lophosiphonia villum β, Z
Martensia sp. α, L
Microcladia borealis β, L
M. coulteri β, L
Nienburgia andersoniana α, L
Odonthalia floccosa β, Z
Polyneura latissima α, L
Polysiphona aquamara β*, Z
P. californica β*, Z
P. collinsii β*, Z*
P. fastigiata[1] β*, Z
Pterosiphonia baileyi β*, L
P. bipinnata β*, Z
P. dendroidea β (Z = L)
Ptilota densa α, L
Rhodomela larix β*, Z
Ricardia saccata β*, Z
Spermothamnion snyderae α, L
Spyridia filamentosa β, L

CRYPTONEMIALES

Bossea corymbifera β, L
B. orbigniana β, L
B. plumosa β, L
Calliarthron cheilosporioides β*, Z
C. setchelliae β*, Z
Corallina chilensis β, L
C. gracilis β, L
C. rosea β, L
Cryptosiphonia woodii (α), L
Endocladia muricata β, L
Farlowia compressa α, L
F. mollis α, L
Grateloupia californica α, L
G. filicina[9] β*, L
G. proteus[9] β, L
G. setchellii α, L
Lithophyllum neofarlowii β, L
Prionitis andersonii α, L
P. australis α, L
P. lanceolata α, L

GELIDIALES

Gelidium coulteri α, L
Pterocladia capillaceae α (L, Z)[4]
P. lucida α, L

GIGARTINALES

Agardhiella coulteri β, L
Erythrophyllum delesseriodes α, L
Gigartina agardhii β, L
G. californica ($\alpha = \beta$), L

TABLE XIII—*continued*

G. canaliculata β, L
G. corymbifera α, L
G. papillata β, L
G. spinosa α, L
Gracilaria edulis β*, L
G. lichenoides[3,4] β*, Z*
G. sjoestedtii β*, Z
Iridophycus flaccidum α, L
I. heterocarpum β, L
Plocamium coccineum α, L
P. hamatum α, Z
P. pacificum α, L
Plocamium sp. β, L
P. violaceum α, L
Rhodoglossum americanum α, L
Solieria robusta α, L
Schizymenia dubyi β, L

RHODYMENIALES

Acantophora spicifera[3,4]
Botryocladia skottsbergii α, L
Fauchia media α, L
Gastroclonium coulteri β, L
Halosaccion glandiforme[5] β, L
Lenormandia prolifera[6,7] α, L

GELIDIALES

Gelidium corneum[9] β, L

Notes and references. [1] Allen *et al.* (1964); [2] Neoxanthin possibly present; [3] Aihara and Yamamoto (1968); [4] β-Cryptoxanthin and antheraxanthin also present; [5] Violaxanthin also present (Strain, 1958); [6] α-Cryptoxanthin also present; [7] Saenger and Rowan (1968); [8] Chapman (1966a); [9] de Nicola and Furnari (1957); possibly also contains taraxanthin (structure unknown).

α- = α-carotene preponderates	Z = zeaxanthin predominates
β- = β-carotene preponderates but some α- present	L = lutein predominates
	Z* = no lutein present
β* = no α-carotene present	L* = no zeaxanthin present

(zooxanthellae) from the sea anemone *A. xanthogrammica* (Strain *et al.*, 1944), clams (*Tridacna* spp., *Hippopus hippopus*) and corals (in particular *Pocillopora* sp.) contained the same pigments with peridinin predominating (Jeffrey and Haxo, 1968). However in two other marine dinoflagellates, *Gymnodinium veneficum* (Riley and Wilson, 1967) and *Glenodinium foliaceum* (Mandelli, 1968), fucoxanthin replaces peridinin as the major pigment. In contrast, it should be noted that, as indicated above, an Australian *Gymnodinium* sp. is reported to contain peridinin and it was this pigment which was used as a control in the clam and coral work.

The carotene fraction (β-carotene) represents about 10% of the total carotenoids in *Prorocentrum micans* (Scheer, 1940) but only about 3% of the total in *Amphidinium* sp. and in the zooxanthellae from the clam *T. crocea* (Jeffrey and Haxo, 1968).

E. CRYPTOPHYTA

The chief characteristics of the carotenoids of members of this group so far examined are (i) that α-carotene predominates over β-carotene (Allen *et al.*, 1964; Chapman and Haxo, 1963), and (ii) the existence of an acetylenic derivative alloxanthin (33) as the major xanthophyll in *Cryptomonas* spp., *Hemiselmis viridis* and *Rhodomonas* sp. D3 (Chapman, 1966; Mallams *et al.*, 1967). Crocoxanthin (34) is also present in *Hemiselmis viridis* and *Rhodomonas* sp. D3 and monadoxanthin (35) in *Rhodomonas* sp. D3 (Chapman, 1966; Mallams *et al.*, 1967).

Cryptomonas ovata also synthesizes ϵ-carotene (30) which is rare amongst algal carotenoids (Chapman and Haxo, 1963).

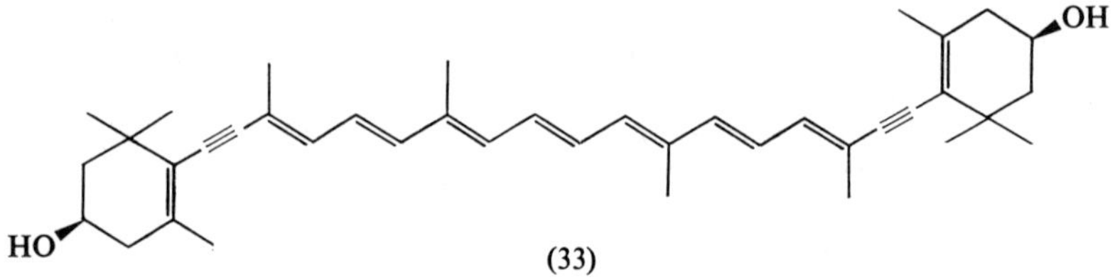

HO OH

(33)

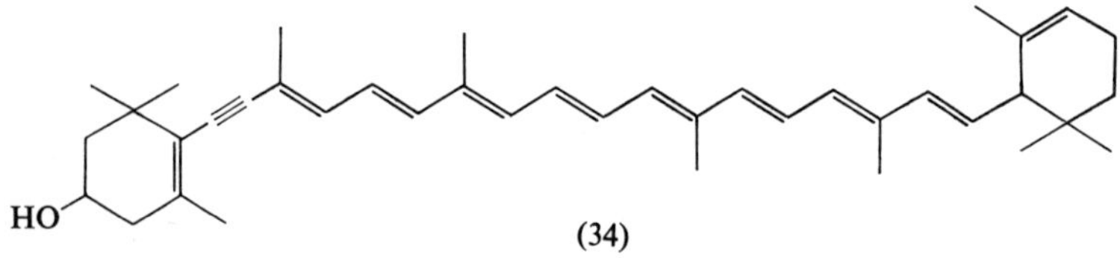

HO

(34)

HO OH

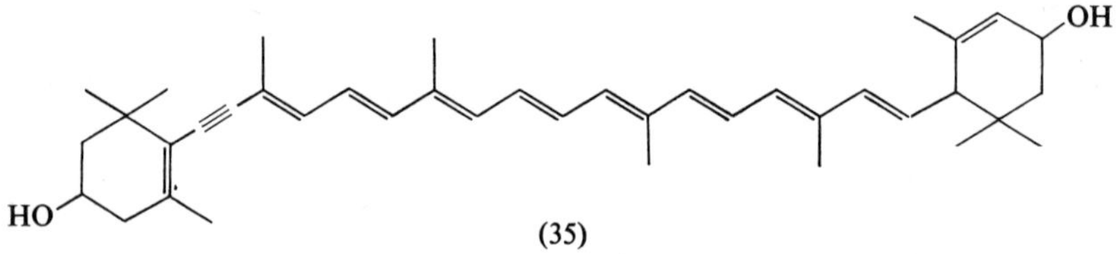

(35)

F. EUGLENOPHYTA

The pattern of carotenoid distribution in various *Euglena* spp., is similar to that in green leaves and most non-siphonalean green algae except that the main xanthophyll is diadinoxanthin (26) and not either antheraxanthin or lutein as previously suggested (Krinsky and Goldsmith, 1960; Goodwin and Jamikorn, 1954b). Trace amounts of ketocarotenoids, echinenone (16), 3-hydroxyechinenone (36) and canthaxanthin (18), are also present (Goodwin and Gross, 1958; Krinsky and Goldsmith, 1960). It is probable that these ketonic carotenoids are located in the eyespot of *Euglena* spp. because chlorotic sub-strains of *E. gracilis*, in which the chloroplasts but not the eyespot have been deleted, still contain echinenone but little if any plastidic carotenoids (Goodwin and Gross, 1958). The pigment distribution in certain

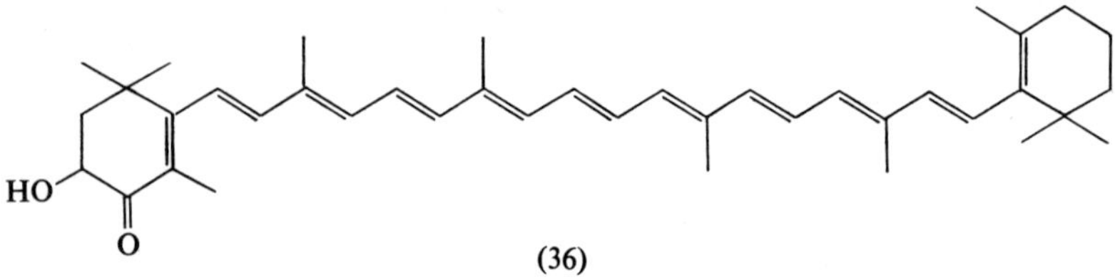

HO
O

(36)

bleached strains is given in Table XIV. However, there is one report that lutein (? = diadinoxanthin) is the main eyespot pigment (Batra and Tollin, 1964).

Astaxanthin once considered to be the eyespot pigment (Heilbron, 1942) has not been detected in recent investigations on *Euglena* pigments; but it has been isolated from the red *E. heliorubescens* and named euglenarhodone (Tischer, 1936) until its identity with authentic astaxanthin was confirmed (Kuhn *et al.*, 1939; Tischer, 1941).

TABLE XIV

Carotenoids in certain bleached strains of *Euglena* [Goodwin and Gross (1958) except where otherwise indicated]

Substrain	Bleaching agent[7]	Carotenoids[1]	
		Carotenes	Xanthophylls
PBZG1 BBZG2 PBZG3	Pyribenzamine[2,3]	Phytofluene β-Carotene ζ-Carotene-like	Normal pigments absent or present in minute traces Echinenone, zeaxanthin, diadinoxanthin[4]
PBZG4	Pyribenzamine[2,3]	No detectable carotenoids	
SML1	Streptomycin[5,6]	Phytofluene β-Carotene ζ-Carotene-like	Normal pigments absent or present in minute traces
SMP SMG	Streptomycin[5,6]	Present in minute traces only	
HBG	High temp. (36°C)[3]	Phytofluene β-Carotene ζ-Carotene-like	Echinenone, zeaxanthin, diadinoxanthin

Notes and references. [1] Concentration is much less than in normal strain (e.g. 0·29 mg/g dry wt in HBG compared with 7·1 mg/g in normal); [2] Gross *et al.* (1955); [3] Gross and Jahn (1958); [4] Originally identified as lutein; [5] Loefer and Guido (1950); [6] Provasoli *et al.* (1948); [7] Other bleaching agents include (i) dihydrostreptomycin (Huzisige *et al.*, 1957) and related antibiotics (Zahalsky *et al.*, 1962); (ii) 3-amino-1,2,4-triazole (Aaronson and Sher (1960); (iii) *O*-methylthreonine (Gray and Hendlin, 1962; Aaronson and Bensky, 1962); (iv) UV radiation (Pringsheim, 1958).

Euglena sp. will grow heterotrophically in the dark without the production of chloroplasts, and carotenoids are reduced in such cultures (Helmy *et al.*, 1967).

G. CYANOPHYTA

Recent investigations have considerably clarified the situation with regard to the nature and distribution of carotenoids in free-living blue-green algae (see Table XV). Characteristics of the pattern are (i) relatively large amounts of β-carotene; (ii) the ubiquity of echinenone and the presence of other ketocarotenoids such as canthaxanthin (18) and 3-hydroxy-4′-keto-β-carotene (37); and (iii) the presence of the unique carotenoid myxoxanthophyll. It is now clear that myxoxanthin (Heilbron and Lythgoe, 1936) and aphanin

(Tischer, 1938, 1939, 1958) are identical with echinenone (Goodwin and Taha, 1951; Goodwin, 1956; Hertzberg and Liaaen-Jensen, 1966a,b), and that aphanicin (Tischer, 1938, 1939) is canthaxanthin (Hertzberg and Liaaen-Jensen, 1966b). Myxoxanthophyll (38) has recently been shown to be a mixed glycoside of a γ-carotene derivative, with rhamnose the major sugar component and a hexose a minor component (Hertzberg and Liaaen-Jensen, 1969a). Oscillaxanthin which was first isolated from *Oscillatoria rubescens* (Karrer and Rutschmann, 1944) has been reisolated from *Athrospira* sp. and shown to be a glycoside derivative of lycopene (39) (Hertzberg and Liaaen-Jensen, 1969b). Myxoxanthophyll and oscillaxanthin are the first glycosides to be reported in algae.

TABLE XV
Blue-green algae examined for carotenoids

Algae	Reference
Anabaena cylindrica	Goodwin (1957)
A. variabilis	Goodwin (1957)
Anabaena sp.	Strain (1958)
Aphanizomenon flosaquae	Hertzberg and Liaaen-Jensen (1966a)
Athrospira sp.	Hertzberg and Liaaen-Jensen (1966a)
Chroococcus sp.	Strain (1958)
Coochloris elabens	Goodwin (1957)
Cylindrospermum sp.	Goodwin (1957)
Hormothamnion enteromorphoides	Strain (1958)
Hydrocoleum sp.	Strain (1958)
Mastigocladus laminosus	Goodwin (1957)
Microcoleus vaginatus	Goodwin (1957)
Nostoc muscorum	Strain (1958)
Nostoc sp.	Goodwin (1957)
Oscillatoria amoena	Tischer (1958)
O. rubescens	Heilbron and Lythgoe (1936)
	Karrer and Rutschmann (1944)
	Hertzberg and Liaaen-Jensen (1966a)
Phormidium autumnale	Strain (1958)
Rivularia atra	Heilbron *et al.* (1935)
R. nitida	Heilbron *et al.* (1935)
Tolypothrix distorta v. *symplocoides*	Manten (1948)

Two symbiotic blue-green algae (cyanomes) which are associated with the colourless algae *Cyanophora paradoxa* and *Glaucocystis nostochinearum* contain only β-carotene and zeaxanthin; no myxoxanthophyll or echinenone was detected (Chapman, 1966a).

An early report of lutein in *O. rubescens* (Heilbron and Lythgoe, 1936) has not been confirmed in later work (Hertzberg and Liaaen-Jensen, 1966a).

In 1927 Kylin described the pigments in *Calothrix scopulorum* as carotene, myxorhodin-α, myxorhodin-β and calorhodin. The relationship between these pigments and those known to occur in other blue-green algae is not known.

The quantitative distribution of carotenoids in some blue-green algae is given in Table XVI; some considerable variations can be expected under different environmental conditions.

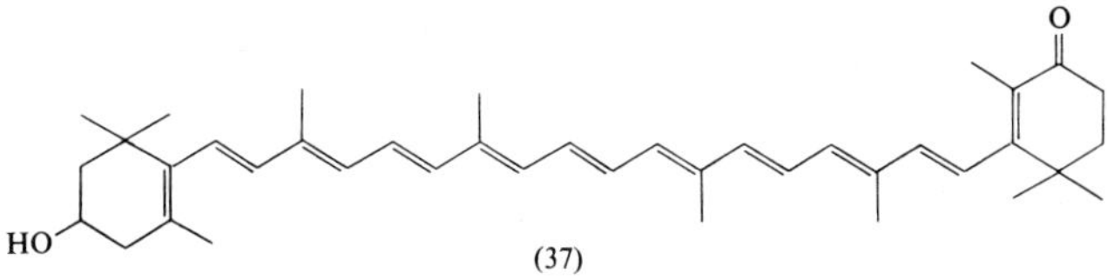

HO O

(37)

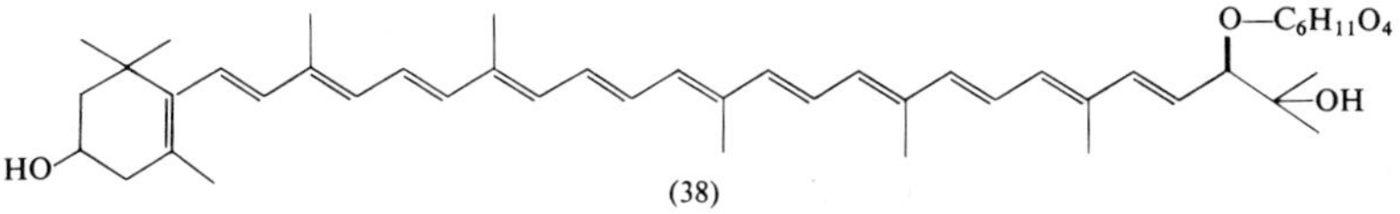

HO $O—C_6H_{11}O_4$ OH

(38)

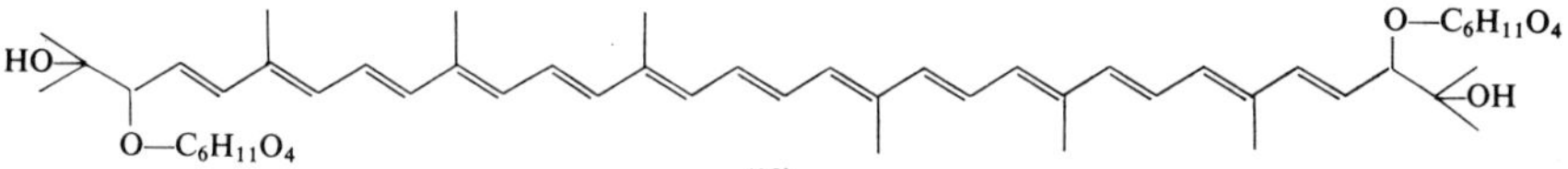

HO $O—C_6H_{11}O_4$ $O—C_6H_{11}O_4$ OH

(39)

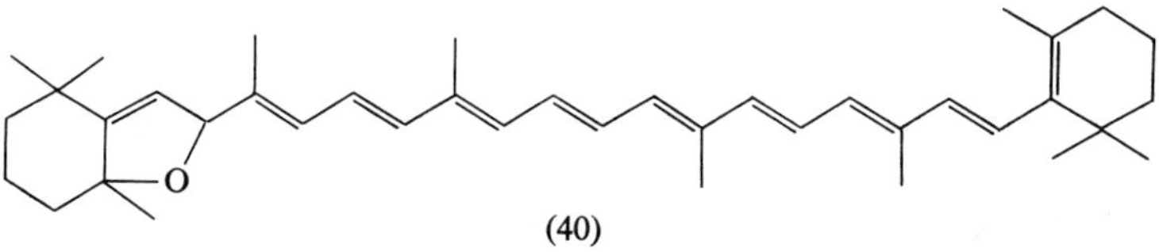

O

(40)

H. CHLOROMONADOPHYTA

Only comparatively recently have the carotenoids of members of this family been examined. *Vacuolaria virescens* and *Gonyostomum semen* seem to have a rather similar pattern to that noted in some Xanthophyceae, i.e. β-carotene, antheraxanthin (= ? diadinoxanthin), lutein epoxide (?) (41) and a hydroxy-lutein monoepoxide-like carotenoid (= ? heteroxanthin) (Chapman and Haxo, 1966).

I. COLOURLESS ALGAE

The heterotrophic phytoflagellate *Polytoma uvella* contains β-carotene and a weakly acidic pigment polytomaxanthin (Links *et al.*, 1960); the xanthophyll: carotene ratio was 4:1 and the total carotenoid concentration between 0·3 and 0·4 mg/g dry wt. In *Astasia ocellata* the major carotene is α-carotene and the xanthophylls are almost entirely ketonic including echinenone, canthaxanthin and what may be 4-keto-α-carotene (phoenicopterone, 42) (Thomas *et al.*, 1967).

TABLE XVI Quantitative distribution o

	Anabaena cylindrica[1]	*Anabaena variabilis*[1]	*Aphanizomenon flos-aquae*[2]	*Athrospira* sp.[3]
β-Carotene	54	43	32	27
Flavacin[4]	—	—	2	—
Cryptoxanthin	—	—	—	2
Echinenone	19	28	14	3
4-Keto-3′-hydroxy-β-carotene	—	—	—	2
Zeaxanthin	trace	4	—	15
Canthaxanthin	—	—	38	—
Myxoxanthophyll	17	33	1	46
Aphanizophyll (structure uncertain)	—	—	13	—
Oscillaxanthin	—	—	—	5

References. [1] Goodwin (1957); [2] Hertzberg and Liaaen-Jensen (1966b); [3] Hertzberg and Liaaen

TABLE XVII Distribution

Pigment	CHLOROPHYTA			PHAEOPHYTA			
	Charo-phyceae[2]	Chloro-phyceae[3]	Prasino-phyceae[4]	Xantho-phyceae[5]	Bacillario-phyceae[6]	Chryso-phyceae[7]	Phaeo-phycea
α-Carotene		+	+			+	
β-Carotene	+	+	+	+	+	+	+
Echinenone							
Lutein	+	+					
Zeaxanthin	+						
Antheraxanthin							
Neoxanthin	+	+	+				
Fucoxanthin					+	+	+
Diatoxanthin					+	+	
Diadinoxanthin				+	+	+	
Peridinin							
Alloxanthin							
Myxoxanthophyll							
Oscillaxanthin							
Violaxanthin	+	+	+				+

Notes

[1] For variations from major pattern see text.

[2] Only one species examined (*Chara fragilis*): lycopene and γ-carotene also present.

[3] γ-Carotene and ε-carotene occasionally encountered; the major xanthophylls in the Siphonale are siphonein and siphonaxanthin.

[4] A keto carotenoid (micronone) and siphonaxanthin also present in some species.

[5] Identity of 2-3 major xanthophylls still in doubt.

[6] ε-Carotene sometimes present.

carotenoids in blue-green algae (% of total)

Coochloris elabens[1]	*Cylindro-spermum* sp.[1]	*Mastigocladus laminosus*[1]	*Microcoleus vaginatus*[1]	*Nostoc* sp.[1]	*Oscillatoria rubescens.*[3]
42	50	42	63	28	29
—	—	—	—	—	—
—	—	—	—	—	4
16	25	39	6	53	19
—	—	—	—	—	1
trace	trace	8	9	trace	8
—	—	—	—	—	—
42	25	11	22	19	30
—	—	—	—	—	—
—	—	—	—	—	10

Jensen (1966a); [4] Probably mutatochrome (40) (Hertzberg and Liaaen-Jensen, 1967).

major carotenoids in algae[1]

RHODOPHYTA Rhodophyceae	PYRROPHYTA Dinophyceae	EUGLENOPHYTA Euglenineae	CYANOPHYTA Cyano-phyceae[11]	CRYPTOPHYTA Cryptophyceae	CHLORO-MONADOPHYTA Chloro-monadineae[5]
+				+	
+	+	+	+		+
		+[10]	+		
+					
+			+		
+[8]		+			
		+			
	+[9]				
	+				? (+)
	+				
				+[12]	
			+		
			+		

[7] γ-Carotene frequently present.
[8] Antheraxanthin only rarely encountered.
[9] Some genera contain fucoxanthin and not peridinin.
[10] Small amounts of other ketocarotenoids noted.
[11] See Table XVI for full details.
[12] Other acetylenic carotenoids present in small amounts.

TABLE XVIII Typical quantitative distribution of major carotenoids in representative a

	CHLOROPHYTA	PHAEOPHYTA			
	Chlorophyceae	Heterokontae	Bacillario-phyceae	Chrysophyceae	Phaeophyce
Pigment	*Chlorella pyrenoidosa*[1]	*Vischeria* sp.[2,d]	*Phaeodactylum bicornutum*[3]	*Ochromonas danica*[1]	*Fucus vesiculosus*
α-Carotene	0·09	—	?	—	—
β-Carotene	0·33	3·27	0·31[a]	0·73	0·14[a]
Echinenone	—	—	—	—	—
Lutein	1·11	—	—	—	—
Antheraxanthin	—	—	—	—	—
Lutein epoxide	—	—	—	—	—
Zeaxanthin	—	—	—	—	—
Violaxanthin	0·22	—	—	—	—
Neoxanthin	0·21	—	—	—	—
Fucoxanthin	—	—	5·35	3·16	0·59
Diadinoxanthin	—	1·74[e]	0·22	—	—
Alloxanthin	—	—	—	—	—
Peridinin	—	—	—	—	—
Myxoxanthophyll	—	—	—	—	—
Unknown pigments	0·25	traces	—	0·40	0·14
TOTAL	2·21	5·01	5·88	4·29	0·87

References. [1] Allen *et al.* (1960); [2] Thomas and Goodwin (1965); [3] Jeffrey (1961); [4] Allen *et* and Haxo (1966). [a] Total carotenes; [b] Dinoxanthin also present; [c] Considerable quantitati total; [e] A mixture.

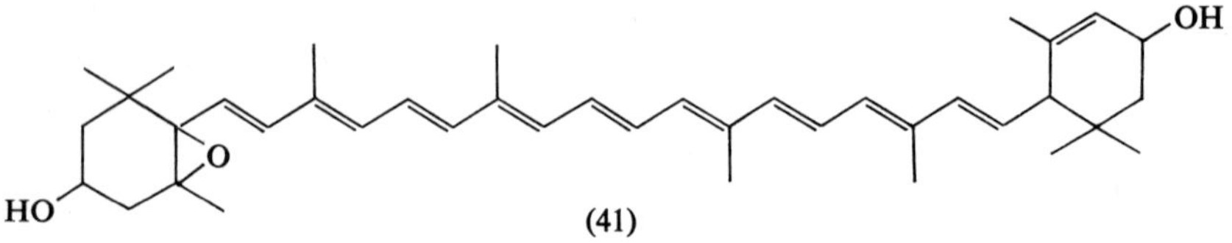

(41)

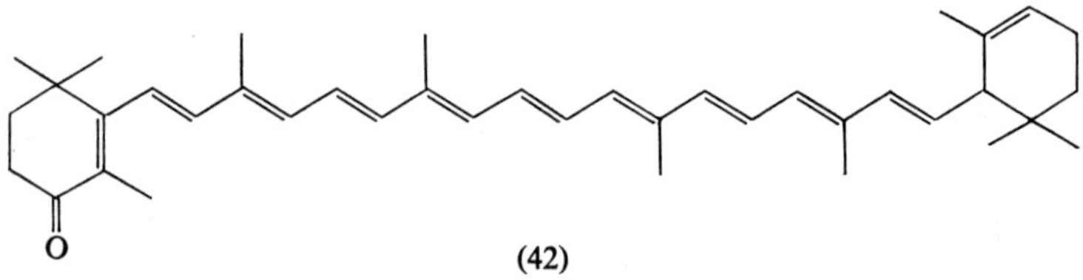

(42)

J. ALGAL SYMBIONTS OF LICHENS

The symbiotic chlorophyte *Trebouxia decolorans* from the lichen *Xanthoria parientina* has the same carotenoids as the free living *T. humicola* although there are relatively more xanthophylls in the phycobiont (de Nicola and Tomaselli, 1961) (see also p. 318).

om different divisions and classes. Concentration (mg/g dry wt)

RHODOPHYTA		PYRRO-PHYTA	EUGLEN-OPHYTA	CYANO-PHYTA	CRYPTO-PHYTA	CHLORO-MONADO-PHYCEAE
ngioideae	Florideae					
hodosomus marinus[4]	*Nemalion multifidum*[4]	*Gymnodinum* sp.[3]	*Euglena gracilis* v. *bacillaris*[5]	*Anabaena variabilis*[6, 7, c]	*Cryptomonas* sp.[4]	*Vacuolaria virescens*[8]
—	0·02	—	—	—	0·28	—
0·09	0·10	0·10[b]	0·79	1·90	—	5·4
—	—	—	—	1·56	—	—
0·03	0·40	—	—	—	—	—
—	—	—	5·86	—	—	—
—	—	—	—	—	—	5·4(?)
0·29	—	—	—	0·22	—	—
—	—	—	—	—	—	—
—	(?)0·05	—	0·52	—	—	—
—	—	—	—	—	—	—
—	—	0·19[b]	—	—	—	23·8
—	—	—	—	—	1·08	—
—	—	0·52	—	—	—	—
—	—	—	—	1·90	—	—
—	0·05	—	trace	—	0·25	6·75
0·41	0·62	0·81	7·15	5·58	1·61	41·3

964); [5] Goodwin and Jamikorn (1954); [6] Goodwin (1957); [7] Glover and Shah (1957); [8] Chapman
riations with cultural conditions; [d] In other Heterokonts β-carotene represents about 10% of

K. *Cyanidium caldarium*

This intriguing alga requires a short section to itself. Although it produces phycocyanin and only chlorophyll *a*, which suggests a close relationship with the blue-green algae, it has usually been considered as belonging to the Chlorophyta (Hirose, 1950; Allen, 1954, 1959), although more recently it has been placed in the Rhodophyta (Chapman, 1965). Its carotenoid picture, β-carotene (55%), zeaxanthin (32%) and lutein (probably) (14%) (Allen *et al.*, 1960), would support its position in the Rhodophyta although the relatively high level of β-carotene is reminiscent of the blue-green algae.

L. GENERAL SUMMARY

The general distribution of the major algal carotenoids in the various orders is given in Table XVII. Specific deviations from this general picture are discussed in the text. A representative series of values for the quantitative distribution of carotenoids in algae are given in Table XVIII. Although this is useful as a general view of the situation considerable variations are likely to be encountered.

II. Factors Controlling Carotenoid Synthesis

A. General Nutritional Conditions

It is clear (p. 328) that the synthesis of so-called secondary carotenoids can be induced in many algae under conditions of nutritional imbalance, in particular when there is a lack of nitrogen relative to assimilable carbon (Lwoff and Lwoff, 1930; Chodat, 1938; Chodat and Haag, 1940; Wenzinger, 1940; Goodwin and Jamikorn, 1954a; Droop, 1954). Under some conditions, e.g. submerged cultures of *Dictyococcus cinnabarinus* on glucose, secondary carotenoids are produced by growing cells (Dentice di Accadia *et al.*, 1966).

The mechanism involved in this phenomenon is not understood. It is claimed that the additional pigments are synthesized in the chloroplast and then diffuse into the fat vacuoles where, if they are hydroxy-carotenoids, they are esterified. As the synthesis parallels loss of chlorophyll it is also suggested that the metabolites of chlorophyll catabolism are used for pigment synthesis (Czygan, 1968b). There is no evidence for this.

B. Mineral Constituents

The generalized claim that lack of Mg^{2+}, SO_4^{2-} or PO_4^{3-} causes an accumulation of carotenoids (Haag, 1941) was not confirmed in the case of *Trentepohlia aurea* (Czygan and Kalb, 1966); a similar lack of effect of deficiency of K^+ or Ca^{2+} was also reported for this alga. In *Ankistrodesmus*, however, deficiency of PO_4^{3-}, SO_4^{2-}, Ca^{2+} and Fe^{3+} stimulated carotenogenesis whilst Mg^{2+} and Mn^{2+} deficiency was without noticeable effect (Dersch, 1960). A low phosphate medium stimulated carotenogenesis three-fold in a bleached strain (SMLI) of *Euglena* (Blum and Bégin-Heick, 1967). Some years ago it was claimed that *Dunaliella salina* synthesizes only β-carotene when cultured in saturated saline, in solutions only $\frac{1}{2}$ or $\frac{2}{3}$ saturated less carotene was synthesized (Fox and Sargent, 1938). In later work it was shown that α-carotene, lutein, violaxanthin and neoxanthin are present as well as β-carotene but that the increase in carotenoid concentration in salinities above 5% NaCl is due almost entirely to β-carotene synthesis (Loeblich, 1969).

C. Light

The carotenoid levels of *Euglena* spp. are very much reduced when the algae are cultured heterotrophically in the dark (Goodwin and Jamikorn, 1954b; Wolken *et al.*, 1955). If the cells are continually sub-cultured in the dark the carotenoid levels become extremely low because chloroplasts cease to be produced. No neoxanthin is present in dark-grown *Euglena* but it appears along with chlorophyll when dark-grown cells are exposed to light and begin to form chloroplasts (Krinsky *et al.*, 1964). On the other hand some algae, for example *Chlorella*, produce chloroplasts when grown autotrophically in the dark and thus also the usual plastid carotenoids; the amount in young dark-grown cultures is relatively less than in light-grown cultures but in old cultures the situation is reversed (Goodwin, 1954). This observation is probably linked with the finding that under intense illumination (100,000 lux) the carotenes of *C. pyrenoidosa* are destroyed more rapidly than the xanthophylls (Aach, 1953; Sironval and Kandler, 1958). Maximum carotenoid synthesis in *Anacystis nidulans* cultured at 30–35°C occurs at a light intensity of 400–800-ft candles (Halldal, 1958).

Under naturally occurring autotrophic conditions changes in light intensity appear to have little effect on carotenoid production, because samples of the same algae taken at various depths in a Mid-West lake differed only slightly in their carotenoid content (Dutton and Juday, 1944). Some early reports on qualitative differences have been published; for example *C. pyrenoidosa* produces more β-carotene than α-carotene under high light intensity and the reverse under low light intensity (Strain and Manning, 1943) and *Nitzschia closterium* produces less diadinoxanthin when cultured under "snowwhite" fluorescent light than under neon tubes although the other pigments were unaffected (Strain *et al.*, 1944).

A number of *Chlorella* mutants are killed by a combination of light and oxygen; the significance of this is considered later (p. 348). Mutant 5/520 synthesizes a series of acyclic colourless polyenes in the dark; on illumination under nitrogen cyclic carotenes are formed and red light (670 nm) is the most effective photostimulator (Claes, 1966a,b). The significance of this observation is not yet fully understood because in the native strain of *Chlorella* light is not required for the synthesis of cyclic carotenoids. Hydroxylation in the mutant strain is also not light-dependent because if the illuminated cells are returned to darkness and oxygenated xanthophylls are formed (Claes, 1954, 1956). The small amounts of prolycopene and proneurosporene present in dark-grown cells are isomerized by light which they themselves absorb, into the all *trans* isomers in the presence of oxygen. In the absence of oxygen light absorbed by the small amounts of chlorophyll formed in the dark will also carry out this photoisomerization (Claes and Nakayama, 1959). Light and oxygen together are also lethal to a pale green mutant of *Chlamydomonas reinhardi* (Sager and Zalokar, 1958).

Light is not essential for the formation of astaxanthin in encysting *Haematococcus pluvialis* (Droop, 1954) but it is stimulatory (Goodwin and Jamikorn, 1954b). Synthesis of extra-plastidic carotenoids is also independent of light intensity in *Trentepholia aurea* (Czygan and Kalb, 1966) and other Chlorophyceae (Czygan, 1968b).

D. OXYGEN—REVERSIBLE EPOXIDATION

It is clear that xanthophyll epoxides can be formed in chloroplasts when they are illuminated strongly in the presence of oxygen. The reverse process, de-epoxidation, takes place in the dark or under dim light. Cell-free preparations have been obtained from *Euglena gracilis* which will de-epoxidize diadinoxanthin in the dark in the presence of a hydrogen donor such as

$FMNH_2$; in intact cells light will substitute for $FMNH_2$ and the reaction is a photoreduction (Bamji and Krinsky, 1965). Epoxidation is considered to be a non-enzymic reaction (Krinsky, 1966).

Similar experiments with *Chlorella pyrenoidosa* show that the light absorbed by chlorophyll is active in the stepwise de-epoxidation of violaxanthin to zeaxanthin via antheraxanthin. The reaction is inhibited by inhibitors of light reaction II in photosynthesis (see Fig. 2) such as DCMU and this inhibition is

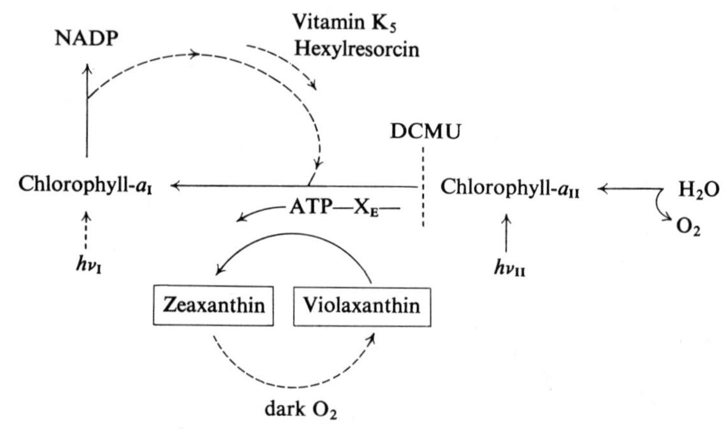

FIG. 2. Outline scheme of the two light hypothesis of photosynthesis indicating where zeaxanthin and violaxanthin may be implicated (after Hager, 1969).

reversed by agents, such as vitamin K_5, which increase cyclic electron transport at chlorophyll a_I. The reaction is also completely suppressed by uncouplers of photophosporylation (Hager, 1967a). The reverse reaction, the epoxidation of zeaxanthin can be observed immediately after a period of strong illumination and is stimulated by exposure to pure oxygen or dim light. Apparent changes observed in freeze dried cells or chloroplasts is said to be merely photooxidative destruction of carotenoids (Hager, 1967b).

E. SEASONAL FACTORS

Different algae vary considerably in the pattern of seasonal variation of carotenoid pigment content. For example in Norwegian waters the values for *Fucus serratus* are reasonably steady throughout the year whilst those for *Pelvetia canaliculata* undergo considerable fluctuations (Fig. 3). In the latter alga the minimum values coincide with the period of maximum fructification (Jensen, 1966).

III. FUNCTION OF CAROTENOIDS

A. IN PHOTOSYNTHESIS

The close association maintained throughout evolution between chlorophylls and carotenoids in chloroplasts and chloroplast-like structures points to a basic function of carotenoids in photosynthesis. In diatoms and brown

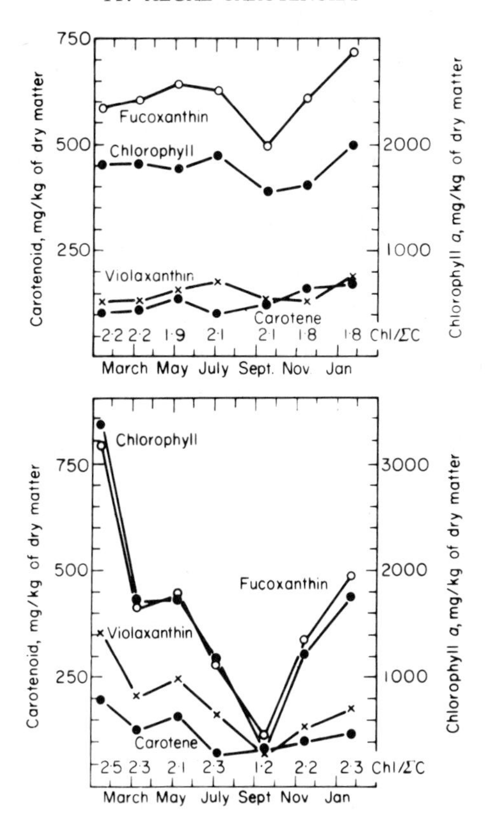

FIG. 3. The annual variation in carotenoid and chlorophyll levels in (A) *Fucus serratus* and (B) *Pelvetia canaliculata* growing in Norwegian waters (after Jensen, 1966).

algae it was early concluded that the light energy absorbed by carotenoids was efficiently transferred to chlorophyll for utilization in photosynthesis (Dutton *et al.*, 1943; Tanada, 1951); rather less efficient transfer was observed with green algae (Emerson and Lewis, 1943). With the emergence of the "two light" theory of photosynthesis this simple concept had to be modified. The generalized pathway of photosynthesis has already been indicated in Fig. 2. Light absorbed by a special form of chlorophyll *a* (Ca_I) causes ejection of an electron at a redox potential sufficiently negative to allow it to be utilized to form NADPH. The second pigment which can be another form of chlorophyll *a*, chlorophyll *b*, a carotenoid or a phycobilin, absorbs light and ejects an electron with a redox potential sufficiently negative to be used to neutralize the charged Ca_I molecule with the simultaneous production of ATP. The charged pigment II is neutralized by an electron from OH^- with the liberation of O_2. Haxo (1960) by measuring action spectra for photosynthesis in *Navicula* showed that fucoxanthin was implicated in the pigment II reaction. However, Goedheer (1969) has concluded from fluorescence action spectra, fluorescence spectra, and absorption spectra at room temperature and at 77°K that in blue-green and red algae energy transfer from β-carotene occurs exclusive in photosystem I, whilst in green algae transfer takes place in both systems. The transfer is almost 100% efficient. Light absorbed by xanthophylls is not utilized in this way (Goedheer, 1969).

The light-induced removal of the epoxide oxygen from violaxanthin is not the process by which oxygen is released in photosynthesis but there is evidence that it is coupled with the electron-transport pathway between reduced plastoquinone and oxidized chlorophyll a_I; the ATP produced during this electron flow might be used to break the C–O bonds in the epoxide (Hager, 1967a). Furthermore the oxygen in violaxanthin in *Chlorella vulgaris* is derived from $^{18}O_2$ and not $H_2{}^{18}O$ (Yamamoto and Chichester, 1965). According to Hager (1969) the light induction of de-epoxidation is basically due to a light-induced drop in pH in the compartment of the chloroplast carrying the enzyme. De-epoxidation in the dark takes place under conditions which produce a low pH, such as in a buffer below 7·0, by NaF addition at low pH (HF formed which is easily permeable), or by aeration with CO_2 (HCO_3^- production within the cell).

B. PROTECTION AGAINST PHOTODYNAMIC SENSITIZATION

The absence of carotenoids but not chlorophylls from chloroplasts or chromatophores of an organism makes it very susceptible to killing by photodynamic sensitization, that is it is killed in the presence of both light and oxygen. This is clearly the case in the *Chlorella* mutants of Claes (p. 321). The phenomenon was first described in detail in mutants of photosynthetic bacteria (Stanier and Cohen-Bazire, 1957) and it also occurs in higher plants (see e.g. Anderson and Robertson, 1960). The mechanism of the effect is still not clear but singlet oxygen may be involved (see Foote, 1969). It is now clear why no naturally occurring plant has been found which contains chlorophylls but no coloured carotenoids; it represents a lethal mutation.

IV. CAROTENOIDS AND ALGAL TAXONOMY AND EVOLUTION

Before considering the implications of carotenoid studies in this aspect of Phycology the basic concepts of carotenoid biosynthesis must be considered (the details are discussed on p. 255 in this volume). The first C-40 compound is phytoene which is stepwise desaturated to the fully unsaturated lycopene (Fig. 4). This or a closely related compound can cyclize to form carotenes with either α- or β-ionone rings which are produced by separate pathways and not by isomerization of the formed rings. Thus two enzymes (yet to be isolated) α-cyclase and β-cyclase are involved. The major xanthophylls present have a hydroxyl group at C-3 and C-3′ and the reaction is that of a typical mixed function oxidase (Walton *et al.*, 1969). 5,6-Epoxidation of xanthophylls take place (see previous section) by a well-established enzymic step. The recent

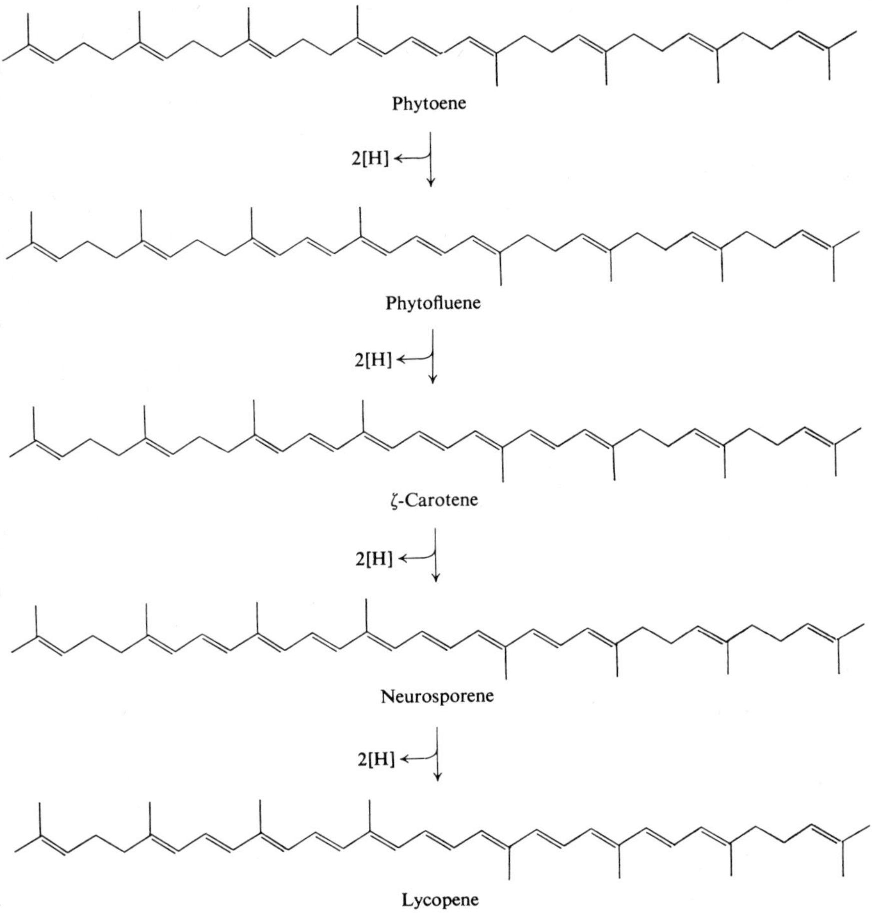

Phytoene

$2[H] \leftarrow$

Phytofluene

$2[H] \leftarrow$

ζ-Carotene

$2[H] \leftarrow$

Neurosporene

$2[H] \leftarrow$

Lycopene

FIG. 4. The biosynthetic pathway from phytoene to lycopene.

elucidation of the structure of neoxanthin allows one to make the reasonable assumption that it arises by isomerization of violaxanthin under the influence of a postulated enzyme, violaxanthin isomerase (Eq. 1).

We now have considered the formation of the major carotenoids in many green algae and incidentally the chloroplasts of all higher plants. Thus apart from the general enzymes which must lead to the immediate acyclic precursor

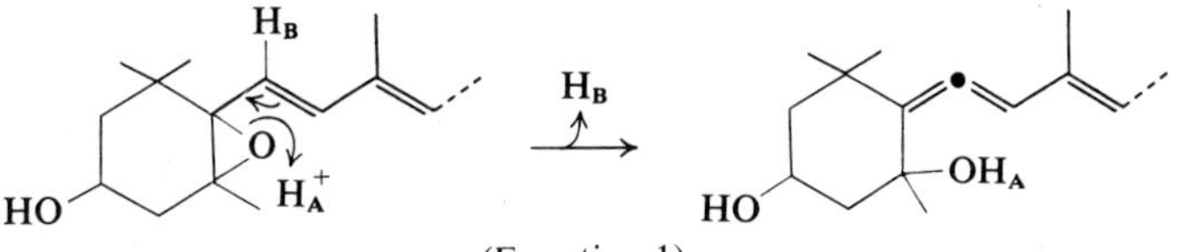

H_B HO O H_A^+ $\xrightarrow{H_B}$ HO OH_A

(Equation 1)

(? lycopene), the following enzymes must have evolved to produce the pigment pattern observed (1) α-cyclase; (2) β-cyclase; (3) α- or β-carotene hydroxylase; (4) epoxidase; (5) violaxanthin isomerase.

With this basic information in mind a taxonomic and evolutionary scheme (Fig. 5) can be constructed which can be compared with other schemes constructed on different bases.

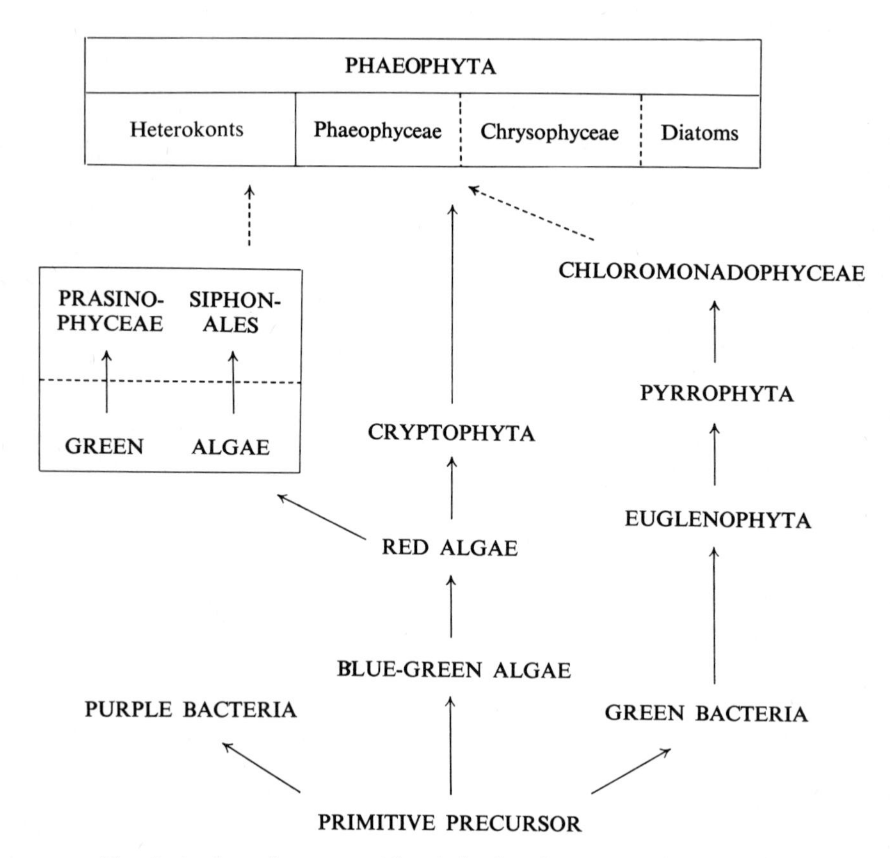

FIG. 5. A scheme for carotenoid evolution based on carotenoid distribution.

If one assumes that the primitive precursor of all algae can synthesize only acyclic carotenoids then the first class to emerge is the blue-green algae. Only β-cyclase is present because only β-carotene derivatives are present (β-carotene, echinenone, canthaxanthin). However, the other major xanthophyll myxoxanthophyll is an acyclic carotenoid glycoside. This suggests a close relationship of the blue-green algae with bacteria because outside the blue-green algae carotenoid glycosides are found only in bacteria, and acyclic carotenoids with hydroxyl groups at C-1 only in photosynthetic bacteria. The red algae could have developed from the blue-greens by evolving an α-cyclase (α-carotene and lutein are frequently present) and a 3-hydroxylase (zeaxanthin and lutein are the major xanthophylls). In so doing they must have lost or repressed the ability to insert oxygen at C-4, or alternatively the blue-green evolved this enzyme after the reds had branched off. It is also likely that in developing the cyclizing enzymes the red algae lost the enzymes concerned with glycosylating acyclic carotenoids because, as just indicated, these do seem to be characteristic of primitive forms.

Xanthophyll epoxides are rare in red algae and neoxanthin is of doubtful occurrence; therefore the main evolutionary development in the appearance of the Chlorophyta is the development of a 5,6-epoxidase to produce antheraxanthin and violaxanthin, and a violaxanthin isomerase to produce neoxanthin. This allows the chlorophyta to produce plastid pigments similar to those found in the higher plants, which have presumably evolved from the Chlorophyta. Some Chlorophyceae have developed a reaction unique to algae, oxidation of an in-chain methyl to produce loroxanthin, which at the moment appears to be unpredictably distributed (p. 318).

This in-chain oxidation of a methyl group is also a characteristic of the unique pigment of the Siphonales, siphonaxanthin. It is not known where this oxidation occurs in the biosynthetic sequence leading to siphonaxanthin but a possible way of inserting the keto group in this pigment is via a triple bond as indicated in Equation 2. However, in *Codium fragile* which has been extensively studied, no acetylenic carotenoids could be detected (Walton *et al.*, 1970). The Siphonales thus separate themselves from other Chlorophyta by having

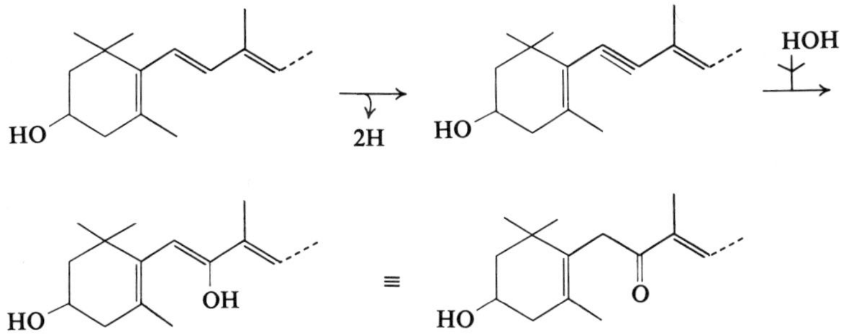

HO … $\xrightarrow{-2H}$ HO … $\xrightarrow{HOH}$

HO … OH ≡ HO … O

(Equation 2)

evolved further specific oxidizing enzymes. Another unique feature of the Siphonales is the fact that siphonein, which is a major component, is esterified siphonaxanthin. The major component fatty acid in *Codium fragile* is a dodecenoic acid (Walton *et al.*, 1970); the position of the double bond and the location of the acid in the molecule is still undecided. If as is supposed, siphonein is in the chloroplast then this is a rare occurrence; the only other esterified carotenoid found in chloroplasts is fucoxanthin, which is acetylated.

One group of the Prasinophyceae (e.g. *Micromonas* sp.) synthesize a pigment, micronone, which has similarities with siphonaxanthin; others synthesize siphonaxanthin and yet others synthesize "normal" plastid pigments (Ricketts, 1967, 1970) (p. 321).

The Cryptophyta could have evolved from the Rhodophyta by elaborating an enzyme which would catalyse the oxidation of an ethylenic bond to an acetylenic bond, that is it would convert zeaxanthin into alloxanthin. The mechanism involved is not known but the simplest possibility is the direct dehydrogenation of the appropriate double bond.

In the Phaeophyta, the Phaeophyceae, Chrysophyceae and Bacillariophyceae all synthesize fucoxanthin, the keto function of which could arise by the reaction indicated in Equation (2) for siphonaxanthin. Appropriate acetylenic precursors are present in the Chrysophyceae and Bacillariophyceae and have recently been detected in some Phaeophyceae. The Heterokontae are the only members of the Phaeophyta which do not synthesize fucoxanthin but *Vaucheria* synthesizes a pigment which is probably of the fucoxanthin/siphonein type.

As the Phaeophyta appear to show little, if any, α-cyclase activity it is possible that they might arise not from the red algae and Cryptomonads as previously proposed (Goodwin, 1964) but by a parallel pathway from blue-green algae via the Euglenophyta and Chloromonadophyceae. *Euglena* spp. synthesize only β-carotene derivatives and their major xanthophyll is the acetylenic diadinoxanthin. Traces of echinenone, the characteristic blue-green xanthophyll are also present, probably in the eye-spot. The only Chloromonadophyceae examined so far have similar pigments to the heterokonts.

The Pyrrophyta are probably closely related to the Phaeophyta, because the characteristic pigment of many members is peridinin which is similar to fucoxanthin: indeed fucoxanthin is the major xanthophyll in some genera (p. 333).

V. General Conclusion

In considering algal carotenoids in relation to taxonomy and evolution both the mandatory requirement for carotenoids in the photosynthetic apparatus and the basic pattern of biosynthesis must be borne in mind. This means that the significant changes to be looked for are variations appearing towards the end of the biosynthetic pathway, that is from lycopene onwards. Most of these changes can be considered as the result of single enzyme changes, mainly those which bring about isomerization or oxidation. The appearance of a particular "terminal" carotenoid is likely to be much more useful than quantitative studies on the pigment components and the detection or otherwise of normal biosynthetic intermediates.

All these thoughts on algal evolution and taxonomy as seen through the eyes of a carotenoid biochemist should be balanced by a consideration of other parameters discussed in detail by Klein and Cronquist (1967).

REFERENCES

Aach, H. G. (1953). *Arch. Mikrobiol.* **19**, 166.
Aaronson, S. and Sher, S. (1960). *J. Protozool.* **7**, 156.
Aaronson, S. and Bensky, B. (1962). *J. gen. Microbiol.* **27**, 75.
Aihara, M. S. and Yamamoto, H. Y. (1968). *Phytochemistry* **7**, 497.
Aitzetmüller, K., Strain, H. H., Svec, W. A., Grandolfo, M. and Katz, J. J. (1969). *Phytochemistry* **8**, 1761.
Allen, M. B. (1954). *Rappt. Comm. 8^e Cong. Int. Bot.* **7**, 41.
Allen, M. B. (1959). *Arch. Mikrobiol.* **32**, 270.
Allen, M. B., Goodwin, T. W. and Phagpolngarm, S. (1960). *J. gen. Microbiol.* **23**, 93.
Allen, M. B., Fries, L., Goodwin, T. W. and Thomas, D. M. (1964). *J. gen. Microbiol.* **34**, 259.
Anderson, I. C. and Robertson, D. S. (1960). *Plant Physiol.* **35**, 531.
Bamji, M. S. and Krinsky, N. I. (1965). *J. biol. Chem.* **240**, 467.
Batra, P. P. and Tollin, G. (1964). *Biochim. Biophys. Acta* **79**, 371.
Berkaloff, C. (1967).
Blum, J. J. and Bégin-Heick, N. (1967). *Biochem. J.* **105**, 821.
Bold, H. C. and Parker, B. C. (1962). *Arch. Mikrobiol.* **42**, 267.
Bunt, J. S. (1964). *Nature, Lond.* **203**, 1261.
Carter, P. W., Heilbron, I. M. and Lythgoe, B. (1939). *Proc. r. Soc.* **128B**, 82.
Carter, P. W., Cross, L. C., Heilbron, I. M. and Jones, E. R. H. (1948). *Biochem. J.* **43**, 349.
Chapman, D. J. (1965). Ph.D. Thesis, University of California.
Chapman, D. J. (1966a). *Arch. Mikrobiol.* **55**, 17.
Chapman, D. J. (1966b). *Phytochemistry* **5**, 1331.
Chapman, D. J. and Haxo, F. T. (1963). *Pl. Cell Physiol.* **4**, 56.
Chapman, D. J. and Haxo, F. T. (1966). *J. Phycol.* **2**, 89.
Chodat, F. (1938). *Arch. Sci. Phys. Nat. Genève* **20**, 96.
Chodat, F. and Haag, E. (1940). *Arch. Sci. Phys. Nat. Genève* **57**, 265.
Claes, H. (1954). *Z. Naturf.* **9b**, 462.
Claes, H. (1956). *Z. Naturf.* **11b**, 260.
Claes, H. (1966b). *In* "Biochemistry of Chloroplasts" (T. W. Goodwin, ed.), p. 441, Vol. II, Academic Press, London and New York.
Claes, H. and Nakayama, T. O. M. (1959). *Nature, Lond.* **183**, 1053.
Czygan, F. C. (1964). *Experientia* **20**, 573.
Czygan, F. C. (1966). *Z. Naturf.* **21b**, 197.
Czygan F. C. (1968b). *Arch. Mikrobiol.* **61** 81.
Czygan F. C. (1968b). *Arch. Mikrobiol.* **62** 209.
Czygan, F. C. and Kalb, K. (1966). *Z. Pflanzenphysiol.* **55**, 59.
Czygan, F. C. and Kessler, E. (1967). *Z. Naturf.* **22b**, 1085.
Dales, R. P. (1960). *J. Marine Biol. Assoc. U.K.* **39**, 693.
de Nicola, M. G. (1961). *Boll. Ist. Bot. Univ. Catania* **2**, 35.
de Nicola, M. G. and di Benedetto, G. (1962). *Boll. Ist. Bot. Univ. Catania*, Sec. IV **3**, 22.
de Nicola, M. and Funari, F. (1954). *Boll. Ist. Bot. Univ. Catania* **1**, 149.
de Nicola, M. G. and Furnari, F. (1957). *Boll. Ist. Bot. Univ. Catania* **1**, 180.
de Nicola, M. G. and Tomaselli, R. (1961). *Boll. Ist. Bot. Univ. Catania* **2**, 22.
Dentice di Accadia, F., Gribanovski-Sassu, O., Romagnoli, A. and Tuttobello, L. (1966). *Biochem. J.* **107**, 735.

Dentice di Accadia, F., Gribanovski-Sassu, O. and Reyes, N. L. (1968). *Experientia* **24**, 1177.
Dersch, G. (1960). *Flora* **149**, 566.
Droop, M. R. (1954). *Arch. Mikrobiol.* **20**, 391.
Dutton, H. J. and Juday, C. (1944). *Ecology* **25**, 273.
Dutton, H. J., Manning, W. M. and Duggar, B. M. (1943). *J. Physical Chem.* **47**, 308.
Egger, K., Nitsche, H. and Kleinig, H. (1969). *Phytochemistry* **8**, 1583.
Emerson, R. and Lewis, C. M. (1943). *Amer. J. Bot.* **30**, 165.
Falk, H. and Kleinig, H. (1968). *Arch. Mikrobiol.* **61**, 347.
Foote, C. (1969). *J. Am. chem. Soc.*
Fox, D. L. and Sargent, M. C. (1938). *Chemy Ind.* **57**, 1111.
Glover, J. and Shah, P. P. (1957). *Biochem. J.* **67**, 15P.
Goedheer, J. C. (1969). *Biochim. biophys. Acta* **172**, 252.
Goodwin, T. W. (1952). "Comparative Biochemistry of the Carotenoids", Chapman & Hall, London.
Goodwin, T. W. (1954). *Experientia* **10**, 213.
Goodwin, T. W. (1956). *Biochem. J.* **63**, 481.
Goodwin, T. W. (1957). *J. gen. Microbiol.* **14**, 467.
Goodwin, T. W. (1964). *In* "Chemistry and Biochemistry of Plant Pigments" (T. W. Goodwin, ed.), Academic Press, London and New York.
Goodwin, T. W. and Taha, M. M. (1951). *Biochem. J.* **47**, 513.
Goodwin, T. W. and Jamikorn, M. (1954a). *Biochem. J.* **57**, 376.
Goodwin, T. W. and Jamikorn, M. (1954b). *J. Protozool.* **1**, 216.
Goodwin, T. W. and Gross, J. A. (1958). *J. Protozool.* **5**, 292.
Gray, R. A. and Hendlin, D. (1962). *Pl. Physiol.* **37**, 223.
Gross, J. A., Jahn, T. L. and Bernstein, E. (1955). *J. Protozool.* **2**, 71.
Gross, J. A. and Jahn, T. L. (1958). *J. Protozool.* **5**, 126.
Haag, E. (1941). *C. r. Phys. Hist. Nat. Genève* **58**, 288, 291.
Hager, A. (1967a). *Planta* **74**, 148.
Hager, A. (1967b). *Planta* **76**, 138.
Hager, A. (1969). *Planta* **89**, 224.
Halldal, P. (1958). *Physiol. Plant.* **11**, 118.
Haxo, F. T. (1960). *In* "Comparative Biochemistry of Photoreactive Pigments" (M. B. Allen, ed.), p. 339, Academic Press, New York and London.
Haxo, F. T. and Clendenning, K. A. (1953). *Biol. Bull.* **105**, 103.
Heilbron, I. M. (1942). *J. Chem. Soc.* 79.
Heilbron, I. M. and Phipers, R. F. (1935). *Biochem. J.* **29**, 1369.
Heilbron, I. M. and Lythgoe, B. (1936). *J. chem. Soc.* 1376.
Heilbron, I. M., Lythgoe, B. and Phipers, R. F. (1935a). *Nature, Lond.* **136**, 989.
Heilbron, I. M., Jackson, H. and Jones, R. N. (1935b). *Biochem. J.* **29**, 1384.
Helmy, F. M., Hack, M. H. and Yaeger, R. G. (1967). *Comp. Biochem. Physiol.* **23**, 565.
Herrmann, R. G. (1968). *Protoplasma* **66**, 387.
Hertzberg, S. and Liaaen-Jensen, S. (1966a). *Phytochemistry* **5**, 557.
Hertzberg, S. and Liaaen-Jensen, S. (1966b). *Phytochemistry* **5**, 565.
Hertzberg, S. and Liaaen-Jensen, S. (1967). *Phytochemistry* **6**, 1119.
Hertzberg, S. and Liaaen-Jensen, S. (1969a). *Phytochemistry* **8**, 1259.
Hertzberg, S. and Liaaen-Jensen, S. (1969b). *Phytochemistry* **8**, 1281.
Hirose, H. (1950). *Bot. Mag.* **63**, 107.
Huzisige, H., Terada, T., Nishimura, M. and Nemura, T. (1957). *Biol. J. Okayama Univ.* **3**, 209.
Jeffrey, S. W. (1961). *Biochem. J.* **80**, 336.
Jeffrey, S. W. and Allen, M. B. (1964). *J. gen. Microbiol.* **36**, 277.

Jeffrey, S. W. and Haxo, F. T. (1968). *Biol. Bull.* **135**, 149.
Jensen, A. (1966). Report No. 31 Norwegian Institute of Seaweed Research.
Karrer, P., Fatzer, W., Favarger, M. and Jucker, E. (1943). *Helv. Chim. Acta* **26**, 2121.
Karrer, P. and Rutschmann, J. (1944). *Helv. Chim. Acta* **27**, 1691.
Kessler, E. and Czygan, F. C. (1965). *Ber. dtsch. Bot. Ges.* **78**, 342.
Kessler, E. and Czygan, F. C. (1966). *Arch. Mikrobiol.* **54**, 37.
Kessler, E. and Czygan, F. C. (1967). *Arch. Mikrobiol.* **55**, 320.
Kessler, E., Czygan, F. C., Fott, B. and Norakova, M. (1968). *Arch. Protistenkd.* **110**, 462.
Klein, R. M. and Cronquist, A. (1967). *Q. Rev. Biol.* **42**, 108.
Kleinig, H. (1967). *Z. Naturf.* **22b**, 977.
Kleinig, H. and Egger, K. (1967a). *Phytochemistry* **6**, 1681.
Kleinig, H. and Egger, K. (1967b). *Z. Naturf.* **22b**, 868.
Kleinig, H. and Czygan, F. C. (1969). *Z. Naturf.* **24b**, 927.
Krinsky, N. (1966). *In* "Biochemistry of Chloroplasts", Vol. 1 (T. W. Goodwin, ed.), p. 423, Academic Press, London and New York.
Krinsky, N. I. and Goldsmith, T. H. (1960). *Archs Biochem. Biophys.* **91**, 271.
Krinsky, N. I. and Levine, R. P. (1964). *Pl. Physiol.* **39**, 680.
Krinsky, N. I., Gordon, A. and Stern, A. I. (1964). *Pl. Physiol.* **39**, 441.
Kuhn, R. and Brockmann, H. (1932). *Hoppe-Seyler's Z. Physiol. Chem.* **213**, 192.
Kuhn, R., Stene, J. and Sörensen, N. A. (1939). *Ber. dtsch. Chem. Ges.* **72**, 1688.
Kylin, H. (1927). *Hoppe-Seyler's Z. Physiol. Chem.* **166**, 39.
Lang, N. J. (1968). *J. Phycol.* **4**, 12.
Larsen, B. and Haug, A. (1956). *Acta Chem. Scand.* **10**, 470.
Liaaen, S. and Sörensen, N. A. (1956). *Proc. 2nd Int. Seaweed Symp. Trondheim.* (T. Braaud and N. A. Sorensen, eds), Pergamon Press, Oxford.
Liaaen-Jensen, S. (1971). *In* "Aspects of Terpenoid Biochemistry" (T. W. Goodwin, ed.), p. 223, Academic Press, London and New York.
Links, J., Verloop, A. and Havinga, E. (1960). *Arch. Mikrobiol.* **36**, 306.
Loeblich, L. A. (1969). *J. Protozool.* **16**, Suppl. p. 22.
Loefer, J. B. and Guido, V. M. (1950). *Texas J. Sci.* **2**, 225.
Lunde, G. (1937). *Tekn. Ukeblad.* **84**, 192.
Lwoff, M. and Lwoff, A. (1930). *C. r. Soc. Biol., Paris* **105**, 454.
McLean, R. (1967). *Physiol. Plant.* **20**, 41.
Mallams, A. K., Waight, E. S., Weedon, B. C. L., Chapman, D. J., Haxo, F. T., Goodwin, T. W. and Thomas, D. M. (1967). *Proc. chem. Soc.* 301.
Mandelli, E. F. (1968). *J. Phycol.* **4**, 347.
Manten, A. (1948). Phototaxis, phototropism and photosynthesis in purple bacteria and blue-green algae. Thesis. Utrecht.
Mattox, K. R. and J. P. Williams (1965). *J. Phycol.* **1**, 191.
Mayer, F. and Czygan, F. C. (1969). *Planta* **86**, 175.
Olson, R. A., Jennings, W. R. and Allen, M. B. (1967). *J. Cell. Physiol.* **70**, 133.
Owen, E. C. (1954). *J. Sci. Food Agric.* **5**, 449.
Parsons, T. R. (1961). *J. Fish. Res. Bd Canada* **18**, 1017.
Parsons, T. R. and Strickland, J. D. H. (1963). *J. Mar. Res.* **21**, 155.
Pinckard, J. H., Kittredge, J. S., Fox, D. L., Haxo, F. T. and Zechmeister, L. (1953). *Archs Biochem. Biophys.* **44**, 189.
Poulton, E. B. (1930). *New Phytol.* **29**, 1.
Pringsheim, E. G. (1958). *Rev. Algol.* **4**, 41.
Provasoli, L., Hutner, S. H. and Schatz, A. (1948). *Proc. Soc. exp. Biol. Med.* **69**, 279.
Richter, G. (1958). *Planta* **52**, 259.

Ricketts, T. R. (1966). *Phytochemistry* **5**, 571.
Ricketts, T. R. (1967). *Phytochemistry* **6**, 1375.
Ricketts, T. R. (1970). *Phytochemistry* **9**, 1835.
Riley, J. P. and Wilson, T. R. S. (1965). *J. Marine Biol. Ass. U.K.* **45**, 583.
Saenger, P. and Rowan, K. S. (1968). *Helgolaendes Wiss. Meeresunters.* **18**, 549.
Sager, R. and Zalokar, M. (1958). *Nature, Lond.* **182**, 98.
Scheer, B. T. (1940). *J. biol. Chem.* **136**, 275.
Sestak, Z. and Baslerova, O. (1963). *In* "Microalgae and Photosynthetic Bacteria", p. 423, Tokyo.
Seybold, A. and Egle, K. (1938). *Jrb. wiss Bot.* **86**, 50.
Seybold, A., Egle, K. and Hülsbruch, W. (1941). *Bot. Arch.* **42**, 239.
Sironval, C. and Kandler, O. (1958). *Biochim. biophys. Acta* **29**, 359.
Stanier, R. Y. and Cohen-Bazire, G. (1957). *In* "Microbial Ecology" (R. E. O. Williams and C. C. Spicer, eds.), p. 56, Cambridge University Press, London.
Strain, H. H. (1949). *In* "Photosynthesis in Plants", Iowa State College Press.
Strain, H. H. (1951). *In* "Manual of Phycology" (G. M. Smith, ed.), Waltham.
Strain, H. H. (1958). 32nd Annual Priestley Lectures. Penn. State University.
Strain, H. H. (1965). *Biol. Bull. Woods Hole* **129**, 366.
Strain, H. H. (1966). *In* "Biochemistry of Chloroplasts" (T. W. Goodwin, ed.), Vol. 1, Academic Press, London and New York.
Strain, H. H. and Manning, W. M. (1943). *J. Am. chem. Soc.* **65**, 2258.
Strain, H. H., Manning, W. M. and Hardin, G. J. (1944). *Biol. Bull. Woods Hole* **86**, 169.
Strain, H. H., Svec, W. A., Aitzetmüller, K., Grandolfo, M. and Katz, J. J. (1968). *Phytochemistry* **7**, 1417.
Strain, H. H., Benton, F. L., Grandolfo, M. C., Aitzetmüller, K., Svec, W. A. and Katz, J. J. (1970). *Phytochemistry* **9**, 2561.
Stransky, H. and Hager, A. (1970). *Arch. Mikrobiol.* **71**, 164.
Sweeney, B. M., Haxo, F. T. and Hastings, J. W. (1959). *J. gen. Physiol.* **43**, 285.
Tanada, T. (1951). *Am. J. Bot.* **38**, 276.
Thomas, D. M. and Goodwin, T. W. (1965). *J. Phycol.* **1**, 118.
Thomas, D. M., Goodwin, T. W. and Ryley, J. F. (1967). *J. Protozool.* **14**, 654.
Tischer, J. (1936). *Hoppe-Seyler's Z. Physiol. Chem.* **239**, 257.
Tischer, J. (1938). *Hoppe-Seyler's Z. Physiol. Chem.* **251**, 109.
Tischer, J. (1939). *Hoppe-Seyler's Z. Physiol. Chem.* **260**, 257.
Tischer, J. (1941). *Hoppe-Seyler's Z. Physiol. Chem.* **267**, 281.
Tischer, J. (1958). *Hoppe-Seyler's Z. Physiol. Chem.* **311**, 140.
Tsukida, K., Yokota, M., Shimamoto, H. and Cho, S. (1968). *Vitamin* **38**, 388.
Villela, G. (1966). *C. r. Acad. Sci.* Ser. D. 1383.
Walton, T. J., Britton, G. and Goodwin, T. W. (1969). *Biochem. J.* **112**, 383.
Walton, T. J., Britton, G. and Goodwin, T. W. (1970). *Phytochemistry* **9**, 2545.
Wassink, E. C. and Kersten, J. A. H. (1946). *Enzymologia* **12**, 1.
Wenzinger, F. (1940). *Bull. Soc. Bot. Genève* **30**, 129.
Whittle, S. J. and Casselton, P. J. (1969). *Br. Phycol. J.* **4**, 55.
Willstätter, R. and Page, H. J. (1914). *Liebigs Ann.* **404**, 237.
Wolken, J. J., Mellon, A. D. and Greenblatt, C. L. (1955). *J. Protozool.* **2**, 89.
Yamamoto, H. Y. and Chichester, C. O. (1965). *Biochim. biophys. Acta* **109**, 303.
Zahalsky, A. C., Hutner, S. H., Keane, M. and Burger, R. M. (1962). *Arch. Mikrobiol.* **42**, 46.

CHAPTER 12

Biosynthesis of Isoprenoid Quinones and Chromanols

D. R. THRELFALL AND G. R. WHISTANCE

Department of Biochemistry and Agricultural Biochemistry,
University College of Wales, Aberystwyth, Wales

I. Introduction

Since the topic was last reviewed for the Phytochemical Society (Threlfall, 1967) significant advances have been made towards an understanding of how isoprenoid quinones and chromanols are biosynthesized by members of the plant kingdom: for example, the identities of most of the distal and in some cases the more immediate precursors are now known. No less significant are the advances made in the fields of isoprenoid quinone biosynthesis by bacteria and animals. Indeed, in some bacteria a complete pathway for the biosynthesis of ubiquinones has been elucidated.

The present review is concerned primarily with the biosynthesis of ubiquinones, plastoquinones, plastochromanols, tocotrienols, tocopherols, tocotrienolquinones, tocopherolquinones and phylloquinones by higher plants, algae and fungi. Some sections also deal with aspects of the biosynthesis of ubiquinones by bacteria and animals and menaquinones by bacteria, since these have an important bearing on current theories regarding the biosynthesis of ubiquinones and phylloquinones by plants. As the quinones and chromanols under review fall into three biogenetically distinct groups their biosynthesis is dealt with in three separate sections: the first section deals with the biosynthesis of ubiquinones, the second with the biosynthesis of plastoquinones, plastochromanols, tocotrienols, tocopherols, tocotrienolquinones and tocopherolquinones, and the third with the biosynthesis of phylloquinones and menaquinones.

II. Ubiquinones

A. Structure and Distribution

The ubiquinones (1) constitute a family of 5,6-dimethoxy-3-methyl-2-all-*trans*-polyprenyl-1,4-benzoquinones which is distributed widely in nature. With the exception of Gram-positive bacteria and blue-green algae, ubiquinones have been detected in all groups of living organisms examined (see Crane, 1965; Pennock, 1966). Intracellular distribution studies have shown that they are localized in the organelles which carry out respiratory chain

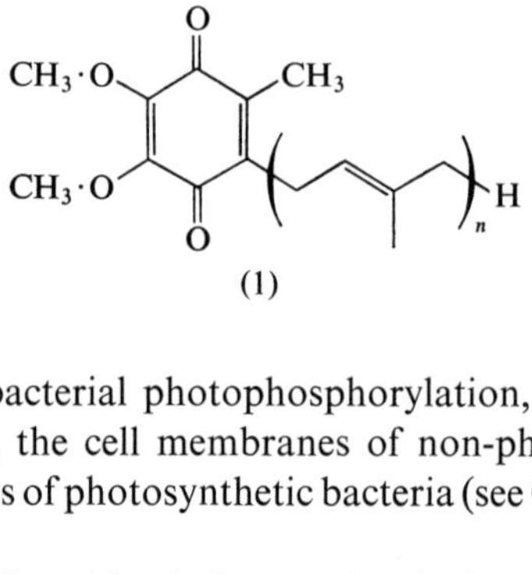

O
$CH_3 \cdot O$ CH_3
$CH_3 \cdot O$ H
O n

(1)

phosphorylation and bacterial photophosphorylation, i.e. the mitochondria of animals and plants, the cell membranes of non-photosynthetic bacteria, and the chromatophores of photosynthetic bacteria (see Crane, 1965; Pennock, 1966).

Ubiquinones possessing side chains varying in length from C_5 (1; $n = 1$) to C_{60} (1; $n = 12$) have been isolated from biological sources (Table I). In general, most organisms contain a series of ubiquinones, with the major homologue [usually ubiquinone-8, -9 or -10 (the numerals refer to the number of isoprene units/mol.)] constituting more than 85% of the total (Lavate and Bentley, 1964; Field, 1969; Whistance *et al.*, 1969b; Threlfall and Whistance, 1970).

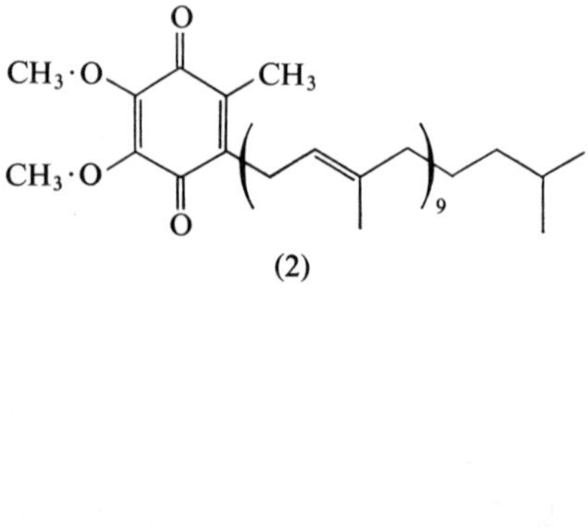

O
$CH_3 \cdot O$ CH_3
$CH_3 \cdot O$
O 9

(2)

Some moulds contain a ubiquinone-10 in which the tenth isoprene unit from the nucleus is saturated (2) (see Pennock, 1966). This compound is known as ubiquinone-10 (X–H_2).

Two aminoquinones, rhodoquinone-9 (3; $n = 9$) and -10 (3; $n = 10$), have been isolated from some members of the Athiorhodaceae (Carr, 1964; Moore and Folkers, 1966; Maroc *et al.*, 1968), strains of the alga *Euglena gracilis* (Powls and Hemming, 1966a; Threlfall and Whistance, 1970) and the parasitic nematodes *Metastrongylus elongatus* and *Ascaris lumbricoides* var. *suis* (Sato and Ozawa, 1969). Biogenetic studies have indicated that in *Rhodospirillum rubrum* and *E. gracilis* they are metabolites of ubiquinones (Parson and Rudney, 1965a; Powls and Hemming, 1966b).

O
NH2 CH3
CH3·O H
n
O
(3)

Monoepoxy derivatives (4) of ubiquinones have been isolated from *R. rubrum*, *Pseudomonas ovalis* Chester, *Proteus mirabilis*, Vibrio 01 (*Moraxella* sp.), bakers' yeast and mouse (Friis *et al.*, 1967a; Nilsson *et al.*, 1968; Whistance *et al.*, 1970; G. R. Whistance and D. R. Threlfall, unpublished observations). At present it is not clear whether these quinones are naturally occurring or whether they are artifacts of the isolation procedures.

O
CH3·O CH3
O
CH3·O H
O x y
(4)

B. BIOSYNTHESIS

1. Para-*Benzoquinone Rings*

The most effective precursors of the *p*-benzoquinone ring of ubiquinones so far discovered are *p*-hydroxybenzaldehyde and *p*-hydroxybenzoic acid. Although *p*-hydroxybenzaldehyde has only been tested in *Rhodospirillum rubrum*, *Azotobacter vinelandii*, *Saccharomyces cerevisiae* (bakers' yeast) and rat kidney (Parson and Rudney, 1964), *p*-hydroxybenzoic acid has been shown to be a precursor of ubiquinones in a wide range of organisms (Table II). It also gives rise to the *p*-benzoquinone rings of ubiquinone-10 (X-H_2) in *Aspergillus flavus* (Ah Law *et al.*, 1970), rhodoquinone-10 in *R. rubrum*

TABLE I
Distribution of ubiquinones

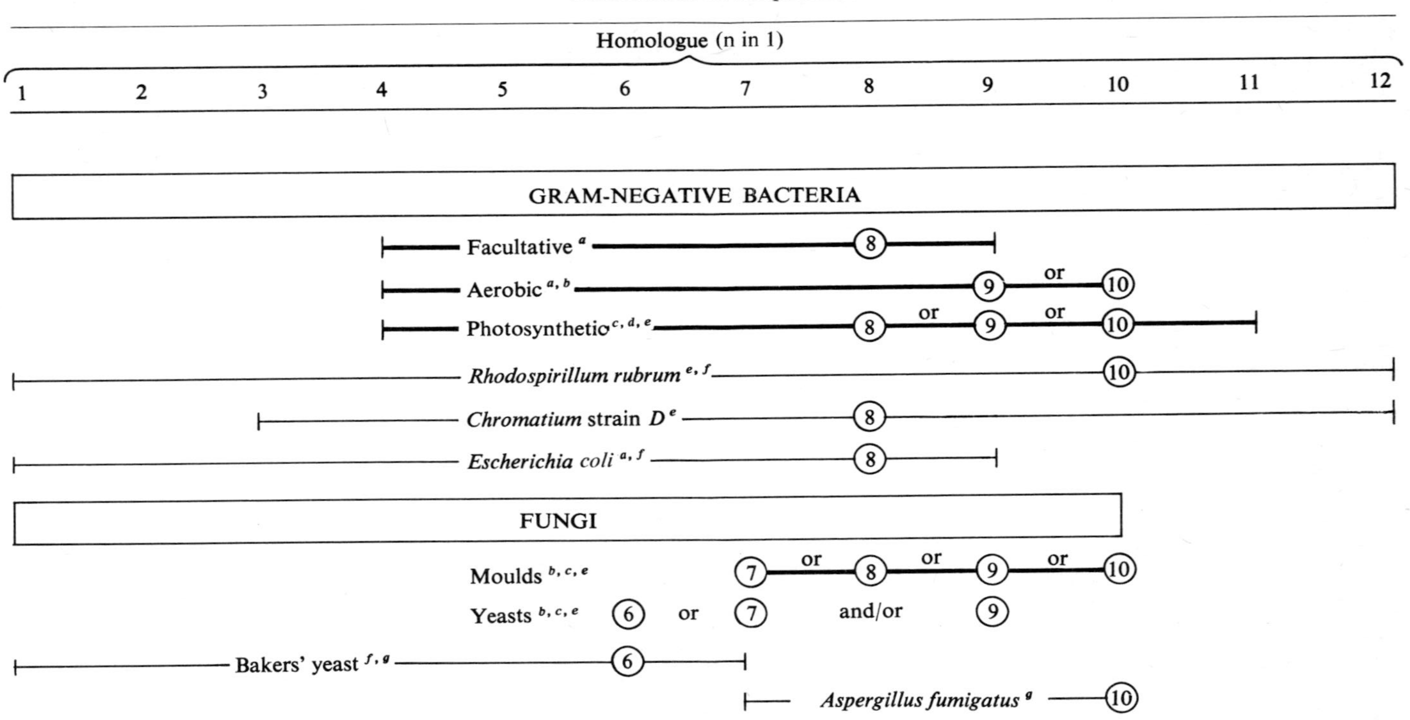

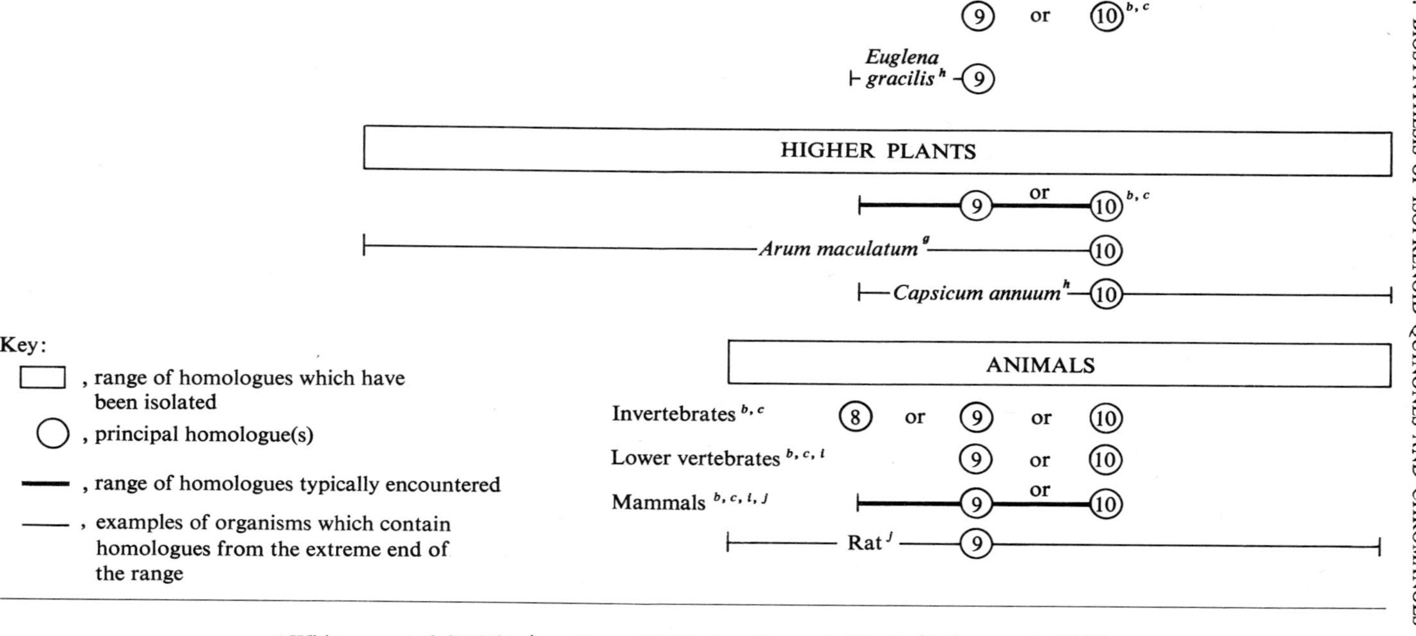

[a] Whistance *et al.* (1969b); [b] see Crane (1965); [c] see Pennock (1966); [d] Maroc *et al.* (1968); [e] G. R. Whistance and D. R. Threlfall, unpublished observations; [f] Daves *et al.* (1967); [g] Ah Law, D. R. Threlfall and G. R. Whistance, unpublished observation; [h] Threlfall and Whistance (1970); [i] Diplock and Haslewood (1967); [j] Field (1969).

TABLE II
Organisms in which *p*-hydroxybenzoic acid is a precursor of the ubiquinone nucleus

Organism	Homologue(s)[a]	Organism	Homologue(s)[a]
PHOTOSYNTHETIC BACTERIA		MOULDS	
Rhodospirillum rubrum	11,⑩,9,8,7,6,5,4[b,c,d]	*Agaricus campestris*	⑨[k]
Rhodopseudomonas spheroides	11,⑩,9,8,7,6,5,4[d,e]	*Phycomyces blakesleeanus*	⑨,8,7[j,k]
R. palustris	⑩[e]		
R. capsulatus	⑩[e]	ALGAE	
Chromatium strain D	10,9,⑧,7,6,5,4[d]	*Euglena gracilis*	⑨,8[l,m]
		Ochromonas danica	⑩,9[d,n]
NON-PHOTOSYNTHETIC BACTERIA			
Azotobacter vinelandii	⑧[f]	HIGHER PLANTS	
Escherichia coli	⑧,7,6,5,4[d,e,g,h,i]	*Phaseolus vulgaris*	⑩,9[d,n]
Organism PC4 (*Achromobacter* sp.)	⑨,8,7,6,5[d,i]	*Zea mays*	11,10,⑨,8[d,m,n]
Pseudomonas denitrificans	⑩[e]		
P. ovalis Chester	10,⑨,8,7,6,5,4[d,i]	PROTOZOA	
P. fluorescens	⑨,8,7,6,5[d,i]	*Plasmodium knowlesi*	9,8,7[o]
Vibrio 01 (*Moraxella* sp.)	10,⑨,8,7,6[d,i]		
		MAMMALS	
YEASTS		Monkey	⑩[o]
Rhodotorula glutinis	⑨,8[j]	Mouse	10,⑨,8[p,q]
Saccharomyces cerevisiae	⑥,5,4[k]	Rat	11,10,⑨,8,7[p,q,r,s]

[a] ◯, principal homologue; [b] Parson and Rudney (1965a); [c] Whistance *et al.* (1966a); [d] G. R. Whistance and D. R. Threlfall, unpublished observations; [e] Rudney and Raman (1966); [f] Parson and Rudney (1964); [g] Cox and Gibson (1964); [h] Jones and Lascelles (1967); [i] Whistance *et al.* (1969b); [j] Ah Law, D. R. Threlfall and G. R. Whistance (1970), in preparation; [k] Spiller *et al.* (1968); [l] Powls and Hemming (1966b); [m] Threlfall and Whistance (1970); [n] Whistance *et al.* (1967); [o] Skelton *et al.* (1969); [p] Nilsson *et al.* (1968); [q] Field (1969); [r] Olson *et al.* (1963); [s] Olson (1966).

(Parson and Rudney, 1965a; Whistance *et al.*, 1966a), rhodoquinone-9 in *Euglena gracilis* (Powls and Hemming, 1966b; Threlfall and Whistance, 1970), epoxyubiquinones-9 in *Pseudomonas ovalis* Chester and epoxyubiquinones-6 in bakers' yeast (G. R. Whistance and D. R. Threlfall, unpublished observations).

In higher plants *p*-hydroxybenzoic acid can be formed from shikimic acid by pathways involving the aromatic amino acids phenylalanine and tyrosine (Fig. 1), and there is now good evidence that these pathways are operative in the biosynthesis of ubiquinones. Thus, ^{14}C-tracer studies have established that in maize (*Zea mays*) shoots and French bean (*Phaseolus vulgaris*) shoots D-shikimic acid, L-phenylalanine, L-tyrosine, cinnamic acid, *p*-coumaric acid, and *p*-hydroxybenzoic acid are all precursors of ubiquinones (Whistance *et al.*, 1967; Threlfall *et al.*, 1970; G. R. Whistance and D. R. Threlfall, unpublished observations). Further, isotope dilution studies have shown that in the above tissues the incorporation into ubiquinones of D-[U-^{14}C] shikimic acid is reduced in the presence of L-phenylalanine and L-tyrosine, that that of L-[U-^{14}C] phenylalanine and L-[U-^{14}C] tyrosine is reduced in the presence of cinnamic acid or *p*-coumaric acid, and that that of [U-^{14}C] cinnamic acid and *p*-[U-^{14}C] coumaric acid is reduced in the presence of *p*-hydroxybenzoic acid (G. R. Whistance and D. R. Threlfall, unpublished observations). It is of interest that the pathway from tyrosine is similar to that which exists in animals (Fig. 2).

So far only a few studies have been carried out on the formation of the *p*-hydroxybenzoic acid utilized for the biosynthesis of ubiquinones in algae and fungi. However, the observations that shikimic acid is a precursor of ubiquinones in bakers' yeast (Spiller *et al.*, 1968) and that cinnamic acid, *p*-coumaric acid and *p*-hydroxybenzoic acid are precursors of ubiquinones in *Euglena gracilis*, *Phycomyces blakesleeanus* and bakers' yeast (Threlfall *et al.*, 1970), provide some evidence that it is formed by pathways similar to those in higher plants. The possibility must also be considered that it can be formed directly from chorismic acid, as it is in bacteria (Fig. 1) (Gibson and Pittard, 1968), although at present there is no evidence for this.

2. Polyprenyl Side Chains

Mevalonic acid is now firmly established as the specific distal precursor of the polyprenyl side chains of ubiquinones in animals, fungi and higher plants (see Glover, 1965; Threlfall, 1967). However, its involvement in ubiquinone biosynthesis in algae and bacteria has still to be demonstrated. The pathway by which mevalonic acid gives rise to isoprenoid units has been fully reviewed by Clayton (1965).

It is generally believed that the polyprenyl units of ubiquinones are synthesized independently of the nuclei as polyprenylpyrophosphates, the pyrophosphate unit of requisite length then coupling with the appropriate nucleus (see Threlfall, 1967). Indirect evidence in support of this belief comes

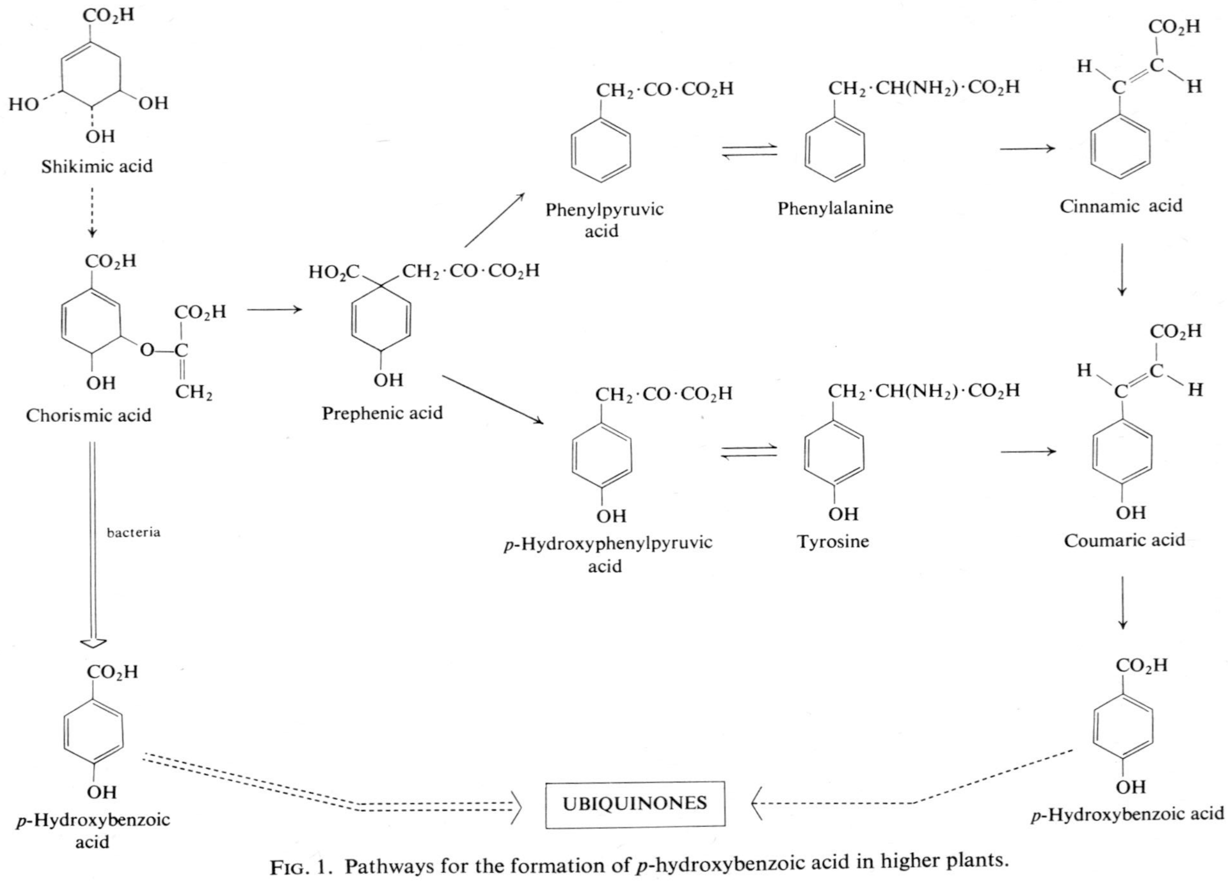

FIG. 1. Pathways for the formation of *p*-hydroxybenzoic acid in higher plants.

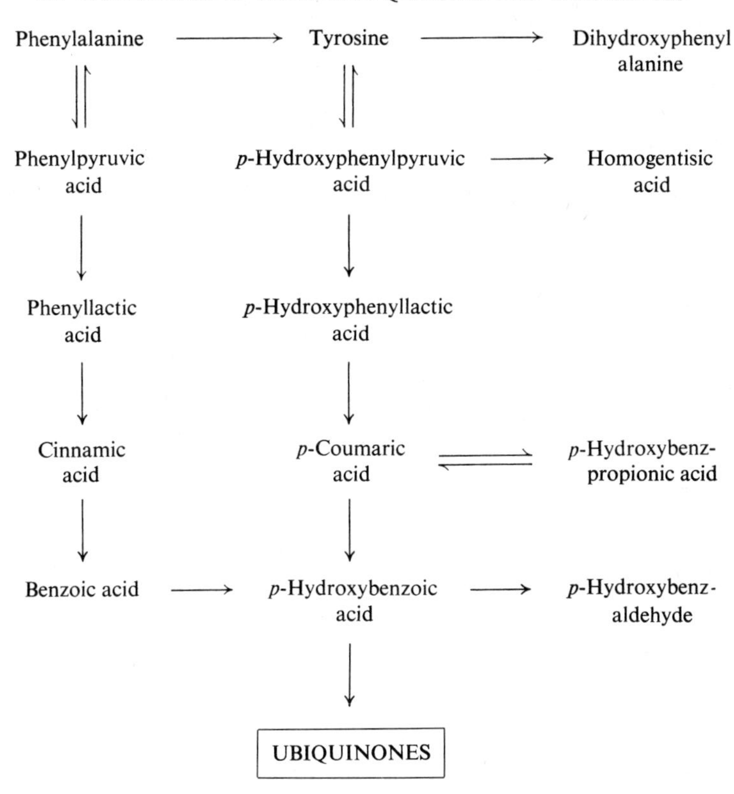

FIG. 2. Pathways for the formation of *p*-hydroxybenzoic acid in animals (after Olson, 1966).

from the fact that polyprenylalcohols of the correct length and configuration, i.e., spadicol (all-*trans*-decaprenol) and solanesol (all-*trans*-nonaprenol) (Hemming, *et al.*, 1963; Kofler *et al.*, 1959) and polyprenylpyrophosphate synthetases occur in nature. [The synthetase from *Micrococcus lysodeikticus* can synthesize from di- to deca-prenylpyrophosphate, with hepta- and octa-prenylpyrophosphate predominating (Allen, C. M. *et al.*, 1967)]. Direct evidence comes from the reports that (a) cell-free extracts of *Rhodospirillum rubrum* (Raman *et al.*, 1969), rat kidney, brain and liver (Winrow and Rudney, 1969), when supplemented with polyprenylpyrophosphate synthetase from *M. lysodeikticus* and isopentenylpyrophosphate, can carry out the prenylation of *p*-hydroxybenzoic acid, i.e. an authenticated step in the biosynthesis of ubiquinones by bacteria (see below), and (b) rat liver mitochondria are capable of synthesizing ubiquinone-9 from ubiquinone-0 and nonaprenyl-pyrophosphate (Stoffel and Martius, 1960).

Through the use of (3*RS*)-[2-^{14}C,(4*R*)-4-^{3}H$_1$] mevalonic acid and (3*RS*)-[2-^{14}C,(4*S*)-4-^{3}H$_1$] mevalonic acid it has been possible to study the stereochemical aspects of the biosynthesis of the nonaprenyl side chain of ubiquinone-9 in *Aspergillus fumigatus* Fresenius and maize (Stone and Hemming, 1967; Dada *et al.*, 1968). The results obtained showed that in both cases the side chain is biogenetically *trans*.

3. *Nuclear* C *and* O-*Methyl Groups*

Numerous investigations have shown that in algae, animals, bacteria, fungi and higher plants the nuclear *C*- and *O*-methyl groups of ubiquinones are derived from the *S*-methyl group of L-methionine (or more precisely *S*-adenosylmethionine) (Table III). Studies with L-[*Me*-^{14}C,$^{3}H_3$] methionine and L-[*Me*-$^{2}H_3$] methionine have shown that in *Escherichia coli* (Jackman *et al.*, 1967) and maize (Threlfall *et al.*, 1968) transmethylation takes place (Fig. 3).

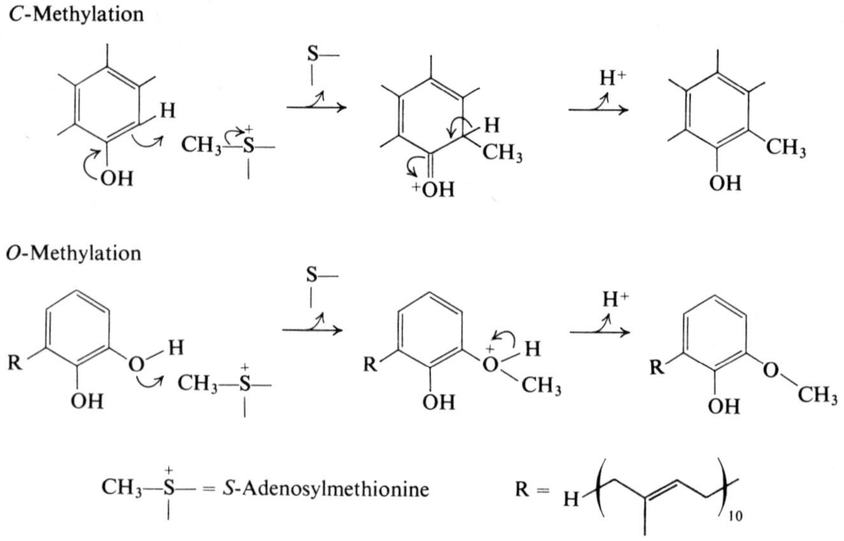

C-Methylation

O-Methylation

CH_3—$\overset{+}{S}$— = *S*-Adenosylmethionine   R = H(—C₅—)₁₀

FIG. 3. Proposed mechanisms of *C*- and *O*-methylation in isoprenoid quinone and chromanol biosynthesis.

4. *Pathways from* p-*Hydroxybenzoic Acid*

The most enlightening findings in respect of this problem have come from studies with Gram-negative bacteria. The initial break-through was made by Parson and Rudney (1965b) who found that when cells of the photosynthetic bacterium *Rhodospirillum rubrum* were incubated under dark, anaerobic conditions with *p*-hydroxy[U-^{14}C] benzoic acid radioactivity accumulated in a lipid, compound X, which on illumination of the cells was converted into ubiquinone-10. Further, through the use of [^{14}C] acetate, *p*-hydroxy [*carboxyl*-^{14}C] benzoic acid and L-[*Me*-^{14}C] methionine these workers showed that compound X possessed a polyprenyl side chain but no nuclear methyl groups, and that during its formation the carboxyl group of *p*-hydroxybenzoic acid

TABLE III

Organisms in which the *S*-methyl group of L-methionine is the source of the nuclear *C*- and *O*-methyl groups of ubiquinones

Organism	Homologue	Organism	Homologue
PHOTOSYNTHETIC BACTERIA		ALGAE	
Rhodospirillum rubrum	10[a,b]	*Euglena gracilis*	9[b]
Chromatium strain D	8[b]		
		HIGHER PLANTS	
NON-PHOTOSYNTHETIC BACTERIA		*Hedera helix*	10[g]
Escherichia coli	8[c]	*Zea mays*	9[h]
Pseudomonas ovalis	9[d]		
		MAMMALS	
YEASTS		Rat	9[i]
Saccharomyces cerevisiae	6[e,f]		

[a] Parson and Rudney (1965b); [b] G. R. Whistance and D. R. Threlfall, unpublished observations; [c] Jackman *et al.* (1967); [d] Whistance *et al.* (1969b); [e] Rudney and Sugimura (1961); [f] Spiller *et al.* (1968); [g] Whistance and Threlfall (1968b); [h] Threlfall *et al.* (1968); [i] Olson (1966).

was lost. Subsequently, Olsen *et al.* (1965) characterized compound X fully and showed it to be 2-decaprenylphenol (5).

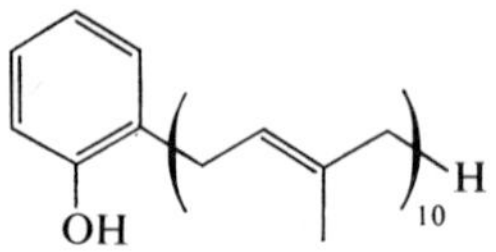

OH H 10

(5)

The discovery of 2-decaprenylphenol prompted Folkers and co-workers (Olsen *et al.*, 1966a,b; Friis *et al.*, 1966; Daves *et al.*, 1966; Friis *et al.*, 1967b) to fractionate a massive amount of lipid from *R. rubrum* in an attempt to find other prenylated precursors of ubiquinones. This led to the isolation of a series of decaprenylphenols and quinones (small amounts of the tetra- through nona-prenyl forms of some of the compounds were also isolated) which are generally believed to be precursors of ubiquinone-10 in *R. rubrum* (Fig. 4). [A similar group of phenols and quinones has also been detected in two other photo-synthetic bacteria, *Rhodopseudomonas spheroides* and *Chromatium* strain D (G. R. Whistance and D. R. Threlfall, unpublished observations)].

Whistance *et al.* (1969b) examined a number of non-photosynthetic bacteria for the presence of polyprenylphenols of the type shown to be precursors of ubiquinones in *Rhodospirillum rubrum*. Of the 22 organisms examined all (10) of the facultative anaerobes and half (6) of the obligate aerobes contained both 2-polyprenylphenols (7, Fig. 4) and 6-methoxy-2-polyprenylphenols (8, Fig. 4), while the other six obligate aerobes contained neither [two (Vibrio 01 and organism PC4) did, however, contain 5-demethoxyubiquinones (Whistance *et al.*, 1969a, 1970; G. R. Whistance and D. R. Threlfall, unpublished observations)]. Whistance *et al.* (1969a, 1970) showed that *Pseudomonas ovalis* Chester and *Proteus mirabilis* contain, in addition to 2-polyprenyl-phenols and 6-methoxy-2-polyprenylphenols, 6-methoxy-2-polyprenyl-1,4-benzoquinones (9, Fig. 4), 2-polyprenyl-1,4-benzoquinones (?) and 5-de-methoxyubiquinones (10, Fig. 4) [the presence of 6-methoxy-2-nonaprenyl-phenol and 5-demethoxyubiquinone-9 in *Pseudomonas ovalis* (species undefined) had been reported previously by Imamoto and Senoh (1968)]. Further, it was shown that in *P. ovalis* Chester 6-methoxy-2-polyprenylphenols, 2-polyprenylphenols, 2-polyprenyl-1,4-benzoquinones (?), 6-methoxy-2-poly-prenyl-1,4-benzoquinones and 5-demethoxyubiquinones are precursors of ubiquinones, which provides good evidence that pathway A (Fig. 4) and an alternative pathway involving 2-polyprenyl-1,4-benzoquinones (?) are opera-tive for the biosynthesis of ubiquinones in this organism.

Recently, investigations have been carried out using bacterial cell extracts. Raman *et al.* (1969) have shown that cell-free extracts of *R. rubrum* (prepared by a combination of lysis with lysozyme and osmotic shock or by sonication) will incorporate radioactivity from *p*-hydroxy[U-^{14}C]-benzoic acid into

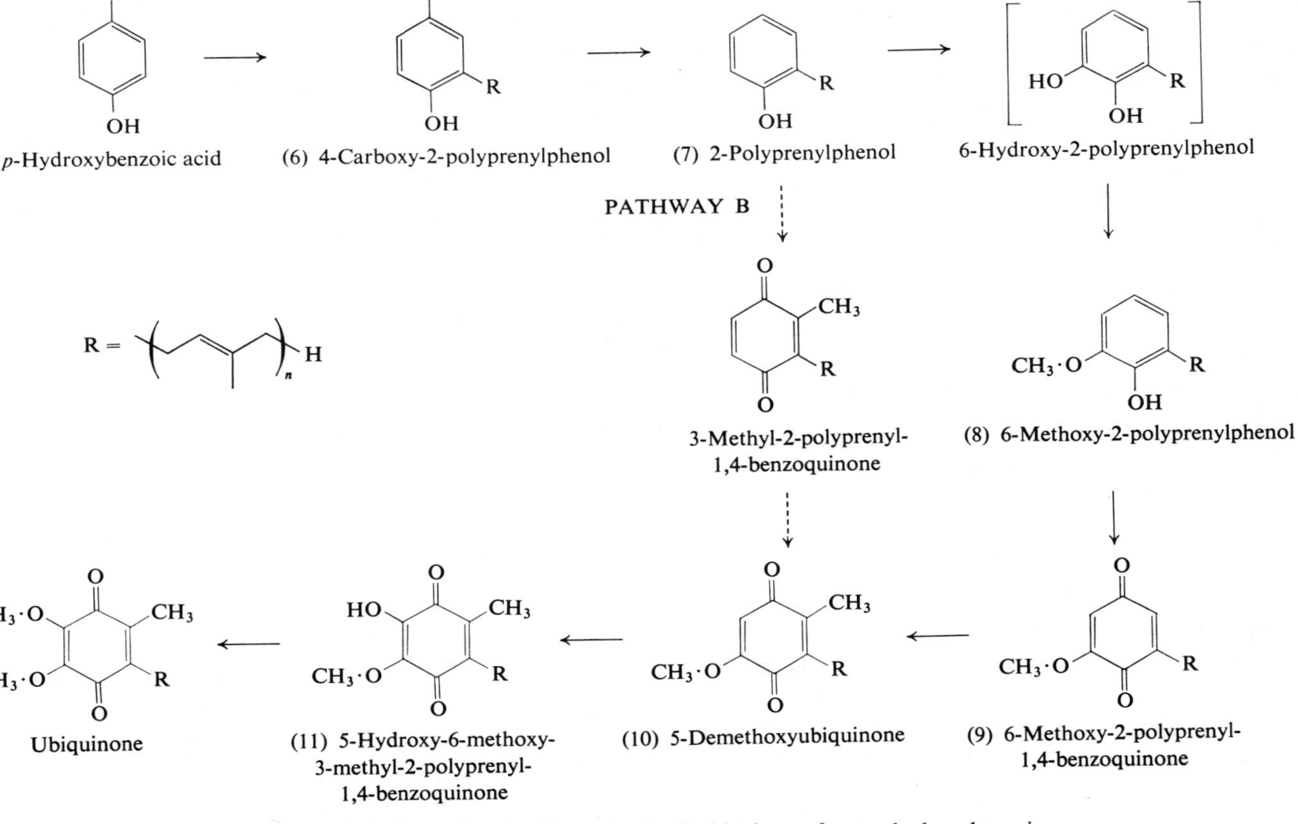

FIG. 4. Proposed pathways for the biosynthesis of ubiquinone from *p*-hydroxybenzoic acid in *Rhodospirillum rubrum* (after Daves *et al.*, 1966; Friis *et al.*, 1967b).

4-carboxy-2-decaprenylphenol (6, Fig. 4; $n = 10$) and 2-decaprenylphenol (7, Fig. 4; $n = 10$), especially when they are supplemented with polyprenylpyrophosphate synthetase from *Micrococcus lysodeikticus* and isopentenyl pyrophosphate. It is of interest that although the polyprenylphosphate synthetase of *M. lysodeikticus* synthesizes predominantly hepta- and octaprenylpyrophosphate (Allen, C. M. *et al.*, 1967) the enzyme which prenylates *p*-hydroxybenzoic acid in *R. rubrum* selects preferentially the relatively small amount of decaprenylpyrophosphate formed. Whistance *et al.* (1970) have obtained incorporation of *p*-hydroxy[U-^{14}C] benzoic acid into 2-polyprenylphenols, 6-methoxy-2-polyprenylphenols, 6-methoxy-2-polyprenyl-1,4-benzoquinones, 2-polyprenyl-1,4-benzoquinones (?), 5-demethoxyubiquinones and ubiquinones using crushed cell preparations of *P. ovalis* Chester and *Escherichia coli*. Although the incorporations obtained (0·1% and 4% respectively) could be attributed in part to intact cells in the preparations (2–5% at the end of the incubations), the number of cells present was not sufficient to account for the total. With cell membranes of *P. ovalis* Chester, on which ubiquinones and their prenylated precursors are known to be localized (Whistance *et al.*, 1970), these workers could obtain no incorporation of radioactivity from *p*-hydroxy-[U-^{14}C]benzoate into any of the above compounds, even when they supplemented the membranes with cytoplasm. This suggests that irreparable damage to the synthesizing centres occurred during preparation of the membranes. Cox *et al.* (1969) have obtained two mutant strains of *E. coli* which are unable to synthesize ubiquinones. One of these accumulates 4-carboxy-2-octaprenylphenol (6, Fig. 4; $n = 8$), while the other accumulates 2-octaprenylphenol (7, Fig. 4; $n = 8$). Studies with crude cell extracts have shown that the mutant which accumulates 2-octaprenylphenol possesses 4-carboxy-2-octaprenylphenol carboxylase activity.

In animals, green algae, moulds and yeasts there is some evidence that the pathways may be similar to those in bacteria (Fig. 4). Thus, 5-demethoxyubiquinones, compounds known to be intermediates in the biosynthesis of ubiquinones in bacteria, have been isolated from all four groups of organisms (Table IV). Further, in the cases of rat, bakers' yeast, *Rhodotorula glutinis* and *Phycomyces blakesleeanus* these quinones are precursors of ubiquinones (Olson, 1966; Olson and Aiyar, 1966; Ah Law, D. R. Threlfall and G. R. Whistance, in preparation). However, with the exception of 6-methoxy-2-nonaprenylphenol which has been found in rat (Nowicki *et al.*, 1969), none of the other phenols or quinones shown in Fig. 4 have been detected in organisms of this type. Recently, Ah Law *et al.* (1970) have isolated the dihydro forms of 6-methoxy-decaprenyl-1,4-benzoquinone (12) and 5-demethoxyubiquinone-10 (13) from the mould *Aspergillus flavus* and have shown that they are precursors of ubiquinone-10 (X-H_2) in this organism. These workers have also isolated 5-demethoxyubiquinone-10 (X-H_2) from two other moulds, *Aspergillus quadrilineatus* and *Neurospora crassa* (Ah Law, D. R. Threlfall and G. R. Whistance, in preparation).

TABLE IV
Occurrence of 5-demethoxyubiquinones

Organism	Homologue	Organism	Homologue
PHOTOSYNTHETIC BACTERIA		YEASTS	
Rhodospirillum rubrum	10,9[a]	*Rhodotorula glutinis*	9[e]
Rhodopseudomonas spheroides	10[b]	*Saccharomyces cerevisiae*	6[d,e]
Chromatium strain D	8[b]	*Torula* sp.	7[d]
NON-PHOTOSYNTHETIC BACTERIA		MOULDS	
Escherichia coli	8[b]	*Aspergillus niger*	9[e]
Organism PC4 (*Achromobacter* sp.)	9[b]	*Phycomyces blakesleeanus*	9[e]
Proteus mirabilis	8[c]		
Pseudomonas ovalis	9,8,7,6,5,4[c,d]		
Vibrio 01 (*Moraxella* sp.)	9[c]	ALGAE	
		Euglena gracilis	9[f]
		MAMMALS	
		Rat	9[g]

[a] Daves *et al*. (1966); [b] G. R. Whistance and D. R. Threlfall, unpublished observations; [c] Whistance *et al*. (1970); [d] Imamoto and Senoh (1968); [e] Ah Law, D. R. Threlfall and G. R. Whistance (1970), in preparation; [f] Threlfall and Whistance (1970); [g] Olson and Aiyar (1966).

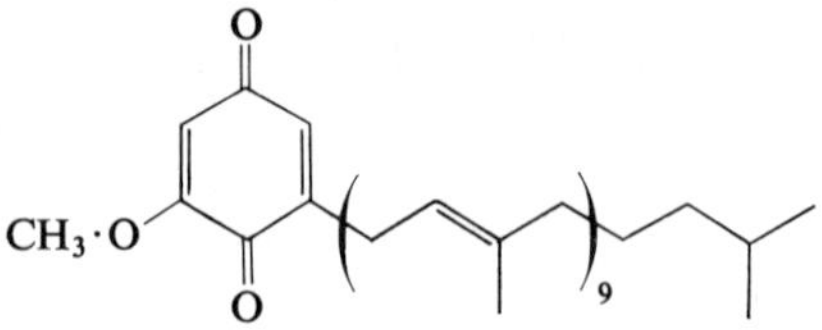

$CH_3 \cdot O$ ... 9

(12)

CH_3 $CH_3 \cdot O$ ... 9

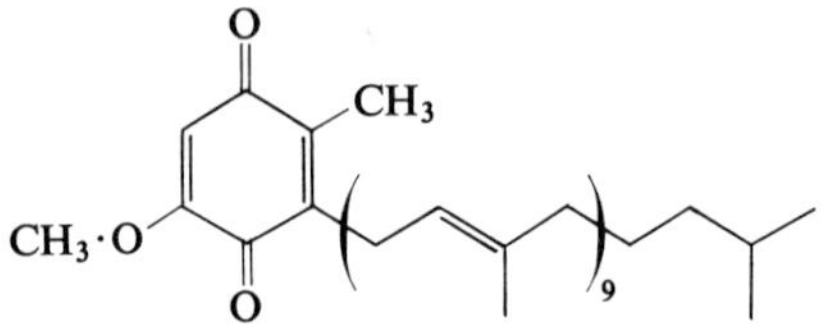

(13)

In higher plants the problem remains completely unresolved. Despite a thorough search no evidence has been found of the presence in any higher plant examined of polyprenylphenols or quinones of the type shown in Fig. 4 (Whistance *et al.*, 1967; Threlfall and Whistance, 1970). Whistance *et al.* (1966b, 1967) detected in maize shoots a "compound" which became highly labelled on administration of *p*-hydroxy[U-^{14}C] benzoic acid and which they believed at the time to be a polyprenylphenol or quinol precursor of ubiquinone-9. However, it has since been shown that this "compound" was a mixture of two compounds, ubiquinone-8 (the radioactive component) and plastochromanol-8 (18) (the phenolic component), neither of which is a precursor of ubiquinone-9 in maize (Threlfall and Whistance, 1970).

III. Plastoquinones, Plastochromanols, Tocotrienols, Tocopherols, Tocotrienolquinones and Tocopherolquinones

A. Structure and Distribution

The plastoquinones, plastochromanols, tocotrienols, tocopherols, tocotrienolquinones and tocopherolquinones constitute six groups of isoprenoid quinones and chromanols which, in addition to having strong chemical links with each other, are biogenetically closely related. They have also in common the fact that their biosynthesis is confined almost exclusively to higher plants and algae.

The plastoquinones are a family of 2,3-dimethyl-5-polyprenyl-1,4-benzoquinones whose members are indigenous to higher plants and algae. The individual plastoquinones differ only in the nature of their polyprenyl side chains, which may be either unsubstituted (plastoquinones-*n*) (14), monohydroxylated (plastoquinones-*C*) (15), monoacylated (plastoquinones-*B*) (16), both monohydroxylated and monoacylated (plastoquinones-*Z*) or partly saturated (phytylplastoquinone) (17). So far plastoquinones with C_{45} [plastoquinone-9 (14; $n = 9$), plastoquinones-*B* (16), plastoquinones-*C* (15) and plastoquinones-*Z*], C_{40} [plastoquinones-8 (14; $n = 8$)], C_{20} [plastoquinone-4 (14; $n = 4$) and phytylplastoquinone (17)] and C_{15} [plastoquinone-3 (14); $n = 3$)] side chains have been found in nature (for key references to the structures of plastoquinones-*B*, -*C* and -*Z*, see Wallwork and Pennock, 1968).

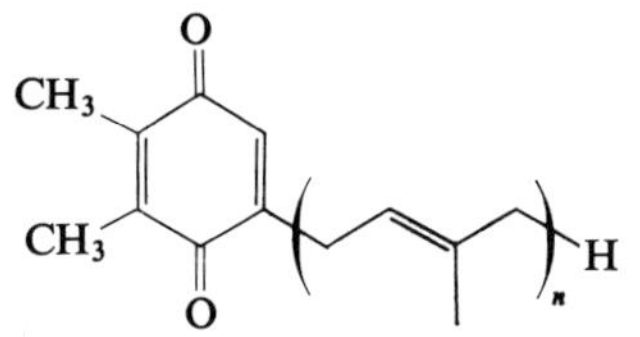

(14)

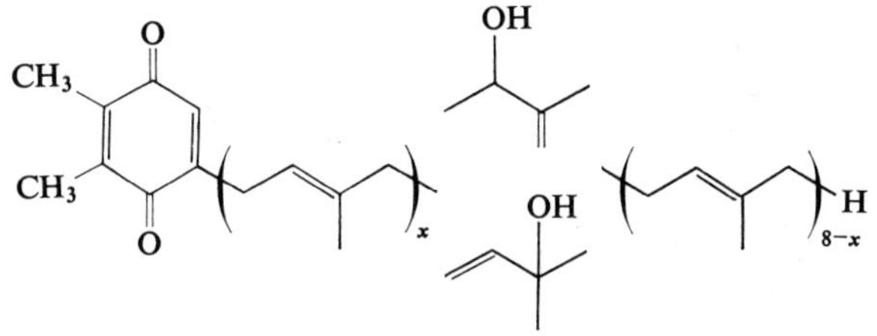

(15)

$x = 1–7$

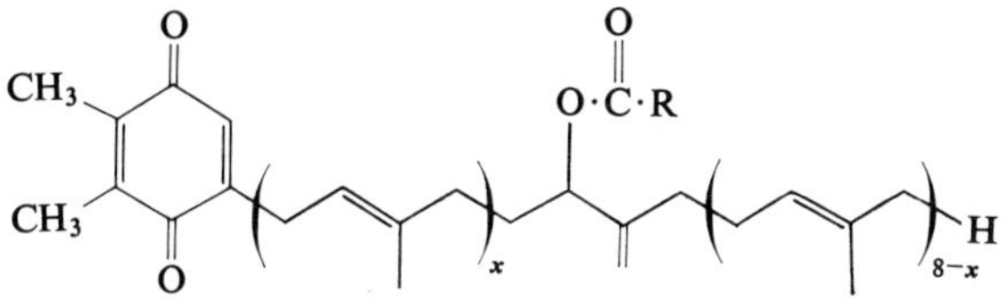

(16)

$x = 1–7$

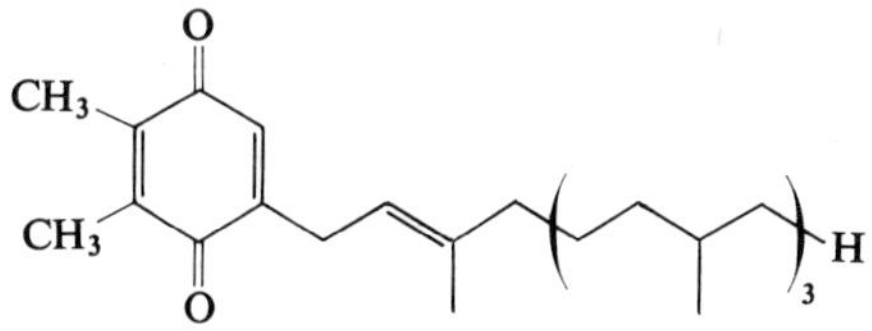

(17)

The most widely distributed plastoquinones are plastoquinone-9 and plastoquinones-*C*. Indeed, these quinones have been detected in all higher plant and algal species examined (see Pennock, 1966; Barr and Crane, 1967; Sun *et al.*, 1968). Plastoquinones-*B* and -*Z* are not nearly so widespread, being confined in the main to higher plants (Barr and Crane, 1967; Sun *et al.*, 1968; Wallwork and Pennock, 1968). Plastoquinone-8 has only been reported in *Aesculus hippocastanum*, *Zea mays* and *Euglena gracilis* (Whistance and

Threlfall, 1970c), while plastoquinone-4, plastoquinone-3 and phytylplastoquinone have yet to be found in organisms other than *Aesculus hippocastanum* (Eck and Trebst, 1963), *Spinacia oleracea* (Misiti *et al*., 1965) and strains of *E. gracilis* and *Tribonema* (Whistance and Threlfall, 1970b) respectively.

In photosynthetic tissues plastoquinones are concentrated in the chloroplasts (see Pennock, 1966; Barr *et al*., 1967; Threlfall and Goodwin, 1967; Lichtenthaler, 1968a). In young chloroplasts they are associated almost entirely with the lamellae (Barr *et al*., 1967). In old chloroplasts, on the other hand, they are associated both with the lamellae and osmiophilic globules (Barr *et al*., 1967). When present in non-photosynthetic tissues they are believed to be localized in "plastids", i.e. amyloplasts, chromoplasts, etioplasts and leucoplasts (Lichtenthaler, 1968b).

A cyclic relative of plastoquinone-9 known as plastochromanol-8 (18) has been isolated from the leaves of many plant species (Dunphy *et al*., 1966). Intracellular distribution studies have shown that in the leaves of *Polygonum cupsidatum* it is localized in the chloroplasts (Dunphy *et al*., 1966). Recently, plastochromanols related to plastoquinones-*B*, -*C* and -*Z* have been isolated from *Lingustrum* sp. and *Polygonum* sp. (Dunphy *et al*., 1969).

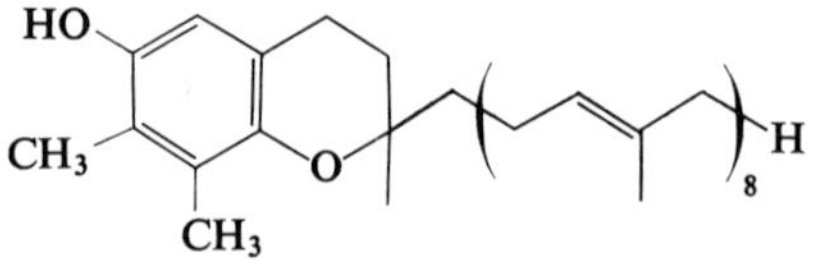

(18)

The tocopherols and tocotrienols constitute two families of isoprenoid chromanols whose members can formally be regarded as the methyl substituted derivatives of tocol[2-methyl-2-(4′,8′,12′-trimethyl-tridecyl)chroman-6-ol] and tocotrienol[2-methyl-2-(4′,8′,12′-trimethyltrideca-3′,7′,11′-trienyl)chroman-6-ol) respectively. So far four members from each family have been found in nature: these are the 8-methyl(δ-), 5,8-dimethyl(β-), 7,8-dimethyl(γ)- and 5,7,8-trimethyl(α-) substituted forms (19 and 20) (Pennock, *et al*., 1964).

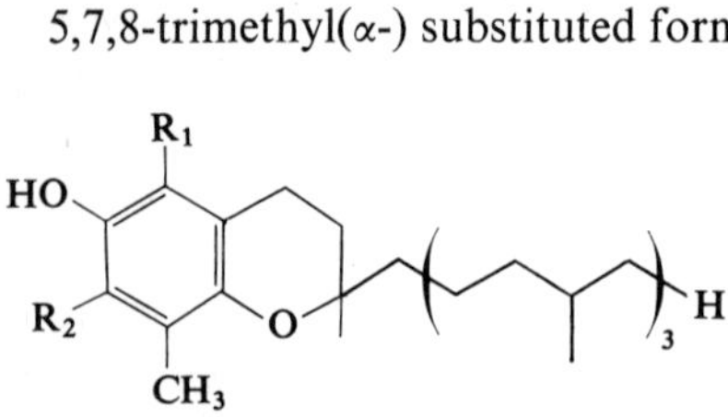

R_1	R_2	
H	H	δ-Tocopherol
CH_3	H	β-Tocopherol
H	CH_3	γ-Tocopherol
CH_3	CH_3	α-Tocopherol

(19)

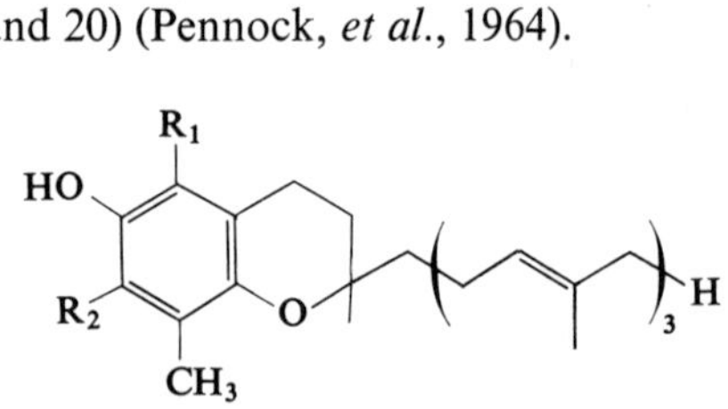

R_1	R_2	
H	H	δ-Tocotrienol
CH_3	H	β-Tocotrienol
H	CH_3	γ-Tocotrienol
CH_3	CH_3	α-Tocotrienol

(20)

Tocopherols are widespread among higher plants and algae (see Green, 1970; Lichtenthaler, 1968a). In addition, they have been reported to occur in some fungi (Kubin and Fink, 1961; Diplock *et al.*, 1961). In higher plants they occur mainly in the photosynthetic regions, natural oils, fruits and seeds. The principal tocopherol found in the photosynthetic regions of plants is α-tocopherol. Often lesser amounts of at least one non-α-tocopherol are also present. In non-photosynthetic regions non-α-tocopherols are much more abundant, particularly in natural oils and fruits, where they sometimes predominate.

In the photosynthetic regions of higher plants and algae α-tocopherol is concentrated mainly in the chloroplasts (see Lichtenthaler, 1966; 1968a). Small amounts are also believed to reside in the mitochondria (Dilley and Crane, 1963; Threlfall and Goodwin, 1967). In young chloroplasts of higher plants α-tocopherol is found in the lamellar regions, but as the chloroplast ages it also appears in osmiophilic globules (Lichtenthaler, 1969). The intracellular distribution of the non-α-tocopherols in photosynthetic tissues has not been clearly defined. Booth (1963), working with *Hedera helix* and *Taxus baccata*, reported that γ-tocopherol and δ-tocopherol are located outside the chloroplasts. Dada (1968), on the other hand, reported that in *Hedera helix* at least half the γ-tocopherol is within the chloroplasts. It seems probable that in non-photosynthetic plant tissues tocopherols are associated with "plastids" (Lichtenthaler, 1968b).

Tocotrienols occur mainly in the natural oils and seeds of higher plants, with palm oil and the seeds of wheat, barley and rice being particularly rich sources (see Pennock *et al.*, 1964). Appreciable amounts have also been found in the latex of *Hevea brasiliensis* (Dunphy *et al.*, 1965).

The tocopherolquinones and tocotrienolquinones are the corresponding γ-hydroxyquinone derivatives of the tocopherols and tocotrienols respec-

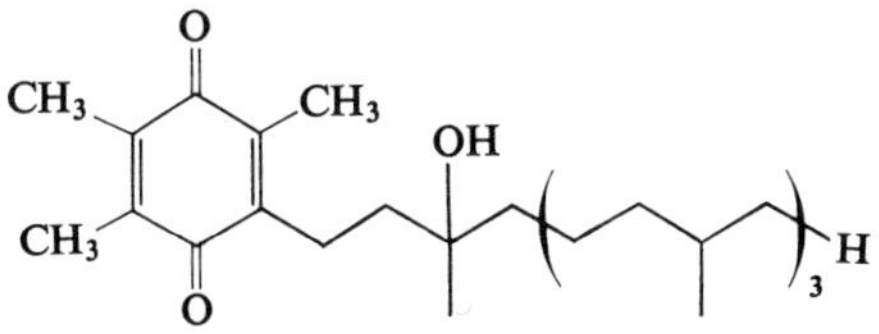

(21)

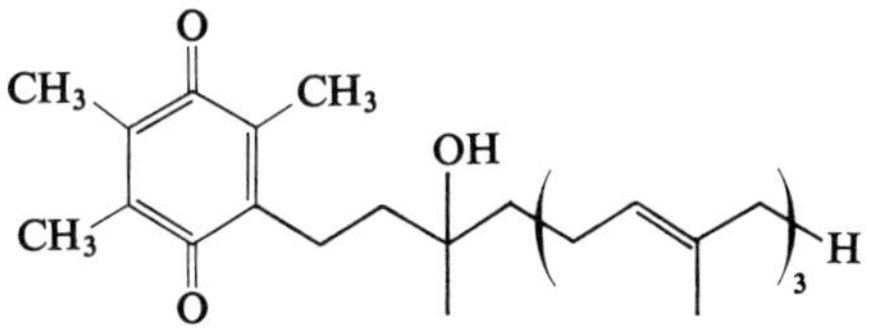

(22)

tively, from which they can readily be formed by mild oxidation. Four tocopherolquinones [α- (21), β-, γ- and δ-] and one tocotrienolquinone [α- (22)] have so far been isolated from natural sources.

α-Tocopherolquinone is distributed widely amongst species of higher plants and algae (see Lichtenthaler, 1968a). β-, γ- and δ-Tocopherolquinones (Henninger *et al.*, 1963; Barr and Arntzen, 1969) and α-tocotrienolquinone (Pennock, 1966), on the other hand, appear to be of limited distribution. Indeed, α-tocotrienolquinone has only been detected in the latex of *Hevea brasiliensis*.

In photosynthetic tissues, where they are normally located, α-, β-, γ- and δ-tocopherolquinones are concentrated in the chloroplasts. Their intrachloroplastidic distribution appears to be similar to that of plastoquinones and tocopherols (see Henninger *et al.*, 1963; Lichtenthaler, 1969; Barr and Arntzen, 1969).

B. BIOSYNTHESIS

1. P-*benzoquinone Rings and Aromatic Nuclei*

Whistance *et al.* (1966, 1967), whilst investigating the biosynthesis of the *p*-benzoquinone rings of ubiquinones in plants, observed that radioactivity from D-[U-^{14}C] shikimic acid, L-[U-^{14}C] phenylalanine and L-[U-^{14}C] tyrosine (but not *p*-hydroxy-[U-^{14}C] benzoic acid) was incorporated into plastoquinone-9, tocopherols and α-tocopherolquinone by greening etiolated maize (*Zea mays*) shoots. Appropriate chemical degradations established that the carbon atoms of the alicyclic ring of shikimic acid and the aromatic rings of tyrosine and phenylalanine had been incorporated into the *p*-benzoquinone rings of plastoquinone-9 and α-tocopherolquinone and the aromatic rings of α-tocopherol and γ-tocopherol. These results provided the first evidence that aromatic compounds formed by the shikimic acid pathway of aromatic biosynthesis are involved in the biosynthesis of plastoquinones, tocopherols and tocopherolquinones by higher plants.

At about this time evidence was obtained from an experiment in which L-[*Me*-^{14}C,^{3}H$_3$] methionine was administered to maize shoots that in plastoquinone-9, α-tocopherol, γ-tocopherol and α-tocopherolquinone one of the nuclear *C*-methyl groups is not derived from the *S*-methyl group of L-methionine (Threlfall *et al.*, 1968; Whistance and Threlfall, 1968b). This led to a consideration of the possibility that the source of this nuclear methyl group might be the β-carbon atom of one of the ring precursors, for example, tyrosine or phenylalanine. To investigate this possibility the incorporation of DL-[β-^{14}C] tyrosine and DL-[β-^{14}C] phenylalanine into plastoquinone-9, tocopherols and α-tocopherolquinone by maize shoots was investigated. Chemical degradation of the ^{14}C-labelled plastoquinone-9, α-tocopherol, γ-tocopherol and α-tocopherolquinone showed that the β-carbon atom of both tyrosine and phenylalanine can indeed give rise to a nuclear *C*-methyl group in each compound (Whistance and Threlfall, 1967, 1968b). In view of these findings the experiments with L-[U-^{14}C] tyrosine were repeated, when it was found that the aromatic ring carbon atoms and β-carbon atom of tyrosine are incorporated as a C_6-C_1 unit into plastoquinone-9, α-tocopherol, γ-tocopherol and α-tocopherolquinone (Whistance and Threlfall, 1968b). Similar results were obtained using ivy (*Hedera helix*) leaves and French bean (*Phaseolus vulgaris*) shoots as the experimental tissues (Whistance and Threlfall, 1968b). One additional observation made in the experiments with ivy leaves was that the carbon atoms of the aromatic ring and nuclear methyl group of δ-tocopherol can also be derived from the nuclear carbon atoms and β-carbon atom of tyrosine.

To explain the incorporation of the β-carbon atom of tyrosine (and phenylalanine) into plastoquinone-9, tocopherols and α-tocopherolquinone it was postulated that one of the biosynthetic steps in their formation is an intramolecular rearrangement of *p*-hydroxyphenylpyruvic acid (Whistance and Threlfall, 1967). This postulate, when considered in conjunction with the knowledge that shikimic acid is a distal precursor of the nuclei of plastoquinone-9, tocopherols and α-tocopherolquinone and with reports of the occurrence in plants of homogentisic acid (23, Fig. 5) (Bertel, 1903), homogentisic acid glucoside (Matsumura and Shibata, 1964) and homoarbutin (24, Fig. 5) (2-methylquinol-4-β-D-glucoside) (Inoue *et al.*, 1958), led to the proposal that plastoquinones, tocopherols, tocotrienols and tocopherolquinones might be synthesized by the pathway shown in Fig. 5.

A series of isotope competition experiments provided good evidence that *p*-hydroxyphenylpyruvic acid and homogentisic acid are precursors of plastoquinone-9, α-tocopherol, γ-tocopherol and α-tocopherolquinone in maize shoots, but they provided no evidence to support the view that homoarbutin has such a role (Whistance and Threlfall, 1968b).

Para-hydroxy[U-^{14}C] phenylpyruvic acid, *p*-hydroxy[β-^{14}C] phenylpyruvic acid, [U-^{14}C] homogentisic acid, [α-^{14}C] homogentisic acid and [*Me*-^{14}C] toluquinol were next synthesized and their incorporation into plastoquinone-9 and biogenetically related compounds by maize shoots investigated (Whistance and Threlfall, 1968a, 1969b, 1970a). As expected, radioactivity from ^{14}C-labelled species of *p*-hydroxyphenylpyruvic acid and homogentisic acid was incorporated, whereas no incorporation of radioactivity from [*Me*-^{14}C] toluquinol took place. Chemical degradations showed that the aromatic ring carbon atoms and β-carbon atom of *p*-hydroxyphenylpyruvic acid and the aromatic ring carbon atoms and α-carbon atom of homogentisic acid gave rise as C_6-C_1 units to the carbon atoms of the nuclei and nuclear *C*-methyl groups (one in each case) of plastoquinone-9, α-tocopherol, γ-tocopherol and α-tocopherolquinone. More recently it has been shown that homogentisic acid (*p*-hydroxyphenylpyruvic acid was not tested) is incorporated in a similar manner into the plastoquinones, tocopherols and tocopherolquinones present in the higher plants lettuce (*Lactuca sativa*) and red veined dock (*Rumex*

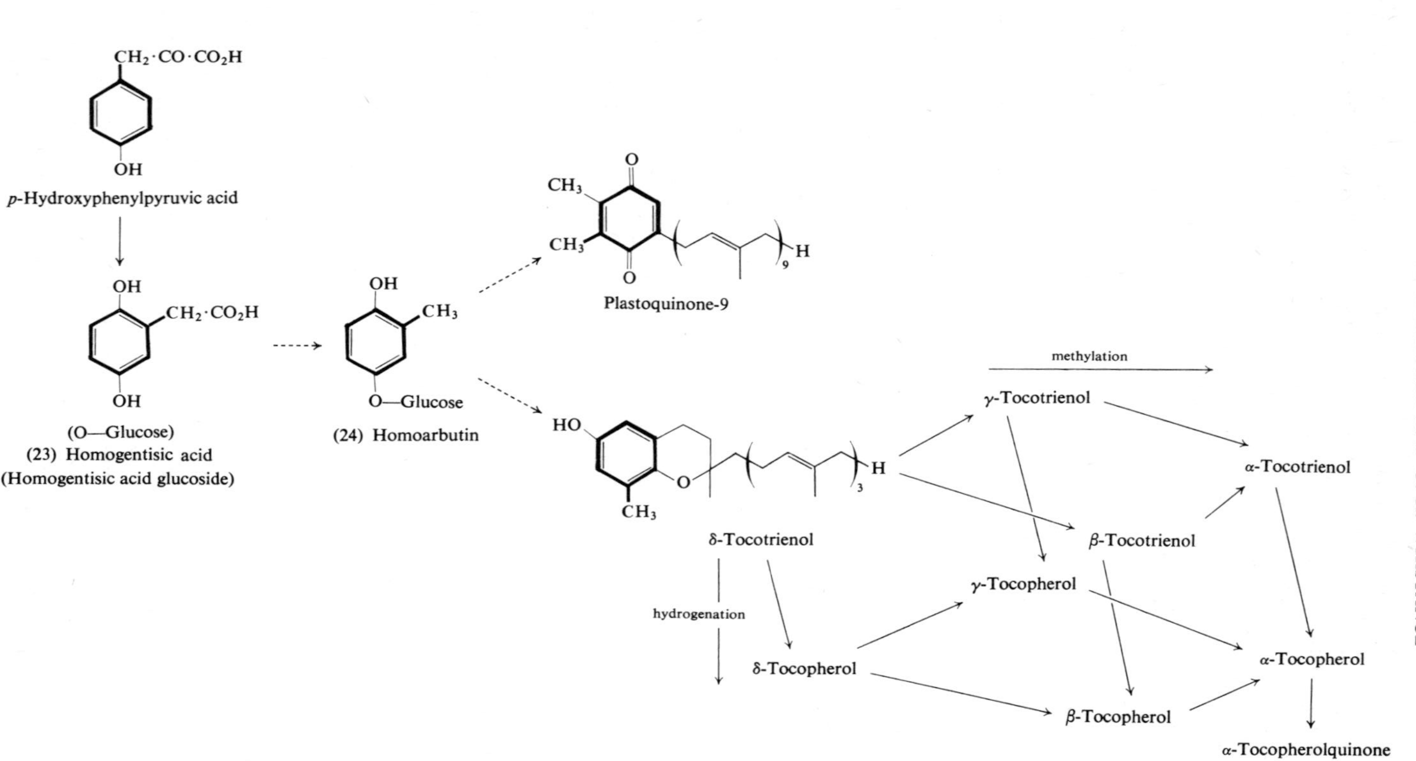

FIG. 5. Proposed pathways for the biosynthesis of plastoquinone-9, α-tocopherolquinone, tocopherols and tocotrienols by higher plants (after Whistance and Threlfall, 1967).

sanguineus), the algae *Euglena gracilis* and *Chlorella pyrenoidosa* and the blue-green alga *Anacystis nidulans* (Whistance and Threlfall, 1970a).

There seems little doubt that in maize shoots homogentisic acid arises from *p*-hydroxyphenylpyruvic acid. Evidence for this are the facts that (a) homogentisic acid is incorporated far more effectively than *p*-hydroxyphenylpyruvic acid into plastoquinone-9, tocopherols and α-tocopherolquinone (Whistance and Threlfall, 1970a), (b) homogentisic acid markedly diminishes the incorporation of radioactivity from *p*-hydroxy[U-^{14}C] phenylpyruvic acid into plastoquinone-9, tocopherols and α-tocopherolquinone (Whistance and Threlfall, 1970a), and (c) phenylacetic acid is not utilized for the biosynthesis of plastoquinone-9, tocopherols and α-tocopherolquinone (Whistance and Threlfall, 1970a) [this excludes the possibility that homogentisic acid arises via *ortho* and *meta* hydroxylation of phenylacetic acid, as it does in some microorganisms (Evans, 1955)]. The mechanism of the conversion remains to be defined. However, the finding that *p*-hydroxyphenylacetic acid (a compound which can be visualized as an intermediate in the conversion of *p*-hydroxyphenylpyruvic acid to homogentisic acid) is not a precursor of plastoquinone-9, tocopherols or α-tocopherolquinone (Whistance and Threlfall, 1970a) suggests that, as proposed by Whistance and Threlfall (1967, 1968b), it is similar to that in animals (see Meister, 1965) and some microorganisms (Evans, 1963), i.e. concomitant hydroxylation of the aromatic ring, shift of the side chains and the formation of carbon dioxide. Direct support for a side-chain shift has come from studies using DL-[1,2-^{14}C] shikimic acid in which the orientation of the non-methionine-derived nuclear *C*-methyl groups of plastoquinone-9, α-tocopherol, γ-tocopherol and α-tocopherolquinone with respect to the polyprenyl side chain has been investigated in maize shoots (see Section IIIB3).

2. *Polyprenyl Side Chains*

As in the case of the ubiquinones it has been established that mevalonic acid is the specific distal precursor of the polyprenyl moieties of plastoquinone-9, plastoquinones-*C*, α-tocopherol, γ-tocopherol and α-tocopherolquinone in higher plants (Threlfall *et al.*, 1967b; Griffiths *et al.*, 1968; Dada *et al.*, 1968). Dada *et al.* (1968) studied the incorporation of (3*RS*)-[2-^{14}C,(4*R*)-4-^{3}H$_1$] mevalonic acid and (3*RS*)-[2-^{14}C,(4*S*)-4-^{3}H$_1$] mevalonic acid into the isoprenoid quinones and chromanols of maize shoots. Their findings proved that, except for the loss of C$_1$, the carbon skeleton of mevalonic acid is incorporated *in toto* into the polyprenyl portions of plastoquinone-9, α-tocopherol, γ-tocopherol and α-tocopherolquinone. In addition they proved that each isoprene residue of the polyprenyl portions of these compounds is biogenetically *trans*. At present the involvement of mevalonic acid in the biosynthesis of algal plastoquinones, tocopherols and tocopherolquinones has still to be demonstrated.

Apart from the information obtained by using ^{14}C and ^{3}H labelled mevalonic acids nothing is known about either the way in which the side chains are assembled or how and at what stage of their formation they are coupled with the appropriate nucleus. However, it seems likely that the steps will be similar to those outlined for the biosynthesis of the ubiquinones (see Section IIB2).

An interesting aspect of the biosynthesis of tocopherols and tocopherolquinones is the mechanism of hydrogenation of the polyprenyl side chain. Wellburn (1968) has recently shed some light on this problem by demonstrating that in seedlings of French bean and oat (*Avena sativa*) radioactivity from [4*R*-$^3H_1$] NADPH (but not [4*S*-$^3H_1$] NADPH) is incorporated into the side chain of α-tocopherol, indicating that a NADPH-dependent reductase is involved.

3. *Nuclear* C-*Methyl Groups*

In higher plants and algae one nuclear *C*-methyl group of plastoquinone-9, plastoquinone-8, phytylplastoquinone, α-tocopherol, γ-tocopherol and α-tocopherolquinone is derived from the α-carbon atom of homogentisic acid (see Section IIIB1). Studies in which L-[*Me*-^{14}C] methionine and L-[*Me*-^{14}C,$^3H_3$] methionine have been administered to maize shoots, ivy leaves and cells of *Euglena gracilis* have established that the remaining nuclear *C*-methyl groups arise by transmethylation from L-methionine (or more precisely *S*-adenosylmethionine) (Threlfall *et al.*, 1967a, 1969; Whistance and Threlfall, 1969b; G. R. Whistance and D. R. Threlfall, unpublished observations). It is to be expected that the mechanism of *C*-methylation will prove to be the same as that proposed for the *C*-methylation of the ubiquinone nucleus (see Section IIB3).

The finding of Whittle *et al.* (1967) that when they administered L-[*Me*-^{14}C] methionine to latex of *Hevea brasiliensis* radioactivity was incorporated into α-tocopherol, α-tocotrienolquinone, β-tocotrienol, γ-tocotrienol but not δ-tocotrienol suggests that the nuclear *C*-methyl groups of these compounds are derived from the same source as those of the plastoquinones, tocopherols and tocopherolquinones.

Recently, an experiment has been carried out using DL-[1,2-^{14}C] shikimic acid to determine which nuclear *C*-methyl groups (one in each case) of plastoquinone-9, α-tocopherol, γ-tocopherol and α-tocopherolquinone are derived from the α-carbon atom of homogentisic acid (G. R. Whistance and D. R. Threlfall, in preparation). In outline this consisted of administering DL-[1,2-^{14}C] shikimic acid to maize shoots, then subjecting the resulting ^{14}C-labelled plastoquinone-9, α-tocopherol, γ-tocopherol and α-tocopherolquinone and the nuclear *C*-methyl derivatives (prepared chemically) of the [^{14}C] plastoquinone-9 and γ-[^{14}C] tocopherol to Kuhn-Roth oxidation.

Reference to Fig. 6 shows that on Kuhn-Roth oxidation of the [^{14}C] plastoquinone-9 and γ-[^{14}C] tocopherol 25% of the radioactivity in the molecules will be recovered in acetic acid, whereas oxidation of their nuclear

C-methyl derivatives, α-[^{14}C] tocopherol and α-[^{14}C] tocopherolquinone will lead to either 25% or 50% of the radioactivity appearing in acetic acid, depending on whether the homogentisic acid-derived nuclear methyl group is *meta* or *para* to the polyprenyl side chain.

In practice 25% of the total radioactivity in each molecule was recovered in acetic acid. This provided conclusive evidence that the α-carbon atom of homogentisic acid gives rise in plastoquinone-9, α-tocopherol, γ-tocopherol and α-tocopherolquinone to the nuclear *C*-methyl substituent which is *meta* to the polyprenyl side chain, a belief (Whistance and Threlfall, 1967) which has been held since the demonstration that the nuclear *C*-methyl substituent of δ-tocopherol in ivy can arise from the β-carbon atom of tyrosine (Whistance and Threlfall, 1967) and that the nuclear methyl group of δ-tocotrienol in *Hevea* latex is not derived from methionine (Whittle *et al.*, 1967).

4. *Pathways from Homogentisic Acid*

The steps in the conversion of homogentisic acid to plastoquinones, plastochromanols, tocopherols, tocotrienols, tocopherolquinones and tocotrienolquinones have still to be elucidated. However, isotope competition experiments and ^{14}C-tracer studies have provided good evidence that in the case of plastoquinone-9, α-tocopherol, γ-tocopherol and α-tocopherolquinone such plausible intermediates as the C_6-C_1 compound gentisic acid (25), gentisaldehyde (26), gentisylalcohol (27), toluquinol (28) and their glucosides are not involved (Whistance and Threlfall, 1968b, 1970a, and unpublished observations). This, together with the knowledge that the α-carbon atom of homogentisic acid gives rise in plastoquinones and biogenetically related compounds to the nuclear *C*-methyl group *meta* to the polyprenyl side chain (see Section IIIB3), suggests that the first step(s) may be the sequential or concomitant prenylation and decarboxylation of homogentisic acid to give 3-all-*trans*-polyprenyltoluquinols (29, Fig. 7).

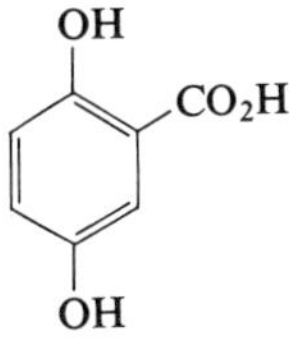

OH
CO_2H
OH
(25)

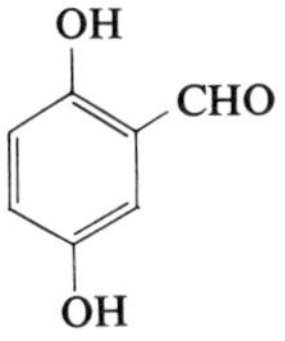

OH
CHO
OH
(26)

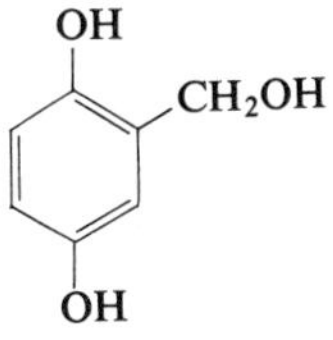

OH
CH_2OH
OH
(27)

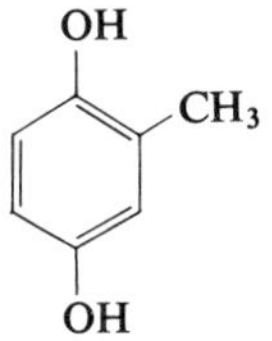

OH
CH_3
OH
(28)

BIOSYNTHETIC STEPS

DL-[1,2-^{14}C]Shikimic acid

p-Hydroxyphenylpyruvic acid

Homogentisic acid

Plastoquinone-9

OR

γ-Tocopherol

OR

α-Tocopherol

OR

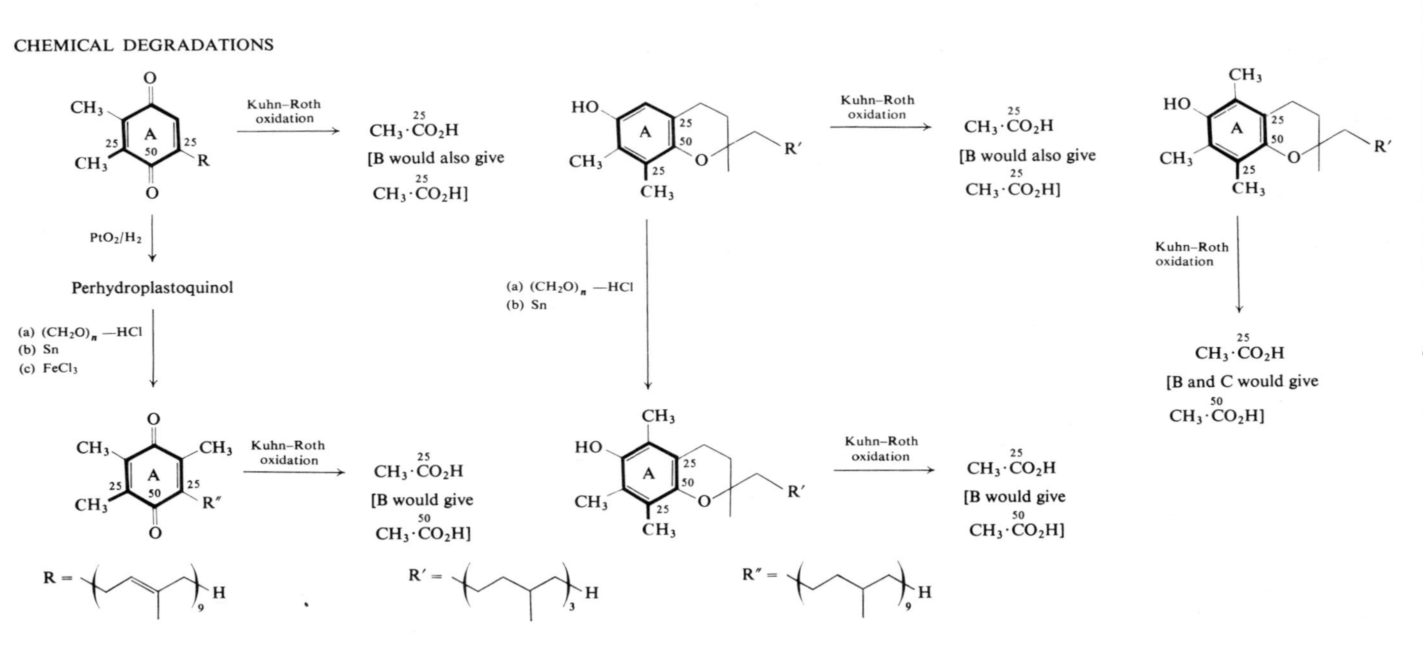

Fig. 6. Manner of incorporation of DL-[1,2-^{14}C] shikimic acid into plastoquinone-9, α-tocopherolquinone, α-tocopherol and γ-tocopherol by maize shoots. The results for α-tocopherolquinone were identical to those obtained for α-tocopherol.

Methylation and oxidation of the nonaprenyl form of 3-all-*trans*-polyprenyltoluquinol (29, Fig. 7; $n = 9$) would give plastoquinone-9 (methylation and oxidation of the octa-, tetra- and triprenyl forms would give plastoquinone-8, plastoquinone-4 and plastoquinone-3 respectively) (Fig. 7), which in turn could give rise to plastoquinones-*B*, -*C* and -*Z* (Fig. 7), possibly by the pathways proposed by Wallwork and Pennock (1968) (Fig. 8). Some evidence to support the view that plastoquinone-9 is a precursor of plastoquinones-*B*, -*C* and -*Z* is the fact that in ^{14}C-tracer experiments with higher plants plastoquinones-*B*, -*C* and -*Z* have always been found to have specific radioactivities appreciably lower than those of plastoquinone-9 (Griffiths *et al.*, 1968; Whistance and Threlfall, 1968b; G. R. Whistance and D. R. Threlfall, unpublished observations). Plastochromanol-8 (18) can also be envisaged as arising from 3-nonaprenyltoluquinol (Fig. 7); however, recent biogenetic studies with [^{14}C] homogentisic acid have provided some evidence that it is more likely to be formed by cyclization of plastoquinol-9 (G. R. Whistance and D. R. Threlfall, unpublished observations). The other known plastochromanols (see Section IIIA) could be formed either from plastochromanol-8 by reactions similar to those involved in the formation of plastoquinones-*B*, -*C* and -*Z* from plastoquinone-9 or by cyclization of the appropriate plastoquinols-*B*, -*C* and -*Z* (Fig. 7).

Cyclization of the tetraprenyl form of 3-all-*trans*-polyprenyltoluquinol (29, Fig. 7; $n = 4$) would give δ-tocotrienol (Fig. 7), a compound which, on paper, can be regarded as the parent member of both the tocotrienol and tocopherol series (Fig. 7). By a suitable combination of methylation and, in the case of the tocopherols, hydrogenation reactions δ-tocotrienol could give rise to all members of both series; these in turn could be converted by oxidation into their corresponding tocotrienolquinones and tocopherolquinones (Fig. 7). At the experimental level specific radioactivity data from biogenetic studies with maize shoots, French bean shoots and ivy leaves have provided some evidence that γ-tocopherol is a precursor of α-tocopherol and that α-tocopherol and α-tocopherolquinone are in equilibrium (Dada *et al.*, 1968; Threlfall *et al.*, 1968; Whistance and Threlfall, 1970a). However, it has provided evidence against δ-tocopherol being a precursor of γ-tocopherol and α-tocopherol (Whistance and Threlfall, 1968b). Similar data from biogenetic studies with *Hevea* latex and leaves and latex of *Ficus elastica* have suggested that in these tissues δ-tocotrienol, γ-tocotrienol, β-tocotrienol and α-tocotrienol may be precursors of α-tocopherol (Whittle *et al.*, 1967; Wellburn, 1970).

Clearly pathways for the biosynthesis of tocotrienols and tocopherols other than those proposed above can be invoked. For example, methylation could precede prenylation of homogentisic acid or cyclization of 3-tetraprenyltoluquinol, as it would if plastoquinol-4 and phytylplastoquinol were precursors of γ-tocotrienol and γ-tocotrienol respectively (Fig. 7). Again, in the case of tocopherol biosynthesis phytylation of homogentisic acid may take

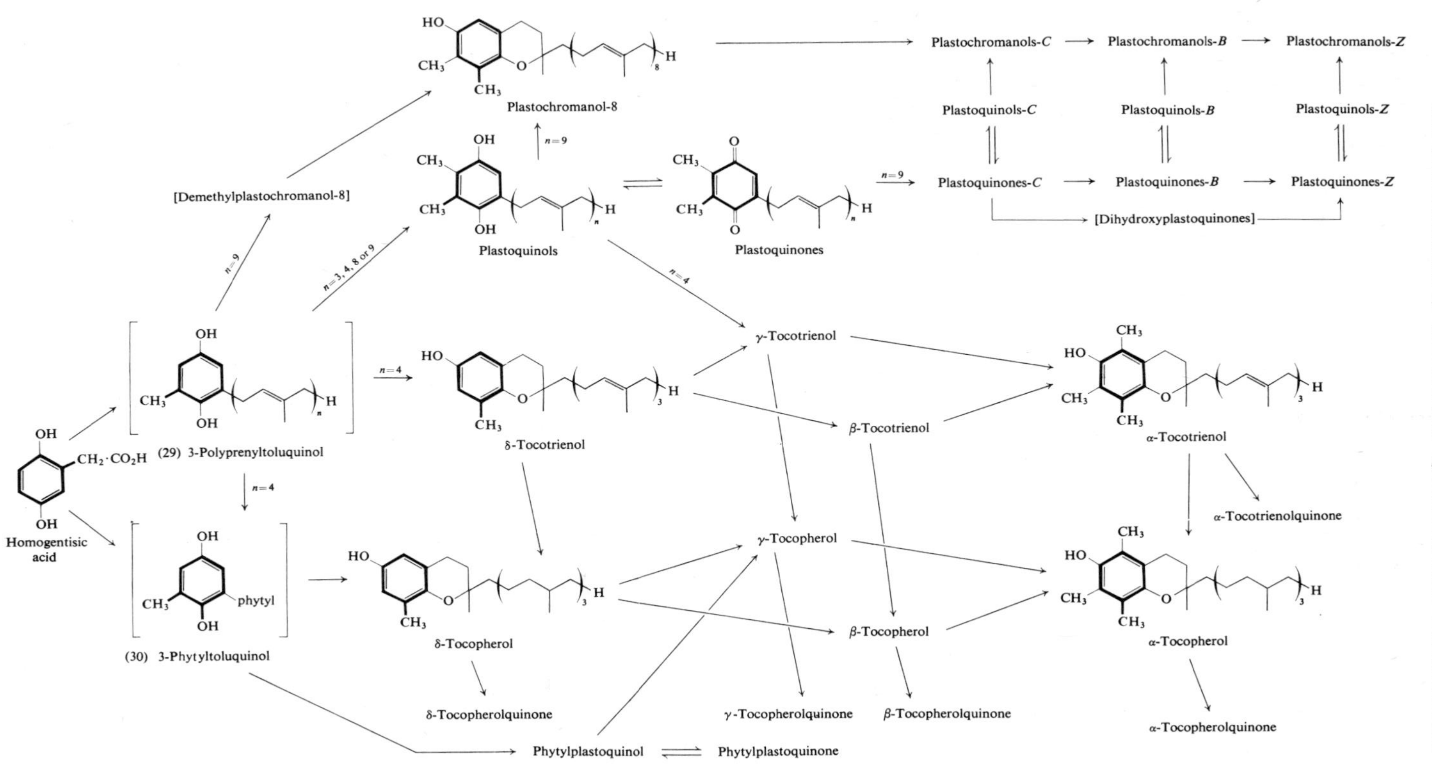

FIG. 7. Possible pathways for the synthesis of plastoquinones, tocotrienolquinones, tocopherolquinones, plastochromanols, tocotrienols and tocopherols from homogentisic acid.

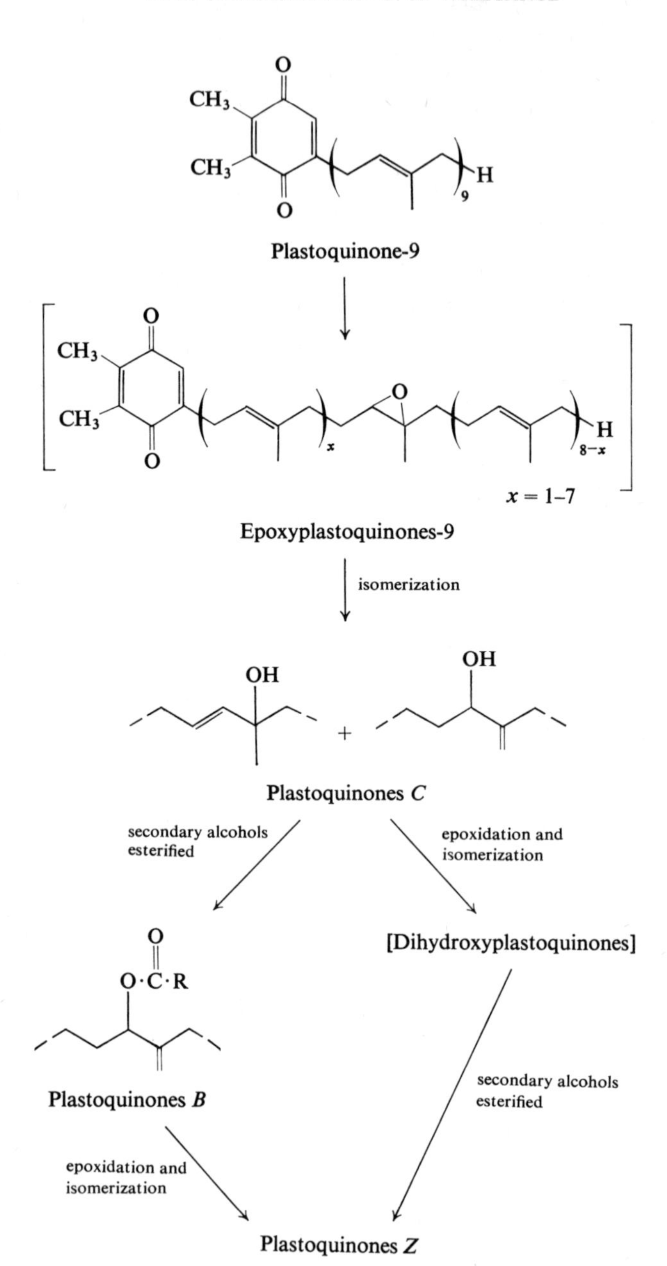

FIG. 8. Proposed biosynthetic interrelationships of plastoquinone-9, plastoquinones-*B*, -*C* and -*Z* (after Wallwork and Pennock, 1968).

place (30, Fig. 7). It should be stressed, however, that at present the evidence for the existence of any of the pathways discussed in this subsection is at best only tentative.

IV. Phylloquinones and Menaquinones

A. STRUCTURE AND DISTRIBUTION

The phylloquinones and menaquinones comprise two families of isoprenoid naphthoquinones which are chemically and biogenetically closely related. The families are readily differentiated, however, by the fact that the biosynthesis of the phylloquinones is confined to higher plants and algae, whereas that of the menaquinones is confined to bacteria and some fungi.

The phylloquinone family, as it is known at present, consists of three members, phylloquinone (31), demethylphylloquinone (32) and hydroxyphylloquinone (the hydroxyl group of this quinone is located at an unidentified position in the phytyl side chain). Phylloquinone is distributed widely amongst species of higher plants and algae (see Pennock, 1966). Demethylphylloquinone and hydroxyphylloquinone, on the other hand, appear to be of limited distribution. Thus, the former has been detected only in spinach chloroplasts (McKenna *et al.*, 1964), while the latter has been found only in photosynthetically grown cells of the blue-green alga *Anacystis nidulans* (Allen, C. F. *et al.*, 1967) and the green alga *Chlorella pyrenoidosa* (Whistance and Threlfall, 1970a).

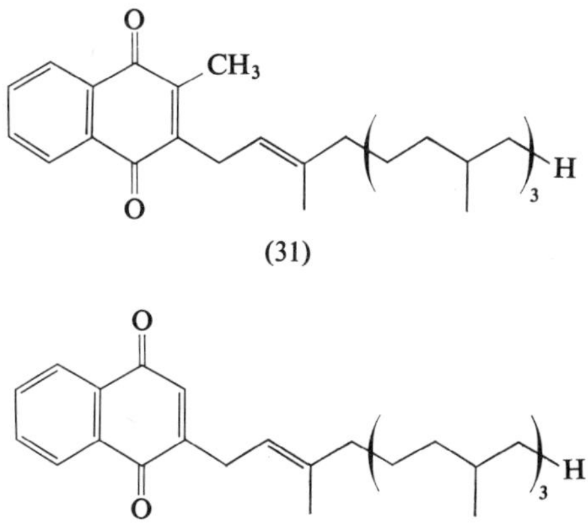

(31)

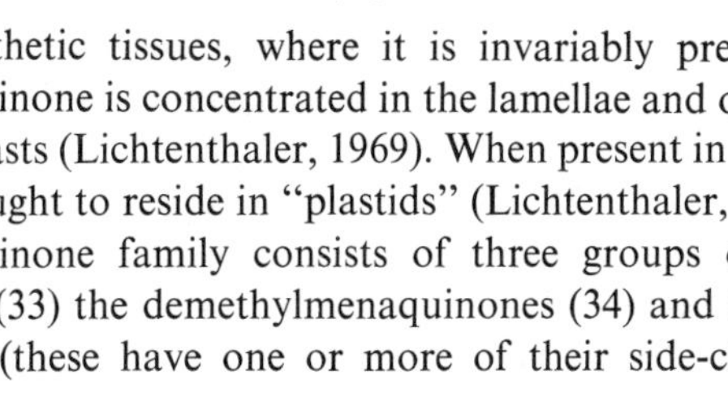

(32)

In photosynthetic tissues, where it is invariably present (see Pennock, 1966), phylloquinone is concentrated in the lamellae and osmiophilic globules of the chloroplasts (Lichtenthaler, 1969). When present in non-photosynthetic tissues it is thought to reside in "plastids" (Lichtenthaler, 1968).

The menaquinone family consists of three groups of compounds, the menaquinones (33) the demethylmenaquinones (34) and the partly saturated menaquinones (these have one or more of their side-chain isoprene units saturated).

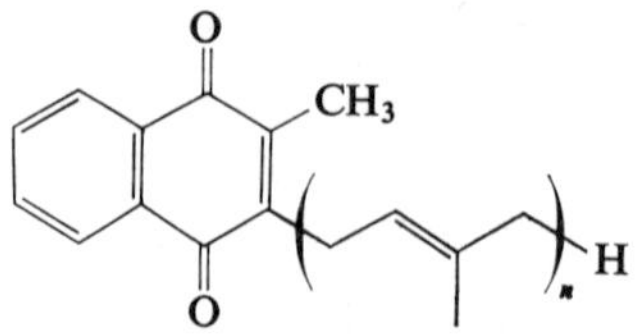

(33) 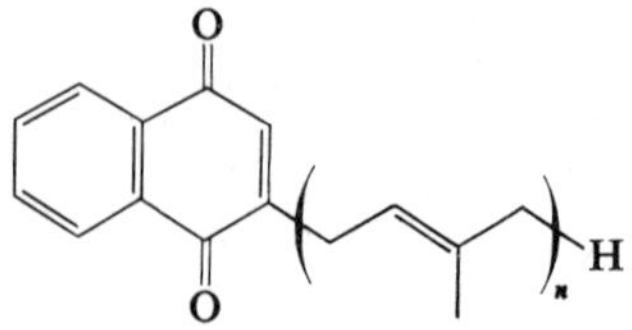 (34)

Menaquinones are of widespread occurrence amongst Gram-positive bacteria (see Pennock, 1966), non-photosynthetic and anaerobic Gram-negative bacteria (Whistance *et al.*, 1969; Maroc *et al.*, 1970; Weber *et al.*, 1970), photosynthetic Gram-negative bacteria (see Pennock, 1966) and animals (see Doisy and Matschiner, 1970). So far menaquinones with side chains varying in length from C_5 to C_{65} have been isolated from biological sources (Table V). Some organisms contain a series of menaquinones; for example, *Staphylococcus aureus* contains menaquinone-1 (33; $n = 1$) through menaquinone-8 (33; $n = 8$) (Hammond and White, 1969a), while *Chromatium* strain *D* contains menaquinone-3 through menaquinone-8 (G. R. Whistance and D. R. Threlfall, unpublished observations). A 1′-oxo derivative of menaquinone-7 called chlorobiumquinone (35) has been isolated from two species of the photosynthetic bacterium *Chlorobium* (Powls *et al.*, 1968); biogenetic studies have indicated that it is a metabolite of menaquinone-7 (Bartlett, 1968). These species of *Chlorobium* also contain a naphthoquinone which is thought to be the 1′-hydroxy derivative of menaquinone-7 (Powls and Redfearn, 1969).

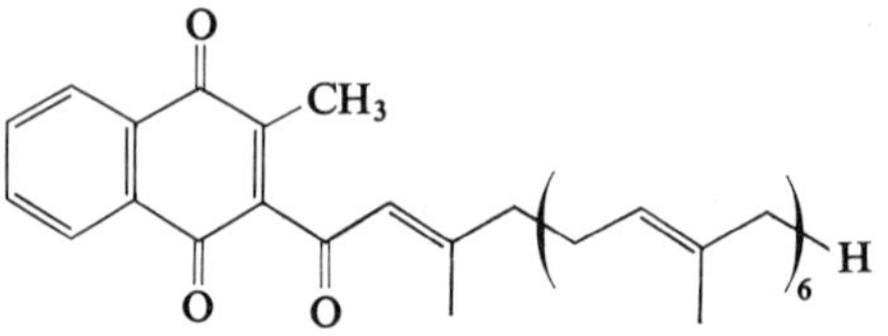

(35)

Demethylmenaquinones occur in some members of the above groups of bacteria but they are not nearly as widely distributed as the menaquinones (Table VI). Usually they are found only in organisms containing menaquinones; however, in *Streptococcus faecalis* (Dolin and Baum, 1965) and *Haemophilus parainfluenzae* (Lester *et al.*, 1964) they are the sole naphthoquinone components. Demethylquinones with side chains varying in length from C_{25} to C_{40} have been found in nature. Often a series of demethylmenaquinones is present in any one organism (Lester *et al.*, 1964).

The partly saturated menaquinones are of limited distribution. The commonest forms, menaquinone-9 (H_2), -8 (H_2) and -7 (H_2), occur in some members of the *Micrococcaceae* family of Gram-positive bacteria and in some strains of the fungus *Streptomyces* [in most cases the position of the saturated isoprene

TABLE V
Distribution of menaquinones

Organism	Principal homologue
GRAM-NEGATIVE BACTERIA	
Photosynthetic	
Rhodospirillum fulvum, *Rhodospirillum molischianum*, *Rhodospirillum viridis*	9[a]
Rhodospirillum sp., *Rhodopseudomonas gelatinosa*, *Chromatium* strain D, *Chromatium vinosum*	8[a,b,c]
Chlorobium ethylicum, *Chlorobium thiosulphatophilum*	7[d]
Non-photosynthetic	
(a) Aerobic	
Halobacterium cutirubrum	8[e]
(b) Facultative	
Aeromonas hydrophila, *Aeromonas punctata*, *Erwinia carotovora*, *Escherichia aurescens*, *Escherichia coli*, *Escherichia freundii*, *Klebsiella aerogenes*, *Proteus mirabilis*, *Proteus vulgaris*	8[f]
(c) Anaerobic	
Desulphatomaculum nigrificans	7[c]
Desulphovibrio gigas, *Desulphovibrio vulgaris*	6[g,h]
GRAM-POSITIVE BACTERIA	
Micrococcus lysodeikticus, *Mycobacterium tuberculosis*	9[i,j]
Sarcina lutea, *Staphylococcus aureus*	8[k,l]
Bacillus cereus, *Bacillus megaterium*, *Bacillus subtilis*, *Sarcina flava*	
Staphylococcus albus	7[k,m]
rumen liquors	10,11,12,13[q]
MOULDS	
Streptomyces griseus, *Streptomyces olivaceus*	9[n,o]
MAMMALS	
Ox liver	10,11,12[p]
Rat	4[r]

[a] Maroc *et al.* (1968); [b] Osnitskaya *et al.* (1964); [c] G. R. Whistance and D. R. Threlfall, unpublished observations; [d] Redfearn and Powls (1967); [e] Tornabene *et al.* (1969); [f] Whistance *et al.* (1969b); [g] Weber *et al.* (1970); [h] Maroc *et al.* (1970); [i] Bishop and King (1962); [j] Noll *et al.* (1960); [k] Bishop *et al.* (1962); [l] Jeffries *et al.* (1967); [m] Jacobsen and Dam (1960); [n] Phillips *et al.* (1969); [o] Ah Law, D. R. Threlfall and G. R. Whistance (1970), in preparation; [p] Matschiner and Amelotti (1968); [q] Matschiner (1970); [r] Martius (1961).

TABLE VI
Distribution of demethylmenaquinones

Organism	Homologue	Organism	Homologue(s)
PHOTOSYNTHETIC BACTERIA		*Escherichia freundii*	8[b]
Chlorobium thiosulphatophilum	7[a]	*Klebsiella aerogenes*	8[b]
		Proteus mirabilis	8[b]
NON-PHOTOSYNTHETIC BACTERIA		*Proteus vulgaris*	8[b]
(a) Gram negative		(b) Gram positive	
Aeromonas hydrophila	8[b]	*Streptococcus faecalis*	9,8,7[c]
A. punctata	8[b]	*Haemophilus parainfluenzae*	7,6,5[d]
Erwinia carotovora	8[b]		
Escherichia aurescens	8[b]	HIGHER PLANTS	
E. coli	8[b,c]	Teak wood	1[e]

[a] Bartlett (1968); [b] Whistance *et al.* (1969b); [c] Dolin and Baum (1965); [d] Lester *et al.* (1964); [e] Sandermann and Simatupang (1962).

unit is unknown (Jeffries *et al.*, 1967; Phillips *et al.*, 1969)]. The latter organisms also contain menaquinone-9 (H_4), -9 (H_6) and -9 (H_8) (Phillips *et al.*, 1969). The Gram-positive bacterium *Mycobacterium phlei* contains in addition to menaquinone-9 (II-H_2) (36) small amounts of 3′-*cis*-menaquinone-9 (II-H_2) (37) (Dunphy *et al.*, 1969).

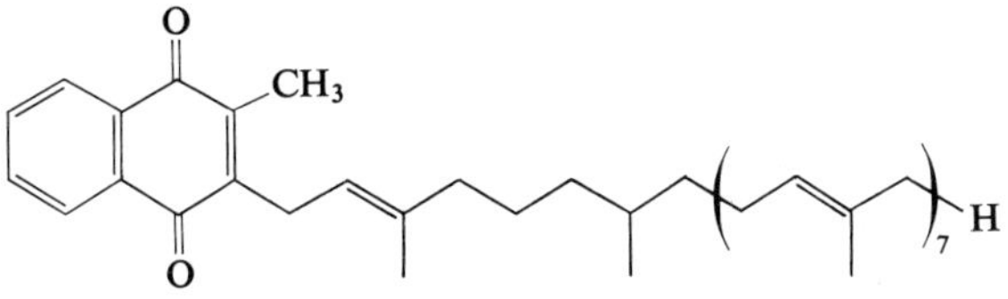

(36)

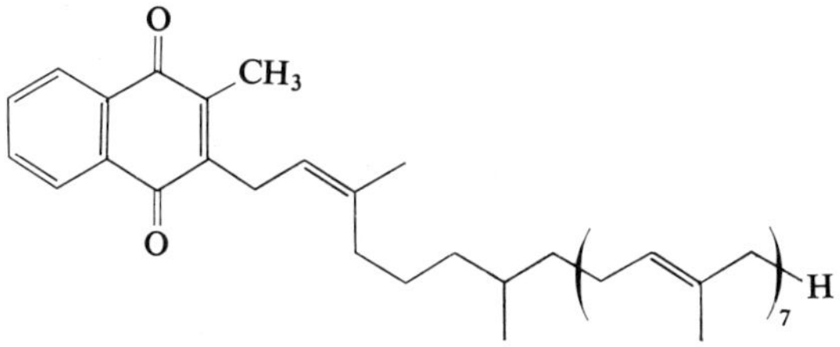

(37)

Intracellular distribution studies have shown that in Gram-positive and non-photosynthetic Gram-negative bacteria (anaerobic Gram-negative bacteria have not been examined) menaquinones, demethylmenaquinones and partly saturated menaquinones are localized on the protoplast membranes (see Pennock, 1966). In photosynthetic Gram-negative bacteria menaquinones are concentrated in the chromatophores (see Pennock, 1966).

B. BIOSYNTHESIS

1. Naphthoquinone Rings

It has been established that shikimic acid is a precursor of the naphthoquinone nuclei of menaquinones in Gram-positive bacteria (*Bacillus megaterium*, *B. subtilis*, *Sarcina lutea*, *Micrococcus lysodeikticus* and *Staphylococcus aureus*) and facultative Gram-negative bacteria (*Escherichia coli* and *Proteus vulgaris*), demethylmenaquinones in facultative Gram-negative bacteria (*E. coli*), menaquinones (H_2) in Gram-positive bacteria (*Mycobacterium phlei*) and phylloquinone in higher plants (*Zea mays*) (Cox and Gibson, 1964, 1966; Leistner *et al.*, 1967; Campbell *et al.*, 1967; Whistance *et al.*, 1966, 1967; Ellis and Glover, 1968). In the case of *E. coli* and *M. phlei* it has been shown that shikimic acid is incorporated as a C_6-C_1 unit into the menaquinone and menaquinone (II-H_2) nuclei (Campbell *et al.*, 1967; Leistner *et al.*, 1967), the alicyclic carbon atoms and carboxyl carbon atom giving rise to the aromatic ring (ring *A*) and one of the carbon atoms (either C-1 or C-4) carrying a quinone function respectively. The manner in which shikimic acid is incorporated into phylloquinone is at present unknown. However, the fact that in the higher plants *Impatiens balsamina*, *Juglans regia* and *Rubia tinctorum* shikimic acid gives rise to ring A and a carbonyl carbon atom of the naphthoquinones lawsone (38) and juglone (39) (Zenk and Leistner, 1967; Leistner and Zenk, 1968a) and the anthraquinones alizarin (40; R = H) and purpurin (40; R = OH) (Leistner and Zenk, 1967, 1968b) suggests that it is similar to that found for menaquinones.

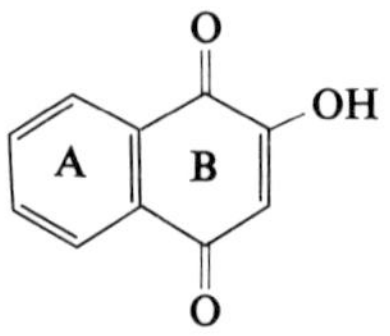

(38)

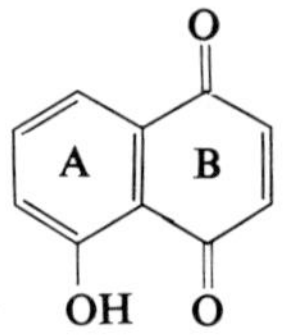

(39)

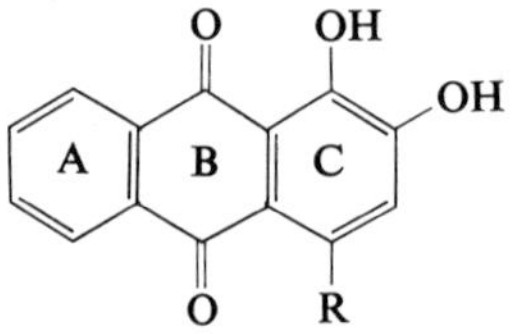

(40)

The chemical degradations of menaquinones, lawsone, alizarin and purpurin labelled from [^{14}C] shikimic acid left two questions unanswered with regard to the absolute relationship between the carbon atoms of the precursor and products. Firstly, do C-2 and C-6 and C-3 and C-5 of shikimic acid retain their identities or do they become equivalent? Secondly, does C-7 of shikimic acid contribute to both or only one of the carbonyl carbon atoms? Recently insight into these problems has been gained by studying the incorporation of ^{14}C-labelled species of shikimic acid into juglone, a compound whose C-5 hydroxyl group confers asymmetry on its chemical degradation products. The results of these studies showed that C-1 and C-6 of shikimic acid give rise to C-9 and C-10, i.e. the carbon atoms forming the bridge between rings A and B, and that C-2 and C-7 contribute to C-5 and C-8 and C-1 and C-4 respectively (Leistner and Zenk, 1968a), indicating that a symmetrical dihydroxy intermediate is involved in the biosynthesis of this compound and perhaps menaquinones and phylloquinones also.

The source(s) of the remaining three carbon atoms of the nuclei of menaquinones and phylloquinones remain(s) to be determined unequivocally. Cox and Gibson (1966) obtained some incorporation of radioactivity from [1,2-^{14}C] acetic acid into the quinone ring of menaquinone-8 in *E. coli.*

Campbell *et al.* (1967) reported similar findings for menaquinone-9 (II-H_2) in *M. phlei*. Campbell (1969), on the basis of the fact that radioactivity from DL-[2-^{14}C] glutamate is incorporated only into C-1 and C-4 of lawsone by *I. balsamina*, has proposed that a thiamine pyrophosphate-succinate semi-aldehyde complex is the immediate source (Fig. 9).

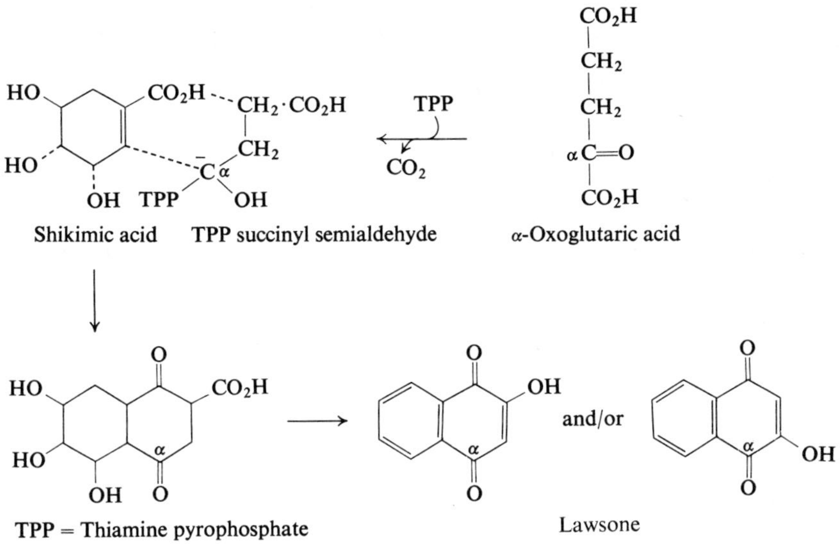

CO_2H | CH_2 | CH_2 | $\alpha C{=}O$ | CO_2H

HO, HO, OH, CO_2H, $CH_2 \cdot CO_2H$, CH_2, $\bar{C}\alpha$, TPP, OH

TPP → CO_2

Shikimic acid   TPP succinyl semialdehyde   α-Oxoglutaric acid

HO, HO, OH, O, O, α, CO_2H → O, O, α, OH and/or O, O, α, OH

TPP = Thiamine pyrophosphate   Lawsone

FIG. 9. Hypothetical scheme for the biosynthesis of lawsone by *Impatiens balsamina* (after Campbell, 1969).

Leistner *et al.* (1967) proposed that α-naphthol is the first naphthalenic intermediate in the biosynthesis of isoprenoid naphthoquinones. This proposal was based on their observation that α-[1-^{14}C] naphthol was incorporated *in toto* into menaquinone-7 by the Gram-negative bacterium *B. megaterium* (Leistner *et al.*, 1967). Since that time there has been both support for and opposition to this proposal. Thus, Hammond and White (1969b) have shown that α-naphthol is incorporated *in toto* into the nucleus of menaquinone-8 by the Gram-positive bacterium *S. aureus*, a finding confirmed by Hurd (1970), whereas Brown *et al.* (1968), Ellis and Glover (1968), Hall (1970) and Hurd (1970) could obtain no incorporation of radioactivity from α-[1-^{14}C] naphthol into menaquinones in the Gram-positive bacteria *B. megaterium* (strain used by Leistner *et al.*, 1967), *Micrococcus lysodeikticus* and *B. subtilis* and the facultative Gram-negative bacteria *E. coli* and *P. mirabilis* and phylloquinone in maize shoots. A possible explanation for these anomalous findings is that α-naphthol is not a true intermediate in the biosynthesis of menaquinones and phylloquinones, as would be the case if the pathway proposed by Campbell (1969) for the formation of the naphthoquinone nucleus involving a thiamine pyrophosphate-succinate semialdehyde complex is operative. It would then follow that those organisms which can utilize α-naphthol for the biosynthesis of naphthoquinones have the ability to convert it into a true intermediate.

2. *Polyprenyl Side Chains*

Mevalonic acid has been shown to be a precursor of the phytyl side chain of phylloquinone in the higher plants tobacco (*Nicotiana tabacum*) and maize (Griffiths *et al.*, 1968; Dada *et al.*, 1968). It has also been shown to give rise to the octaprenyl side chain of menaquinone-8 in the Gram-positive bacterium *Staphylococcus aureus* (Hammond and White, 1969).

Investigations with (3*RS*)-[2-^{14}C,(4*R*)-4-$^{3}H_1$] mevalonic acid have established that in maize shoots all four isoprene units of the phylloquinone side chain are biogenetically *trans* (Dada *et al.*, 1968).

Nothing is known about either the way in which the side chains are assembled or how and at what stage of their formation they are coupled with the appropriate nucleus. However, it seems likely that the steps will be similar to those outlined for the biosynthesis of ubiquinones (see Section IIB2), plastoquinones, tocopherols, etc.(see Section IIIB2).

3. *Nuclear* C-*Methyl Groups*

The nuclear *C*-methyl groups of menaquinones in Gram-positive and Gram-negative bacteria and of phylloquinone in higher plants and algae are derived from the *S*-methyl group of L-methionine (Table VII). Studies with L-[*Me*-$^{2}H_3$] methionine and L-[*Me*-^{14}C,$^{3}H_3$] methionine have shown that in *Escherichia coli* (Jackman *et al.*, 1967), *Mycobacterium smegatis* (Jauréguiberry *et al.*, 1966) and maize shoots (Threlfall *et al.*, 1968) transmethylation takes place.

4. *Terminal Steps*

The natures of the terminal steps involved in the biosynthesis of menaquinones and phylloquinones have still to be elucidated.

Leistner *et al.* (1967) proposed that the terminal steps in the biosynthesis of menaquinones are as shown in Fig. 10. This proposal was based on their observation that α-[1-^{14}C] naphthol was incorporated *in toto* into the nucleus of menaquinone-7 by *Bacillus megaterium* and the observation of Martius and Leuzinger (1964) that 1,4-[^{3}H] naphthoquinone was incorporated into menaquinone-9 by *Fusiformis nigrescens*.

As stated previously there is some doubt whether α-naphthol is a true intermediate in the biosynthesis of menaquinones (see Section IVB1). Similarly, controversy exists regarding the role of menadione. Thus, Hammond and White (1969) and Hurd (1970) have shown that in *Staphylococcus aureus* and *B. megaterium* menadione is incorporated into the nuclei of menaquinones, whereas Hurd (1970) has shown that in *Escherichia coli* and *Micrococcus*

TABLE VII

Organisms in which the *S*-methyl group of L-methionine has been shown to be the source of the nuclear *C*-methyl groups of menaquinones (MK) and phylloquinone (K)

Organism	Quinone	Organism	Quinone
PHOTOSYNTHETIC BACTERIA		ALGAE	
Chlorobium thiosulphatophilum	MK-7, chlorobiumquinone	*Euglena gracilis*	K[b]
Chromatium strain D	MK-8[b]		
		HIGHER PLANTS	
NON-PHOTOSYNTHETIC BACTERIA		*Hedera helix*	K[g]
Escherichia coli	MK-8[c]	*Zea mays*	K[h]
Fusiformis nigrescens	MK-9[d]		
Mycobacterium phlei	MK-9 (II-H_2)[e]		
M. smegatis	MK-9 (II-H_2)[f]		

[a] Bartlett (1968); [b] G. R. Whistance and D. R. Threlfall, unpublished observations; [c] Jackman *et al.* (1967); [d] Martius and Leuzinger (1964); [e] Guerin *et al.* (1965); [f] Jauréguiberry *et al.* (1966); [g] Whistance and Threlfall (1968b); [h] Threlfall *et al.* (1968).

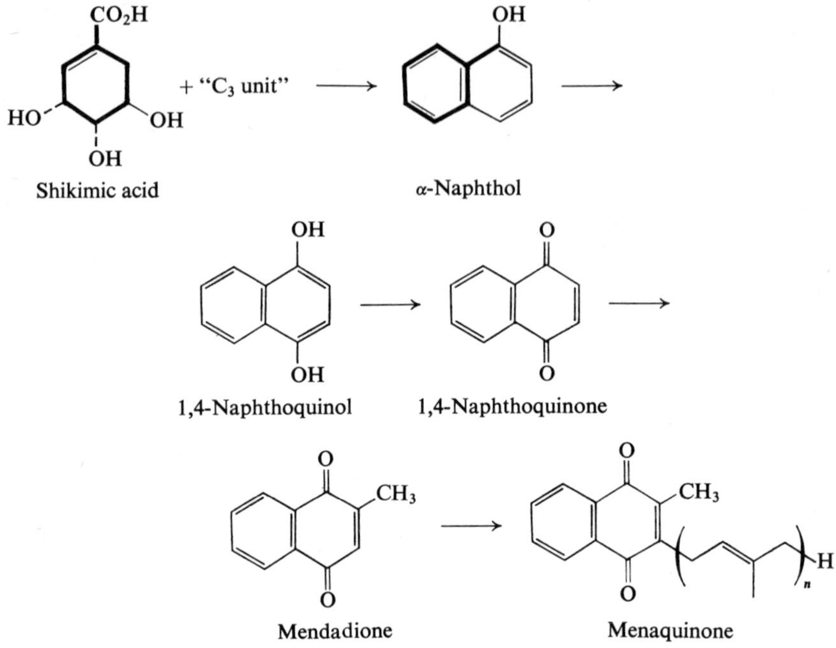

Shikimic acid + "C_3 unit" → α-Naphthol →

1,4-Naphthoquinol → 1,4-Naphthoquinone →

Mendadione → Menaquinone

FIG. 10. Proposed pathways for the biosynthesis of menaquinones in bacteria (after Leistner *et al.*, 1967).

lysodeikticus it is not. Again, Azerad *et al.* (1967) and Samuel and Azerad (1969) working with cell-free extracts of *Mycobacterium phlei* have provided some evidence that demethylmenaquinones rather than menadione are the immediate precursors of menaquinones. [It is noteworthy in this context that Bartlett (1968) and Ellis and Glover (1968) could find no evidence that demethylmenaquinones are precursors of menaquinones in *Chlorobium thiosulphatophilum* and *E. coli*.]

Other than the observation of Martius and Leuzinger (1964) that 1,4-[^{3}H] naphthoquinone was incorporated into menaquinone-9 by *F. nigrescens* there is no experimental support for the view that 1,4-naphthoquinones or 1,4-naphthoquinols are involved in menaquinone biosynthesis.

In the case of the partly saturated menaquinones there exists the problem of whether they are synthesized from menaquinones or from partly saturated precursors. Azerad *et al.* (1965) have provided some evidence that the former may be the case. Thus, these workers have shown that *M. phlei* possesses a menaquinone reductase which in the presence of NADH or NADPH converts menaquinone-9 to menaquinone-9 (II-H_2).

Virtually nothing is known about the final steps in the biosynthesis of phylloquinone, although it has been shown that in maize shoots α-naphthol and menadione are not involved (Hall, 1970). Leistner and Zenk (1968a), on the basis of the incorporation of radioactivity from 1,4-[^{14}C] naphthoquinone into the naphthoquinone nucleus of juglone by leaves of *Juglans regia* (Leistner and Zenk, 1968a), have proposed that 1,4-naphthoquinone (or 1,4-naphthoquinol) is a key intermediate in the biosynthesis of naphthoquinones by higher plants. Support for this proposal has come from the observation that 1,4-naphthoquinone can also be utilized by *Rubia tinctorum* for the biosynthesis of rings *A* and *B* of the naphthoquinone-derived anthraquinone alizarin (Leistner and Zenk, 1968b). At present, however, there is no experimental evidence to suggest that this compound (or 1,4-naphthoquinol) is a precursor of phylloquinone, although it is apparent that to be converted to phylloquinone it would only require methylation and either phytylation or prenylation followed by side chain saturation. The occurrence of demethylphylloquinone in spinach chloroplasts (McKenna *et al.*, 1964) perhaps provides some evidence that in higher plants the last step is one of methylation.

V. Intracellular Sites of Synthesis of Isoprenoid Quinones and Chromanols in Plants

Goodwin and Mercer (1963) proposed that there are two sites for the biosynthesis of terpenoids in the photosynthetic plant cell, an intrachloroplastidic site which would be responsible for the synthesis of carotenoids and the isoprenoid side chains of chlorophylls, phylloquinones, plastoquinones, tocopherols and tocopherolquinones, and an extrachloroplastidic site where 4-demethylsterols, pentacyclic triterpenes and the side chains of ubiquinones would be synthesized.

To test this hypothesis in relation to the biosynthesis of the isoprenoid side chains of phylloquinone, plastoquinone-9, plastoquinones-*C*, α-tocopherol, α-tocopherolquinone, ubiquinone-9 and ubiquinone-10 Threlfall (1967), Threlfall *et al.* (1967b) and Griffiths *et al.* (1968) compared the incorporation of radioactivity from $^{14}CO_2$ and [2-^{14}C] mevalonic acid into these compounds, β-carotene and 4-demethylsterols by greening etiolated maize (*Zea mays*) and French bean (*Phaseolus vulgaris*) shoots and young seedlings of tobacco (*Nicotiana tabacum*). In each set of experiments radioactivity from $^{14}CO_2$ was incorporated well into the intrachloroplastidic compounds β-carotene, phylloquinone, plastoquinone-9, plastoquinones-*C*, α-tocopherol and α-tocopherolquinone, but only poorly into the extrachloroplastidic compounds ubiquinones and 4-demethylsterols. Conversely, radioactivity from [2-^{14}C] mevalonic acid was incorporated well into ubiquinones and 4-demethylsterols, but only poorly into β-carotene, plastoquinone-9, plastoquinones-*C*, α-tocopherol and α-tocopherolquinone. These findings led Threlfall (1967), Threlfall *et al.* (1967b) and Griffiths *et al.* (1968) to conclude that, in agreement with the proposals of Goodwin and Mercer (1963), the side chains of phylloquinone, plastoquinone-9, plastoquinone-*C*, α-tocopherol and α-tocopherolquinone are synthesized intrachloroplastidically and that those of ubiquinones are synthesized extrachloroplastidically.

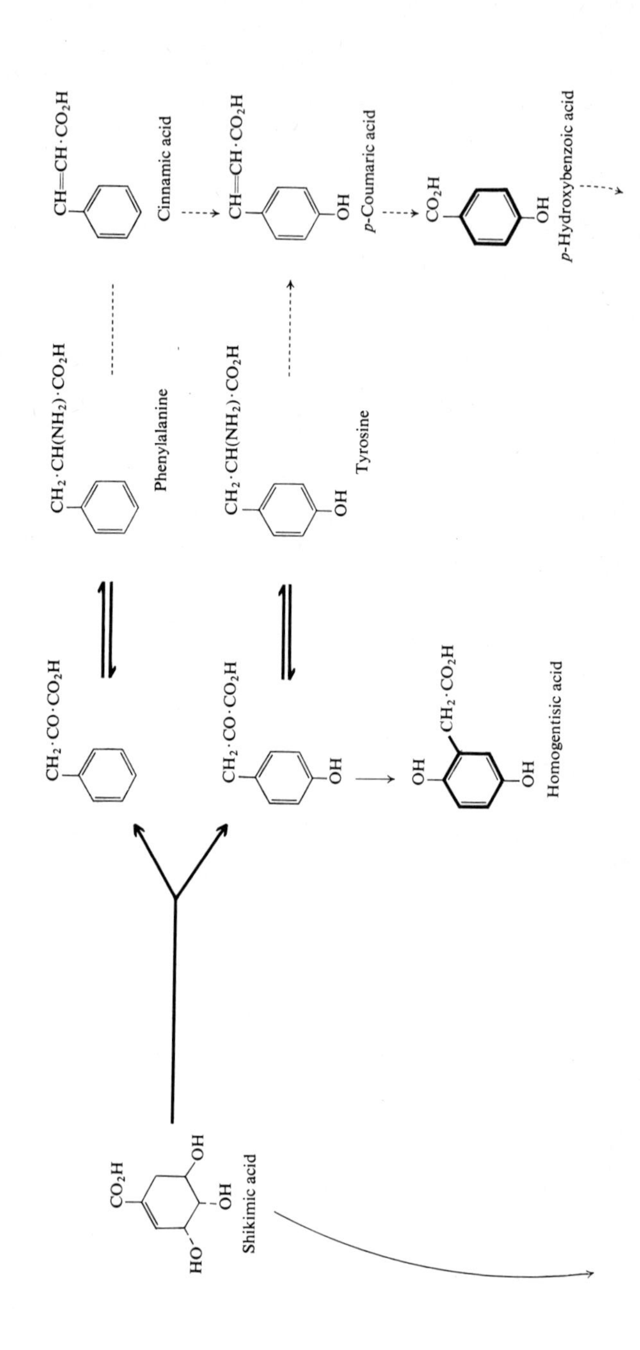

$CH=CH \cdot CO_2H$
Cinnamic acid
$CH=CH \cdot CO_2H$
OH
p-Coumaric acid
CO_2H
OH
p-Hydroxybenzoic acid
$CH_2 \cdot CH(NH_2) \cdot CO_2H$
Phenylalanine
$CH_2 \cdot CH(NH_2) \cdot CO_2H$
OH
Tyrosine
$CH_2 \cdot CO \cdot CO_2H$
$CH_2 \cdot CO \cdot CO_2H$
OH
OH
$CH_2 \cdot CO_2H$
OH
Homogentisic acid
CO_2H
HO
OH
OH
Shikimic acid

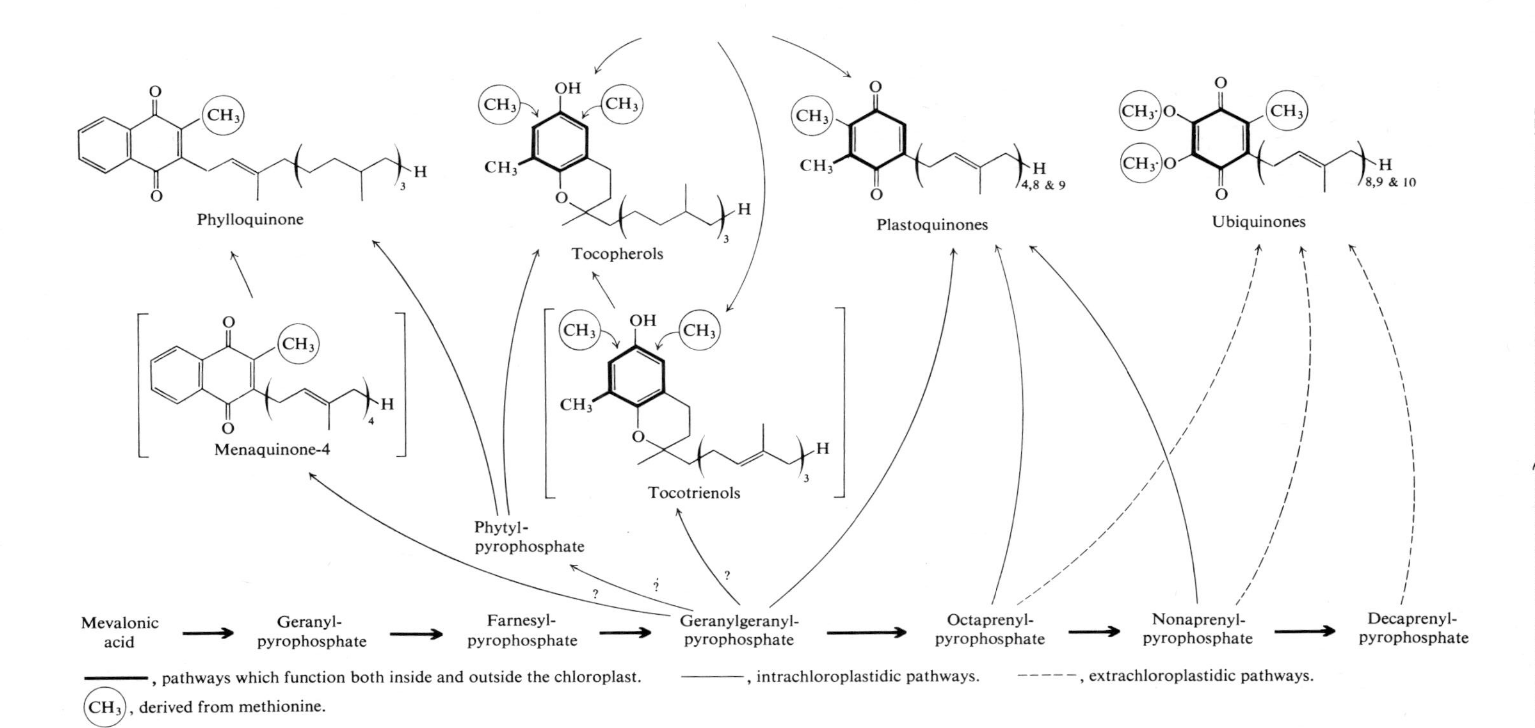

FIG. 11. Compartmentation of isoprenoid quinones and chromanols biosynthesis in the photosynthetic plant cell.

It is currently believed that the overall biosynthesis of phylloquinone, plastoquinones, tocopherols and tocopherolquinones takes place within the chloroplast and that that of ubiquinones occurs elsewhere in the cell (mitochondria?) (Fig. 11) (Threlfall and Griffiths, 1967; Dada *et al.*, 1968; Threlfall *et al.*, 1968; Whistance and Threlfall, 1968b). The factors which have led to this belief are threefold. Firstly, the synthesis of plastoquinone-9 and α-tocopherolquinone closely parallels chloroplast formation, whereas that of ubiquinones does not, (Threlfall and Griffiths, 1967). Secondly, the comparative specific radioactivity data obtained from experiments in which L-[*Me*-^{14}C] methionine, D-[U-^{14}C] shikimic acid, L-[U-^{14}C] phenylalanine and L-[U-^{14}C] tyrosine were administered to maize and French bean shoots is similar to that obtained in experiments with [2-^{14}C] mevalonic acid, the specific radioactivities of ubiquinone-9 and -10 being in most cases appreciably greater than those of phylloquinone, plastoquinone-9, tocopherols and α-tocopherolquinone (Whistance *et al.*, 1967; Threlfall *et al.*, 1968). Thirdly, chloroplasts isolated from leaves of horsechestnut (*Aesculus hippocastanum*) have been shown to be capable of incorporating [2-^{14}C] mevalonic acid into plastoquinone-9 (Wellburn and Hemming, 1967).

In yeasts there is some evidence that ubiquinones are biosynthesized in the mitochondria (their intracellular locality). Thus, when anaerobically grown cells of bakers' yeast (which contain promitochondria) are aerated ubiquinones are synthesized in step with mitochondrial development (Sugimura and Rudney, 1960).

ACKNOWLEDGEMENTS

We wish to thank the members, both past and present, of the Department of Biochemistry and Agricultural Biochemistry, University College of Wales, Aberystwyth who have collaborated with us in our studies.

REFERENCES

Ah Law, Threlfall, D. R. and Whistance, G. R. (1970). *Biochem. J.* **117**, 799.
Allen, C. F., Franke, H. and Hirayama, O. (1967). *Biochem. biophys. Res. Commun.* **26**, 562.
Allen, C. M., Alworth, W., MacRae, A. and Bloch, K. (1967). *J. biol. Chem.* **242**, 1895.
Azerad, R., Bleiler-Hill, R., Catala, F., Samuel, O. and Lederer, E. (1967). *Biochem. biophys. Res. Commun.* **27**, 253.
Barr, R. and Arntzen, C. J. (1969). *Pl. Physiol.* **44**, 591.
Barr, R. and Crane, F. L. (1967). *Pl. Physiol.* **42**, 1255.
Barr, R., Magree, L. and Crane, F. L. (1967). *Am. J. Bot.* **54**, 365.
Bartlett, K. (1968). B.Sc. Thesis: University College of Wales, Aberystwyth.
Bertel, R. (1903). *Chem. Zbl.* **1**, 178.
Bishop, D. H. L. and King, H. K. (1962). *Biochem. J.* **85**, 550.
Bishop, D. H. L., Pandya, K. P. and King, H. K. (1962). *Biochem. J.* **83**, 606.
Booth, V. H. (1963). *Phytochemistry* **2**, 421.
Brown, B. S., Whistance, G. R. and Threlfall, D. R. (1968). *FEBS Letters* **1**, 323.

Campbell, I. M. (1969). *Tetrahedron Letters* **54**, 4777.
Campbell, I. M., Coscia, C. J., Kelsey, M. and Bentley, R. (1967). *Biochem. biophys. Res. Commun.* **28**, 25.
Carr, N. G. (1964). *Biochem. J.* **91**, 28P.
Clayton, R. B. (1965). *Q. Rev. chem. Soc.* **19**, 168, 201.
Cox, G. B. and Gibson, F. (1964). *Biochim. biophys. Acta* **93**, 204.
Cox, G. B. and Gibson, F. (1966). *Biochem. J.* **100**, 1.
Cox, G. B., Young, I. G., McCann, L. M. and Gibson, F. (1969). *J. Bact.* **99**, 450.
Crane, F. L. (1965). *In* "Biochemistry of Quinones" (R. A. Morton, ed.), p. 183, Academic Press, London and New York.
Dada, O. A. (1968). M.Sc. Thesis: University College of Wales, Aberystwyth.
Dada, O. A., Threlfall, D. R. and Whistance, G. R. (1968). *Europ. J. Biochem.* **4**, 329.
Daves, G. D., Jun., Friis, P., Olsen, R. K. and Folkers, K. (1966). *Vitams Horms* **24**, 427.
Daves, G. D., Muraca, R. F., Whittick, J. S., Friis, P. and Folkers, K. (1967). *Biochemistry, Easton* **6**, 2861.
Dilley, R. A. and Crane, F. L. (1963). *Pl. Physiol.* **38**, 452.
Diplock, A. T. and Haslewood, G. A. D. (1967). *Biochem. J.* **104**, 1004.
Diplock, A. T., Green, J., Edwin, E. E. and Bunyan, J. (1961). *Nature, Lond.* **187**, 749.
Doisy, E. A., jun. and Matschiner, J. T. (1970). *In* "Fat Soluble Vitamins" (R. A. Morton, ed.), Vol. 9, p. 293, Pergamon Press Ltd., Oxford.
Dolin, M. I. and Baum, R. H. (1965). *Biochem. biophys. Res. Commun.* **18**, 202.
Dunphy, P. J., Whittle, K. J., Pennock, J. F. and Morton, R. A. (1965). *Nature, Lond.* **207**, 521.
Dunphy, P. J., Whittle, K. J. and Pennock, J. F. (1966). *In* "Biochemistry of Chloroplasts" (T. W. Goodwin, ed.), Vol. I, p. 165, Academic Press, London and New York.
Dunphy, P. J., Gutnick, D. L., Phillips, P. G. and Brodie, A. F. (1968). *J. biol. Chem.* **243**, 398.
Dunphy, P. J., Peake, I. R. and Pennock, J. F. (1969). *Biochem. J.* **113**, 35P.
Eck, H. and Trebst, A. (1963). *Z. Naturf.* **186**, 446.
Ellis, J. R. S. and Glover, J. (1968). *Biochem. J.* **110**, 22P.
Evans, W. C. (1963). *J. gen. Microbiol.* **32**, 177.
Field, F. E. (1969). B.Sc. Thesis: University College of Wales, Aberystwyth.
Friis, P., Daves, G. D., jun. and Folkers, K. (1966). *J. Am. chem. Soc.* **88**, 4754.
Friis, P., Daves, G. D., jun. and Folkers, K. (1967a). *Biochemistry, Easton* **6**, 3618.
Friis, P., Nilsson, J. L. G., Daves, G. D., jun. and Folkers, K. (1967b). *Biochem. biophys. Res. Commun.* **28**, 234.
Gibson, F. and Pittard, J. (1968). *Bacteriol. Rev.* **32**, 465.
Glover, J. (1965). *In* "Biochemistry of Quinones" (R. A. Morton, ed.), p. 207, Academic Press, London and New York.
Goodwin, T. W. and Mercer, E. I. (1963). *Biochem. Soc. Symp.*, No. 24, p. 37.
Green, J. (1970). *In* "Fat Soluble Vitamins" (R. A. Morton, ed.), Vol. 9, p. 71, Pergamon Press Ltd., Oxford.
Griffiths, W. T., Threlfall, D. R. and Goodwin, T. W. (1968). *Europ. J. Biochem.* **5**, 124.
Guerin, M., Azerad, R. and Lederer, E. (1965). *Bull. Soc. Chim. biol.* **47**, 2105.
Hall, R. A. (1970). B.Sc. Thesis: University College of Wales, Aberystwyth.
Hammond, R. K. and White, D. C. (1969a). *J. Chromat.* **45**, 446.
Hammond, R. K. and White, D. C. (1969b). *J. Bact.* **100**, 573.

Hemming, F. W., Morton, R. A. and Pennock, J. F. (1963). *Proc. r. Soc.* B **158**, 291.
Henninger, M. D., Dilley, R. A. and Crane, F. L. (1963). *Biochem. biophys Res. Commun.* **10**, 237.
Hurd, D. M. (1970). B.Sc. Thesis: University College of Wales, Aberystwyth.
Imamoto, S. and Senoh, S. (1968). *J. chem. Soc. Japan, Pure Chem. Sect.* **89**, 316.
Inoue, H., Arai, T. and Takano, Y. (1958). *Chem. Pharm. Bull. Tokyo* **6**, 653.
Jackman, L. M., O'Brien, I. G., Cox, G. B. and Gibson, F. (1967). *Biochim. biophys. Acta* **141**, 1.
Jacobsen, B. K. and Dam, H. (1960). *Biochim. biophys. Acta* **40**, 211.
Jauréguiberry, G., Lenfant, M., Das, B. C. and Lederer, E. (1966). *Tetrahedron, Suppl.* 8, **1**, 27.
Jeffries, L., Cawthorne, M. A., Harris, M., Diplock, A. T., Green, J. and Price, S. A. (1967). *Nature, Lond.* **215**, 257.
Jones, R. G. W. and Lascelles, J. (1967). *Biochem. J.* **103**, 714.
Kofler, M., Langemann, A., Rüegg, R., Chopard-dit-Jean, L. H., Raymond, O. and Isler, O. (1959). *Helv. chim. Acta* **42**, 1283.
Kubin, H. and Fink, H. (1961). *Fette Seifen Anstrichmittel* **63**, 280.
Lavate, W. V. and Bentley, R. (1964). *Archs Biochem. Biophys.* **108**, 287.
Leistner, E. and Zenk, M. H. (1967). *Z. Naturf.* **22b**, 865.
Leistner, E. and Zenk, M. H. (1968a). *Z. Naturf.* **23b**, 259.
Leistner, E. and Zenk, M. H. (1968b). *Tetrahedron Letters* **7**, 861.
Leistner, E. and Zenk, M. H. (1968c). *Tetrahedron Letters* **11**, 1395.
Leistner, E., Schmitt, J. H. and Zenk, M. H. (1967). *Biochem. biophys. Res. Commun.* **28**, 845.
Lester, R. L., White, D. C. and Smith, S. L. (1964). *Biochemistry, Easton* **3**, 949.
Lichtenthaler, H. K. (1966). *Ber. dtsch. bot. Ges.* **79**, 111.
Lichtenthaler, H. K. (1968a). *Planta, Berlin* **81**, 140.
Lichtenthaler, H. K. (1968b). *Z. Pflanzenphysiol. Bd.* **59**, 195.
Lichtenthaler, H. K. (1969). *Z. Naturf.* **24b**, 1462.
McKenna, M., Henninger, M. D. and Crane, F. L. (1964). *Nature, Lond.* **203**, 524.
Maroc, J., De Klerk, H. and Kamen, M. D. (1968). *Biochim. biophys. Acta* **162**, 621.
Maroc, J., Azerad, R., Kamen, M. D. and Le Gall, J. (1970). *Biochim. biophys. Acta* **197**, 87.
Martius, C. (1961). *Am. J. Clin. Nutr.* **9**, 97.
Martius, C. and Leuzinger, W. (1964). *Biochem. Z.* **340**, 304.
Matschiner, J. T. (1970). *J. Nutr.* **100**, 190.
Matschiner, J. T. and Amelotti, J. (1968). *J. Lipid Res.* **9**, 176.
Matsumura, U. and Shibata, Y. (1964). *Chem. Abstr.* **61**, 2104d.
Meister, A. (1965). *In* "Biochemistry of Amino Acids", Vol. 2, pp. 894–908, Academic Press, London and New York.
Misiti, D., Moore, H. W. and Folkers, K. (1965). *J. Am. chem. Soc.* **87**, 1402.
Moore, H. W. and Folkers, K. (1966). *J. Am. chem. Soc.* **88**, 567.
Nilsson, J. L. G., Farley, T. M. and Folkers, K. (1968). *Analyt. Biochem.* **23**, 422.
Noll, H., Ruegg, R., Gloor, U., Ryser, G. and Isler, O. (1960). *Helv. chim. Acta* **43**, 433.
Nowicki, H. G., Dialameh, G. H., Trumpower, B. I. and Olson, R. E. (1969). *Fedn Proc. Fedn Am. Socs exp. Biol.* **28**, 884.
Olsen, R. K., Smith, J. L., Daves, G. D., jun., Moore, H. W., Folkers, K., Parson, W. W. and Rudney, H. (1965). *J. Am. chem. Soc.* **87**, 2298.
Olsen, R. K., Daves, G. D., jun., Moore, H. W., Folkers, K. and Rudney, H. (1966a). *J. Am. chem. Soc.* **88**, 2346.

Olsen, R. K., Daves, G. D., jun., Moore, H. W., Folkers, K., Parson, W. W. and Rudney, H. (1966b). *J. Am. chem. Soc.* **88**, 5919.

Olson, R. E. (1966). *Vitams Horms* **24**, 551.

Olson, R. E. and Aiyar, A. S. (1966). *Fedn Proc. Fedn Am. Socs exp. Biol.* **25**, 217.

Olson, R. E., Bentley, R., Aiyar, A. S., Dialameh, G. H., Gold, P. H., Ramsey, V. G. and Springer, C. M. (1963). *J. biol. Chem.* **238**, PC 3146.

Osnitskaya, L. K., Threlfall, D. R. and Goodwin, T. W. (1964). *Nature, Lond.* **204**, 80.

Parson, W. W. and Rudney, H. (1964). *Proc. natn Acad. Sci. U.S.A.* **51**, 444.

Parson, W. W. and Rudney, H. (1965a). *J. biol. Chem.* **240**, 1855.

Parson, W. W. and Rudney, H. (1965b). *Proc. natn Acad. Sci. U.S.A.* **53**, 599.

Pennock, J. F. (1966). *Vitams Horms* **24**, 307.

Pennock, J. F., Hemming, F. W. and Kerr, J. (1964). *Biochem. biophys. Res. Commun.* **17**, 542.

Phillips, P. G., Dunphy, P. J., Servis, K. L. and Brodie, A. F. (1969). *Biochemistry, Easton* **8**, 2856.

Powls, R. and Hemming, F. W. (1966a). *Phytochemistry* **5**, 1235.

Powls, R. and Hemming, F. W. (1966b). *Phytochemistry* **5**, 1249.

Powls, R. and Redfearn, E. R. (1969). *Biochim. biophys. Acta* **172**, 429.

Powls, R., Redfearn, E. R. and Trippet, S. (1968). *Biochem. biophys. Res. Commun.* **33**, 408.

Raman, T. S., Rudney, H. and Buzzelli, N. K. (1969). *Archs Biochem. Biophys.* **130**, 164.

Redfearn, E. and Powls, R. (1967). *Biochem. J.* **106**, 56P.

Rudney, H. and Raman, T. S. (1966). *Vitams Horms* **24**, 531.

Rudney, H. and Sugimura, T. (1961). *In* "Ciba Foundation Symposium on Quinones in Electron Transport" (G. E. W. Wolstenholme and C. M. O'Connor, eds.), p. 211, J. & A. Churchill, London.

Samuel, O. and Azerad, R. (1969). *FEBS Letters* **2**, 336.

Sandermann, W. and Simatupang, H. M. (1962). *Angew. Chem.* **74**, 782.

Sato, M. and Ozawa, H. (1969). *J. Biochem., Tokyo* **65**, 861.

Skelton, F. S., Lunan, K. D., Folkers, K., Schrell, J. V., Siddiqui, W. A. and Geiman, Q. M. (1969). *Biochemistry, Easton* **8**, 1284.

Spiller, G. H., Threlfall, D. R. and Whistance, G. R. (1968). *Archs Biochem. Biophys.* **125**, 786.

Stoffel, W. and Martius, C. (1960). *Biochem. Z.* **333**, 440.

Stone, K. J. and Hemming, F. W. (1967). *Biochem. J.* **104**, 43.

Sugimura, T. and Rudney, H. (1960). *Biochim. biophys. Acta* **37**, 560.

Sun, E., Barr, R. and Crane, F. L. (1968). *Pl. Physiol.* **43**, 1935.

Threlfall, D. R. (1967). *In* "Terpenoids in Plants" (J. P. Pridham, ed.), p. 191, Academic Press, London and New York.

Threlfall, D. R. and Goodwin, T. W. (1967). *Biochem. J.* **103**, 573.

Threlfall, D. R. and Griffiths, W. T. (1967). *In* "Biochemistry of Chloroplasts" (T. W. Goodwin, ed.), Vol. 2, p. 254, Academic Press, London and New York.

Threlfall, D. R. and Whistance, G. R. (1970). *Phytochemistry* **9**, 355.

Threlfall, D. R., Whistance, G. R. and Goodwin, T. W. (1967a). *Biochem. J.* **102**, 49P.

Threlfall, D. R., Griffiths, W. T. and Goodwin, T. W. (1967b). *Biochem. J.* **103**, 831.

Threlfall, D. R., Whistance, G. R. and Goodwin, T. W. (1968). *Biochem. J.* **106**, 107.

Threlfall, D. R., Ah Law and Whistance, G. R. (1970). *Biochem. J.* (In press).

Tornabene, T. G., Kates, K., Gelpi, E. and Oro, J. (1969). *J. Lipid Res.* **10**, 294.
Wallwork, J. C. and Pennock, J. F. (1968). *Chemy Ind.* 1571.
Weber, M. M., Matschiner, J. T. and Peck, H. D. (1970). *Biochem. biophys. Res. Commun.* **38**, 197.
Wellburn, A. R. (1968). *Phytochemistry* **7**, 1523.
Wellburn, A. R. (1970). *Phytochemistry* **9**, 743.
Wellburn, A. R. and Hemming, F. W. (1967). *Biochem. J.* **104**, 173.
Whistance, G. R. and Threlfall, D. R. (1967). *Biochem. biophys. Res. Commun.* **28**, 295.
Whistance, G. R. and Threlfall, D. R. (1968a). *Biochem. J.* **109**, 482.
Whistance, G. R. and Threlfall, D. R. (1968b). *Biochem. J.* **109**, 577.
Whistance, G. R. and Threlfall, D. R. (1970a). *Biochem. J.* **117**, 593.
Whistance, G. R. and Threlfall, D. R. (1970b). *Phytochemistry* **9**, 213.
Whistance, G. R. and Threlfall, D. R. (1970c). *Phytochemistry* **9**, 737.
Whistance, G. R., Threlfall, D. R. and Goodwin, T. W. (1966a). *Biochem. J.* **101**, 5*P*.
Whistance, G. R., Threlfall, D. R. and Goodwin, T. W. (1966b). *Biochem. biophys. Res. Commun.* **23**, 849.
Whistance, G. R., Threlfall, D. R. and Goodwin, T. W. (1967). *Biochem. J.* **104**, 145.
Whistance, G. R., Brown, B. S. and Threlfall, D. R. (1969a). *Biochim. biophys. Acta* **176**, 895.
Whistance, G. R., Dillon, J. F. and Threlfall, D. R. (1969b). *Biochem. J.* **111**, 461.
Whistance, G. R., Brown, B. S. and Threlfall, D. R. (1970). *Biochem. J.* **117**, 119.
Whittle, K. J., Audley, B. G. and Pennock, J. F. (1967). *Biochem. J.* **103**, 21C.
Winrow, M. J. and Rudney, H. (1969). *Biochem. biophys. Res. Commun.* **37**, 833.
Zenk, M. H. and Leistner, E. (1967). *Z. Naturf.* **22b**, 460.

Author Index

Numbers in italics are those pages on which references are listed

H

I

J

L

Y

Genus and Species Index

Subject Index

(*Where a page number is* **in italics**, *the compound referred to is first shown structurally*)